33,51,57   191
58,   415

,27

FUNDAMENTALS OF
EXPERIMENTAL DESIGN

FUNDAMENTALS OF EXPERIMENTAL DESIGN

THIRD EDITION

JEROME L. MYERS
University of Massachusetts at Amherst

Allyn and Bacon, Inc.
Boston • London • Sydney • Toronto

In memory of David A. Grant

Production Editor: Jane Dahl
Interior Designer: Libby Griffiths

Library of Congress Cataloging in Publication Data

Myers, Jerome L.
    Fundamentals of experimental design.

    Includes bibliographies and index.
    1. Experimental design. 2. Psychometrics.
I. Title.
QA279.M93   1979   001.4'34   78-12815

ISBN 0-205-06615-1

Printed in the United States of America.

CONTENTS

PREFACE

This edition of *Fundamentals of Experimental Design* contains much that did not appear in the two previous editions. There are five new chapters, 12 through 15, and 18; two chapters that have been largely rewritten, 7 and 11 (formerly 13); and sections added or expanded in most of the remaining chapters. Some of the revision is an attempt to say things more clearly than in past editions and to update the coverage of those earlier editions. Largely, however, the changes in this edition are centered about certain common issues in research design and analysis, issues that were treated either inadequately or not at all in previous editions.

One such issue is the appropriate analysis of data from factorial designs in which cell frequencies are disproportionate. Chapter 15 provides a detailed account of several alternative regression analyses and tries to clarify several questions raised in recent articles on this topic. Chapters 12 and 14, on simple and multiple regression, were added to the text both because they served as logical prerequisites to Chapter 15 and because of my growing conviction that we too often use factorial designs and analysis of variance when random sampling and regression analysis would be more appropriate. Chapter 13 provides some elementary but useful matrix operations. The contents are a prerequisite for much of Chapters 14, 15, and 18. Even readers who already know matrix algebra should read Section 13.5, "Some Statistical Applications," in which some important properties of regression coefficients are derived.

Chapter 16 (formerly 12) deals with analysis of covariance, and Chapter 17 (formerly 14) with orthogonal polynomial analysis. Because both chapters rely explicitly on concepts associated with regression analysis, they fit neatly in this section of the book. Chapters 13 through 15 are not prerequisite to these two chapters but Chapter 12 will help in trying to understand them. Chapter 18 is a new chapter and treats such topics as Hotelling's T^2 and multivariate analysis of variance. It is introductory and therefore limited; it does, however, treat the relative merits of univariate and multivariate analyses for repeated-measurement designs.

The added chapters are not the only departure from the previous edition. Chapter 11 formerly dealt with such applications of expected mean squares as pooling, quasi F ratios, omega squared, and measures of reliability. That material is now distributed throughout the book, integrated at appropriate points with other analyses of data.

Chapter 7, although not new, has been extensively revised. The presentation of models for repeated-measurement designs, and the discussion of the implications of these models, is more detailed than previously. The distinction between fixed and random effects, and its nature and the implications for analyses of data, also have received considerable attention.

Chapter 11 (formerly 13) also has been largely rewritten. I have tried to clarify the issues that arise in choosing among multiple comparison procedures and have considered such issues as the appropriate technique when ns are unequal, the proper error term in repeated-measurement designs, and the relation between the F test of the overall null hypothesis and various multiple comparison procedures. Readers should find this chapter better organized, more clearly written, and more responsive to their concerns than it was in previous editions.

Most other chapters have been somewhat revised; at least, references have been updated and exercises added. A few of the changes are here detailed. The discussion of the implications of violations of the analysis of variance model in Chapter 4 has been elaborated on to include the robustness of the F test when the dependent measure is restricted to a very few possible values. Chapter 6 presents a recently published variation of the extreme-groups design that is better than the approaches detailed earlier. The coverage in Chapter 9 of nesting of within-subject variables has been expanded; this should prove helpful whenever subjects are exposed to several classes (for example, level of difficulty, type of problem) of randomly sampled stimuli. Chapter 16 (formerly 12) on analysis of covariance now includes a section on interpretative problems; misuses in the research literature and a series of often contradictory recommendations on how to use covariance properly have led me to include the new section.

I wish to thank Professor Alphonse Chapanis of the Johns Hopkins University, Professor James Bowen of the University of Texas, Professor Larry Jones of the University of Illinois, Professor Fred Hornbeck of San Diego State College, and Professor David Lyon of Western Michigan University, all of whom reviewed the manuscript at some stage of its development.

I am grateful to the Literary Executor of the late Sir Ronald A. Fisher, F. R. S., to Dr. Frank Yates, F. R. S., and to Longman Group Ltd., London, for permission to reprint Tables III, IV, and V from their book *Statistical Tables for Biological, Agricultural and Medical Research* (6th edition, 1974).

I am indebted to the University of Massachusetts for the sabbatical leave that allowed time to write this edition, and to the University of Houston, which graciously made available the office space and library resources that I needed. I particularly want to thank Professor Bart Osburn, of the Psychology Department of the University of Houston, who called to my attention several important references and helped clarify my thinking on several topics. Finally, and as always, I am indebted to my wife and colleague, Professor Nancy A. Myers. Her contributions have included proofreading the manuscript, discussing the material at length, and tolerating — even encouraging — a preoccupied spouse.

Jerome L. Myers

TO THE INSTRUCTOR

This, the third edition of *Fundamentals of Experimental Design,* has five added chapters — those on regression (Chapters 12, 14, and 15), multivariate analysis (Chapter 18), and the matrix algebra appropriate to the level of presentation of this material (Chapter 13). I have noted in the preface other additions in this third edition. The added material should enhance the value of the book as a textbook and a reference. Given these additions, however, alternative approaches in using the current edition as a textbook may merit consideration.

It may help to note that Chapters 16 and 17 (analysis of covariance, trend analysis) do not depend on the material in Chapters 12 through 15, although the student should know something about simple linear regression to be able to understand fully the material in Chapters 16 and 17. In any event, the instructor who wants to teach only the material in the first two editions can do so by dropping the added chapters — 12 through 15, and 18 — from the syllabus.

My own bias is that regression, and its relation to analysis of variance, should be stressed more than is usual in experimental design courses. I should therefore prefer including some of the new material at the sacrifice of certain advanced topics in design and analysis. One possibility is presenting Chapters 1 through 5, 7 and 8, and 11 through 15. Time permitting, the instructor could add any of the remaining chapters that are of particular interest. My students have generally been through the first 11 or 12 chapters of Hays's *Statistics for the Social Sciences* in a first-semester course. In my courses, therefore, I plan to omit Chapters 1 through 3 (and possibly 4), which will allow me to add Chapters 16 and 17 to my syllabus.

It should be evident that the two approaches I have just sketched do not exhaust the possibilities. There are many possible subsets of chapters that can be given in a one-semester course. The path any one instructor takes must depend finally on the expertise and interests of the instructor and the emphasis desired, and the background and interests of the students.

Jerome L. Myers

1 PLANNING THE EXPERIMENT

1.1 INTRODUCTION

We undertake psychological experiments to determine what factors influence a certain behavior, and the extent and direction of the influence. We seek answers to such questions as: What are the relative effects of these three drugs on the number of errors made in learning a maze? Which of these training methods is most effective? What changes in auditory acuity occur as a function of certain changes in sound intensity? If an experiment is to answer questions adequately, we must first specify the factors whose effects are to be studied (*independent variables*); minimize the effects of factors not of current interest (*irrelevant variables*); carefully select a measure, or measures, of the behavior we are investigating (*dependent variables*); and choose the beings whose behavior is to be measured (*subjects*). Planning these four basic aspects of an experiment is the first and most critical step in getting answers about behavior. Although much of our discussion about these topics may seem obvious to the well-trained and experienced researcher, we hope that the student of experimental design will profit from this review of the many things that must be considered in planning an experiment.

1.2 THE INDEPENDENT VARIABLE

Once we have decided what independent variable or variables must be studied, we must choose the actual treatments: the levels—specific types or amounts—of the independent variable that will be tested in the experiment. We must decide which drugs, which training methods, or which sound intensities will be compared. For this class of decisions, it is helpful to distinguish two types of independent variables, quantitative and qualitative.

1.2.1 QUANTITATIVE VARIABLES.   A quantitative independent variable is a variable whose levels differ in amount. Examples of such a variable are amount of reward, intensity of shock, and number of practice trials. Generally, we are not interested in the specific numerical levels chosen for inclusion in the experiment.

For example, in a study of the effects of intertrial interval on the speed of learning lists of words, we might choose 2, 4, and 6 seconds as levels of interval length. Probably 1.8, 3.8, and 5.8 seconds would be equally adequate for our purpose, but we tend to think of whole numbers. The levels of a quantitative independent variable are usually of interest only in that they allow an experimenter to determine whether any change in the quantity manipulated results in a change in behavior, and if so, what the characteristics (for example, shape, slope, and position) of the function relating the independent and dependent variables are. Therefore, the levels of the independent variable should be chosen to cover a wide enough range to detect any behavioral change that might result, and in sufficient number and proximity so that the shape of the function will be clearly defined.

In any single experiment, it may be hard to achieve the ideal of broadly covering the continuum of the independent variable with many levels, close together. It is not always possible to decide without some pilot experimentation how many levels we should have and how close they should be. Furthermore time, money, and subjects may be limited, thus making the ideal difficult to realize. It will often be best in initial experiments to determine generally whether any behavioral change occurs as the independent variable is manipulated, and to attain a rough description of the shape of the function relating the independent and the dependent variables. If desirable, further experiments can be designed to yield a more precise definition of the function.

Suppose that we are interested in analyzing the relation between x and y of Figure 1–1. Assuming that there are other independent variables that we also want to investigate, we may not be able to include as many levels of x as we deem

Figure 1–1   A function relating dependent and independent variables.

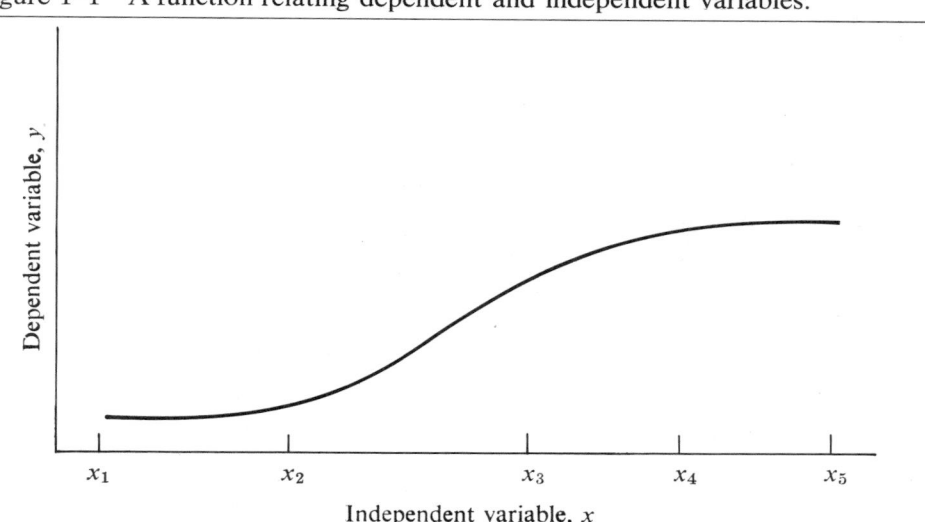

ideal. In our first experiment, we might include the levels x_1, x_3, and x_5, thus learning that variations in x do result in variations in y and obtaining some idea of the slope of the function and the minimum range of x over which y varies. In subsequent experiments, we might choose levels between x_2 and x_4, getting a more specific picture of the function.

Note that selecting levels depends on the results of previous research. Such decisions may also be influenced, however, by theoretical considerations. Suppose that the experimenter is trying to decide which of two theories is correct; one predicts gradual changes in y (in Figure 1–1) with changes in x, and the other predicts a "staircase" effect, or stepwise changes in the function. In the latter case, an examination of a broad range of levels of x might be sacrificed to allow concentrating more levels within a narrow range.

Some inferential pitfalls occur particularly when only a few levels of the quantitative independent variable are included in an experiment. Suppose that the variable x of Figure 1–1 is under investigation and that the relation between independent variable and dependent variable is as depicted in the figure. If no other information were available, and if only x_1 and x_2 or x_4 and x_5 were selected for investigation, the data might lead us to conclude that the independent variable x does not influence behavior. If we happened to choose x_1, x_3, and x_4, we might conclude that performance and x are related by a linear function. The first instance exemplifies extrapolation beyond the range of levels selected, and the second instance interpolation between the levels selected. It is wise to recognize the tentative status of inferences that go beyond the levels included in the experiment.

1.2.2   QUALITATIVE VARIABLES.   A qualitative independent variable has "levels" that differ in type.[1] Examples of such a variable are type of punishment, method of training, and type of instruction. The particular levels—the specific types—of the variable that are included in the experiment are usually of direct interest to us. They contrast with the numerical levels of quantitative variables, which are often less important in themselves than for information they provide about some function relating the independent and dependent variables. Qualitative treatments are chosen because previous research, theoretical predictions, or practical considerations dictate their choice. Two particular methods of teaching reading, for example, may be chosen for experimental comparison because (1) previous experimentation suggests that these are the two most efficient procedures available, (2) the results of the comparison should differentiate between two educational theories in which the experimenter is interested, or (3) other methods take so much time and money that they are impractical and would not be used in practice even if they were effective.

---

[1] The word *levels* actually does not apply to the qualitative variable, since all definitions of the word suggest an ordering, or ranking. Since it is a useful concept for later chapters, however, we have used it for both the quantitative and qualitative variables.

1.2.3   CONTROL LEVELS.   Common to both qualitative and quantitative variables is the selection of *control* levels. These are treatments that may not be of interest in themselves but that give additional information about the effects of one or more other treatments. Suppose we wanted to compare the effects of teaching reading versus not teaching it in kindergarten on reading scores obtained at the end of first grade. The experiment demands two groups of first graders, both with kindergarten experience, but only one with reading experience in kindergarten. Adding a control group, children of no kindergarten experience at all, would permit evaluating the possibility that organized activity alone is helpful preparation for grade school performance.

1.2.4   FIXED AND RANDOM VARIABLES.   An implicit premise in our consideration of variables has been that we are dealing with *fixed* variables in our research, variables whose levels we arbitrarily choose. A second, less frequent but equally important, possibility is that the variable is *random,* that is, that its levels were chosen from some larger population of levels on the principle that all members of the population have an equal opportunity of being chosen. An experimenter chooses levels in this way when the main consideration is getting a reliable estimate of the variability in the population. To determine whether newly manufactured calculating machines perform similarly, for example, we should test a sample representing the entire population of such machines. Another important random variable in psychological research is the subject. Individual subjects are rarely chosen for unique personal attributes. Generally, subjects in psychological experimentation are a random sample from some population, perhaps all the students taking basic psychology at some college.[2]

The distinction between fixed and random independent variables is important to our inferential processes. When the levels of the variable have been arbitrarily chosen, any inferences about differences among the effects of the levels are limited to the particular levels chosen. Having compared the effects of three arbitrarily chosen methods of teaching reading, we can draw inferences only about these three methods, not about any broader population of methods. On the other hand, the observed variability among performances on five calculators chosen randomly from a factory's output leads to conclusions about the variability in the population of machines (the factory's output). Our inferential statements extend beyond the actual levels sampled (the five machines in the experiment) to the broader population from which they were randomly sampled. This distinction also has important implications for analyzing data.

1.3   IRRELEVANT VARIABLES

Drawing conclusions about a certain behavior would present no problems if the only variables affecting the behavior were those the experimenter selected for

---

[2] The issue of what population has been sampled is much more complex than indicated here. A fuller discussion is presented in Chapter 7. We shall also develop in detail there the distinction between fixed and random variables and the implications of the distinction for data analyses and for drawing inferences from our data.

study. Unfortunately, behavior is never so simply caused; it may be a function of such variables as the intelligence, prior experience, attitude, and age of the subject, the time of day at which the data are collected, and even the tone of voice in which instructions are read. If we are not interested in the effects of these variables in some particular experiment, they are irrelevant in the limited sense that the experiment has not been performed to investigate them. Nevertheless, in designing our experiments we must consider which irrelevant variables might influence the data and how to deal with such influences in our experimental design. There are two reasons for attending to such variables in planning our experiment. First, we wish to avoid systematic biases, the *confounding* of one or more irrelevant variables with some variable of interest to us. Suppose we want to compare memory for a set of facts organized in two different ways. Assume that, as is often the case, our subjects will be introductory psychology students who volunteer to serve for course credit. Perhaps subjects who volunteer later in the semester will be more in need of course credit; in general, they may be less able or motivated than those who were subjects earlier. If these variables affect performance in the memory task, and if we were to run the first half of our subjects with one organization and the second half with the other, the effects of organization would be confounded with effects of ability and motivation. To what do we attribute any difference in performance between the two groups of subjects? Is organization responsible—or ability, motivation, some combination of these, or even some other consequence of the method by which we established our experimental group? If the two groups perform equally well, are the two types of organization equally effective, or was the more effective treatment handicapped by being applied to less able or less motivated subjects?

Note that the problem is not that subjects assigned to one type of organization are generally more intelligent or more motivated; it is rather that the procedure for assigning subjects to conditions ensures that this particular condition would again have the advantage if we were to repeat the experiment. This would not be the case if each subject had an equal chance of being assigned to either type of organization. If there is this equal chance, the bias is not systematic and the variables are not confounded. Under such random assignment of subjects to conditions, we can assess the likelihood that differences in the abilities of our subjects were sufficient to account for differences in average performance of the two experimental groups. If this likelihood is very small, we shall conclude that the two levels of the independent variable differ in their effects on performance.

Even if confounding is not a problem, there is a second reason for focusing on irrelevant variables in planning the experiment. The presence of such variables may be reflected in *error variance*, the variability among scores that cannot be attributed to the effects of the independent variables. There will always be some error variability, even among scores that have been obtained under the same experimental treatment, for these scores will come either from different individuals who differ in such variables as intelligence, attitudes, and age or from the same individuals at different times, for which they will show change in such variables as attentiveness, practice, and fatigue. The greater the error variability, the more difficult the determination of the effects of independent variables. The following example will show why this is so.

Assume that the two sets of data below have been obtained under two treatments *A* and *B*, each applied to three subjects. The mean performance under treatment *B* is better, reflected by a mean of 6 as against the mean 5 for treatment *A*. The individuals within a group, however, differ by at least as much as the group means do, and we cannot be sure whether the difference between the *A* and *B* means comes from individual or treatment differences.

A	*B*
4	5
5	6
6	7

Now suppose that all the scores for *A* were 5 and all those for *B* were 6. The treatment means are the same as before; but if people treated alike do not differ and people treated differently do, it seems reasonable to conclude that the treatments have different effects.

How can we control irrelevant variables, eliminating systematic biases and reducing error variance? The approaches to this control problem can be grouped in three categories, as follows.

1.3.1   UNIFORM APPLICATION OF THE IRRELEVANT VARIABLE.   If only one level of the irrelevant variable is present in the experiment, it contributes no variability at all and therefore cannot give advantage to any one level of the independent variable, nor contribute to error variance. An important aid in uniformly applying the irrelevant variable is automation. Thus, electronic timers allow us to keep intertrial intervals constant over trials, subjects, and experimental conditions; tape-recorded instructions are presented in the same words and tone to every subject; and automated animal test cages (which also house the animals) eliminate handling and provide a uniform environment for all animals. While automation is helpful, the careful experimenter can do much to minimize variability without it. Care can be taken to read the same instructions to all subjects or to provide similar living conditions and a minimal amount of handling for all animal subjects. Every experiment will have its own potential sources of irrelevant variation, but by careful analysis of a situation, one can eliminate or minimize many of these.

1.3.2   RANDOMIZATION.   Randomization guards against the danger of systematic biases in the data. Suppose we are interested in comparing the effects of two sets of instructions on problem-solving performance. One obvious potential source of bias is problem-solving ability; it is necessary to guard against a systematic bias from applying one set of instructions to the better problem solvers. To avoid the hazard of systematic bias, we can assign the subjects randomly to treatments, that is, by some method that ensures that each subject is equally likely to be assigned to either set of instructions. Each subject could draw a number from a hat; then the odd-numbered subjects could be assigned to one treatment and the even-numbered subjects to the second. Note that randomization does not ensure two experimental groups perfectly matched for those variables that might influence problem solving. Randomization does ensure that over many replications of the experiment neither treatment will have an advantage. In any one experiment, one group could have an advantage (the odd-numbered subjects might, by chance, have higher intelligence), but statistical procedures that assume randomization take these biases into account.

Randomization does not apply only to assigning subjects to the levels of independent variables. It can also apply, for example, to selecting orders of presentation of treatments when each subject is tested at several levels of the independent variable. In fact, there is a vast array of different schemes for collecting data. Some of these will be considered next, with particular emphasis on the ways in which error variance may be further reduced.

1.3.3   EXPERIMENTAL DESIGN AND ANALYSIS.   An experimental design is a plan for running the experiment. One such plan is complete randomization, in which each subject is randomly assigned to only one combination of levels of the independent variables. We could extend such a design by including irrelevant variables as independent variables. In the experiment on instructions and problem solving, for example, the subject pool might be divided into three levels of intelligence—low (IQ under 85), medium (86–115), and high (above 115). Low IQ subjects would then be randomly assigned to the two instructional sets, and likewise the medium and high subjects, giving six combinations of IQ and instructions with an equal number of subjects in each group. The advantage of this plan over the original completely random assignment of subjects to the two treatments is that it permits a more accurate assessment of the effects of instructions. One can now remove, through statistical analysis, variability in the data from differences in problem-solving ability among the three levels of intelligence. This design, in which the levels of the independent variable are matched on some irrelevant variable, is said to be more *efficient,* or to result in less error variability, than the completely randomized design first considered. (A more extensive discussion of the matching design appears in Chapter 6.)

Treatments can be matched on the basis of other variables besides the personal attributes of the subjects. If it were necessary, say, to divide the subject testing between two experimenters, subjects could be randomly assigned to experimenters regardless of treatment group, or one could ensure that half of each treatment group would be run by each experimenter. The second method is similar to the matching of intelligence previously suggested and would be recommended if there were any reason to suspect that experimenter differences might be a source of variability in the data.

There are advantages to running each subject through all levels of the independent variable. A subject might do one problem-solving task under one set of instructions, for example, and a second equivalent task under a second set of instructions. Since all subjects experience both treatments, there can be no systematic bias due to personal attributes like intelligence. Furthermore, computations exist for removing variability due to individual differences, permitting evaluation of the effects of instructions against a smaller error variability. One big drawback of this design is the possibility of a systematic bias from temporal effects. If one set of instructions is always given first, the second set might profit from practice or be handicapped because of fatigue. For this reason, the order of presentation should be random; each treatment should have an equal chance of being assigned to each position in the sequence of treatment presentations. (See Chapter 7 for repeated-measurement designs of this sort.)

The randomization just described might be restricted to ensure that each treatment appeared equally often in each position in the sequence of presentations. In the problem-solving experiment, each of 20 subjects could be assigned a number from 1 to 20; each number would be assigned exactly once. If the number were even, one instructional set would be given first; otherwise the other set would come first. Such a design further reduces error variance by permitting removal of variability due to temporal effects and also variability due to individual differences. This type of design is generally referred to as a Latin square design (more fully discussed in Chapter 10).

With any design, further control of the effects of irrelevant variables is possible if one can somehow analyze what part of each score is caused by the irrelevant variable. Rather than match subjects on intelligence test scores, for example, one could use the design described earlier in which subjects were randomly assigned to treatments regardless of intelligence level. Subjects' intelligence scores could then be used as a basis for adjusting their problem-solving scores. (The method of adjustment, called analysis of covariance, is discussed in Chapter 16.)

If control of irrelevant variables were the only concern of the experimenter, there would be greater uniformity in selecting experimental designs and possibly more extensive use of analysis of covariance. There are, however, at least three other factors to consider in selecting a design—the information desired, the model for the data analysis, and the practical requirements of the situation.

1.4   FACTORS IN SELECTING EXPERIMENTAL DESIGNS

1.4.1   INFORMATION.   Different kinds of designs yield different sorts of information. If information about the effects of time is wanted, a repeated-measurement design is required. If there is interest in what effects order of presentation of treatments has, a Latin square design is indicated. If the experimenter hypothesizes that the effect of the independent variable is a function of certain characteristics of his subjects, some effort might be made to match these characteristics. If certain single or joint effects of variables are of more interest than others, there are designs that permit a more efficient evaluation of these effects at the expense of losing information on others. These are but a few examples of how designs can differ in the information they provide.

1.4.2   THE MODEL.   The validity of any inference drawn from a statistical analysis rests on the validity of an underlying model, a set of assumptions about the data. Violating any one of the assumptions may result in an incorrect inference about the effects of the independent variable. Since the model is a function of the design used, selecting the experimental design means thinking about which assumptions are implied, whether they are likely to be met, and how failing to meet them will affect the validity of the inferences drawn from the statistical analysis. Using the Latin square design, for example, assumes that the size of the treatment effects is not a function of the order of presenting the treatments. This assumption will often be false, with the usual result an increased probability of concluding that the treatment has no effect, when it really has.

1.4.3   PRACTICAL REQUIREMENTS.   The selection of designs must often be dictated, or at least narrowed, by such considerations as the number of available subjects or the time available per subject. In animal research, for example, it may be more convenient, and certainly less expensive, to run a few subjects through many conditions than to run many groups of subjects each under a different treatment. On the other hand, in research with children, where teachers may object if any one child loses too much class time, it may be more reasonable to use a completely randomized design, with fewer measurements on more children.

1.5   THE DEPENDENT VARIABLE

1.5.1   CHOOSING THE DEPENDENT VARIABLE.   Choosing an appropriate dependent variable, or measure, may seem to be a trivial problem—after all, we are interested in leadership, or aggressive behavior, or learning, or visual acuity, and this is what will be measured. Unfortunately, there are always several measures that can be reasonably interpreted as indices of the behavioral process under investigation. Consider, for example, learning a list of words. Shall we measure the number of trials to attain some predetermined criterion? Shall we measure the speed of each response? Is the basic information the number of errors in each block of five trials? ten trials? all trials? Should errors of commission (incorrect responses) and errors of omission (failures to respond) be analyzed separately? Certainly these alternatives are not mutually exclusive, but there are practical limits to how many measures can be considered in analyzing any single experiment. What, then, is involved in selecting the dependent variable or variables?

There are several characteristics of dependent variables on which the accuracy of inferences depends. Ideally, the dependent variable should be reliable, sensitive, and distributed in a way that conforms to the assumptions of the data analysis model. Reliability will be a factor to the extent that measures equivalent in all other respects differ in variability. The measure that is least variable under constant experimental conditions is preferred.

Sensitivity means that certain measures show greater differential effects than other measures do as a function of changes in the independent variable. In a study of escape from conflict, for example, animals under conflict might not differ from control subjects who were not under conflict in the number of escape responses (presses of a platform that result in removal of the conflictual stimuli). The two groups might differ in mean duration of the escape responses, however.

Statistical analysis is often complicated because the distribution of the measure does not conform to the assumptions of available statistical models. If all other considerations are equal, we want that measure whose distribution is consistent with the model that is associated with the statistical analysis to be applied. This implies a thorough knowledge of previous research and the kinds of results got with different measures, and also consideration of the statistical model.

As in every other phase of experimental planning, practical considerations are involved in choosing measures. All else being equal, we want measures we can get easily. In research on personality, for example, if a paper and pencil test and a projective test are equally reliable and sensitive and conform similarly to the statistical model, the paper and pencil test is preferable; it is administered and scored much more rapidly. Of course, all other things are rarely equal; the experimenter may consider the projective test more sensitive, while the paper and pencil test is probably more reliable and possibly more likely to result in a normal distribution of data. The moral is that in all phases of experimental planning the ideal is rarely attained; the experiment is a compromise among the factors that have been indicated.

Theory may also be an issue in selecting measures. Consider an experiment in which the subject guesses which of two events will occur on each trial. The measure generally taken in such experiments is the percentage of each type of guess, partly because it is easily obtained and partly because pertinent theories of the behavior under investigation yield predictions of this measure. With the recent advent of theories that generate exact quantitative predictions of response latency for choice behavior, latency has become more common as a measure.

1.5.2   CHOOSING THE MEASURING TECHNIQUE.   In many instances, the choice of a measure still leaves unanswered how the measure is to be recorded. In deciding this point, the experimenter should consider the ease of obtaining the data as well as the probable degree of reliability of the recording technique. Again, imagine that a subject must guess which of two events will occur on each trial. There are at least three methods of recording which response occurs on each trial: (1) the experimenter can manually record the subject's choice on each trial, (2) responses can be automatically recorded by some kind of event-pen system, and (3) responses can be automatically punched out onto IBM cards. The manual technique is the least reliable, since it is subject to recording errors by the experimenter. Manually recorded data are generally easier to score and tabulate than ink records, however, and this nonautomated recording system is less subject to breakdowns. The event-pen method frees the experimenter's time during experimental sessions, involves an initial expenditure of several hundred dollars, and is generally reliable in recording data, but involves record-reading labor and possible related error. The data punchout method is efficient in both collecting and analyzing data and is highly reliable, but may involve an initial expenditure of several thousand dollars. The automatic system may also be less reliable from day to day in the sense that breakdowns may be more frequent than with the alternative procedures. Clearly, the choice of a system for recording the data is, like all other decisions made in planning an experiment, a compromise among numerous considerations.

1.6 SUBJECTS

There are two main classes of decisions the experimenter must make about subjects: From what population should they be drawn? How many should be run?

1.6.1 THE SUBJECT POPULATION.   The purpose of the research will often dictate the kind of subject population, or at least considerably narrow that choice. This is obviously the case when we are concerned with such processes as schizophrenic performance, development of memory skills, marital interactions, or migration in birds. Less obviously, theoretical considerations may affect what subjects are chosen. For example, adult human subjects, when asked to guess which of two events will occur, tend to guess one event if the other event has had a long run, that is, has come up several times in succession. This behavior conflicts with the predictions of some prominent theories of choice behavior; it has been hypothesized that the subject behaves like this having experienced only short event runs in his pre-experimental history, and he therefore expects short runs in the laboratory. A test of this hypothesis might involve using subjects with limited previous exposure to event sequences—for example, young children or rats.

Certain basic processes do not, at least superficially, seem to need a particular subject population. Simple conditioning, both classical and operant, is one possibility. Here, the need for minimizing error variance may be the important consideration. The abundance of data on animals may be largely due to the control we have of the prior experience of such subjects. Often, the experimenter can control their eating habits, genetic history, and environmental influences from the time of their birth. As Tolman has remarked, "Rats live in cages; they do not go on binges the night before one has planned the experiment...."[3]

1.6.2 HOW MANY SUBJECTS?   A primary consideration in deciding on the number of subjects is the *power* desired. Roughly, power is the probability of correctly concluding that differences among the effects of treatments exist. Power depends on the direction and size of the effect to be detected, how large a risk of wrongly concluding that the treatments have different effects one is willing to take, and the error variance expected. The error variance in turn depends on the experimental design used. For the time being, we merely note that decisions about the factors influencing power will influence the number of subjects required for attaining a desired degree of power.

[3] E. C. Tolman, "A Stimulus-Expectancy Need-Cathexis Psychology," *Science* 101:160–166 (1945).

1.7   CONCLUDING REMARKS

We have implied that experimental plans are partly related to the data analysis; choosing designs and dependent variables is a function of the demands of the statistical model. This point is now made explicit: the statistical analysis should be planned *in detail* before a single subject is run. The alternative is a post-experimental search for an analysis that is consistent with the design and the distribution of the measure; and that analysis may not exist. One can be reasonably sure that an appropriate analysis exists only if one considers design and analysis together, before the data are collected. When alternative analyses exist, the choice depends on factors like those previously cited in discussing the choice among experimental designs (Section 1.4). Thus one should take into account the relative *efficiencies* of analyses, the resulting *information*, and the *computational labor* involved, as well as *assumptions* and whether or not the assumptions are likely to be met.

Sometimes the considerations involved in planning experiments will point to a single decision; more often they will conflict. Thus, the simplest measure to obtain may be the least reliable, and the most efficient design may imply a statistical model to which our data will not conform. Experimental planning means weighing such considerations and compromising among them. It is impossible to state a single set of rules for weighing these considerations, since we should need different rules for each experiment. However, we can determine for ourselves what the important factors influencing our decisions should be.

2 NOTATION

We must have a common language to talk about the derivations and computational formulas that relate to psychological experimentation. Such a language exists in the notational system here presented. If you try to master it now, your efforts will be amply repaid. You will find first a few simple rules, which are then applied to some elementary statistical quantities.

2.1 A SINGLE GROUP OF SCORES

2.1.1 SOME BASIC RULES. In a group of scores like $Y_1, Y_2, Y_3, Y_4, \ldots, Y_n$, the subscript has no purpose except to distinguish among the individual scores. The quantity n is the total number of scores in the group. Suppose that $n = 5$ and we want to show that all five scores are to be added together. We could write

$$Y_1 + Y_2 + Y_3 + Y_4 + Y_5$$

or more briefly,

$$Y_1 + Y_2 + \cdots + Y_5$$

Still more briefly, we write

$$\sum_{i=1}^{5} Y_i$$

This expression is read "sum the values of Y for all i from 1 to 5." In general, $i = 1, 2, \ldots, n$ (that is, i takes on the values 1 to n), and the summation of a group of n scores is indicated by

$$\sum_{i=1}^{n} Y_i$$

The quantity i is the *index*, and 1 and n are the *limits* of summation. When the context of the presentation permits no confusion, the index and limits are often dropped. Thus we may often indicate by $\sum Y$ that a group of scores are to be summed.

Three rules for summation follow.

RULE 1.   *The sum of a constant times a variable equals the constant times the sum of the variable; or*

(2.1)
$$\sum CY = C \sum Y$$

The term C is a constant in the sense that its value does not change as a function of i; the value of Y depends on i, and Y is therefore a variable relative to i. Equation (2.1) is easily proved.

$$\sum CY = CY_1 + CY_2 + CY_3 + \cdots + CY_n$$
$$= C(Y_1 + Y_2 + Y_3 + \cdots + Y_n)$$
$$= C \sum Y$$

RULE 2.   *The sum of a constant equals n times the constant, where n equals the number of quantities summed; or*

(2.2)
$$\sum C = C + C + \cdots + C = nC$$

RULE 3.   *The summation sign operates like a multiplier on quantities within parentheses.*

EXAMPLE 1.

$$\sum_{i}^{n} (X_i - Y_i) = \sum_{i}^{n} X_i - \sum_{i}^{n} Y_i$$

PROOF.

$$\sum (X - Y) = (X_1 - Y_1) + (X_2 - Y_2) + \cdots + (X_n - Y_n)$$
$$= (X_1 + X_2 + \cdots + X_n) - (Y_1 + Y_2 + \cdots + Y_n)$$
$$= \sum X - \sum Y$$

EXAMPLE 2.

$$\sum (X - Y)^2 = \sum X^2 + \sum Y^2 - 2 \sum XY$$

PROOF.

$$\sum (X - Y)^2 = (X_1 - Y_1)^2 + \cdots + (X_n - Y_n)^2$$
$$= (X_1^2 + Y_1^2 - 2X_1 Y_1) + (X_2^2 + Y_2^2 - 2X_2 Y_2) + \cdots$$
$$+ (X_n^2 + Y_n^2 - 2X_n Y_n)$$
$$= (X_1^2 + X_2^2 + \cdots + X_n^2) + (Y_1^2 + Y_2^2 + \cdots + Y_n^2)$$
$$- 2(X_1 Y_1 + X_2 Y_2 + \cdots + X_n Y_n)$$
$$= \sum X^2 + \sum Y^2 - 2 \sum XY$$

These three rules apply to some commonly computed statistics. The mean of a group of n scores is given by

(2.3)
$$\bar{Y} = \frac{\sum Y}{n}$$

The mean is at the center of our data in the sense that the sum of the distances of all scores from the mean is zero. Our three rules for summation let us show this easily. First, we represent the sum of the distances, or deviations, from the mean by $\sum (Y - \bar{Y})$. Applying Rule 3, we have

(2.4)
$$\sum (Y - \bar{Y}) = \sum Y - \sum \bar{Y}$$

However, $\bar{Y}$ is a constant; its value remains the same regardless of the value of the index i. It should be clear that we are summing over i from 1 to n throughout this presentation, even though the index and the limits are not explicitly presented in each expression. Applying Rule 2, we rewrite Equation (2.4) as

(2.5)
$$\sum (Y - \bar{Y}) = \sum Y - n\bar{Y}$$

At this point we substitute Equation (2.3) into Equation (2.5), getting

(2.6)
$$\sum (Y - \bar{Y}) = \sum Y - n\left(\frac{\sum Y}{n}\right) = \sum Y - \sum Y = 0$$

It has been proved that the sum of deviations of scores about their mean is zero, an elementary but important result. Of course, our purpose has been primarily to show how applying the summation rules can result in proofs of basic statistical properties.

2.1.2   RAW SCORE FORMULAS.   The summation rules can be applied to simplify computations. Two examples are used here, the sample variance and the sample correlation. The sample variance, when used as an estimator (as it will be in most of this book), is defined as follows:

(2.7)
$$\hat{\sigma}^2 = \frac{\sum (Y - \bar{Y})^2}{n - 1}$$

The Greek *sigma*, σ, represents the standard deviation of a population of scores; the diacritical mark ($\hat{}$) indicates that we have an estimate of that standard deviation. Computations, particularly with a desk calculator, are greatly simplified by obtaining a raw score formula for $\hat{\sigma}^2$; for this formula, the mean is eliminated and only the numerator of the right side of Equation (2.7) will be manipulated. To get the raw score formula, expand the quantity within the summation sign. Thus,

(2.8)
$$\sum (Y - \bar{Y})^2 = \sum (Y^2 + \bar{Y}^2 - 2Y\bar{Y})$$

Rule 3 is applied, permitting elimination of the parentheses:

(2.9)
$$\sum (Y - \bar{Y})^2 = \sum Y^2 + \sum \bar{Y}^2 - \sum 2Y\bar{Y}$$

Applying Rule 2 leads to a further change, when we note that $\bar{Y}^2$ is a constant:

(2.10)
$$\sum (Y - \bar{Y})^2 = \sum Y^2 + n\bar{Y}^2 - \sum 2Y\bar{Y}$$

The quantity $2\bar{Y}$ is a constant and can, by Rule 1, be placed before the summation sign. Thus,

(2.11)
$$\sum (Y - \bar{Y})^2 = \sum Y^2 + n\bar{Y}^2 - 2\bar{Y}\sum Y$$

Now replace $\bar{Y}$, using Equation (2.3).

(2.12)
$$\sum (Y - \bar{Y})^2 = \sum Y^2 + n\frac{(\sum Y)^2}{n^2} - 2\left(\frac{\sum Y}{n}\right)\sum Y$$

The final step is to obtain the simplest form of the above expression, which is

(2.13)
$$\sum (Y - \bar{Y})^2 = \sum Y^2 - \frac{(\sum Y)^2}{n}$$

Dividing the right-hand side of Equation (2.13) by $n - 1$ gives the raw score formula for $\hat{\sigma}^2$ that was sought.

A term that we shall want to know subsequently is the covariance:

(2.14)
$$C_{xy} = \frac{\sum (X - \bar{X})(Y - \bar{Y})}{n - 1}$$

A little insight enables us to bypass the previous derivational steps. We first note that Equation (2.13) could be rewritten as

(2.13′)
$$\sum (Y - \bar{Y})^2 = \sum (Y - \bar{Y})(Y - \bar{Y}) = \sum YY - \frac{(\sum Y)(\sum Y)}{n}$$

By analogy, the numerator of C_{xy} has the raw score formula

(2.15)
$$\sum (X - \bar{X})(Y - \bar{Y}) = \sum XY - \frac{(\sum X)(\sum Y)}{n}$$

Dividing the above expression by $n - 1$, we now have a raw score formula for the covariance of X and Y, which is C_{XY}.

Obtaining raw score formulas from those formulas originally used to define a statistic is a common enough problem to warrant a brief summary of the steps involved in the preceding two examples. They are expansion, application of Rule 3 (removing parentheses), application of Rule 2 (replacing summation signs that precede constants by appropriate multipliers), application of Rule 1 (placing constants before summation signs), substitution for quantities not currently in raw score form, and algebraic simplifications.

2.1.3   VARIANCE OF A SUM.   Notational rules are also helpful in deriving many statistical relations. One such derivation, besides being an additional example of how to manipulate statistical symbols, also gives a result that is closely related to later developments.

Suppose that we have a test that has p parts. Then the total score for the ith individual, X_{it}, represents the sum of the individual's scores on the p parts. That is,

$$X_{it} = X_{i1} + X_{i2} + \cdots + X_{ij} + \cdots + X_{ip} = \sum_{j=1}^{p} X_{ij}$$

For n individuals who have taken the test, we can represent their average total score as $\bar{X}_{.t}$; the dot subscript indicates that we are averaging over values of i, in other words, over subjects. We can readily show that the average total test score is the sum of the averages on the p parts; by definition of a mean,[1]

$$\bar{X}_{.t} = \frac{\sum_{i=1}^{n} X_{it}}{n}$$

Replacing X_{it} in terms of the scores on the p parts, we have

$$\frac{\sum X_{it}}{n} = \frac{\sum X_{i1}}{n} + \cdots + \frac{\sum X_{ip}}{n}$$

or

$$\bar{X}_{.t} = \bar{X}_{.1} + \cdots + \bar{X}_{.p}$$

The variance of the n total test scores bears a more complicated relation to the variances of the n scores on the p subtests. From the definition of a variance, we know that

$$\hat{\sigma}_t^2 = \left(\frac{1}{n-1}\right) \sum (X_{it} - \bar{X}_{.t})^2$$

Replacing X_{it} and $\bar{X}_{.t}$ in terms of scores and means for the p subtests, we have

$$\hat{\sigma}_t^2 = \left(\frac{1}{n-1}\right) \sum_{i=1}^{n} [(X_{i1} + \cdots + X_{ij} + \cdots + X_{ip})$$

(2.16)
$$- (\bar{X}_{.1} + \cdots + \bar{X}_{.j} + \cdots + \bar{X}_{.p})]^2$$

$$= \left(\frac{1}{n-1}\right) \sum_{i=1}^{n} [(X_{i1} - \bar{X}_{.1}) + \cdots + (X_{ij} - \bar{X}_{.j}) + \cdots + (X_{ip} - \bar{X}_{.p})]^2$$

Expanding the squared term, we have

$$\hat{\sigma}_t^2 = \left(\frac{1}{n-1}\right) \sum_{i=1}^{n} [(X_{i1} - \bar{X}_{.1})^2 + \cdots + (X_{ij} - \bar{X}_{.j})^2 + \cdots$$

(2.17)
$$+ (X_{ip} - \bar{X}_{.p})^2 + 2(X_{i1} - \bar{X}_{.1})(X_{i2} - \bar{X}_{.2}) + \cdots$$
$$+ 2(X_{ij} - \bar{X}_{.j})(X_{ij'} - \bar{X}_{.j'}) + \cdots$$
$$+ 2(X_{i,p-1} - \bar{X}_{.,p-1})(X_{ip} - \bar{X}_{.p})]$$

---

[1] To conserve space, whenever indices of summation are presented in a line of text or in a fraction, we write them in the form $\sum_{i=1}^{n}$. This is equivalent to the notation

$$\sum_{i=1}^{n}$$

Note that any term of the form $\sum_i (X_{ij} - \bar{X}_{.j})^2/(n-1)$ is the variance of the n scores on part j. Furthermore, the cross-product term,

$$\frac{\sum_i (X_{ij} - \bar{X}_{.j})(X_{ij'} - \bar{X}_{.j'})}{n-1}$$

is $C_{jj'}$, the covariance of scores on parts j and j'. The correlation of scores on any two parts j and j' is expressed by

$$r_{jj'} = \frac{C_{jj'}}{\hat{\sigma}_j \hat{\sigma}_{j'}}$$

Substituting in Equation (2.17), we have

(2.18)
$$\hat{\sigma}_t^2 = \sum_{j=1}^{p} \hat{\sigma}_j^2 + 2 \sum_{\substack{j,j'' \\ j \neq j'}} r_{jj'} \hat{\sigma}_j \hat{\sigma}_{j'}$$

In words, the variance of the total equals the sum of the variances of the parts plus twice the sum of all possible covariances; the covariance is the product of the two standard deviations and the correlation coefficient. If for all parts, the scores are independent of those obtained on the other parts, the covariances are zero and $\hat{\sigma}_t^2 = \sum_j \hat{\sigma}_j^2$.

2.2   SEVERAL GROUPS OF SCORES

The simplest possible experimental design involves several groups of scores. Thus one might have a groups of n subjects each, which differ in the amount of reward they receive for their performance on some learning task. In setting the data down on paper, there would be a column for each level of amount of reward, that is, for each experimental group. The scores for a group could be written in order within the appropriate column. In referring to a score, we should designate it by its position in the column (or experimental group) and by the position of the column. Table 2–1 illustrates this procedure. Note that the first subscript refers to the position in the group (row), the second to the position of the group (column). Thus Y_{22} is the second score in group 2, and in general, Y_{ij} is the ith score in the jth group.

Table 2–1   A two-dimensional matrix

	GROUPS				
	Y_{11}	Y_{12}	$\cdots$ Y_{1j}	$\cdots$	Y_{1a}
	Y_{21}	Y_{22}	$\cdots$ Y_{2j}	$\cdots$	Y_{2a}
	$\vdots$	$\vdots$	$\vdots$		$\vdots$
Subjects	Y_{i1}	Y_{i2}	$\cdots$ Y_{ij}	$\cdots$	Y_{ia}
	$\vdots$	$\vdots$	$\vdots$		$\vdots$
	Y_{n1}	Y_{n2}	$\cdots$ Y_{nj}	$\cdots$	Y_{na}

Suppose we want to refer to the mean of a single column. The term used previously, $\bar{Y}$, is obviously inadequate since it does not designate the row or column that we want. Even $\bar{Y}_1$ is not clear, since it might as easily refer to the mean of the first row as to the mean of the first column.[2] The appropriate designation is $\bar{Y}_{.1} = [(1/n) \sum_i^n Y_{i1}]$; the dot represents summation over i, the index that ordinarily appears in that position. Similarly, the mean of row i would be designated by $\bar{Y}_{i.} = [(1/a) \sum_j^a Y_{ij}]$; summation is over the index j. The mean of all an scores would be designated by $\bar{Y}_{..} = [(1/an) \sum \sum Y_{ij}]$, or merely $\bar{Y}$.

Some examples using the double summation $(\sum_i \sum_j)$ may be helpful. Suppose we have

$$\sum_{j=1}^{a} \sum_{i=1}^{n} Y_{ij}^2$$

This is an instruction to set i and j initially at 1; the resulting score Y_{11} is then squared. Holding j at 1, we step i from 1 to n, squaring each score thus obtained and adding it to those previously squared. When n scores have been squared and summed, we reset the index i at 1 and step j to 2; the squaring and summing is then carried out for all Y_{i2}. The process continues until all an scores have been squared and summed. The process just described can be represented by

$$(Y_{11}^2 + Y_{21}^2 + \cdots + Y_{na}^2)$$

If we have

$$\sum_{j=1}^{a} \left(\sum_{i=1}^{n} Y_{ij} \right)^2$$

the notation indicates that a sum of n scores is to be squared. We again set j at 1, and after adding together all the Y_{i1}, square the total. The index j is then stepped to 2 and i is reset at 1; we get another sum of n scores, which is squared and added to the previous squared sum. We again continue until all an scores have been accounted for. The process can be represented by

$$(Y_{11} + Y_{21} + \cdots + Y_{n1})^2 + \cdots + (Y_{1a} + Y_{2a} + \cdots + Y_{na})^2$$

A third possibility is

$$\left(\sum_{j=1}^{a} \sum_{i=1}^{n} Y_{ij} \right)^2$$

which indicates that the squaring operation is carried out once on the total of an scores; we then have

$$[(Y_{11} + Y_{21} + \cdots + Y_{n1}) + \cdots + (Y_{1a} + Y_{2a} + \cdots + Y_{na})]^2$$

Note that the indices within the parentheses show how many scores are to be summed prior to squaring, and the indices outside the parentheses show how

[2] In the design we used for an example, the mean of the first row would not be a quantity of interest, since we stipulated that the order within each column was arbitrary. There are designs, however, giving rise to tables like Table 2–1 for which it is as interesting to obtain row means as it is to obtain column means.

Table 2–2   Some sample data

	GROUP 1	GROUP 2	GROUP 3
	4	1	6
	1	7	4
	3	2	5
	2	4	4
$\sum_i Y_{ij} = 10$		14	19
$\sum_i Y_{ij}^2 = 30$		70	93

many squared totals are to be summed. When no parentheses appear, as in $\sum\sum Y^2$, we treat the notation as if it were $\sum\sum(Y^2)$. When no indices appear outside the parentheses, it is understood that we are dealing with a single squared term, as in $(\sum\sum Y)^2$. When several indices appear together, whether inside or outside the parentheses, the product of their upper limits tells us the number of terms involved. Thus, $(\sum_{j=1}^a \sum_{i=1}^n Y)^2$ indicates that *an* scores are summed before the squaring.

Our three illustrations of the double summation can be further clarified if we use some numbers. Let us use the three groups of four scores each shown in Table 2–2. Now,

$$\sum_j \sum_i Y_{ij}^2 = 30 + 70 + 93 = 193$$

and

$$\sum_j \left(\sum_i Y_{ij}\right)^2 = (10)^2 + (14)^2 + (19)^2 = 657$$

and

$$\left(\sum_j \sum_i Y_{ij}\right)^2 = (10 + 14 + 19)^2 = 1849$$

As another example of how to use double summation, we might derive a raw score formula for the average group variance, often referred to as the *within-group mean square*. This is the sum of the group variances divided by a, the number of groups, or

$$\frac{1}{a}\left[\frac{\sum_{i=1}^n (Y_{i1} - \bar{Y}_{.1})^2}{n-1} + \cdots + \frac{\sum_{i=1}^n (Y_{ia} - \bar{Y}_{.a})^2}{n-1}\right]$$

More briefly, this average is indicated by

$$\frac{1}{a(n-1)}\sum_j^a \sum_i^n (Y_{ij} - \bar{Y}_{.j})^2$$

Now, expanding the numerator (or "sums of squares") of the above quantity, we get

(2.19) $$\sum_{j=1}^{a}\sum_{i=1}^{n}(Y_{ij}-\bar{Y}_{.j})^2=\sum_{j=1}^{a}\sum_{i=1}^{n}(Y_{ij}^2+\bar{Y}_{.j}^2-2Y_{ij}\bar{Y}_{.j})$$

We "multiply through" by $\sum_i$, noting that $\bar{Y}_{.j}$ varies only with j; it is constant when i is the index of summation. Terms are also rearranged so that sums are premultiplied by constants.

(2.20) $$\sum_{j}\sum_{i}(Y_{ij}-\bar{Y}_{.j})^2=\sum_{j}\left(\sum_{i}Y_{ij}^2+n\bar{Y}_{.j}^2-2\bar{Y}_{.j}\sum_{i}Y_{ij}\right)$$

Note that $\sum_i\bar{Y}_{.j}=n\bar{Y}_{.j}$. While $\bar{Y}_{.j}$ is a variable relative to the index j, it is a constant relative to i, the index over which we are currently summing; therefore Rule 2 applies.

Substituting raw score formulas for the group means gives

(2.21) $$\sum_{j}\sum_{i}(Y_{ij}-\bar{Y}_{.j})^2=\sum_{j}\left[\sum_{i}Y_{ij}^2+n\frac{(\sum_i Y_{ij})^2}{n^2}-2\left(\frac{\sum_i Y_{ij}}{n}\right)\sum_{i}Y_{ij}\right]$$

Simplifying gives

(2.22) $$\sum_{j}\sum_{i}(Y_{ij}-\bar{Y}_{.j})^2=\sum_{j}\left[\sum_{i}Y_{ij}^2-\frac{(\sum_i Y_{ij})^2}{n}\right]$$

which can also be written

$$\sum_{j}\sum_{i}Y_{ij}^2-\frac{\sum_j(\sum_i Y_{ij})^2}{n}$$

To simplify notation, we shall generally use T (for "total") to replace $\sum Y$. The sum of scores, for example, for group j is

$$T_{.j}=\sum_{i}Y_{ij}$$

and the raw score expression just derived can be rewritten

$$\sum_{j}\sum_{i}Y_{ij}^2-\frac{\sum_j T_{.j}^2}{n}$$

2.3   MORE THAN TWO INDICES

The principles thus far developed can be extended to any number of summations and variables. We might want to represent operations on a data set obtained from children varying in both age and socioeconomic level. Let A_j represent the jth level of age and B_k the kth socioeconomic level. If we let i take on values from 1 to n indicating children within each AB classification, the typical score is Y_{ijk}, representing the value obtained for the ith of n children at the jth age level and the kth socioeconomic level. Table 2–3 presents a sample data matrix.

Table 2–3   Some sample data involving three indices

	B_1	B_2
A_1	64	70
	78	67
	67	83
	70	81
	$T_{.11} = 279$	$T_{.12} = 301$
A_2	68	84
	82	76
	79	81
	74	70
	$T_{.21} = 303$	$T_{.22} = 311$
A_3	75	91
	84	87
	88	78
	83	86
	$T_{.31} = 330$	$T_{.32} = 342$

During many of the statistical analyses we shall make, a variety of operations will be performed on data matrices like the one in Table 2–3. We might meet the notation

$$\sum_{j=1}^{a} T_{.j.}^2$$

This implies a squared total for each of a levels of A; each such total is based on bn scores. The dot subscripts remind us that we have summed over the indices i and k. By the data of Table 2–3, we should have

$$(279 + 301)^2 + (303 + 311)^2 + (330 + 342)^2$$

Another common term is

$$\sum_{j=1}^{a} \sum_{k=1}^{b} T_{.jk}^2$$

This indicates six (ab) squared totals, each total based on four (n) scores. In our example, we have

$$(279)^2 + (303)^2 + \cdots + (342)^2$$

It is just as simple to go from verbal instructions to symbolic representation. For example, the instruction, "Add all an scores at each level of B, square the totals, and sum the squared terms," implies:

1.   There are b squared quantities to be summed, so we have $\sum_{k=1}^{b} (\quad)^2$.

2. To obtain each squared total, we have summed over the indices i and j (that is, *subjects* and levels of A at each level of B). Therefore, replace these indices with dot subscripts and get $\sum_{k=1}^{b} T_{..k}^2$.

EXERCISES

Answers are provided at the end of the book. It is extremely important to try to do each problem before turning to the answers for corroboration or help.

2.1 Write the summation of the Ys encircled below, indicating the limits.

$$Y_1 + Y_2 + \boxed{Y_3 + Y_4 + Y_5 + Y_6 +} Y_7 + Y_8$$

2.2 Simplify the following expression (k is a constant):

(a) $\sum_{i=1}^{n} (Y_i - \bar{Y})$

(b) $\sum_{i=1}^{n} (Y_i + k - n)$

2.3 Show that

$$\frac{1}{k} \sum_{i=1}^{k} (k + kY_i) = k + \sum_{i=1}^{k} Y_i$$

indicating which summation rules have been applied.

2.4 Prove that the variance remains the same if a constant is added to all scores.

2.5 Let the variance of a set of n scores be S_y^2. Prove that if each score is multiplied by C, the new variance is $C^2 S_y^2$.

2.6 Let $z_i = (Y_i - \bar{Y})/S_y$. Prove that (a) $\bar{z} = 0$; (b) $S_z^2 = 1$.

2.7 Let $D_i = X_i - Y_i$. Prove that
(a) $\bar{D} = \bar{X} - \bar{Y}$
(b) $S_D^2 = S_X^2 + S_Y^2 - 2r_{XY}S_X S_Y$

2.8 In an experiment on how several variables affect the running times of rats, there are a levels of amount of reward A, and d levels of delay of reward D, and t levels of intertrial interval T. There are n subjects in each of the adt combinations of treatments, yielding a total of $nadt$ subjects. Write the following verbal instructions in a complete and specific notational form, first clearly specifying your indices and limits of summation:
(a) Sum all scores obtained under A_1. Square the sum. Do the same for all other levels of A. Add the squared quantities.
(b) Add all the scores obtained under $D_1 T_1$. Square this sum. Do the same for each DT combination. Add the squared quantities.

2.9 Assume that each of n subjects is tested under each of a conditions. Prove that

$$\sum_{j=1}^{a} \sum_{i=1}^{n} (\bar{Y}_{i.} - \bar{Y}_{..})^2 = \frac{\sum_{i=1}^{n} T_{i.}^2}{a} - \frac{T_{..}^2}{an}$$

2.10 There are a levels of A, b levels of B, and n subjects in each of the ab combinations. Let

$$i = 1, 2, \ldots, n; \quad j = 1, 2, \ldots, a; \quad k = 1, 2, \ldots, b$$

Prove that

$$na \sum_{k=1}^{b} (\bar{Y}_{..k} - \bar{Y}_{...})^2 = \frac{\sum_k T_{..k}^2}{na} - \frac{T_{...}^2}{nab}$$

2.11

	C_1			C_2		
	A_1	A_2	A_3	A_1	A_2	A_3
B_1	14	3	4	2	1	12
	12	8	11	8	9	5
B_2	3	4	5	4	1	7
	6	7	3	2	8	6
B_3	4	⑦	2	1	2	6
	5	1	3	④	7	3

Y_{1231} Y_{2132}

Let $i = 1, 2$ (scores in cells)
  $j = 1, 2, 3$ (levels of A)
  $k = 1, 2, 3$ (levels of B)
  $m = 1, 2$ (levels of C)
Thus $Y_{2132} = 4$, and $Y_{1231} = 7$. Find

(a) $\sum_j \sum_k T_{.jk.}^2$ (b) $\sum_j \sum_k \sum_i T_{ijk.}^2$

(c) $\bar{Y}_{...2}$ (d) $\bar{Y}_{.2.}$

2.12 We have response time X and error frequencies Y for each of 3 subjects (Ss) in each of 12 experimental groups. The data are:

X Data

	C_1				C_2			
	B_1	B_2	B_3		B_1	B_2	B_3	
	4	1	8	13	6	4	1	11
A_1	3	9	6	18	8	9	12	29
	4	5	2	11	7	3	10	20
	11	15	16	42	21	16	23	60
	12	8	9	29	16	9	14	39
A_2	9	14	11	34	8	6	7	21
	13	10	7	30	15	13	11	39
	34	32	27	93	39	28	32	99

Y Data

	C_1				C_2			
	B_1	B_2	B_3		B_1	B_2	B_3	
	6	4	8	18	3	7	14	24
A_1	12	3	6	21	11	6	7	24
	9	12	10	31	8	9	12	29
	27	19	24	70	22	22	33	77

	7	6	14	27		16	10	5	31
A_2	14	2	6	22		11	12	13	36
	9	12	10	31		8	7	9	24
	30	20	30	80		35	29	27	91

Let $i = 1, 2, 3$ (the level of subject)
  $j = 1, 2$ (the level of A)
  $k = 1, 2, 3$ (the level of B)
  $m = 1, 2$ (the level of C)

Set up the calculations appropriate for the following formulas:

(a) $\sum_m T_{...m(x)} T_{...m(y)}$

(b) $\sum_k \sum_m T^2_{..km(y)}$

(c) $\sum_j \sum_k T_{.jk.(x)} T_{.jk.(y)}$

(d) $\sum_i T^2_{i...(x)}$

3 STATISTICAL INFERENCE

3.1 INTRODUCTION

Since it is generally impractical to measure the behavior of all individuals in a population, the experimenter usually confines the investigation to a small sample of subjects. An experimenter interested in the effects of amount of reward on the running times of rats might randomly assign 40 rats to two groups of 20 subjects each, one group receiving one food pellet at the end of a six-foot runway and the other receiving a reward of two food pellets. The experimenter does not primarily care whether these two groups of rats differ in performance. Rather, he is interested in assessing the relative performances of two populations of rats, systematically differing only in amount of reward for running. If the subjects can be considered random samples from the populations in which the experimenter is interested, it is reasonable to use the sample data as a basis for conclusions about the populations. Conclusions from sample data about such things as the mean and variance of the running times of large populations of rats are statistical inferences. There are several processes by which such inferences are drawn.

The fundamental problem in statistical inference is that samples are not miniature replicas of the populations from which they have been drawn. Individuals tested under the same experimental conditions will perform differently owing to differences in ability, personality, or motivation. The same individuals tested at different times will show variability in their scores because of changes in set or motivation. The consequence of such error, or chance, variability is that the measures computed from the sample—say the mean, the variance, the proportion of latencies longer than five seconds—will vary among samples and no single sample will accurately specify the characteristics of the population in which we are interested.

Consider the example in which two sizes of reward are tested for their effect on the running speed of rats. Because of error variability, it is possible that the experimental results will misrepresent the relative performances of two populations of rats differing in reward size. For large numbers of animals, the smaller reward may, on the average, result in slower running. In our experiment, however, we could have by chance placed the faster rats in the one-pellet sample.

Thus our experiment might detect no difference between the reward magnitudes; it might even lead us to conclude that the smaller reward results in faster running times.

Fortunately, there are patterns to error variability, and knowing these patterns gives us the basic tool of statistical analysis. If we have the correct statistical model—that is, if we know how scores are distributed in the population—we can state the probabilities of various sampling results under various hypotheses about the populations. Certain basic concepts underlie these inferential processes.

3.2   RANDOM VARIABLES

Our data analysis begins with sets of numbers; the numbers might be latency scores or trials to criterion for each subject in an experiment, or number of subjects falling into each of several categories, such as personality type or political preference. We designate any particular one of these numbers by the symbol Y. Since the various values of Y occur with different frequencies, we can assign a probability to each value of Y. Then Y is called a random variable.

3.2.1   DISCRETE RANDOM VARIABLES.   A random variable is discrete if it assumes only a finite, or denumerable, number of values. Suppose we have 20 indivduals classified for whether they solve a particular problem; then Y, the number of solvers, can take only the values from zero to 20. The probability that Y is one of some subset of values—that Y is, for example, greater than 15—is the sum of the probabilities of the values composing the subset; in other terms,

$$P(Y > 15) = P(Y = 16) + \cdots + P(Y = 20)$$

We shall generally designate as $p(y)$ the probability that the random variable Y takes on some specific numerical value y; this function is a *discrete density function* if it happens for all y that $\sum p(y) = 1$.

If we know the density function of Y, we know its probability distribution, the theoretical probability of each possible value of Y. Consider a set of four coins that are repeatedly tossed. If the tosses are independent (the probabilities of heads and tails do not change as a function of the outcome of any other toss), and if each coin is equally likely to come up heads or tails on any one toss, then the probability of y heads is

$$P(y) = \frac{4!}{y!\,(4-y)!} \left(\frac{1}{2}\right)^4$$

and the tabled probability distribution is

y	$p(y)$
0	$\frac{1}{16}$
1	$\frac{1}{4}$
2	$\frac{3}{8}$
3	$\frac{1}{4}$
4	$\frac{1}{16}$

These values of $p(y)$ are the proportions of sets of four tosses having y heads *in the long run* if our assumptions (independence, equal likelihood of heads and tails on a single toss) are correct. If the four coins have been tossed relatively few times, the observed distribution of Y will deviate from the theoretical distribution tabled above; I should be surprised if 16 sets of tosses yielded exactly one set with zero heads, four sets with one head, and so on. On the other hand, 1600 sets of tosses should yield observed proportions of heads within 1 or 2 percent of the theoretical values of $p(y)$.

3.2.2   CONTINUOUS RANDOM VARIABLES.   A continuous random variable is one that can take on any value within a given interval. A common example in psychological research is response time, which theoretically takes on any value between zero and infinity. Of course, observed response times usually fall between some boundaries, such as 200 milliseconds and 10 seconds. Even within such boundaries, continuity is more theoretical than real, since the best laboratory timing devices rarely record in units smaller than thousandths of a second. Generally, we shall not be able to record all the values that a continuous random variable will take on. Nevertheless, many of the statistics that figure prominently in our inferential processes are continuous variables, so the concept is important.

A logical problem arises when we try to deal with the probabilities of values of Y. This can be seen by considering a relatively crude clock capable of registering response times to within .1 sec. Times longer than .95 sec but less than 1.05 sec will be recorded as 1 sec. Suppose we now substitute a more accurate clock capable of registering time to the nearest hundredth of a second. Latencies in the interval .995–1.005 now will be recorded as 1 sec. The proportion of such latencies will clearly be smaller than the proportion in the .95–1.05 interval. Extending the argument, it should become apparent that the probability of a response time of exactly 1-sec duration is essentially zero.

Usually when Y is a continuous random variable, the concept of the probability of some exact value of Y is useless; if we are to distinguish among continuous random variables having different distributions, we need some other function of Y besides its probability for determining the distribution of Y. Such a function is $f(y)$, the *probability density function* of Y at some specific value y. To define this function, we first consider the probability that $Y < y$, which we denote by $F(y)$. Similarly, $F(y + \Delta y)$ is the probability that $Y < y + \Delta y$, where Δy is an increment to y. Then, $F(y + \Delta y) - F(y)$ is the probability that Y lies in the interval Δy. As we noted in our clock example of the preceding paragraph, this probability approaches zero as Δy becomes very small; however, the ratio $[F(y + \Delta y) - F(y)]/\Delta y$ approaches a limiting value that varies as a function of the way Y is distributed and of the particular value of Y under consideration. This limiting value is $f(y)$, the continuous density function. It measures not the probability of y, but the rate of change in the probability relative to some small change in y. By plotting $f(y)$ for a wide range of values of y, we can get a picture of the probability distribution.

When we considered discrete random variables, we noted that the probability of any subset of values of Y was given by the sum of the probabilities of the members of the subset; furthermore, the sum of all $p(y)$ had to be 1 if $p(y)$ was truly a density function. When the random variable is continuous, the probability that it lies within some subset of values, say between $Y = a$ and $Y = b$, is obtained by integration:

$$P(a \leq Y \leq b) = \int_a^b f(y) \, dy$$

For those unfamiliar with calculus, imagine dividing the area under the curve determined by $f(y)$, between the points $Y = a$ and $Y = b$, into narrow rectangular strips of height $f(y)$ and width Δy. The product of $f(y)$ and Δy roughly gives the area in the strip above Δy—that is, the probability that Y lies in the interval Δy. Integrating between the limits a and b is essentially obtaining the limiting sum of these probabilities as the interval widths become very small. To carry the analogy to the discrete case a step further, the integral over the entire range of values of Y must equal 1 if $f(y)$ is a probability density function.

3.3   EXPECTED VALUES

Frequently we shall have occasion to refer to the *expected value* of some random variable. The expected value of Y, denoted by $E(Y)$, is an average theoretically computed over all possible values of Y. To put $E(Y)$ in perspective, consider the more familiar arithmetic mean $\bar{Y}$. Suppose that $\bar{Y}$ is a discrete random variable that in a sample of nine scores takes on the values 3, 5, 5, 8, 9, 9, 9, 12, and 12. Then, $\bar{Y}$ is computed as the sum of scores, 72, divided by the number of scores, 9. We could obtain the same result by multiplying each distinct value by the proportion of its occurrences, and then summing these terms; that is,

$$\bar{Y} = (3)(\tfrac{1}{9}) + (5)(\tfrac{2}{9}) + (8)(\tfrac{1}{9}) + (9)(\tfrac{3}{9}) + (12)(\tfrac{2}{9})$$

It is this representation of the mean that the concept of expected value embodies; we generally have

(3.1) $$E(Y) = \sum y p(y)$$

where $\sum p(y)$ must equal 1. If the random variable is continuous,

(3.2) $$E(Y) = \int y f(y) \, dy$$

where the summation sign of Equation (3.1) is replaced by the integral sign and $p(y)$ is replaced by $f(y) \, dy$.

Why do we need this concept of an average? The usual arithmetic mean is appropriate when we wish to describe the average of an observed finite set of numerical values. In drawing statistical inferences, however, we are often concerned not with the average of a set of values that have been available for observation but with what the average would be in the long run if we were to collect many observations. The game roulette provides a simple example of this application of the expected value concept. The roulette wheel contains 16 black numbers, 16 red numbers, and "0" and "00," which are neither black nor red. Thus, if we bet on either color on each spin of the wheel, the probability of winning is $\frac{16}{34}$ and of losing is $\frac{18}{34}$. If each bet is one dollar, the value of the bet is either plus or minus one dollar and the average value per trial in the long run will be

$$E(Y) = (1)(\tfrac{16}{34}) + (-1)(\tfrac{18}{34}) = -\tfrac{1}{17}$$

Due to the presence of "0" and "00," the gambling house has a slight edge, and in the long run, we shall lose a little less than six cents (one-seventeenth of a dollar) on each spin of the wheel.

Other aspects of the expected distribution are also interesting. In particular, the variance over the long run can be expressed in expected values. Again, it helps to begin with the variance for an observed sample of values. This can be viewed as an average squared deviation, $\sum (Y - \bar{Y})^2/n$; its raw score formula is $\sum Y^2/n - (\sum Y/n)^2$. Note that $\sum Y^2/n$ is the arithmetic mean of Y^2 and $(\sum Y/n)^2$ the square of the arithmetic mean of Y. If we replace arithmetic means by expected values, the variance in the long run would be

(3.3)
$$\sigma^2 = E[Y - E(Y)]^2$$
$$= E(Y^2) - [E(Y)]^2$$

The quantity $E(Y^2)$ is generally referred to as the second raw moment, and for a discrete random variable, it is $\sum y^2 p(y)$. For a continuous random variable the expression is $\int y^2 f(y)\, dy$.

In Chapter 4, we shall consider some derivations involving expected values. These will be straightforward if we note that the expectation operator E follows the same rules that the summation operator $\sum$ does. For example,

$$E(X + Y) = E(X) + E(Y)$$
and
$$E(cY) = cE(Y)$$
and
$$E(X + Y)^2 = E(X^2) + E(Y^2) + 2E(XY)$$

Furthermore, E and $\sum$ are manipulated in the same way that two summation signs are. For example,

$$E\left(\sum X\right) = \sum E(X)$$
and

$$E\left[\sum (X + Y)^2\right] = E\left(\sum X^2\right) + E\left(\sum Y^2\right) + 2E\left(\sum XY\right)$$
$$= \sum E(X^2) + \sum E(Y^2) + 2\sum E(XY)$$

3.4 SOME IMPORTANT PROBABILITY DISTRIBUTIONS

One particular distribution, the F distribution, is frequently encountered in data analysis. We shall briefly examine the nature of this distribution. However, the F is closely related to two other distributions, the normal and the chi square (χ^2); in fact, the normal distribution logically precedes the χ^2, and the χ^2 logically precedes the F. Therefore we shall take up the normal and the χ^2 distributions first.

3.4.1 THE NORMAL DISTRIBUTION.   A theoretical continuous distribution that is important in statistical inference is the normal distribution, characterized by the density function

(3.4)
$$f(y) = \frac{1}{\sigma\sqrt{2\pi}} e^{-(y-\mu)^2/2\sigma^2}$$

where μ and σ are the mean and the standard deviation of the population and π and e are mathematical constants. The random variable Y can take on any value between $-\infty$ and $+\infty$ and the curve is symmetric about the mean. Not all symmetric distributions over the range of all real numbers are normal distributions, although the normal distribution may very well approximate such symmetric distributions.

The location of the distribution is determined by μ and the spread by σ; thus there are infinitely many possible normal distributions. Fortunately, all these can be reduced to a single function. Given a random variable Y, we can transform it by $z = (Y - \mu)/\sigma$; if Y is normally distributed, z will be also. The mean of the z distribution will be zero, however, and its standard deviation will be 1 (see Exercise 2.6). The density function is now

(3.5)
$$f(z) = \frac{1}{\sqrt{2\pi}} e^{-z^2/2}$$

This transformation entails no loss of information about the relative position of an individual score and is extremely helpful. We can plot or table a single normal distribution, that for z, and use it to get information about any normally distributed random variable. Table A–2 presents values of $F(z)$ (labeled $1 - \alpha$ in the table), the probability that a normally distributed random variable with mean zero and standard deviation equal to 1 will be exceeded by a given value z. Because the distribution is symmetric, only positive values of z have been tabled. Thus, if we desire the probability that z is less than -1.0, we must note that this equals the probability that z is greater than 1.0, which is approximately $1.0 - .84$. Typically, we want information about the distribution of Y; we might want to know, for example, what proportion of individuals have IQ scores greater than 130. If we can assume that such scores are normally distributed in the population, and if we know the value of the population mean and variance, Table A–2 solves our problem. If the mean is 100 and the standard deviation is 15, a score of 130 can be transformed into a z score of 2.00. Less than 2.5 percent of the populations have z scores larger than this, or if we return to the original random variable, IQ scores larger than the corresponding value of 130.

The normal distribution is prominently involved in testing hypotheses about population parameters and in estimating these parameters, since many random variables (not all) are at least approximately normally distributed. Furthermore, regardless of the distribution of Y, the distribution of the sample mean $\bar{Y}$ will rapidly approach normality as the sample size increases if the observations are independently sampled. The normal also provides a good approximation to several other distributions when there are many cases; the binomial, which is a discrete distribution, provides one example. There are several other reasons for the central importance of the normal distribution in statistical inference; of primary concern to us, however, is that if we assume that some random variable Y is normally distributed, we can derive the distribution of other random variables functionally related to Y. The chi square distribution is one illustration.

3.4.2   THE CHI SQUARE DISTRIBUTION.   Suppose that we randomly sample n values from a normally distributed population. Each value of the random variable Y is then transformed into a z score, which is then squared; the set of n squared z scores are then summed. This sum will be referred to as χ^2; that is,

(3.6)
$$\chi_n^2 = \sum_{i=1}^{n} \left(\frac{Y_i - \mu}{\sigma} \right)^2$$

If we independently draw many such samples of size n from a normal population, each time calculating χ_n^2, we have a distribution of values of χ_n^2 with a characteristic density function. The formula for the density function, $f(\chi_n^2)$, is not particularly revealing and will be omitted; the important point is that the density depends not only on the value of χ_n^2 but also on the *degrees of freedom* (df), the number of independent observations that were summed in obtaining χ_n^2. We actually have a different-shaped distribution for each value of df.

Suppose that we have one df; that is,

$$\chi_1^2 = \left(\frac{Y - \mu}{\sigma} \right)^2$$

The relation between the chi square and normal distributions is, in this instance, quite direct. Generally if some proportion P of the z scores lie between $-z$ and $+z$, then P of the values of χ_1^2 lie between zero and z^2. For example, approximately 50 percent of the z scores lie between $-.67$ and $.67$; then approximately 50 percent of the values of χ^2 lie between zero and $.45$. The first row of Table A–4, which contains values of χ^2 that will be exceeded with certain selected probabilities, verifies this observation. If the χ^2 is based on more observations, it stands to reason that larger values are more likely to occur. Thus a chi square of .45 is exceeded more than 70 percent of the time when there are two df, and more than 90 percent of the time when there are three df. These values are also based on Table A–4.

The χ^2 distribution is of interest for several reasons. A discrete random variable, generally referred to as χ^2 but actually distributed only approximately as χ^2 (which is itself continuously distributed), is useful in testing hypotheses about frequency distributions—for example, whether the proportions of cases in certain categories conform to some theory held by the investigator. The χ^2 distribution also forms the basis for drawing inferences about the population variance. This application presupposes that the ratio of the numerator of the sample variance to the population variance is distributed as χ^2 on $n-1$ *df*; that is,

(3.7)
$$\frac{\sum_{i=1}^{n}(Y_i-\bar{Y})^2}{\sigma^2}=\chi^2_{n-1}$$

This is readily proved. We begin with the identity

$$(Y_i-\mu)=(Y_i-\bar{Y})+(\bar{Y}-\mu)$$

Squaring both sides and summing over the n values, we have

$$\sum_i(Y_i-\mu)^2=\sum_i(Y_i-\bar{Y})^2+n(\bar{Y}-\mu)^2-2(\bar{Y}-\mu)\sum_i(Y_i-\bar{Y})$$

Note the application of the rules of Chapter 2. Since we proved in that chapter that $\sum_i(Y_i-\bar{Y})=0$, we can proceed with

$$\frac{\sum(Y-\mu)^2}{\sigma^2}=\frac{\sum(Y-\bar{Y})^2}{\sigma^2}+\frac{n(\bar{Y}-\mu)^2}{\sigma^2}$$

Rewriting the last term as $(\bar{Y}-\mu)^2/(\sigma^2/n)$, and noting that σ^2/n is the sampling variance of $\bar{Y}$, we find that this term is a squared z score; $(\bar{Y}-\mu)/(\sigma/\sqrt{n})$ is the deviation of a quantity $\bar{Y}$ from its expected value, divided by the standard deviation of the quantity, which meets the general definition of a z score. Furthermore, since the mean of scores that come from a normal distribution is itself normally distributed, the squared z score is a χ^2 variable on one *df*. Of course, the term to the left of the equals sign is a χ^2 variable on n *df*. Then

$$\frac{\sum(Y-\bar{Y})^2}{\sigma^2}=\chi^2_n-\chi^2_1$$

$$=\chi^2_{n-1}$$

because of the additive property of χ^2 variables. Equation (3.7) is thus verified. This result is critical in defining the F distribution.

3.4.3 THE F DISTRIBUTION. We typically use the sample statistic S^2 to estimate the value of the population variance σ^2; S^2 is defined as $\sum(Y-\bar{Y})^2/(n-1)$. In Section 3.6, we shall consider the properties of this estimator and justify the use of $n-1$ in the denominator. For the present, note that $(n-1)S^2=\sum(Y-\bar{Y})^2$, and according to the results of the preceding section,

$$\frac{(n-1)S^2}{\sigma^2}=\chi^2_{n-1}$$

or with terms transposed,

(3.8)
$$\frac{S^2}{\sigma^2} = \frac{\chi^2_{n-1}}{n-1}$$

Thus, the ratio of the sample variance to the population variance is equivalent to a χ^2 variable divided by its *df*.

Suppose that we independently drew two samples from the population and gave two independent estimates of σ^2, S_1^2, and S_2^2. The ratio of these two sample variances F is then related to χ^2; that is, from Equation (3.8),

(3.9)
$$\frac{S_1^2}{S_2^2} = \frac{S_1^2/\sigma^2}{S_2^2/\sigma^2} = \frac{\chi^2_{n_1-1}/(n_1-1)}{\chi^2_{n_2-1}/(n_2-1)} = F$$

where n_1 and n_2 are the two sample sizes. If we repeatedly draw pairs of independent samples, we can get a frequency distribution of the ratio of two χ^2s divided by their *df*. Such a distribution is called the F distribution.

Note the conditions under which we can conclude that the ratio of sample variances is distributed as F. First, we assume that the random variable Y is distributed normally; then we can prove that $\sum (Y - \bar{Y})^2/\sigma^2$ is distributed as χ^2_{n-1}, or that S^2/σ^2 is distributed as $\chi^2_{n-1}/(n-1)$. Second, we assume that we have two independent estimates of the variance of the normal population. Then the ratio of the two sample variances is an F statistic, distributed on n_1-1 and n_2-1 *df*.

Since we are dealing with ratios of variances, F can take values between zero and plus infinity. The actual density function is complicated and will not be presented. The distribution is generally asymmetric with low probabilities of large values, however, and depends on the values of df_1 and df_2, the numerator and denominator *df*s. Table A–5 presents values of F that will be exceeded with certain selected probabilities. If df_1 equals four and df_2 equals eight, for example, 1 percent of the values of F will exceed 7.01 and 10 percent will exceed 2.81.

There is little point in discussing the applications of the F statistic in this chapter. Most of the remainder of this book will do exactly that.

3.5   SAMPLING DISTRIBUTIONS

Three aspects of statistical inference are closely related—point estimation, interval estimation, and hypothesis testing. The concept of a sampling distribution is fundamental to all. To have some preliminary sense of why this is so, and of what we mean by the term *sampling distribution*, let us examine a simple example.

Assume that we randomly sample 50 individuals from some well-defined population. We ask them to rate a new brand of breakfast cereal. The rating scale ranges from 1 ("strongly dislike") to 11 ("strongly like") with 6 representing neutrality. The mean of our 50 ratings is 8.9. The matter of interest is what μ, the mean of the sampled population, is. We cannot know the answer to this. If we do know that $\bar{Y}$, the mean rating in the sample, will vary little from one sample to another however, we can be reasonably confident that our obtained estimate of

μ, which is 8.9, is close to the actual value of μ, and that the population at large did indeed clearly like the product.

The critical point is that it is useful to picture many random replications of our 50-subject sampling experiment, with each replication giving rise to a value of $\bar{Y}$. This hypothetical distribution of $\bar{Y}$ is called the sampling distribution of the mean. As we see by the above example, knowing the properties of the sampling distribution enables us to assess inferences made on the basis of a single sample. Every statistic has a sampling distribution. For the time being, however, we shall focus on the mean and variance of the sampling distribution of $\bar{Y}$, the sample mean. These two properties of the sampling distribution of the mean will be useful to know subsequently.

First, consider the mean of the sampling distribution of $\bar{Y}$, which is $E(\bar{Y})$.

$$E(\bar{Y}) = E\left(\frac{\sum Y}{n}\right)$$

$$= \left(\frac{1}{n}\right) E\left(\sum Y\right)$$

$$= \left(\frac{1}{n}\right) \sum (E(Y))$$

(3.10)

$$= \left(\frac{1}{n}\right) \sum \mu$$

$$= \left(\frac{1}{n}\right) n\mu$$

$$= \mu$$

Thus, the mean of the sampling distribution of $\bar{Y}$ is the population mean μ.

The derivation of the variance of the sampling distribution of $\bar{Y}$, namely $\sigma_{\bar{Y}}^2$, closely follows the derivation of the variance of a sum presented in Section 2.1.3. From Equations (3.3) and (3.10),

$$\sigma_{\bar{Y}}^2 = E(\bar{Y} - \mu)^2$$

$$= E\left(\frac{Y_1 + \cdots + Y_n)}{n} - \frac{n\mu}{n}\right)^2$$

$$= \frac{1}{n^2} E[(Y_1 - \mu) + \cdots + (Y_n - \mu)]^2$$

$$= \frac{1}{n^2} E[(Y_1 - \mu)^2 + \cdots + (Y_n - \mu)^2 - 2(Y_1 - \mu)(Y_2 - \mu) - \cdots$$
$$- 2(Y_{n-1} - \mu)(Y_n - \mu)]$$

Consider $E(Y_i - \mu)^2$. It is as though we have taken many samples, selected Y_i from each, and then computed $\sigma_{Y_i}^2$. The sampling distribution of the ith score should be the same for all i; more to the point, $\sigma_{Y_i}^2 = \sigma^2$ for all i. Therefore, we now have

$$\sigma_{\bar{Y}}^2 = \frac{1}{n^2}\{n\sigma^2 - 2E(Y_1 - \mu)(Y_2 - \mu) - \cdots - 2E(Y_{n-1} - \mu)(Y_n - \mu)\}$$

Terms like $E(Y_1 - \mu)(Y_2 - \mu)$ are population covariances. Such an average cross-product is the numerator of a correlation coefficient. It is as though we were to draw many random samples, retaining only the ith and i'th score from each, and then computing a correlation of Y_i and $Y_{i'}$. If scores are independently distributed (meaning that the probability that Y_i has any particular value is independent of the value of $Y_{i'}$), this correlation will be zero. More precisely, the covariance will be zero, and we now have

(3.11) $$\sigma_{\bar{Y}}^2 = \frac{\sigma^2}{n}$$

The above derivation underlines the importance of assuming that random variables are independent. Many of the formulas we use are special cases of more general forms, resulting when cross-product terms go to zero because of the independence assumption. If that assumption is false, the inferences based on our formulas may not be valid.

Equation (3.11) feels right intuitively. If $n = 1$, the sampling distribution of $\bar{Y}$ is the population distribution of Y, since each sample mean corresponds to a single score. In this case, $\sigma_{\bar{Y}}^2$ should equal σ^2, as Equation (3.11) implies. On the other hand, as n becomes very large, the composition of the sample should not change much with repeated samplings; thus $\bar{Y}$ should vary little, and consequently, $\sigma_{\bar{Y}}^2$ should approach zero, as Equation (3.11) implies.

3.6 POINT ESTIMATION

A quantity that can be computed from the population data, such as the population mean or the population variance, is generally referred to as a *population parameter*. Throughout this book, Greek letters will be used to denote parameters. The population mean will be indicated by μ (*mu*) and the population standard deviation by σ (*sigma*). Similar quantities computed from the sample data are referred to as sample *statistics*, and Latin letters will be used to denote these. Thus the sample mean is represented by $\bar{Y}$ and the standard deviation by S. One purpose of experimenting is to get estimates of the magnitude of population parameters from sample statistics. This kind of inference is often referred to as *point estimation*, to distinguish it from the estimation of an interval containing the parameter.

3.6.1 PROPERTIES OF ESTIMATORS.

An infinite number of possible estimators of any single population parameter exist. The population mean might, for example, be estimated by the sample mean, the sample median, or even the first score drawn from the sample. Which quantity best estimates the parameter can be decided by establishing criteria for good estimators and then examining how closely various estimators meet these criteria. The criteria that are generally agreed on are based on knowledge of the sampling distribution of the estimator. If the value of the sample mean were computed for each of 20 samples independently drawn from the same population, the result would be a distribution of values. The same would be true of the median or mode or any other sample statistic. It seems best to choose an estimator whose sampling distribution has a small variance. Any one estimate, then, will have a high probability of being close to the parameter value being estimated. It also seems desirable that this requirement should have greater probability of being met as the size of the sample is increased; increased information should result in increased reliability. The following criteria embody the properties just described.

   Lack of bias. If the mean of the sampling distribution of a statistic equals the parameter being estimated, the statistic is said to be an *unbiased estimator* of the parameter. A familiar example of such an estimator is the sample mean, $\bar{Y}$. Equation (3.10) expresses this fact. The statement that $\hat{\theta}$, an estimator of some parameter θ, is unbiased can usually be expressed by

(3.12) $E(\hat{\theta}) = \theta$

For a finite population, the population variance σ^2 is defined as $\sum (Y - \mu)^2/N$, the sum of the squared deviations of scores about the population mean divided by the number of scores in the population. A logical estimator of σ^2 seems to be $\sum (Y - \bar{Y})^2/n$, the sum of the squared deviations about the sample mean divided by the number of scores in the sample. This estimate is biased, however, since it can be proved that

(3.13) $E\left[\dfrac{\sum (Y - \bar{Y})^2}{n}\right] = \left(\dfrac{n-1}{n}\right)\left[\dfrac{\sum (Y - \mu)^2}{N}\right] = \left(\dfrac{n-1}{n}\right)\sigma^2$

Multiplying both sides of Equation (3.13) by $n/(n-1)$ yields an unbiased estimator, since

(3.14) $E\left[\dfrac{\sum (Y - \bar{Y})^2}{n-1}\right] = \sigma^2$

meeting the condition set forth in Equation (3.12). Because there is no bias, the estimator with $n-1$ in the denominator is most frequently recommended.

   Consistency. An estimator is said to be consistent if

(3.15) $P(\hat{\theta} \to \theta) \to 1$ as $n \to \infty$

This expression is read, "The probability approaches 1 that the estimator approaches the parameter as the sample size approaches infinity." An alternative statement is that the probability of obtaining an estimate closer to the estimated parameter increases as the sample size increases. An example of a consistent estimator is the sample mean $\bar{Y}$. It does not follow, however, that an estimator must be unbiased to be consistent. If an estimator is to approach the population parameter as n increases, it must be unbiased only for very large values of n. The quantity $\sum (Y - \bar{Y})^2/n$ is an example of a biased estimator that is unbiased when the sample size is very large, and that is consistent. It follows from Equation (3.13) that this estimator is unbiased for large n, since $(n-1)/n \to 1$ as $n \to \infty$; for such values of n, the expression states, the expected value of the estimator equals the parameter. Nor does it follow that all unbiased estimators are consistent. Mood (1950) points out that $E(Y_1) = \mu$ where Y_1 is the first score drawn from the sample. However, Y_1 is not a consistent estimator of μ. The value of Y_1 is not necessarily equal to μ even when the entire population is sampled.

Efficiency. In any choice between two estimators of a parameter, that estimator whose sampling distribution shows less spread about the parameter is preferred. Assume, for example, that numerous samples are obtained from a normally distributed population, a mean and a median are computed for each sample, and sums of squared deviations about μ are then computed, one for the sample means and one for the sample medians. The variability of the sample means about μ will be 64 percent of the variability of the sample medians about μ. This is expressed by saying that the *relative efficiency* of the median to the mean (as estimators of μ) is 64 percent. Conversely, the relative efficiency of the mean to the median is $1/.64$, or 157 percent. The relative efficiency of $\hat{\theta}_1$ to $\hat{\theta}_2$ when these are two different estimators of θ is expressed by

$$\frac{E(\hat{\theta}_2 - \theta)^2}{E(\hat{\theta}_1 - \theta)^2}$$

Thus, relative efficiency is defined as the ratio of two averages of squared deviations of estimators about the same population parameter. Note that this is a measure of the efficiency of the estimator in the denominator relative to that in the numerator.

Since all the properties that have been described are not always present in one estimator, some conclusion about the relative importance of these properties is required. Although lack of bias is intuitively appealing, a biased estimator may be very satisfactory if its variability about the parameter is slight compared with the variability of other estimators and decreases with increasing sample size. The relative importance of efficiency and bias are shown in Figure 3–1; although the frequency distribution of $\hat{\theta}_1$ centers about θ, the probability of getting an estimate closer to θ is greater when θ_2 is the estimator, even though θ_2 is somewhat biased.

3.6.2   PRINCIPLES OF ESTIMATION.   Once it is agreed that the properties lack of bias, consistency, and efficiency are desirable, estimators having such properties must be derived. Two principles lead to such derivations.

Maximum likelihood estimation. Assume that we want to estimate the probability of obtaining heads on the toss of a coin. This is essentially the problem of estimating the parameter π, the proportion of heads in an infinite population of tosses. Now suppose that a sample of 10 independent tosses has resulted in 7 heads and 3 tails. The maximum likelihood principle states that the best estimate is the one giving the highest probability, the maximum likelihood, of obtaining the

Figure 3–1   Sampling distributions of two estimators of θ.

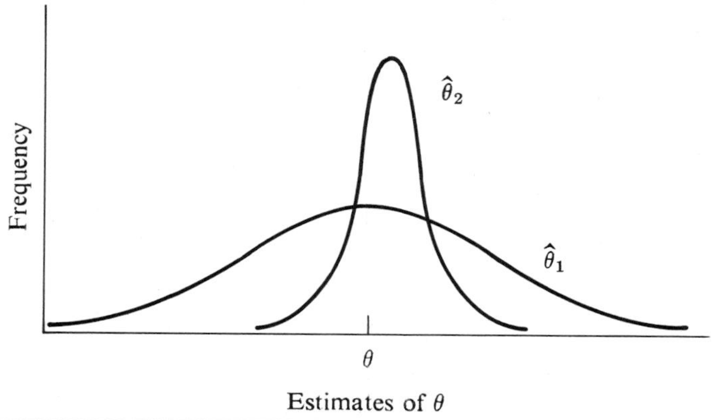

Estimates of θ

observed data. Clearly $\pi = 1$ or $\pi = 0$ is an exceedingly poor estimate according to this principle, for if $\pi = 1$, the sample should have consisted of all heads, and if $\pi = 0$, all tails would have resulted. Somewhere between zero and one is an estimate that makes the occurrence of the observed sequence of results most likely. The probability of the obtained sequence of outcomes P is related to π by

(3.16) $$P = \pi^7(1 - \pi)^3$$

Table 3–1 is the result of substituting different estimates of π into Equation (3.16). When $\pi = .7$, the likelihood of obtaining the observed data is maximal, since P then takes on its largest value. The maximum likelihood estimate is therefore .7.

It is not necessary to work with an actual set of data or to enumerate likelihoods for various estimates. Differential calculus is applied to the equation describing the distribution of the data, and a general formula for the maximum

Table 3–1   The probability P of obtaining seven heads and three tails in the observed sequence for various values of $\hat{\pi}$

$\hat{\pi}$	P
03	0
.1	.0000000729
.2	.0000065535
.3	.0000750133
.4	.00035390
.5	.00097656
.6	.0017886
.7	.0022236
.8	.0016778
.9	.0004783
1.0	0

likelihood estimator is obtained. In our example, Y/n would be the general solution, where Y is the observed number of heads and n is the number of tosses.

Maximum likelihood estimators have the important property of being as efficient as any other estimator, or more efficient, and consistent also. Although a maximum likelihood estimator is not necessarily unbiased, the center of its distribution is generally close to the value of the parameter being estimated.

Least-squares estimation. Assume that any score in a sample is related to μ, the mean of the population, by

(3.17) $$Y_i = \mu + e_i$$

where Y_i is the score of subject i, and e_i is subject i's error component, the deviation of the score from μ. We can obtain a measure of variability by computing $\hat{e}_i (= Y_i - \hat{\mu})$ for all individuals in the sample, squaring these quantities, and summing them. The least-squares principle states that the appropriate estimate of μ is the value that makes the sum of squared deviations as small as possible, that is, that minimizes the variability about $\hat{\mu}$. For example, the least-squares estimate is $\bar{Y}$ and is arrived at by applying differential calculus to find the minimum point of the function relating the sum of the $\hat{e}_i^2$ to various values of $\hat{\mu}$. The same approach can be used to find estimators for other parameters. If the e_i are uncorrelated and their frequency distribution is normal, the method yields results essentially like those of maximum likelihood estimation. The least-squares computations are usually much simpler, however.

Figure 3–2 illustrates the two estimation procedures just discussed. On the top is plotted the probability of obtaining the observed set of data as a function of various estimates of the population parameter. The estimation problem reduces to finding the maxima of the function, the value of $\hat{\theta}$ for which P is greatest. On the bottom is plotted the error variance, the variance of the data points about the estimated parameter, as a function of various estimates. Now the estimation problem reduces to finding the minima of the function, the value of $\hat{\theta}$ for which the error variance is least. The functions drawn are hypothetical and will vary depending on the parameter to be estimated and the distribution of the data. The principles implied are general ones, however.

3.7 INTERVAL ESTIMATION

A point estimate will almost never be correct in the sense of equaling the parameter. Nor can the extent of error of estimate be judged from the estimate itself, since this reflects no information about sampling distribution. To evaluate the adequacy of a point estimate, it is desirable to have an estimate of certain limits within which the parameter falls and a quantitative statement of confidence that the parameter does fall within these limits. Such confidence intervals can be established for many different parameters, but at present establishing such limits will be exemplified by obtaining limits for μ.

Figure 3–2   Two approaches to parameter estimation.

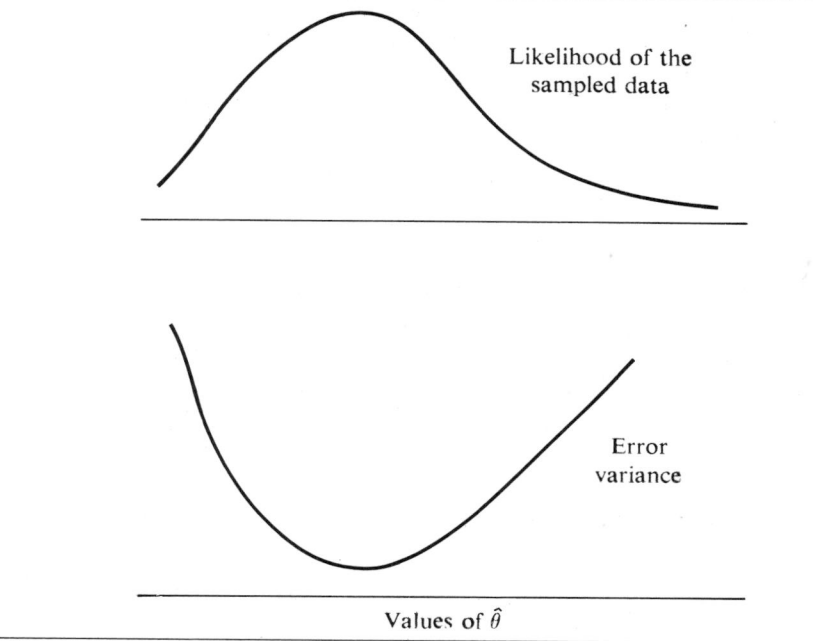

Likelihood of the
sampled data

Error
variance

Values of $\hat{\theta}$

Determining a confidence interval for μ requires the following information:

1. Value of the sample mean.
2. Value of the population standard deviation or some estimate of it.
3. Equation for the frequency distribution of scores in the population.
4. Size of the sample.
5. Degree of confidence required.

The experimenter decides the last item, which will always be a number between zero and 1, generally .90 or above. For our example, we shall assume that:

1. $\bar{Y} = 25$.
2. $\sigma = 5$.
3. The distribution is known to be normal.
4. $n = 100$.
5. 95 percent confidence is desirable.

By Section 3.4.1, if Y is normally distributed, the quantity

(3.18)
$$z = \frac{\bar{Y} - \mu}{\sigma/\sqrt{n}}$$

is also normally distributed, with mean equal to zero and standard deviation equal to 1. The distribution of z scores, with percentages lying between the mean and

Figure 3–3   Normal distribution of *z* scores.

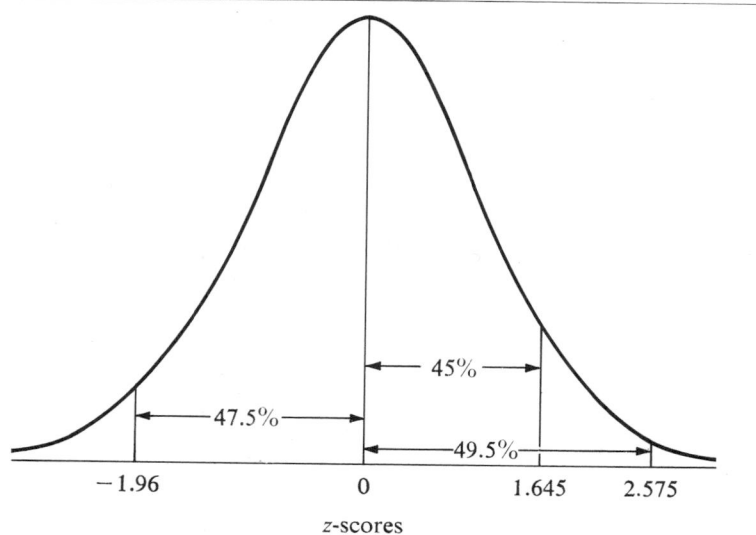

several selected values, is presented in Figure 3–3. That 95 percent of the *z* scores lie between −1.96 and 1.96 can be interpreted as follows; if 100 samples of scores are collected from a normal distribution, and if the means of these samples are transformed into *z* scores according to Equation (3.18), 95 of them would be expected to fall between −1.96 and 1.96. Equation (3.19) expresses this more succinctly.

(3.19)
$$P\left(-1.96 < \frac{\bar{Y} - \mu}{\sigma/\sqrt{n}} < 1.96\right) = .95$$

which is read, "The probability that the *z* transform of the sample mean lies between −1.96 and 1.96 is .95." Algebraic manipulation of Equation (3.19) results in Equation (3.20), which states the probability that μ lies within the designated limits.

(3.20)
$$P\left(\bar{Y} - \frac{1.96\sigma}{\sqrt{n}} < \mu < \bar{Y} + \frac{1.96\sigma}{\sqrt{n}}\right) = .95$$

Plugging in the appropriate values from our example results in the 95 percent confidence interval, $24.02 < \mu < 25.98$. We have 95 percent confidence in the sense that while 100 samples might give rise to 100 different sets of limits, 95 percent of such intervals should contain μ. We therefore have high confidence that the one interval actually derived does contain μ.

Three variables that affect the width of the confidence interval, and consequently the precision of the estimate, are σ, *n*, and the level of confidence. It can be seen in Equation (3.20) that as variability σ decreases, the upper limit of the interval is lowered and the lower limit is raised (becomes less negative). This reduced interval width reflects the increased precision accompanying decreased variability. To see the relation between interval size and σ, suppose that σ is 2.5, rather than 5 as in the original example. The limits are now $25 \pm (1.96)(.25)$, or 24.51 and 25.49.

The width of the interval containing μ also decreases as n increases. If n were 200 (σ, $\bar{Y}$, and the confidence level are again 5, 25, and .95, respectively), the limits would be $25 \pm (1.96)(.354)$, or 24.31 and 25.69.

Increased confidence is paid for by wider interval widths. If the 99 percent confidence interval were desired, 2.575 would be substituted for 1.96 in the previous calculations. (Note in Figure 3–3 that 99 percent of the z scores fall between 2.575 and -2.575.) By the original values of σ, $\bar{Y}$, and n, the interval is now 25 ± 1.29, or 23.71 and 26.29. We have greater assurance than originally that μ falls within the specified interval, but the interval is wider than before and the interval estimate is therefore less precise.

3.8 HYPOTHESIS TESTING

Psychologists have generally been more concerned with evaluating specific hypotheses about the value of the parameter than with estimating the parameter (point or interval). Thus, they have usually asked questions in the form, Does μ equal 100? instead of, What does μ equal? Or, Do these treatments differ in effect? instead of, What is the difference in effect among these treatments? In looking now at an example of how a specific hypothesis would be tested, we are concerned with those steps common to various statistical tests (such as t, χ^2, or F), although we have had to confine ourselves to one particular test to exemplify the inferential process.

Twenty rats are trained to run to a box containing food; the food box is white for half the rats, black for the other half. On the day following the last training day, all rats are tested for their preference between a white and a black box, neither of which contains food. The purpose is to determine whether a preference has been established for the box previously associated with food. To reach a conclusion, it is first necessary to state explicitly two hypotheses, the *null hypothesis*, H_0, and the *alternative hypothesis*, H_1. An appropriate null hypothesis is that in the population from which these subjects are a sample, the percentage of correct responders (subjects who prefer the box that previously contained food) equals the percentage of incorrect responders. This might be expressed by

$$H_0 : \pi = .5$$

where π is the proportion of correct responders in the population. The alternative hypothesis might be that in the population sampled, the proportion of correct responders exceeds the proportion of incorrect responders. This may be stated

$$H_1 : \pi > .5$$

This statement of H_1 ignores the possibility that there is a majority of incorrect responders in the population. Such a statement implies that the experimenter is willing to allow that there are by chance less than 50 percent incorrect responders in the sample, that such an occurrence will be viewed as support for H_0. This situation is described as a test of H_0 against a one-tailed alternative, or simply, a one-tailed test.

Suppose the experimenter considered it possible that a majority of incorrect responders does exist in the population. Possibly, the box that did not contain food is reinforcing because it is a novel stimulus; the subject has not experienced it before. In this situation, H_0 should be tested against a two-tailed alternative, namely

$$H_1 : \pi \neq .5$$

In discussing various aspects of hypothesis testing in this example, we shall first assume a one-tailed procedure. Changes in the testing procedure that arise when H_1 is two-tailed will then be noted.

The choice between H_0 and H_1 requires that a test statistic be computed from the data. Such a statistic should have the following properties:

1. The statistic should reflect the relative merits of the two hypotheses.
2. Its distribution should be obtainable under the assumption that H_0 is true.
3. The assumptions necessary to derive the distribution should be reasonably valid for the data at hand.

Of the various statistics that meet these criteria for our example, the number of correct responders seems the least complicated and therefore the best for illustration. How does the number of correct responders meet our three requirements?

1. If the null hypothesis is true, then the average number of correct responders (over many replications of the experiment) will be 10. If the true value of π is greater than .5, the average number of correct responders will be greater than 10. This monotone-increasing relation between π and the expected value of the test statistic meets requirement 1 above.
2. If the performances of the 20 rats are independent, the probability that the number of correct responders equals x is

$$(3.21) \qquad P = \frac{n!}{x!\,(n-x)!}\,\pi^x (1-\pi)^{n-x}$$

where n is the number of subjects, x is the number of correct responders, and π is the proportion of correct responders in the population. In our example, $n = 20$; and under the null hypothesis, $\pi = .5$. The probability distribution can now be easily obtained by inserting these values of n and π into Equation (3.21) and letting x take on the values $0, 1, \ldots, 20$. The resulting probabilities of various values of x can be found in the column headed "$\pi = .5$" in Table 3–2 and in the solid-line histogram of Figure 3–4.

Table 3–2   Probability of the number of correct responders when $n = 20$ for this value of π

NUMBER CORRECT	$\pi = .5$	$\pi = .75$
0	.000	.000
1	.000	.000
2	.000	.000
3	.001	.000
4	.005	.000
5	.015	.000
6	.037	.000
7	.074	.000
8	.120	.001
9	.160	.003
10	.176	.010
11	.160	.027
12	.120	.061
13	.074	.112
14	.037	.169
15	.015	.202
16	.005	.190
17	.001	.134
18	.000	.067
19	.000	.021
20	.000	.003

3. Since each subject can be considered an independent replication of the same experiment, yielding only two classes of responses (correct, incorrect), the data seem appropriate to the assumptions underlying the application of Equation (3.21).

3.8.1 TYPE I ERRORS.

The null and alternative hypotheses have been stated, and a test statistic has been selected to provide a basis for choosing between them. The next step is determining those values of the statistic that will result in rejection of H_0 in favor of H_1. Such values constitute a critical region, a set of possible values of the test statistic, consistent with H_1, that are so improbable if H_0 is assumed true that their occurrence leads us to reject H_0. An arbitrarily chosen value, alpha (α), defines exactly how improbable "so improbable" is. More precisely, α, the *level of significance*, is the probability of obtaining a test statistic that falls within the critical region when H_0 is true. For example, if α equals .05, the test statistic can be expected to fall within the critical region in 5 of 100 replications of an experiment for which the null hypothesis is true, thus resulting in rejections of true null hypotheses in 5 percent of the replications. The rejection of a true null hypothesis is generally referred to as a *Type I, or α, error*.

The size of the critical region will clearly depend on the size of α. We have stated that this region is also a set of values consistent with H_1. This means that

Figure 3–4   Distributions for $\pi = .5$ and $\pi = .75$.

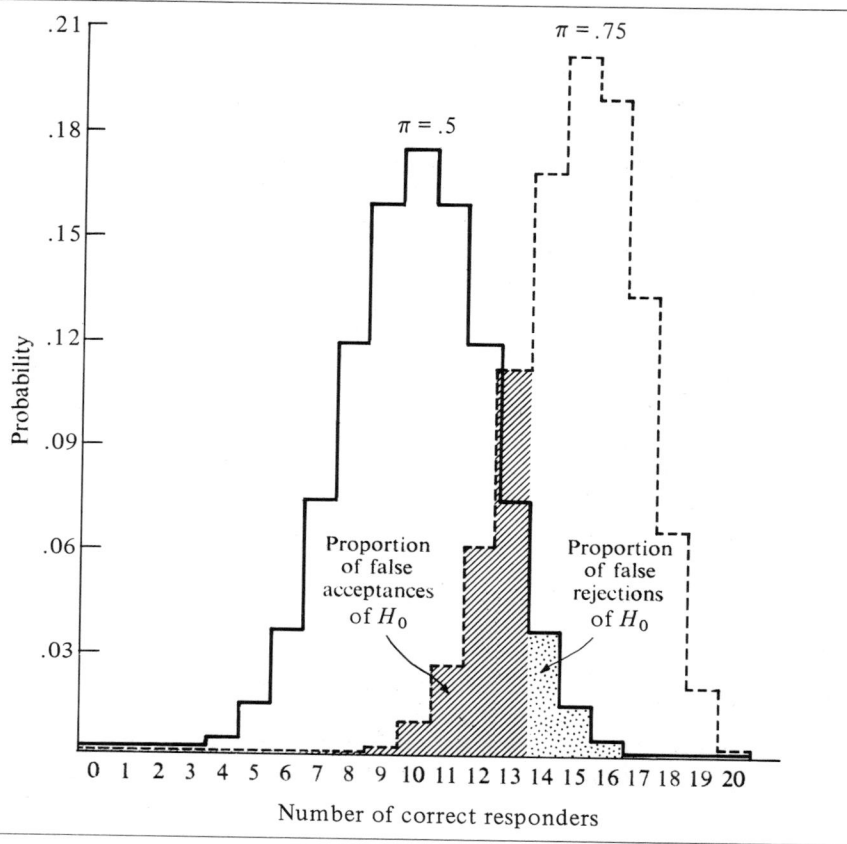

the critical region is so chosen that its location is consistent with H_1. If the alternative hypothesis is, for example, that π is greater than .5, the critical region should include only numbers of correct responders greater than half the sample size. If the alternative is two-tailed, the critical region should include numbers of responders both greater than half the sample size and less.

In the example of the 20 rats faced with a black-white preference test, assume that α has been set equal to .06. From the null hypothesis that the two choices are equally likely and the alternative that the correct side is preferred, the critical region can now be determined for the distribution of Table 3–2 and Figure 3–4 ($\pi = .5$). Since the probability of 14 or more responders is almost 6 percent $(.037 + .015 + .005 + .001 = .058)$ in the right-hand tail of the distribution, this is the critical region. If 14 or more subjects respond correctly, H_0 will be rejected. This region of rejection is indicated in Figure 3–4.

What if the alternative were two-tailed—if the proportion of correct responders in the population were greater than .5 or less? If it is equally important to detect deviations in both directions, and if it is still required that the proportion of cases in the critical region should not exceed 6 percent, 5 or fewer or 15 or more correct responders would result in rejection of H_0.

To summarize briefly, tests of the null hypothesis require explicitly stating the null and alternative hypotheses, choosing a statistic, and knowing the distribution of the statistic if H_0 is assumed true. These things, together with a stated significance level, result in the selection of a critical region. All these decisions must be made before collecting data; the availability of experimental data could influence the statement of hypotheses or the selection of a critical region. Furthermore, if the test statistic and the assumptions underlying its distribution are not considered before the experiment is carried out, the experiment may produce data that are difficult to analyze properly. Once the critical region has been decided, the data can be collected, the appropriate statistic computed, and its value compared with those falling within the critical region.

3.8.2 TYPE II ERRORS.   Since α is the probability of rejecting true null hypotheses, why not set it extremely low? The answer is that decreases in α decrease the probability of rejecting both true and false null hypotheses. In the extreme case where α is zero, Type I errors would never be made; and also, false null hypotheses would never be rejected. Obviously, the failure to reject a false null hypothesis is also an error. This failure to establish a treatment effect when it actually exists is generally referred to as a *Type II*, or *beta* (β), *error*. The example of the 20 rats in the preference test will be used to show the principles involved in computing β, the probability of a Type II error, and also to examine the relations between β and such variables as α and n.

Once again, assume that the null hypothesis $\pi = .5$ is being tested against the alternative that $\pi > .5$. Assume also that unknown to the experimenter, H_0 is false and the proportion of correct responders in the population is actually .75, or $\pi = .75$. The probability distribution for $\pi = .75$ and $n = 20$ is presented in Table 3–2 and in Figure 3–4 (dashed-line histogram). If $\alpha = .06$, as previously stipulated, the experimenter will accept H_0 whenever 13 or fewer rats respond correctly. If π really equals .75, the probability of 13 or fewer rats responding correctly is $.001 + .003 + .010 + .027 + .061 + .112 = .214$. Thus β, the probability of accepting a false null hypothesis (false because $\pi = .75$ is the true state of affairs in the population), is .214. To summarize, the procedure for determining β against a specific alternative value of the parameter consists in dividing the baseline under the H_0 function into a region of acceptance and a region of rejection, and then determining the probability that the statistic will fall in the acceptance region, for an alternative that is true.

Interpreting β by Figure 3–4 may also be helpful. Beta equals that proportion of the total area under the $\pi = .75$ function labeled "proportion of false acceptances of H_0." This region is that part of the $\pi = .75$ distribution lying above abscissa values that also underlies the $\pi = .5$ function but excludes the critical region of the $\pi = .5$ distribution.

Several sources (Cohen, 1969, for instance) contain tables or graphs for *operating characteristics* (OC), or *power* functions, for some of the common statistical tests. The OC function consists of values of β for the range of possible alternative values of the population parameter. The quantity $1 - \beta$ often appears instead of β. This is commonly referred to as the power of the test and is the probability of rejecting H_0 when it is false. Power, or OC, functions are important because several tests of the same H_0 may exist but have different power for the same critical region and sample size. If all other factors (like validity of the model for the data, computational ease, and availability of tables) are equal, the more powerful test is preferred. Furthermore, power and OC functions yield information on the relation of power to α, H_1, and sample size, thus making it easier to decide about these variables.

Concerning the relation between α and β, if the critical region of Figure 3–4 were reduced, the region of false acceptances of H_0 would consequently increase. Contrarily, if α were increased, enlarging the critical region, there would be a corresponding decrease in the size of the region of false acceptances of H_0, and therefore in β. Because of this inverse relation between α and β, the choice of α should always reflect a compromise between the relative importance of Type I and Type II errors. Sometimes Type I errors will be more undesirable. When the consequences of a significant finding (and such consequences may involve applying the findings or future experiments) will be costly in time, effort, and money, the experimenter wants to be sure that an effect really does exist before exploring further. In such an instance, α may be set extremely low, guarding against Type I errors even at the increased risk of Type II errors. The relation between one's own experimental findings and findings of previous studies is also a consideration. If a significant result will conflict with an established body of knowledge, more stringent significance levels (say .01) might be required. On the other hand, in research in which the variables influencing behavior are less well understood, the experimenter might be willing to take a greater risk of a Type I error, reducing β to avoid missing some promising lead.

These comments are guides more than rules for setting α. There are no nice, neat formulas for arriving at the appropriate level of α. Since our inferences are so arbitrary, it follows that judging experimental effects should not depend solely on statistical significance; it is just an important piece of information. Inferences should not be ground out by a computer; they should be thought out by an experimenter. (We shall return to this problem in the last part of this chapter.)

The choice between one-tailed and two-tailed tests also has implications for β. If α is kept constant, the critical region of Figure 3–4 must be reduced to establish a similar region in the left-hand tail of the $\pi = .5$ distribution. The shift from a one-tailed to a two-tailed alternative permits the detection of π values less than .5, but only at the cost of reduced power to detect π values greater than .5.

Figure 3–5 presents distributions of correct responders for the null hypothesis $\pi = .5$ and for the alternative $\pi = .75$ with α equal to .06 and n equal to 15. Comparing these distributions with the distributions of Figure 3–4, one notes that the decrease in n from 20 to 15 has brought about an increase in the

Figure 3–5   Probability distributions of correct responders when $n = 15$.

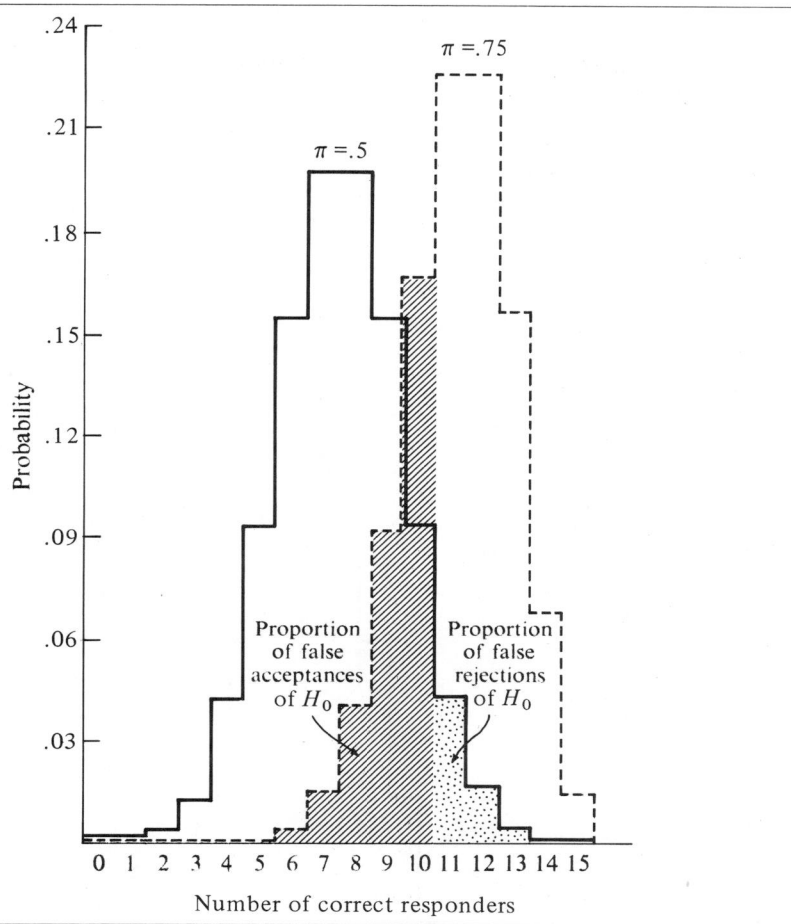

proportion of area under the $\pi = .75$ function, which consists of false acceptances of H_0. Beta has shifted from 21.4 percent to 31.4 percent. A reduction in the amount of data increases the overlap between the two distributions, consequently reducing the ability to detect false null hypotheses. The contrary is also true; increases in n result in decreases in β and increases in power.

You can get some concept of how β varies as a function of the true value of the parameter by looking at Figure 3–6, which shows probability distributions for $\pi = .5$ and for $\pi = .9$. In both cases, α again equals .06 and n equals 20. Note that the proportion of false acceptances of H_0 constitutes only .2 percent of the area under the $\pi = .9$ function, compared with 21.4 percent for the $\pi = .75$ function in Figure 3–4. This result makes good sense; the further apart the null-hypothesized and true parameter values, the more easily detected the false null hypotheses.

Figure 3–6   Distributions of correct responders for $\pi = .5$ and $\pi = .9$.

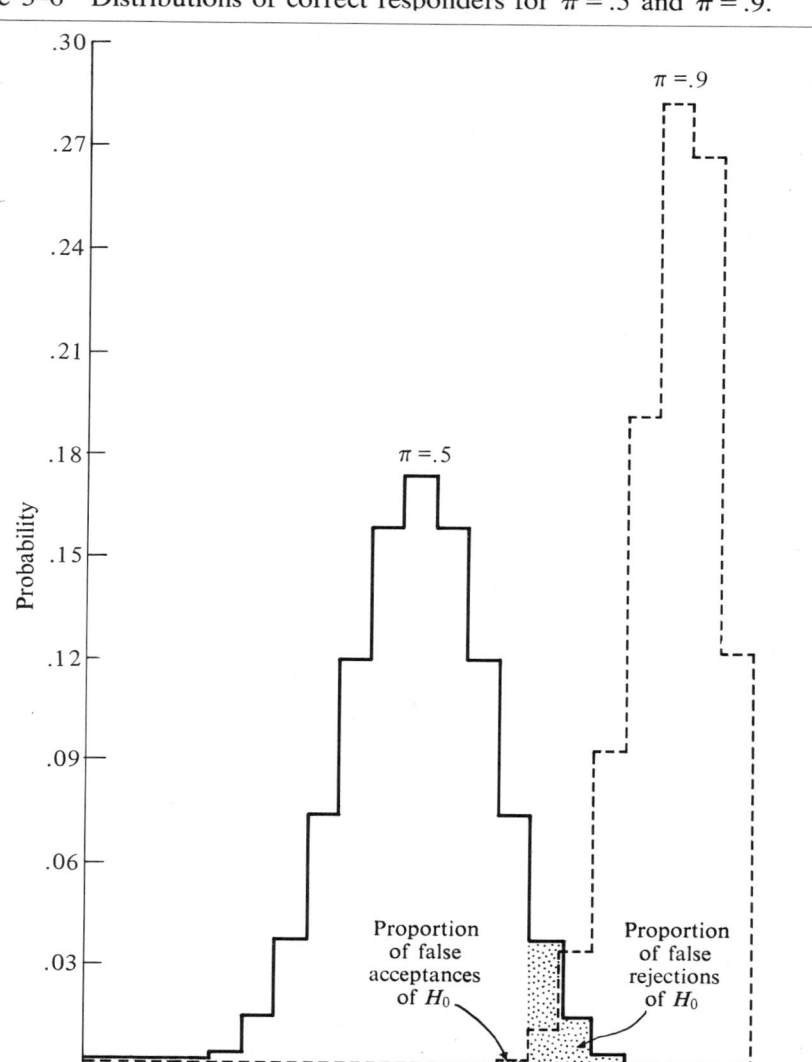

Number of correct responders

To summarize, β is reduced (and power consequently increased) as the difference between the value of the parameter under H_0 and the true value of the parameter increases, as α increases, and as n increases. These relations are negatively accelerated; that is, the reduction in β becomes less marked as the value of the parameter increases. Thus, for example, the reduction in β is much smaller when n is increased from 90 to 100 than when n is increased from 10 to 20. Besides these effects just cited, a one-tailed test will always be more powerful than a two-tailed test against alternatives in that tail, but less powerful against alternatives in the second direction. *In fact it has zero power there*

Beta differs from α in the sense that β can be selected only for some minimal effect. To say that one will set β equal to .10 (as we might set α equal to .10) is meaningless; one can only require that β equal .10 for some stated minimal degree of falsity of the null hypothesis. It is not enough to select a level of β (or power); one must also decide what deviations from the parameter value assumed under H_0 it is important to detect. In the example of the preference experiment with rats, we might ask how much greater than .5 the value of π must be before it becomes important to reject the null hypothesis. It may seem that it is important to detect any deviation from the value of the parameter assumed under H_0. It is rare, however, that the null hypothesis is precisely true. Since that is so, we wish to design the experiment so that the rejection of H_0 will not be a trivial result, merely reflecting a very large amount of data. Our interest lies not in whether H_0 is false (it usually is) but in whether the extent of the falsity is important. Suppose that we want to compare how traditional classroom procedures and teaching machines affect the rate of learning arithmetic. The null hypothesis might be that the two methods are equally effective, and the alternative hypothesis that using teaching machines is more effective. It might be considered important to reject H_0 only if the teaching machine were so much more effective that the added expense of developing teaching machine programs and of building machines was judged to be outweighed.

Now that we know about β, we have a rationale for deciding on how many subjects to run. We need a large enough n to achieve a desired degree of power for rejecting H_0 against some specified alternative, given that the probability of a false rejection is α. The specific steps in deciding on n are the following:

1. State the null and the alternative hypotheses. Continuing with our example of the rat experiment, we have

$$H_0 : \pi = .5$$

$$H_1 : \pi > .5$$

2. Decide on the level of α. Assume that α is set equal to .06.
3. Decide on the minimal degree of falsity of H_0 that is important to detect. In our example, assume that if $\pi \geq .75$, it is important to reject H_0.
4. Decide on the level of power, the probability of rejecting H_0 if the true parameter value deviates to the extent specified under step 3. Assume that if $\pi \geq .75$, we want at least an 80 percent chance of rejecting H_0. Consequently, power $= .80$ and $\beta = .20$.

From the outcomes of the above decisions, the required n can be determined. Turning to Figure 3-4, we note that if $\alpha = .06$, $\pi = .5$ under H_0, $\pi = .75$ under H_1, and $n = 20$, then $\beta = .214$. To have $\beta = .20$, approximately 21 subjects would be needed. The steps indicated are essentially the same for all statistical tests. The main difference in the steps between the simple binomial test developed in this chapter and others such as the F and t tests is that the last two require some estimate of the population variance.[1]

3.9   CONFIDENCE INTERVALS AND SIGNIFICANCE TESTS

After having treated confidence intervals and significance tests separately, let us now look at some similarities and differences in the two procedures. Both techniques involve the same assumptions, manipulations of the same quantities, and the same definitions of probability—the probability of an event is the relative frequency of occurrence of the event. Despite the similarities, the two procedures do lead to somewhat different presentations of the information obtained from the data. In a detailed comparison of the two procedures, the following example is used.

Twenty-five rats are run for 500 trials in a **T** maze. One goal box contains food in 60 percent of the trials; the other goal box contains food in the remaining 40 percent. One theory of maze behavior leads the experimenter to predict that the rats will run to each goal box with the same relative frequency that the goal box contains reward: 60 percent of the responses will be made to the more frequently rewarding goal box and 40 percent to the less frequently rewarding box. To assess this "matching" hypothesis, the experimenter decides to test the null hypothesis that for the last block of 20 trials there are an average number of 12 runs (60 percent) to the more frequently rewarding goal box in a population of rats from which the 25 subjects have been sampled. The t test is the statistic chosen, the alternative hypothesis is that $\mu \neq 12$, and α is chosen as .05. The corresponding critical region consists of values of t less than -2.06 and greater than $+2.06$.

The mean number of runs to the more frequently rewarding goal box is 11, and the standard deviation is 3. Inserting these values, together with the sample size, 25, into the following equation

$$(3.22) \qquad\qquad t = \frac{\bar{Y} - \mu}{S/\sqrt{n}}$$

gives $(11 - 12)/\frac{3}{5} = 1.67$, which does not fall within the critical region. The experimenter concludes that the mean percentage of responses to a goal box does match the percentage of trials in which it contains food.

The experimenter could also approach this problem by establishing a

---

[1] To deal with a power function that could easily be computed by the reader, the foregoing discussion has omitted one factor, variability. Power for many tests (for example, z, t, F) is plotted as a function of a ratio of the parameter value to the population variance. As might be expected, power varies inversely with variability. (A specific example is presented for the F test in Chapter 4.)

confidence interval for μ. First, from Equation (3.23)

(3.23)
$$P\left(-2.06 < \frac{\bar{Y} - \mu}{S/\sqrt{n}} < 2.06\right) = .95$$

Then algebraic manipulation and insertion of the values for $\bar{Y}$, S, and n result in the 95 percent confidence interval, $9.76 < \mu < 12.24$. There is 95 percent confidence that the population mean lies between these values, and again there is support for the belief that the true mean is 12.

With the two analyses completed, certain aspects can be compared. In both techniques the same information has been used: the values of the mean, standard deviation, and the size of the sample. Both techniques rest on the properties of the t distribution and consequently on the assumption that the scores constitute a random sample from a normally distributed population. The t test of $H_0: \mu = 12$ provides evidence for the matching hypothesis. The confidence interval does also, since the sample mean, 11, will not differ significantly (at the 5 percent level, if confidence is 95 percent) from any value that falls in the obtained interval, $9.76 < \mu < 12.24$. The confidence interval goes further in permitting the experimenter to note that certain other hypotheses are also tenable; for example, the hypothesis that the rats develop no preference is consistent with the occurrence of the value 10 within the interval. Briefly, the confidence interval approach permits us to consider all possible null hypotheses simultaneously.

Suppose that the confidence interval had been $4 < \mu < 18$. While it is true that the value predicted by the matching hypothesis falls within this interval, the interval is so wide that it also contains support for numerous alternative theories. Since no theory that predicts means from 5 through 12 can be rejected, this interval would result in a less firm statement that the matching hypothesis was correct. The width of the confidence interval indicates how firmly an experimenter can state an inference about any hypothesized population mean.

Tests of the null hypothesis can also provide the experimenter with an index of assurance when he accepts the null hypothesis. If there is high power of detecting even small deviations from the value assumed under H_0, an experimenter can be quite firm in his support of the matching hypothesis when the statistical test does not yield significant results. If he had little probability of detecting differences of the size observed, the experimenter might be more hesitant about drawing conclusions from the test. Thus the width of the confidence interval and the power of the statistical test provide similar types of information. There is a big difference however, which is to the advantage of the confidence interval approach. The computed confidence interval must provide information about its width, whereas null hypothesis tests can be, and usually are, made without reference to the appropriate power, or OC, functions.

To summarize, the confidence interval allows the experimenter to consider simultaneously all the possible null hypotheses, and it immediately yields an index of the strength of an inference about any one null hypothesis. Also, the confidence interval consists of a set of numbers on the same scale as the original data.

In the null hypothesis test, the experimenter is further removed from his original measures, since the test statistic is on a different scale. It may therefore be easier to digest the information the confidence interval provides.

Tests of the null hypothesis are firmly entrenched in the statistical methodology of psychology. We have not intended, in comparing null hypothesis tests and confidence intervals, to promote one at the expense of the other, but to make you aware of one useful approach to data analysis that researchers have often neglected. There is an overwhelming prevalence of hypothesis testing in the psychological literature, which our approach reflects. However, techniques will also be given for establishing confidence intervals for several parameters. The preceding discussion will at least show that alternative techniques are available, to help you to choose intelligently between estimation and hypothesis testing. Too often, psychologists have tested in ignorance of the alternative.[2]

Be warned against too literally translating statistical results results into scientific conclusions. Certainly the significance or nonsignificance of a test statistic should be a big factor in drawing conclusions about treatment effects, particularly when decisions about such things as α, β, and n have been made before the experiment. But assuming nonsignificance, how sure can we be that an important effect does not exist when the probability of the test statistic falls .1 percent above α, or when we find that the observed variability is greater than the estimate used to decide n? What should we conclude about results that are barely significant or fall just short of significance when the assumptions underlying the test are not met by our data? Should we draw the same inference about two nonsignificant results when the qualitative trends in one set of data are consistent with expectations based on available data and theory, yet no recognizable pattern exists in the second data set? How should published results of others be interpreted, when it is apparent that most experimenters give no thought to β in planning their experiment, that few even preselect α?

Under such circumstances, the test statistic can be at best a rough indicator of population effects rather than a sharp inferential tool. There are no simple answers, but we reject any one-to-one relation between the significance or nonsignificance of a test statistic and the existence or nonexistence of treatment effects, or the tenability or nontenability of a theory under investigation. In drawing inferences, the scientist has the responsibility of adding to the test statistic his a priori expectations, and his knowledge of the literature and of the particular experimental conditions (is there, for example, reason to suspect that some variable whose effects are not analyzable obscured the effects of independent variables?), and of the size and direction of effects; he must then subjectively weight these factors. When the results of the data analysis (regardless of the type) conflict with those factors not built into the test, then the experimenter should

---

[2] References to additional works on confidence intervals and null hypothesis tests are at the end of the chapter. There are also several references that present objections to both interval estimation and null hypothesis tests and in which alternative inferential procedures are proposed. Although this material is beyond the scope of the present textbook, you will find the source material stimulating.

reserve judgment. The ultimate criterion of the credibility of experimental conclusions is whether or not these conclusions are supported by subsequent replications of the experiment or by differently designed investigations of the hypotheses in question.

EXERCISES

3.1 Define the following terms:

(a) parameter   (b) consistent   (c) efficient
(d) unbiased   (e) expected value   (f) power
(g) Type I error   (h) Type II error   (i) alpha
(j) beta   (k) critical region   (l) one-tailed test
(m) confidence interval

3.2 One hundred random samples, each consisting of 225 scores, are drawn from a population of normally distributed scores, with $\mu = 0$, $\sigma = 1$. The 95 percent confidence interval, $\bar{Y} - 1.96(\frac{1}{15}) < \mu < \bar{Y} + 1.96(\frac{1}{15})$, is computed for each sample.

(a) Verify that the formula used is correct.

(b) Only 92 of the 100 computed confidence intervals contain the value of the true mean, 0. Does this fact conflict with the statement that these are 95 percent confidence intervals? If a 95 percent confidence interval does not mean that 95 percent of the obtained intervals will include the parameter value, what does it mean?

3.3 I desire a 95 percent confidence interval for μ, the mean of a normally distributed population whose standard deviation is 15. I should like the interval to be no wider than 10 units. How large should my sample be to achieve this criterion?

3.4 In our discussion of hypothesis testing, the critical region was always selected to be in one or both tails of the distribution. For example, against the alternative hypothesis $\mu_1 > \mu_0$, if we use the z test with $\alpha = .05$, the critical region would consist of z values equal to 1.645 or greater. Consider this rule for rejecting H_0: reject H_0 if the test statistic falls between $z = -.065$ and $z = .065$, this area also being approximately 5 percent of the area under the normal curve. What would happen to the power of the test? Make diagrams similar to those in Figures 3–4 to 3–6 for the usual z test and for the suggested alternative to illustrate your argument.

3.5 Consider $H_0: \mu = 100$; $H_1: \mu = 110$. Assume $n = 25$, $\sigma = 20$, $\alpha = 05$. (a) What is the power of a one-tailed test of H_0 against H_1? Use the normal probability (z) test. (b) What is the power if the test is two-tailed?

3.6 Prove

$$E\left(\frac{\sum (Y - \bar{Y})^2}{n}\right) = \left(\frac{n-1}{n}\right)\sigma^2$$

(*Hint:* Begin with $Y - \mu = (Y - \bar{Y}) + (\bar{Y} - \mu)$.)

REFERENCES

Cohen, J. *Statistical Power Analysis for the Behavioral Sciences.* New York: Academic Press, 1969.

Mood, A. M. *Introduction to the Theory of Statistics.* New York: McGraw-Hill, 1950.

SUPPLEMENTARY READING

For the student with a background in calculus, a more mathematical treatment of the
material in this chapter can be found in the following:

Mood, A. M., and Grayhill, F. A. *Introduction to the Theory of Statistics.* 2d. ed. New York:
McGraw-Hill, 1963.

Some articles written by psychologists treating the problems of null hypothesis testing are:

Baken, D. The test of significance in psychological research. *Psychological Bulletin* 66:
423–37 (1966).

Binder, A. Further considerations on testing the null hypothesis and the strategy and
tactics of investigating theoretical models. *Psychological Review* 70: 101–9 (1963).

Edwards, W. Tactical note on the relation between scientific and statistical hypothesis.
Psychological Bulletin 63: 400–2 (1965).

Edwards, W., Lindman, H., and Savage, L. J. Bayesian statistical inference for psychological research. *Psychological Review* 70: 193–242 (1963).

Grant, D. A. Testing the null hypothesis and the strategy and tactics of investigating
theoretical models. *Psychological Review* 69: 54–61 (1962).

Greenwald, A. G. Consequences of prejudice against the null hypothesis. *Psychological
Bulletin* 82: 1–20 (1975).

LaForge, Rolfe. Confidence intervals or tests of significance in scientific research.
Psychological Bulletin 6: 446–47, (1967).

Lykken, David T. Statistical significance in psychological research. *Psychological Bulletin*
3: 151–59 (1970).

Overall, John E. Classical statistical hypothesis testing within the context of Bayesian theory.
Psychological Bulletin 71: 285–92 (1969).

Rozeboom, W. W. The fallacy of the null hypothesis significance test. *Psychological
Bulletin* 57: 416–28 (1960).

Wilson, W., Miller, H. L., and Lower, J. S. Much ado about the null hypothesis.
Psychological Bulletin 3: 188–96 (1967).

Wilson, W. R., and Miller, H. L. A note on the inconclusiveness of accepting the null
hypothesis. *Psychological Review* 71: 238–42 (1964).

Many of the above include some discussion of significance tests and confidence intervals. A
more detailed comparison is presented in the following.

Natrella, M. G. The relation between confidence intervals and tests of significance—A
teaching aid. *American Statistics* 14: 20–22, 38 (1960).

Several of the above references argue for Bayesian analysis as an alternative to hypothesis
testing. An introduction to this approach is contained in the final chapter (Chapter
19: Some elementary Bayesian methods) of

Hays, W. L. *Statistics for the Social Sciences.* 2d ed. New York: Holt, Rinehart, & Winston,
1973.

An excellent statement of the considerations involved in drawing inferences from data is to
be found in

Tukey, J. W. Conclusions versus decisions. *Technometrics* 2: 423–33 (1960).

4 COMPLETELY RANDOMIZED ONE-FACTOR DESIGNS

4.1 INTRODUCTION

Completely randomized designs are characterized by the random assignment of each subject to only one level of the independent variable, or to only one combination of levels if more than one independent variable is under investigation. The term *random* represents the requirement that each subject has an equal probability of assignment to any level or combination of levels.

Not really correct

For an example of a completely randomized design involving one independent variable, consider the following experiment. Each subject is required to learn the appropriate response to each of 12 stimulus words; the dependent variable is the number of trials (times through the list of 12 stimulus-response pairs) required to attain a criterion of two errorless trials. The independent variable is level of noise intensity, and it is planned to have 40 subjects, 10 tested at each of 4 levels of noise. The subjects are college students, volunteers from a basic psychology course, who indicate on a sign-up sheet their willingness to participate.

One way to assign these subjects randomly to noise levels would be to write the numbers from 1 to 40 on separate slips of paper, place these in a hat, mix well, and then note the order in which the slips are selected out of the hat. If the slip of paper numbered 17 were selected first, then the 17th subject to volunteer would be placed in the first treatment group, for example, the one corresponding to the lowest level of noise. The first 10 selected pieces of paper designate the subjects who go into the first group; the second 10, those who are placed in the second treatment group; and so on.

A simpler way to accomplish this would be to use a table of random numbers such as Table A–1 in the Appendix. Two adjacent columns of digits are chosen at random within a five-column set, and a row within the columns is then selected randomly. Beginning with this row, one proceeds down the columns, noting the order of appearance of the numbers from 01 to 40 and ignoring all other numbers. This order provides the basis for assignment to groups in the same way that the order of drawing numbered slips did. Other numbers besides 01 to 40 might be used by making 41 equivalent to 01, 42 equivalent to 02, and so on.

The numbers 81 to 00 would still be omitted, to give all numbers an equal chance of being selected.

The procedures described for assigning subjects to experimental treatments ensure that all subjects have an equal probability of assignment to each treatment, and consequently, that there exists no systematic source of error. Contrast with such procedures the possible consequences when the first 10 subjects to volunteer are assigned to one treatment, the second 10 volunteers to a second treatment, and so on. Under that procedure, it is difficult to defend any inference based on the data analysis. If the low noise group learns most quickly, is this a function of differences in the effects of noise on performance, or were the earlier volunteers more interested in the experiment, more motivated to do well?

The completely randomized design has one big advantage over other designs—simplicity. This characteristic of such designs extends to the experimental layout, the model underlying the data analysis, and the computations involved in the data analysis. Let us briefly consider these advantages.

1. *The experimental layout.* The experimental layout is a statement of who is tested and the conditions for the test. If the design requires that each subject be tested under several experimental conditions, the order of presentation of conditions for each subject is part of the layout. For the completely randomized design, only the random assignment described previously is required to lay out the experiment. In contrast, stratified designs (see Chapter 6) require grouping the subjects into strata or levels based on some additional measure; then random assignment to the levels of the independent variable is carried out within these strata. Other designs involve still other complications in obtaining the experimental layout.

2. *The model.* The model underlying the completely randomized design has the important feature that it involves fewer assumptions than those made for any other experimental design. Consequently, the derivations of parameter estimates are simpler than in any other instance. An even more important consequence of the parsimony of assumptions is that there is less that can go wrong with the inferential machinery than in the case of other, more complex models. Each additional assumption underlying the derivation of the test statistic is one more assumption that can be violated, undermining the validity of the statistical inference.

3. *The computations.* The analysis of variance for the completely randomized design involves fewer terms than are required for other designs, and it is often easier to compute the terms. Furthermore, related analyses like the analysis of covariance and the estimation of missing data will also generally be simpler for the completely randomized design.

The chief disadvantage of the completely randomized design is its relative inefficiency. The error variance will usually be large compared with that resulting from the use of other designs. This is partly offset since no design yields as many degrees of freedom for the error variance as the completely randomized design does, if we assume some fixed amount of data. The power of a test increases

But this would results in more chance of misspecification.

monotonically with degrees of freedom (see Section 4.6.2 for more on degrees of freedom).

   To help you in learning the subject matter of this book, we shall first consider only limited aspects of a topic, then gradually develop additional concepts and computations. Thus, in this chapter, the following limitations have been placed on the presentation:

1.   We consider only that subset of completely randomized designs that involve only one independent variable; these are one-factor designs.
2.   We consider only independent variables whose levels are fixed; that is, we define the population of levels as consisting only of those that have been selected for the experiment.
3.   We consider only the test of the general null hypothesis that $\mu_1 = \mu_2 = \cdots = \mu_j = \cdots = \mu_a$, where μ_j is the mean of a population of individuals tested under A_j, the jth level of the independent variable A.

In addition to tests of the general null hypothesis, we shall consider here estimation of the variance of the μ_j and what that variance contributes to the total population variance.

4.2   A MODEL FOR THE COMPLETELY RANDOMIZED ONE-FACTOR DESIGN

Consider an infinitely large population of individuals. Assume that each individual in this parent population is randomly assigned to exactly one of a possible treatments (levels of the independent variable). There are now a very large *treatment populations*, systematically differing from each other only in levels of the independent variable A. Next consider a completely randomized one-factor experiment in which *an* subjects are randomly distributed among a treatments in the manner described earlier or by some other procedure that equally assures randomness of assignment. The a experimental groups of n subjects can be considered a random samples, one from each of the treatment populations hypothesized previously. We are generally interested in drawing inferences about $\mu_1, \mu_2, \ldots, \mu_j, \ldots, \mu_a$, the means of the treatment populations. Our specific current concern is with developing a test of the null hypothesis that the μ_j are equal. To develop such a test and to answer other questions about the population parameters, we need some statement about how the data and the parameters of the population are related. A simple possibility is the following:

$$(\mathbf{4.1}) \qquad\qquad Y_{ij} = \mu + \alpha_j + \varepsilon_{ij}$$

where Y_{ij} is the score of the ith subject in the treatment group j; μ is the mean of the $\cdot\mu_j$, or equivalently, the mean of the parent population before treatment populations were established; $\alpha_j = \mu_j - \mu$, the *effect* of treatment A_j; and $\varepsilon_{ij} = Y_{ij} - \mu - \alpha_j = Y_{ij} - \mu_j$.

The quantity ε_{ij} is the error associated with the ith score in the jth group and is a unique contribution of the individual, a deviation of the total score from μ, which cannot be accounted for by the treatment effect. The variability in ε_{ij} may be due to differences among subjects in such factors as ability, set, motivation, the tone in which instructions are read, or the temperature of the room.

Before the consequences of Equation (4.1) are developed, it may be helpful to consider the rationale underlying the equation. If our subjects were identical individuals, identically treated, we should have

$$Y_{11} = Y_{21} = \cdots = Y_{ij} = \cdots = Y_{na} = \mu$$

But each group of n individuals has been treated differently. Assuming n identical individuals in each treatment population, but now allowing for the possibility that the treatments do not have identical effects, we obtain

$$Y_{11} = Y_{21} = \cdots = Y_{i1} = \cdots = Y_{n1} = \mu_1$$
$$\vdots$$
$$Y_{1j} = Y_{2j} = \cdots = Y_{ij} = \cdots = Y_{nj} = \mu_j$$
$$\vdots$$
$$Y_{1a} = Y_{2a} = \cdots = Y_{ia} = \cdots = Y_{na} = \mu_a$$

Finally, we take into account that even individuals treated alike with respect to the independent variable will rarely perform in an identical manner. Because of individual differences, the score of subject i in group j will deviate from μ_j by an amount ε_{ij}. Consequently

$$Y_{11} = \mu_1 + \varepsilon_{11}$$
$$\vdots$$

(4.2)
$$Y_{ij} = \mu_j + \varepsilon_{ij}$$
$$\vdots$$
$$Y_{na} = \mu_a + \varepsilon_{na}$$

Equation (4.2) is unchanged if we add and subtract the constant μ, resulting in

$$Y_{ij} = \mu + \mu_j - \mu + \varepsilon_{ij}$$

Since $\mu_j - \mu = \alpha_j$ by definition, Equations (4.2) and (4.1) are equivalent. Thus the model underlying the data analysis asserts that the score of the ith individual in the jth group is the sum of the following three components:

1. *The parent population mean, μ.* This quantity is a constant component of all scores in the data matrix.

2. *The effect of treatment A_j, namely α_j,* is a constant component of all scores obtained under A_j, but may vary over treatments (levels of j). If the a levels of A are arbitrarily selected, as assumed, they exhaust the population of levels of A, and therefore $\sum_j \alpha_j = 0$, since the sum of all deviations of scores μ_j about their mean μ is zero. The null hypothesis asserts that $\alpha_1 = \alpha_2 = \cdots = \alpha_j = \cdots = \alpha_a = 0$.

3. *The deviation, ε_{ij}, due to uncontrolled variability, of the ith score in group j from the jth treatment population mean.* This component of Y_{ij} is the only source of

variance among scores in the jth group, and if the null hypothesis is true, the only source of variance in the data matrix.

   Equation (4.1) is not sufficient for deriving parameter estimates and statistical tests. In addition, the following assumptions about the distribution of the ε_{ij} are required:

1. The ε_{ij} are independently distributed. This means that the probability of sampling some value of ε_{ij} does not depend on any other values of ε_{ij} in the sample. An important consequence of this is that the ε_{ij} are uncorrelated.
2. The distribution of the ε_{ij} is normal, with zero mean, in each of the a treatment populations.
3. The distribution of the ε_{ij} has variance σ_e^2 in each of the a treatment populations; that is, $\sigma_1^2 = \sigma_2^2 = \cdots = \sigma_j^2 = \cdots = \sigma_a^2 = \sigma_e^2$.

   Since the ε_{ij} are solely responsible for the variability in the jth treatment population, it follows that the Y_{ij} should also be normally and independently distributed, with mean μ_j and variance equal to σ_e^2. The fact provides a basis for testing the above assumptions. (A more detailed analysis of the assumptions will be presented later; for the present, some consequences of the model are shown.)

4.2.1   ESTIMATES OF THE POPULATION PARAMETERS.

Under the normal distribution assumption, the least-squares and maximum likelihood procedures (described in Chapter 2) result in identical values of $\hat{\mu}$ and $\hat{\alpha}$, the estimates of μ and α. Both procedures ordinarily require differential calculus to obtain the estimates; we therefore omit the derivation but offer an algebraic proof that the least-squares estimate of μ is $\bar{Y}_{..}$ and of α_j is $\bar{Y}_{.j} - \bar{Y}_{..}$.

   The simplest approach is first to derive an expression for $\hat{\mu}_j$, the least-square estimate of μ_j. Since μ is an average of the μ_j, then $\hat{\mu}$ can readily be obtained by averaging the $\hat{\mu}_j$. Furthermore, since $\alpha_j = \mu_j - \mu$, then $\hat{\alpha}_j = \hat{\mu}_j - \hat{\mu}$. A simple relation between μ_j and the error component ε_{ij} is provided by Equation (4.2); transposing terms and replacing the parameters by their estimators, we have

$$(\textbf{4.2}') \qquad\qquad \hat{\varepsilon}_{ij} = Y_{ij} - \hat{\mu}_j$$

We desire a value of $\hat{\mu}_j$ that minimizes the quantity $\sum_i \sum_j \hat{\varepsilon}_{ij}^2$, the sum of squared errors. We therefore square both sides of Equation (4.2'), and after summing, we have

$$(\textbf{4.3}) \qquad\qquad \sum_j \sum_i \hat{\varepsilon}_{ij}^2 = \sum_j \sum_i (Y_{ij} - \hat{\mu}_j)^2$$

The least-squares estimate $\hat{\mu}_j$ can be represented as a value some distance from the group mean $\bar{Y}_{.j}$; that is, $\hat{\mu}_j = \bar{Y}_{.j} + C$. Replacing $\hat{\mu}_j$ in Equation (4.3) gives us

$$\sum_j \sum_i \hat{\varepsilon}_{ij}^2 = \sum_j \sum_i (Y_{ij} - \bar{Y}_{.j} - C)^2$$

$$(\textbf{4.4}) \qquad = \sum_j \sum_i (Y_{ij}^2 + \bar{Y}_{.j}^2 + C^2 - 2Y_{ij}\bar{Y}_{.j} - 2Y_{ij}C + 2\bar{Y}_{.j}C)$$

$$= \sum_j \sum_i (Y_{ij}^2 + \bar{Y}_{.j}^2 - 2Y_{ij}\bar{Y}_{.j}) + anC^2 - 2C \sum_j \sum_i Y_{ij} + 2Cn \sum_j \bar{Y}_{.j}$$

Note the application of the summation rules (Chapter 2). Equation (4.4) can be simplified if we replace $\sum_j \sum_i Y_{ij}$ by $T_{..}$ and also replace $\sum_j \bar{Y}_{.j}$ by $\sum_j (T_{.j}/n)$, which in turn is replaced by $T_{..}/n$. Then the rightmost two terms cancel each other and we now have

(4.5)
$$\sum_j \sum_i \hat{\varepsilon}_{ij}^2 = \sum_j \sum_i (Y_{ij}^2 + \bar{Y}_{.j}^2 - 2 Y_{ij} \bar{Y}_{.j}) + anC^2$$

If C is not zero, anC^2 is a positive term. Therefore, the error variability is smallest when C equals zero, or when $\hat{\mu}_j = \bar{Y}_{.j}$. Estimates for the remaining parameters follow readily. We have

(4.6)
$$\begin{aligned}
\hat{\mu}_j &= \bar{Y}_{.j} \\
\hat{\mu} &= \bar{Y}_{..} \\
\hat{\alpha}_j &= \bar{Y}_{.j} - \bar{Y}_{..} \\
\hat{\varepsilon}_{ij} &= Y_{ij} - \bar{Y}_{.j}
\end{aligned}$$

If the ε_{ij} are, as assumed, normally distributed, then the above estimates are efficient and consistent.

4.2.2   THE *F* RATIO.   If the assumptions of the model are met, an appropriate test of the null hypothesis that the μ_j are equal is provided by

(4.7)
$$F = \frac{n \sum_j \hat{\alpha}_j^2/(a-1)}{\sum_i \sum_j \hat{\varepsilon}_{ij}^2/a(n-1)} = \frac{n \sum_j (\bar{Y}_{.j} - \bar{Y}_{..})^2/(a-1)}{\sum_i \sum_j (Y_{ij} - \bar{Y}_{.j})^2/a(n-1)}$$

The numerator of F is generally referred to as the *between-groups mean square*. We shall use the notation MS_A; the subscript designates the independent variable. The MS_A is an estimate of a variance common to the a treatment populations; it is assumed that the treatment populations have identical means (the null hypothesis) and variances (generally referred to as the assumption of *homogeneity of variance*). It is also assumed that observations are independently distributed. Under these assumptions,

(4.8)
$$\frac{\sum_j (\bar{Y}_{.j} - \bar{Y}_{..})^2}{a-1} = \hat{\sigma}_{\bar{Y}}^2 = \frac{\hat{\sigma}^2}{n}$$

and with the terms in Equation (4.8) multiplied by n, it is apparent that MS_A estimates σ_e^2, the variance of any one population.

The denominator of the F ratio is generally referred to as the *within-group mean square*, and we shall use the notation $MS_{S/A}$; here the subscript is read "subjects within levels of A." The $MS_{S/A}$ is an arithmetic average of a estimates of the common (under the assumption of homogeneity of variance) treatment population variance σ_e^2. Thus, under the assumptions of H_0, homogeneity of variance, and independence, numerator and denominator of the F ratio are independent estimates of σ^2. If we add the assumption that the populations have normal distributions, it follows (from Section 3.4.3) that the ratio of mean squares is distributed as F; that is,

$$\frac{MS_A}{MS_{S/A}} = \frac{MS_A/\sigma_e^2}{MS_{S/A}/\sigma_e^2} = \frac{X_{a-1}^2/(a-1)}{X_{a(n-1)}^2/a(n-1)} = F_{a-1,a(n-1)}$$

where $a-1$ is the numerator degree of freedom (df), and $a(n-1)$ is the denominator df.

4.2.3 EXPECTED MEAN SQUARES. Let us now consider why the ratio of mean squares should be sensitive to violations of the null hypothesis. The numerator of the F ratio is n times the variance of the treatment group means. One reason that these means will differ is that each is based on a different set of n individuals; error variance is contributing to the variance of group means. There is another *possible* source of the variance of group means. If the treatments really differ in their effects, the μ_j are not all equal; we should expect this to be reflected in the spread among the group means. Thus, there are two possible sources of the variability among the group means and therefore of the magnitude of the MS_A—individual differences and treatment effects.

Next consider the denominator of the F ratio, $MS_{S/A}$. The variance among individuals in the jth treatment group is $\sum_{i=1}^{n} (Y_{ij} - \bar{Y}_{.j})^2/(n-1)$. Summing over groups and dividing by a yields the variance of individuals averaged over groups. Treatment effects do not contribute to this mean square, for if a constant (say α_j) is added to all the scores in a group, the variance of the scores is unchanged. The $MS_{S/A}$ is a function only of σ_e^2, the error variance.

By the preceding line of reasoning, we should expect the average values of MS_A and $MS_{S/A}$ over many replications of the experiment to equal each other *if the μ_j were equal*. In other words, if the null hypothesis is true, both numerator and denominator of the F ratio reflect only error variance and $E(MS_A) = E(MS_{S/A}) = \sigma_e^2$. In any single experiment, we should not expect the numerator and denominator mean squares to be identical, since by chance, there might be more (or less) variability among individuals in different groups than among individuals in the same group. If MS_A were considerably larger than $MS_{S/A}$, however, there would be grounds for suspecting that treatment effects were being added to error variance in the numerator of the F ratio.

If these intuitive arguments can be proved, we shall have met one important criterion for a test statistic; it should reflect the relative merits of the null and alternative (that the μ_j are not equal) hypotheses. Such a proof follows. Specifically, it will be proved that if H_0 is true, and if the assumptions that the ε_{ij} are independently distributed with variance σ_e^2 are true, on the average over many replications of the experiment the two mean squares are equal; if H_0 is false, the average value of MS_A is greater than the average value of $MS_{S/A}$.

Our approach is to derive separately the expectations of $n \sum_j (\bar{Y}_{.j} - \bar{Y}_{..})^2$ and $\sum_i \sum_j (Y_{ij} - \bar{Y}_{.j})^2$. These quantities are the numerators of the mean squares; the first is referred to as the *sum of squares for A* (SS_A), and the second, the *sum of squares within groups* $(SS_{S/A})$. First the SS_A is redefined by the analysis of variance model. From Equation (4.1),

$$\bar{Y}_{.j} = \frac{\sum_i (\mu + \alpha_j + \varepsilon_{ij})}{n}$$

(**4.9**)

$$= \frac{n\mu + n\alpha_j + \sum_i \varepsilon_{ij}}{n}$$

$$= \mu + \alpha_j + \bar{\varepsilon}_{.j}$$

and

$$\bar{Y}_{..} = \frac{\sum_i \sum_j (\mu + \alpha_j + \varepsilon_{ij})}{an}$$

$$= \frac{an\mu + n \sum_j \alpha_j + \sum_i \sum_j \varepsilon_{ij}}{an}$$

Since $\sum_j \alpha_j = 0$,

(**4.10**) $$\bar{Y}_{..} = \mu + \bar{\varepsilon}_{..}$$

Since $SS_A = n \sum_j (\bar{Y}_{.j} - \bar{Y}_{..})^2$, substituting on the basis of Equations (4.9) and (4.10) and taking the expected value of SS_A over many replications of the experiment, we have

(**4.11**) $$E(SS_A) = E\left[n \sum_j (\alpha_j + \bar{\varepsilon}_{.j} - \bar{\varepsilon}_{..})^2 \right]$$

Expanding Equation (4.11), we obtain

(**4.12**) $$E(SS_A) = nE\left(\sum_j \alpha_j^2 + \sum_j \bar{\varepsilon}_{.j}^2 + \sum_j \bar{\varepsilon}_{..}^2 - 2 \sum_j \bar{\varepsilon}_{.j} \bar{\varepsilon}_{..} \right.$$
$$\left. + 2 \sum_j \alpha_j \bar{\varepsilon}_{.j} - 2 \sum_j \alpha_j \bar{\varepsilon}_{..} \right)$$

Since the expectation of a sum equals the sum of the expectations, the expectation of each term on the right side of Equation (4.12) can be evaluated separately.

(**4.12a**) $$E(n \sum_j \alpha_j^2) = n \sum_j \alpha_j^2$$

since α_j is assumed to be constant over replications of the experiment, and the expectation of a constant is the constant.

(**4.12b**) $$nE\left(\sum_j \bar{\varepsilon}_{.j}^2 \right) = n \sum_j E(\bar{\varepsilon}_{.j}^2) = a\sigma_e^2$$

where σ_e^2 is the variance of the ε_{ij} in the jth treatment population. Note that the expectation of a squared deviation about its average is, by definition, a variance. Thus, $E(\bar{\varepsilon}_{.j}^2)$ is $\sigma_{\bar{\varepsilon}_{.j}}^2$ or $(1/n)\sigma_{\varepsilon_{ij}}^2$, assuming that the ε_{ij} are independently distributed. The final result, $a\sigma_e^2$, follows if we also assume homogeneity of variance; that is,

$$\sigma_{\varepsilon_{ij}}^2 = \sigma_e^2 \qquad \text{for all } j$$

Similarly,

(4.12c)
$$nE\left(\sum_j \varepsilon_{..}^2\right) = n\sum_j \left(\frac{1}{an}\right)\sigma_e^2 = \sigma_e^2$$

and noting that $\sum_j \bar{\varepsilon}_{.j} = a\bar{\varepsilon}_{..}$, we get

(4.12d)
$$nE\left(-2\sum_j \bar{\varepsilon}_{.j}\bar{\varepsilon}_{..}\right) = nE(-2a\bar{\varepsilon}_{..}^2) = -2an\sigma_{\bar{\varepsilon}_{..}}^2 = -2\sigma_e^2$$

The next two terms vanish. Since α_j is a constant independent of ε_{ij}, the covariance of the two terms is zero; that is,

(4.12e)
$$E\left(\sum_j \alpha_j\bar{\varepsilon}_{.j}\right) = 0$$

Since $\sum_j \alpha_j = 0$,

(4.12f)
$$E\left(\sum_j \alpha_j\bar{\varepsilon}_{..}\right) = 0$$

Combining terms, we get

(4.13)
$$E(SS_A) = (a-1)\sigma_e^2 + n\sum_j \alpha_j^2$$

and for the expected mean square of A, which is $E(MS_A)$, we have

(4.14)
$$E(MS_A) = E\left(\frac{SS_A}{a-1}\right) = \sigma_e^2 + n\theta_A^2$$

The symbol θ_A^2 is used instead of σ_A^2 to represent the variability among the μ_j for two reasons: (1) in a strict sense $\sum (\mu_j - \mu)^2/(a-1)$ is not a population variance because of the divisor, and (2) the use of θ^2 is a reminder that the effects of A are fixed instead of random.

The derivation of $E(MS_{S/A})$ is similar to the derivation for $E(MS_A)$. Applying Equation (4.2) gives

(4.15)
$$E\left[\sum_i \sum_j (Y_{ij} - \bar{Y}_{.j})^2\right] = E\left[\sum_i \sum_j (\varepsilon_{ij} - \bar{\varepsilon}_{.j})^2\right]$$

Expansion yields

(4.16)
$$E\left[\sum_i \sum_j (Y_{ij} - \bar{Y}_{.j})^2\right] = E\left[\sum_i \sum_j \varepsilon_{ij}^2 - n\sum_j \bar{\varepsilon}_{.j}^2\right]$$
$$= a(n-1)\sigma_e^2$$

Dividing both sides of Equation (4.16) by $a(n-1)$ yields

(4.17)
$$E(MS_{S/A}) = \sigma_e^2$$

We now can state that

(4.18)
$$\frac{E(MS_A)}{E(MS_{S/A})} = \frac{\sigma_e^2 + n\theta_A^2}{\sigma_e^2}$$

If H_0 is true, $\theta_A^2 = 0$ and the above ratio equals 1. If H_0 is false, $\theta_A^2 > 0$ and $E(MS_A)/E(MS_{S/A}) > 1$. Thus the ratio of mean squares satisfies one requirement of a test statistic; its magnitude can be expected to reflect the validity of H_0. The second requirement of the test statistic is that its distribution under H_0 be known. The distribution of F must be known to determine the critical region, those values whose probability of occurrence is less than α, if H_0 is true. If the F obtained from an experiment is so large that it falls within the critical region, we reject H_0 in favor of the alternative hypothesis that $\theta_A^2 > 0$.

If the conditions of normality, independence, and homogeneity of variance hold, and if H_0 is true, the ratio of mean squares is distributed as F (Section 4.2.2). (Note that the normality assumption was not invoked in deriving the $E(MS)$; it is necessary, however, if the sums of squares are to have the χ^2 distribution.) Consequently, the probability of obtaining various F values can be computed. Values of F required for significance are presented for several combinations of df and α levels in Table A–5 in the Appendix. In an example of using Table A–5, assume an experiment with five levels of the independent variable A and six subjects at each level. Then the df of the numerator (df_1) are 4 and the df of the denominator (df_2) are 25. If α equals .05, then values of F greater than 2.76 will result in the rejection of H_0. If the selected α equaled .01, then Fs greater than 4.18 would be required for significance.

4.3 ASSUMPTIONS UNDERLYING THE F TEST

The third requirement for a statistic to be an appropriate test of H_0 is that the data conform to the assumptions underlying the test. The ratio of mean squares will be distributed as F if:

1. The ε_{ij} are independently distributed.
2. The ε_{ij} are normally distributed.
3. The variance of the ε_{ij} is the same for all treatment populations.
4. The null hypothesis is true.

If the first three assumptions are valid, then significant Fs can be attributed to the falsity of assumption 4. If any of the first three assumptions are false, however, the distribution of mean squares may be such that the true probability of obtaining a result in the critical region is actually not α; whether the true probability of a Type I error is greater or less than α will depend on the form of the violation, although an excess of Type I errors is the more usual occurrence. Figure 4–1 exemplifies one potential problem. The solid line represents the F distribution, the distribution of the ratio of mean squares, when all four assumptions are valid. Under some departures from assumptions 1 through 3, the distribution of the ratio of mean squares might resemble that described by the dotted line. In this case, the probability of obtaining an F in the critical region is less than α. An alternative result of violations of any of the first three assumptions is represented by the dashed line in Figure 4–1. The distribution of the ratio of

Figure 4–1   An example of the distribution of the ratio of mean squares when assumptions are valid (solid line), when violations result in loss of power (dotted line), and when violations result in increased risk of Type I errors (dashed line).

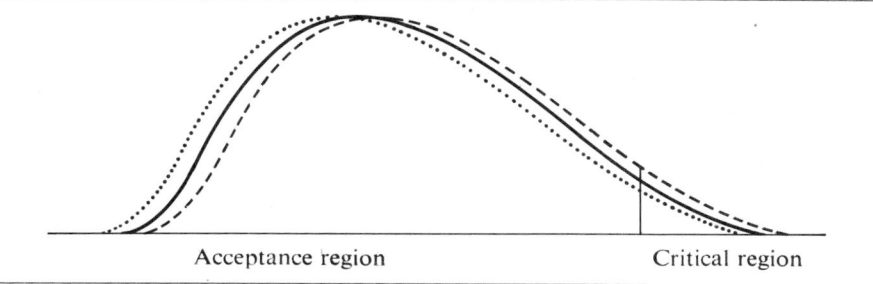

Acceptance region   Critical region

mean squares is now such that the probability of obtaining an *F* in the critical region is greater than α, and the risk of a Type I error is therefore greater than what the experimenter wishes to assume.

4.3.1   INDEPENDENCE.   When only one observation is obtained from each subject, and subjects are randomly assigned to treatment combinations, the assumption of independence of scores will generally be met. In repeated measurement designs, however, the assumption generally is not valid. If subjects are tested, for example, on several days, we can expect that scores for any pair of days will be correlated. This may have serious consequences on the distribution of the ratio of mean squares. (We shall elaborate on this in Chapter 7.) Positive correlations (the usual case) can result in serious inflation of the Type I error rate; negative correlations will result in its deflation (Cochran, 1947; Scheffé, 1959).

4.3.2   NORMALITY.   The validity of the normality assumption, No. 2, will depend on the measure chosen. Projective test scores, for example, tend to show a skewed distribution, and percentage scores will be binomially distributed. The χ^2 test can be used to evaluate the significance of departures of the obtained distribution from the assumed normal distribution; the test would be applied to each of the *a* sets of *n* scores in turn. Generally, such a procedure will be unnecessary, since the distribution of the ratio of mean squares seems little affected by departures from normality. Both mathematical proofs (Scheffé, 1959) and empirical studies (Boneau, 1960; Bradley, 1964; Donaldson, 1968; Lindquist, 1953, pp. 78–90), attest to this conclusion.[1] In the empirical studies, large treatment populations of scores were created, samples were drawn repeatedly from each population with *F* ratios computed for each set of *a* samples, and the empirical distribution of the ratio of mean squares was then compared to the theoretical *F* distribution (Table A–5). When the treatment populations are

[1] An additional, unpublished, computerized sampling experiment has been carried out by J. W. Clinch and the author, producing results consistent with the conclusions presented in this section.

markedly skewed, but homogeneous in shape and variance, the probabilities of large values of F are slightly deflated; when n is 4, the Fs required for significance at the .01 and .05 levels are exceeded in about .003–.009 and .03–.04 of the samples, respectively. Symmetric nonnormal distributions seem to result in slight inflation of the Type I error probabilities; the uniform distribution, in which the density is constant for all values of y, typically yields operative α values of .01–.018 and .05–.065 when the theoretical values are .01 and .05. The worst result I have noted in any of the studies of nonnormality involved three leptokurtic distributions with samples of size three; the operative α levels were .023 and .078. Clearly, even with samples considerably smaller than experimenters typically use, nonnormality has a limited effect on Type I error rates. Furthermore, the situation rapidly improves as n increases. This optimistic picture of the role of the normality assumption should be qualified by noting that substantial errors can occur in estimating intervals for variance components (for example, confidence intervals for σ_e^2) if the treatment populations are not normally distributed.

A particularly flagrant violation of the normality assumption occurs when the dependent variable is discretely distributed. This is quite common in psychological and educational research, occurring whenever rating scale or response frequency data are analyzed. Lunney (1970) has considered the case in which the dependent variable has only two possible values for ns, ranging from 3 to 31 (in steps of 4), as from 2 through 5, and p (probability of a success on a single trial) from .1 to .9. Except for the two most extreme p values, 20 error df sufficed to yield a close approximation of obtained to theoretical α levels; 40 error df sufficed at ps of .1 and .9. These findings have been corroborated and extended by Hsu and Feldt (1969) and Bevan, Denton, and Myers (1974); both of these sets of investigators varied number of rating points and shapes of distributions. Even with a equal to 2, n equal to 4, and only three scale points, theoretical values of α were closely approximated in all but the most skewed distributions. In most cases, the empirical α was no more deviant than the α obtained by sampling error when the parent populations were normally distributed.

Thus far, we have concentrated on how nonnormality affects Type I error rate. Some information about its consequences for Type II error rate (or its complement, power) is also available. Both sampling (Boneau, 1962; Donaldson, 1968; Games and Lucas, 1966) and analytic (Srivastava, 1959) investigations suggest that moderate departures from normality do not much affect the power function. When the parent distributions are flatter than the normal, power is reduced; when distributions are negatively skewed, power is actually increased, except for those values of power above .90.

4.3.3 HOMOGENEITY OF VARIANCE.   The validity of the homogeneity of variance assumption, No. 3, will depend on how careful the experimenter is in administering the experimental treatments to all subjects consistently, on the type of measure taken, and on the treatment levels used. Certainly, the uniform application of treatments to subjects will tend to stabilize variances from group to group.

Despite proper experimental methodology, however, heterogeneity of variance can occur. For example, some measures (response frequency is a common instance) are Poisson-distributed; in this case, the mean and the variance are equal, and either both are homogeneous or both are heterogeneous over treatments. Several methods exist for detecting heterogeneity of variance. Some of these are overly sensitive to departures from normality (Bartlett, 1973; Cochran, 1941; Hartley, 1950). A more robust test of heterogeneity of variance has been proposed by Levene (1960); also, see Layard (1973). You will find this useful when there is interest in how treatment affects variability. It is not clear, however, that any of the available tests are useful as preliminary tests preceding analysis of variance. Here the issue is not whether population variances differ but whether they differ enough for the ratio of mean squares in the ordinary analysis of variance not to be any longer distributed as F. As noted below, so long as ns are equal, variances must be notably heterogeneous to have severe consequences for the F test of equality of treatment population means.

Both mathematical derivations by Box (1954) and the empirical studies cited earlier indicate that the α level is inflated by heterogeneity of variance. If all treatment populations are approximately normally distributed, and if all groups are of the same size, the inflation is slight; with two groups, n equal to five, and a 20:1 variance ratio, the Fs required for significance at the .01 and .05 levels were exceeded in .02 and .07 of the computerized experiments we ran. Similar results were obtained with more groups. Over several investigations, variance ratios of about 4:1 yielded still better results, the operative α levels consistently falling below .02 and .07. Increases in sample size yield slight decreases in the degree of inflation, but some distortion of Type I error rates will occur even with very large n. Unequal ns are more troublesome. Boneau's results (1960), obtained with normally distributed populations, are typical; with variances of 1 and 4 and ns of 5 and 15, the Fs required for significance at the .01 and .05 levels were exceeded in .001 and .01 of the cases; with the ns reversed so that the group with the smaller variance had the larger n, the operative α levels were .06 and .16. Usually when there is a negative correlation of variance and sample size, or when the ns are equal, the true probability of a Type I error is above that assumed; if the correlation is positive, heterogeneity of variance may deflate the Type I error rate.

Clearly, the best protection against distortion of the Type I error rate is using equal ns. If experimental groups are not of equal size, however, or if the error variances differ extremely over treatment levels, it is still possible to carry out the significance test at the desired α level, *provided that the treatment population distributions are approximately normal*. When a equals 2, a reasonable approximation is obtained (Welch, 1937) by computing the quantity

$$(4.19) \qquad t' = \frac{\bar{Y}_{.1} - \bar{Y}_{.2}}{\sqrt{S_1^2/n_1 + S_2^2/n_2}}$$

which is then evaluated for significance in Table A–3 with

$$(4.20) \qquad df = \frac{(n_1-1)(n_2-1)}{(n_2-1)c^2 + (n_1-1)(1-c)^2}$$

where

(4.21)
$$c = \frac{S_1^2/n_1}{S_1^2/n_1 + S_2^2/n_2}$$

and S_j^2 and n_j are the variance and sample size for the jth treatment group. For an example of this approach, suppose that $S_1^2 = 1$, $S_2^2 = 4$, $n_1 = 5$, $n_2 = 15$, $\bar{Y}_{.1} - \bar{Y}_{.2} = 1.8$ and the required α level is .05. The usual t test (which is equivalent to the F test since $t^2 = F$ when two treatment levels are being compared) would fail to reject the null hypothesis; we should use the formula

(4.22)
$$t = \frac{\bar{Y}_{.1} - \bar{Y}_{.2}}{\sqrt{\left[\frac{(n_1 - 1)S_1^2 + (n_2 - 1)S_2^2}{n_1 + n_2 - 2}\right]\left[\frac{1}{n_1} + \frac{1}{n_2}\right]}}$$

which gives

$$t = \frac{1.8}{.943}$$

which is less than 2.101, the value required for significance at the .05 level with 18, $n_1 + n_2 - 2$, df. We arrive at a different conclusion using Welch's solution for the unequal variance problem, however. The adjusted df, by Equations (4.20) and (4.21), is 14.5; thus the critical value of t' lies between 2.131 and 2.145. The value obtained using Equation (4.19) is

$$t' = \frac{1.8}{.683} = 2.64$$

which is clearly significant. Because of the positive correlation of variance and n, the usual significance test, which wrongly assumed homogeneity of variance, failed to detect a result that proved significant when heterogeneity of variance was taken into account. If the variances and sample sizes had been negatively correlated, the usual test might have resulted in significance when Welch's test did not.

When a is greater than 2, other adjustments for heterogeneity of variance are possible. One approximate technique Box (1954) proposed uses the usual ratio of mean squares as the test statistic. The df require adjustment, however, and if the ns are unequal, the F required for significance must be multiplied by a factor b. To test for significance at a given level of α, we turn to Table A–5, find the F required for significance on h' and h df, multiply that value by b if the ns are not equal, and then note whether the ratio of mean squares computed from our data exceeds this value. The appropriate formulas are

(4.23)
$$b = \frac{(N - a)\sum_j (N - n_j)S_j^2}{N(a - 1)\sum_j (n_j - 1)S_j^2}$$

(4.24)
$$h' = \frac{[\sum_j (N - n_j)S_j^2]^2}{(\sum_j n_j S_j^2)^2 + N\sum_j (N - 2n_j)S_j^2)^2}$$

(4.25)
$$h = \frac{[\sum_j (n_j - 1)S_j^2]^2}{\sum_j (n_j - 1)(S_j^2)^2}$$

where $N = \sum_j n_j$.

Thus, if

$$S_1^2 = 1 \qquad S_2^2 = 2 \qquad S_3^2 = 3$$

and

$$n_1 = 7 \qquad n_2 = 5 \qquad n_3 = 3$$

then

$$b = 1.28 \qquad h' = 1.86 \qquad h = 10.00$$

Assuming $\alpha = .05$, we shall reject H_0 if the ratio of mean squares is larger than 1.28 times the F value of F in Table A–5 that is associated with .05 as well as 1.86 and 10 df. Of course, there is some difficulty in finding a column labeled 1.86 df in Table A–5; it is clear, however, that the required F lies between 4.10 and 4.96, presumably closer to 4.10. Linear interpolation yields 4.22, a fair approximation. The critical region therefore consists of those values of F equal to or greater than 1.28×4.22, or 5.40.

The combination of two violations, heterogeneity of variance and non-normality, is not especially worse than heterogeneity of variance alone, provided that all treatment populations have the same distribution function. Increasing n is helpful and n should be constant over groups.

When treatment population distribution functions and variances are both heterogeneous, Type I error rates are again inflated. The degree of inflation will vary with the set of distribution functions as well as with the variances; it is consequently difficult to provide a definitive description of the effects of these violations. Some feeling for the consequences may be obtained from the results of several computerized investigations. Boneau (1960), using samples from two of these three—normal, exponential, uniform—and variances of 1 and 4, found that the Fs required for significance at the .01 level were exceeded in about 1–2 percent of his cases, and at the .05 level, in 5–7 percent. Norton (Lindquist, 1953), with a 45:1 ratio of largest to smallest variance and four degrees of skewness varying from a normal to a J-shaped distribution, obtained .036 and .10 for the nominal .01 and .05 levels with $n = 3$. The error rates dropped to .029 and .081 with $n = 10$. Bradley (1964) used a normal and a skewed distribution and a 4:1 variance ratio. With $a = 2$ and ns equal, the results were similar to Boneau's. With a equal to 3 and 4, the operative α levels varied greatly as a function of the number of groups sampled from each population and the particular variance-distribution combinations used; when normal populations with small variances were pitted against skewed populations with larger variances, error rates tended to be more inflated than when the normal populations had the larger variance; in the former case, Fs required for significance at the .01 and .05 levels were sometimes exceeded in .05 and .12 of the computerized "experiments" whereas in the latter case, the operative α levels were about .02–.03 and .06–.07. Nor does increased sample size necessarily improve the results.

If the parent distributions are all normal and the variances differ, power is not too greatly affected so long as the ns are equal (Scheffé, 1959; Boneau, 1962). My impression is that heterogeneity of variance leads to a slight elevation of the power function to a degree related to the size of the differences among variances, but rarely by more than 5 percent for any value of $\theta_A^2 / \bar{\sigma}_j^2$.

Donaldson (1968) considered exponential distributions, which are skewed and have $\sigma_j^2 = \mu_j^2$. In this case, power to detect small effects was slightly reduced but power to detect larger effects was increased. Again, group sizes were equal.

4.3.4 SUMMARY.   Even extreme nonnormality does not affect Type I or Type II error rates much; in those rare cases in which the departure from the theoretical value is more than a few hundredths, slight increases in error *df* improve the situation. When sample sizes and population distribution functions are identical, heterogeneity of variance is of little consequence, generally resulting in a slight inflation of Type I error rate. Problems arise when heterogeneous variances are accompanied by unequal *n*s or distributions that vary with the experimental treatment. If the population distributions are normal, adjustments for heterogeneous variances are possible despite variations in group size. If distribution functions vary, however, we lack adjustment procedures comparable to the Welch and Box solutions for the normal distribution case. Consequently, it is important to try to assess before the experiment how the independent variable might affect the distribution of the dependent variable. Occasionally, a choice of dependent variables is possible and we can aim for homogeneity of population shape, if not variance. Unequal *n*s should certainly be avoided wherever possible. If the data do reflect heterogeneity of both shape and variance, the experimenter must recognize that the nominal α level is at best an approximation to the actual probability of a Type I error, probably an underestimate ranging from 1 percent to 7 percent, depending on the nominal α level and the particular combination of violations of assumptions; if the *n*s are not equal, the distortion may be far greater.

4.4   TRANSFORMATIONS OF THE DATA

If there is evidence of some systematic relation between treatment population mean and variance, homogeneity of variance can be obtained through an appropriate transformation of the data. Bartlett (1947) has presented a formula for deriving such transformations provided that the relation between μ_j and σ_j^2 is known. Often when the nature of the relation is unclear, the experimenter can, through trial and error, find a transformation that will stabilize the within-group variances. Three of the most useful transformations are noted here.

4.4.1   THE SQUARE ROOT TRANSFORMATION.   The square root transformation is applicable when $\sigma_j^2 = k\mu_j$, that is when the means and variances are proportional for each treatment. This situation is not unusual when the data are in the form of frequency counts, for example, when the dependent variable is number of bar presses. In such cases, the analysis of variance is carried out not on Y but on Y', where

(4.26) $$Y'_{ij} = \sqrt{Y_{ij}}$$

and Y_{ij} is the score originally obtained from the ith subject in the jth group. If some values of Y_{ij} are less than 10, homogeneity of variance is more likely to be produced by the transformation,

(**4.27**) $$Y'_{ij} = \sqrt{Y_{ij}} + \sqrt{Y_{ij} + 1}$$

4.4.2   THE ARC SINE TRANSFORMATION.   The arc sine transformation is applicable when $\sigma_j^2 = k\mu_j(1 - \mu_j)$, that is, when the scores are proportions; an example is percentage correct or percentage predictions of some event. In such cases, the appropriate transformation is

(**4.28**) $$Y'_{ij} = \arcsin \sqrt{Y_{ij}}$$

The transformed score is the angle whose sine equals the square root of the original score. If the original score is .50, its square root is approximately .707. Turning to a table of natural trigonometric functions (available in most books of mathematical or statistical tables), we find that the sine of 45° is .707; Y'_{ij} is 45. The transformation can be made directly, without computing the square root, if tables of $\arcsin \sqrt{Y}$ are available. Such a table is in Snedecor and Cochran (1967) or Fisher and Yates (1957). If the proportion Y_{ij} is close to 0 or 1, replace it in Equation (4.28) by

$$Y^*_{ij} = \frac{kY_{ij} + \frac{3}{8}}{k + \frac{3}{4}}$$

where k is the number of observations on which the proportion is based.

Smith (1976) has suggested applying the arc sine transformation to rating scale data if cell means fall close to scale end points. In this case, each rating would be transformed to a proportion of the distance between end points. A 2 on a scale from 1 to 5 would correspond to .25, which would then be transformed as described above.

4.4.3   THE LOGARITHMIC TRANSFORMATION.   The logarithmic transformation is applicable when $\sigma_j^2 = k\mu_j^2$, that is, when the treatment standard deviation is proportional (k is a constant of proportionality) to the treatment mean. This situation will sometimes arise when the distribution of scores is markedly skewed; thus, reaction time scores may be amenable to this transformation. This transformation is also applicable when the scores are standard deviations. Equation (4.29) characterizes the transformation.

(**4.29**) $$Y'_{ij} = \log Y_{ij}$$

If some of the measures are small, the recommended transformation is

(**4.30**) $$Y'_{ij} = \log(Y_{ij} + 1)$$

4.4.4   INTERPRETING TRANSFORMED DATA.   Transformations can cause interpretative problems. Life is reasonably simple if we are concerned solely with the general null hypothesis for a one-factor design. In this case, a significant result for the transformed data implies rejection of the null hypothesis on the original data scale. Other data analyses will cause difficulties however, even in the one-factor case. First, point or interval estimates on the transformed scale may not mean anything. Most of us should find a point estimate of $\mu'_1 - \mu'_2$, where μ'_j is based on the arc sin $\sqrt{Y_{ij}}$, relatively unintelligible. One solution is to retransform $\bar{Y}'_{.j}$, the mean of the transformed data, to the original scale. Thus, assuming an arc sine transformation, if $\bar{Y}'_{.j}$ is 45 we take as our estimate of μ_j, .50, the value whose transform is 45. Note that this value, .50, will generally not equal $\bar{Y}_{.j}$, the mean on the original scale.

A second problem arises in testing hypotheses about specific functional relations between the independent and the dependent variables. A linear relation, for example, on the original data scale will be nonlinear on a transformed scale. Unless we can translate our original linearity hypothesis using a different specific function on the transformed scale, any gain due to variance stabilization may be more than offset by the ambiguity of the results of the analysis.

Other interpretative problems arise when designs involve several factors. Suppose, for example, we want to select one of two methods of teaching reading. We have equal numbers of male and female subjects, using two subjects in each cell; this is neither a plausible nor a recommended n, but it is enough for the example. The scores on a test of reading ability are:

$$
\begin{array}{ccc}
 & \text{Male} & \text{Female} \\
\text{Method} \quad \begin{matrix} 1 \\ 2 \end{matrix} & \begin{bmatrix} 4, 9 \\ 1, 4 \end{matrix} & \begin{matrix} 9, 16 \\ 16, 25 \end{bmatrix}
\end{array}
$$

Homogeneity of variance can be achieved by a square root transformation. On this scale, the data are:

$$
\begin{array}{ccc}
 & \text{Male} & \text{Female} \\
\text{Method} \quad \begin{matrix} 1 \\ 2 \end{matrix} & \begin{bmatrix} 2, 3 \\ 1, 2 \end{matrix} & \begin{matrix} 3, 4 \\ 4, 5 \end{bmatrix}
\end{array}
$$

Let us assume that it is uneconomical, and perhaps undesirable, to use a different method with each sex. Then, we should want to pick the method that is better when we average over the two sexes. The original data suggest that this is method 2; the transformed data indicate no difference between the methods. Although the transformed data have more desirable distributional properties, it is the original test score that school systems use for such purposes as student placement. The point is that the difference in the average over some second variable may yield conflicting conclusions on two scales. When this is the case, theoretical implications or intended application of results should take precedence over the desire to have homogeneous variances.

There is a second potential problem with transformations in multifactor research. We are frequently interested in interactions, the relative effectiveness of one variable over several levels of another variable. It might be of interest, for example, to ask whether the difference in performance under two methods of teaching reading is constant over several levels of IQ. If not, we might decide to use different methods with different IQ levels. The answer to our question is again likely to be scale-dependent. If differences among effects of methods are constant over IQ levels on one scale, they generally will not be constant on a transformed data scale. Again, substantive considerations should take priority in choosing the data scale.

Besides reducing heterogeneity of variance, transformations sometimes result in a closer approximation to the normal distribution. Furthermore, they are most useful in complex designs, where it is desirable to eliminate certain effects of combinations of variables that are unaccounted for by the usual analysis of variance models. (We reserve discussion of this problem for Chapter 7, when the relevant designs will first be met. For further information on transformations, including additional transformations and estimates of what the variance will be on the new scale, see Bartlett's paper (1947).)

The use of nonparametric statistics has been recommended as an alternative approach to data analysis when assumptions have been violated (Siegel, 1956). However, these techniques are also sensitive to differences in other parameters besides the mean and are not necessarily more powerful than the F test when the assumptions of that test are violated (Boneau, 1962). We feel that the use of nonparametric statistics should usually be limited to instances in which the data are originally in the form of ranks or frequency counts, as in contingency table problems, or when we need a quick approximate indication of significance. The pros and cons of nonparametric statistics and of their advantages and disadvantages relative to those of analysis of variance have been covered by Siegel (1956), Gaito (1959), and Anderson (1961), among others. The main reason for my restraint in using nonparametric statistics is simply that they are not versatile enough, that researchers who use the nonparametric approach are limited in the designs they can use and in what they can ask of their data.

4.5   *F* RATIOS LESS THAN 1

If the null hypothesis is correct, the ratio of expected mean squares is 1, and on the basis of sampling variability, one may expect occasional F ratios less than 1. Such occurrences are regarded merely as support for the null hypothesis. The occurrence of Fs so small that their reciprocals are significant or the occurrence of many Fs less than 1 in a single analysis of variance merits further consideration, however. Such findings suggest that the model underlying the analysis of variance has somehow been violated. A frequent occurrence is the presence of some systematic effect that is not described by the analysis of variance model and consequently is not accounted for in the analysis of the data. If all experimental groups consist of some subjects run by one experimenter and some subjects run by another experimenter, for example, within-group variability may be increased without a concomitant increase in between-groups variability. If the systematic factor can be designated, its contribution to the error variance can often be removed with the loss of a few *df*s in the error term.

4.6   THE ANALYSIS OF VARIANCE

Having considered the relevant theory, we now turn to the actual data analysis. It will be shown that SS_A and $SS_{S/A}$ account for the total variability in the data; then raw score formulas for these quantities will be developed. Finally, these formulas will be applied in analyzing a sample set of data.

4.6.1   COMPONENTS OF VARIABILITY.   Consider the following identity, which states that the deviation of a score from the grand mean consists of two components: (1) the deviation of a score from the mean of its experimental group, and (2) the deviation of the group mean from the grand mean.

(**4.31**) $$Y_{ij} - \bar{Y}_{..} = (Y_{ij} - \bar{Y}_{.j}) + (\bar{Y}_{.j} - \bar{Y}_{..})$$

Squaring both sides of Equation (4.31) gives

(**4.32**) $$(Y_{ij} - \bar{Y}_{..})^2 = (Y_{ij} - \bar{Y}_{.j})^2 + (\bar{Y}_{.j} - \bar{Y}_{..})^2 + 2(Y_{ij} - \bar{Y}_{.j})(\bar{Y}_{.j} - \bar{Y}_{..})$$

If we sum over i and j for both sides of Equation (4.32), remembering the notational rules presented earlier, the result is

(**4.33**) $$\sum_i^n \sum_j^a (Y_{ij} - \bar{Y}_{..})^2 = \sum_i^n \sum_j^a (Y_{ij} - \bar{Y}_{.j})^2 + n \sum_j^a (\bar{Y}_{.j} - \bar{Y}_{..})^2$$
$$+ 2 \sum_i^n \sum_j^a (Y_{ij} - \bar{Y}_{.j})(\bar{Y}_{.j} - \bar{Y}_{..})$$

The cross-product term, $\sum_i^n \sum_j^a (Y_{ij} - \bar{Y}_{.j})(\bar{Y}_{.j} - \bar{Y}_{..})$, equals zero. This can be proved by rearranging terms.

(**4.34**)
$$\sum_i^n \sum_j^a (Y_{ij} - \bar{Y}_{.j})(\bar{Y}_{.j} - \bar{Y}_{..}) = \sum_j \left[(\bar{Y}_{.j} - \bar{Y}_{..}) \sum_i (Y_{ij} - \bar{Y}_{.j}) \right]$$
$$= \sum_j (\bar{Y}_{.j} - \bar{Y}_{..})(0)$$
$$= 0$$

Consider Equation (4.33) again, ignoring the cross-product term, which has been proved equal to zero. The term $\sum_i \sum_j (Y_{ij} - \bar{Y}_{..})^2$ is the numerator of the variance of all scores about the grand mean and will henceforth be referred to as the total sum of squares (SS_{tot}). The term $n \sum_j (\bar{Y}_{.j} - \bar{Y}_{..})^2$ is n times the numerator of the variance of the group means about the grand mean. This is usually referred to as the between-groups sum of squares (SS_A). The term $\sum_i \sum_j (Y_{ij} - \bar{Y}_{.j})^2$ is the within-groups sum of squares ($SS_{S/A}$, the subscript representing "subjects within levels of A").

Table 4–1   Analysis of variance for the completely randomized one-factor design

SV	df	SS	MS	EMS[a]	F
Total	$an-1$	$\sum\limits_{i}^{n}\sum\limits_{j}^{a} Y_{ij}^2 - \dfrac{T_{..}^2}{an}$			
A (between groups)	$a-1$	$\sum\limits_{j}^{a} \dfrac{T_{.j}^2}{n} - \dfrac{T_{..}^2}{an}$	$\dfrac{SS_A}{a-1}$	$\sigma_e^2 + n\theta_A^2$	$\dfrac{MS_A}{MS_{S/A}}$
S/A (within groups)	$a(n-1)$	$\sum\limits_{i}^{n}\sum\limits_{j}^{a} Y_{ij}^2 - \sum\limits_{j}^{a} \dfrac{T_{.j}^2}{n}$	$\dfrac{SS_{S/A}}{a(n-1)}$	σ_e^2	

[a] $\theta_A^2 = \dfrac{\sum(\mu_j - \mu)^2}{a-1}$

4.6.2   SUMMARIZING THE ANALYSIS OF VARIANCE.   Table 4–1 summarizes much of the material presented thus far. The first column on the left contains the *sources of variance SV*. These follow from Equation (4.1), which states that the deviation of a score from the population mean consists of a treatment component A and a component due to individual differences S/A.

Degrees of freedom. The second column from the left in Table 4–1 contains the *df* associated with each *SV*. In deriving the expected mean squares *EMS*, we noted that the *df* associated with the A effects are $a-1$, and those associated with S/A are $a(n-1)$. Let us consider why this is so.

Suppose that we are asked to choose four numbers that sum to 100. The first three numbers chosen can be any three finite numbers. However, the fourth number must be 100 minus the sum of the first three numbers; only three numbers are chosen freely. If the first three numbers chosen are 41, 3, and -18, the fourth number, k_4, must be $100 - 41 - 3 - (-18) = 74$. We characterize this situation by saying that there is one restriction on the data, causing us to lose one *df*. If it is required that the first two numbers chosen sum to 50 and all four sum to 100, there are two restrictions on the data and two *df* are thus lost. Only two numbers may be freely chosen. Analogous geometrical examples exist. Suppose that one is told to draw a triangle, the only restriction being that the figure must be a closed three-sided one. This restriction causes the loss of one *df*; while the first two sides may be of any length and form any angle, for example,

the third side must be a line connecting the points A and C. From these examples, we draw the generalization:

(**4.35**)
$$df = \text{number of independent observations}$$
$$= \text{total number of observations minus number}$$
$$\text{of restrictions on the observations}$$

Next, this line of reasoning is extended to statistical tests by noting that the estimation of a population parameter places a restriction on the data. Computing the variance of a group of scores involves summing squared deviations about the group mean, and the restriction is imposed that $\sum_i Y_i = n\bar{Y}$, or $\sum_i (Y_i - \bar{Y}) = 0$. Consequently, the variance of a single group of scores is based on $(n-1)$ *df*. In computing $MS_{S/A}$, a such quantities are calculated, resulting in $a(n-1)$ *df*. In computing the MS_A, we meet the restriction that $\sum_j \bar{Y}_{.j} = a\bar{Y}_{..}$. We therefore have $(a-1)$ *df*.

The role of *df* may be clearer if the above development for the F test is contrasted with the development for the normal deviate, or z test, where $z = (\bar{Y} - \mu)/(\sigma/\sqrt{n})$. Note that the population standard deviation is *known, not estimated*, and consequently there is no restriction on the data. The concept of *df* is not relevant to this test. An alternative view of the relevance or irrelevance of *df* to a statistical test is obtained by looking at the distribution of the statistic. The shape of the normal curve is not a function of the number of observations, whereas the shape of the F distribution is affected by the numerator and denominator *df*.

The importance of *df* lies not only in that they are necessary components of the F ratio and determiners of the F distribution. These quantities provide a check on the *SV* in more complex designs where a source may be overlooked or the variance wrongly analyzed in some other way. The check assumes that we have the correct *df* for each listed source, in which case they must sum to the total number of scores minus one. Furthermore, the *df* provide an alternative basis for arriving at raw score formulas for the *SS*.

Sums of squares. In view of the general access to high-speed computers, it is tempting to delete, or at least deemphasize, computational formulas and numerical examples. We have resisted this particular temptation for two reasons. First, some data manipulation is essential for a gut-level appreciation of analysis of variance. Second, with the advent of very small and inexpensive electronic calculators, calculation by hand will frequently be not only feasible, but even simpler than by computer. The computer has a decided advantage, however, when many analyses are planned for a single data set or when many terms are to be calculated.

The *SS* formulas of Table 4–1 were derived as shown in Section 2.2. For more complex designs, such procedure is tedious, and we need some simple rule that makes it possible to set down *SS* formulas quickly. The rule should generalize to all the terms of the analysis, and to analyses for any design. Such a rule follows if one notes the isomorphism existing between *df* and *SS*. There is a *df* for each squared quantity involved in the computations of *SS*, and any operation applied to the *df* is applied to the squared quantities. Consider the $SS_{S/A}$ first:

1. Expand the *df* expression.

$$df_{S/A} = a(n-1) = an - a$$

2. Each *df* in the expanded term corresponds to a squared quantity. Thus, we

now have

$$\sum_{j=1}^{a} \sum_{i=1}^{n} Y_{ij}^2 - \sum_{j=1}^{a} T_{.j}^2$$

Note that each squared quantity must have subscripts that correspond exactly to the indices of summation immediately to its left. Any indices that are not subscripted following application of this rule are replaced by dots, indicating that summation has taken place for that index before squaring. The designation $T_{.j}^2$ would imply that at the jth level of A, we have summed the n scores and then squared the total.

3.   Divide each squared quantity by the number of values on which it is based. Since Y_{ij} is a single score, Y_{ij}^2 is divided by 1. The value $T_{.j}$, however, is a total of n scores so

$$SS_{S/A} = \sum_j \sum_i Y_{ij}^2 - \sum_j \frac{T_{.j}^2}{n}$$

The SS_A and SS_{tot} are left as an exercise for the student.

The entries in the MS column of Table 4–1 are the ratios of SS to df. The MS_A is n times the variance of the group means about the grand mean, and the $MS_{S/A}$ is the average over groups of the variances of scores about group means.[2] The EMS have been derived previously (Section 4.2.3) and have been shown to justify using the F ratio as a test statistic.

4.6.3   NUMERICAL EXAMPLES.   Table 4–2 contains speeds of traversing a runway, in feet per second, for four groups of eight rats. All rats were allowed access to a solution of sucrose for 20 sec at the end of each run. The groups differed in percentage of sucrose in the solution offered.

The first step in analyzing is to get $\sum_i Y_{ij}$ and $\sum_i Y_{ij}^2$ for each group, that is, for each value of j. The two sums can be obtained simultaneously on most desk calculators. They provide the basic ingredients for the F test of the null hypothesis that the μ_j are equal. They are entered in Table 4–2 together with group means and variances; the variances are computed according to Equation (4.36):

(4.36) $$\hat{\sigma}_j^2 = \frac{\sum_i Y_{ij}^2 - T_{.j}^2/n}{n-1}$$

Now we can do the actual analysis of variance. Since terms like $T_{..}^2/an$ appear in many components of all analyses of variance, a special notation is used. Such terms are called *correction terms* and designated by C, where

(4.37) $$C = \frac{(\text{sum of all scores in the data matrix})^2}{\text{total number of scores}}$$

---

[2] Several electronic calculators have a key that on being pressed gives the standard deviation for some previously entered set of values. If you are working with such a calculator, enter the $\bar{Y}_{.j}$, press the standard deviation key, square, and multiply by n; the result is MS_A as defined in Equation (4.7). The quantity $MS_{S/A}$ is merely an arithmetic average of the $a\hat{\sigma}_j^2$ if the ns are equal; if they are not, multiply each $\hat{\sigma}_j^2$ by $(n_j - 1)/df_{S/A}$ and add the products to obtain a weighted average of the variances.

Table 4–2   Data for four groups of rats in a runway study

	PERCENTAGE OF SUCROSE IN WATER			
	8	16	32	64
	1.4	3.2	6.2	5.8
	2.0	6.8	3.1	6.6
	3.2	5.0	3.2	6.5
	1.4	2.5	4.0	5.9
	2.3	6.1	4.5	5.9
	4.0	4.8	6.4	3.0
	5.0	4.6	4.4	.5.9
	4.7	4.2	4.1	5.6
$T_{.j} =$	24.0	37.2	35.9	45.2
$\bar{Y}_{.j} =$	3.00	4.65	4.49	5.65
$\sum_i Y_{ij}^2 =$	86.54	186.78	171.67	264.24
$\hat{\sigma}_j^2 =$	2.08	1.97	1.51	1.27

In our example,

$$C = \frac{(24.0 + 37.2 + 35.9 + 45.2)^2}{32} = 632.79$$

and

$$SS_{\text{tot}} = 86.54 + 186.78 + 171.67 + 264.24 - 632.79$$
$$= 709.23 - 632.79$$
$$= 76.44$$

According to Equation (4.37),

$$SS_A = \frac{\sum_j T_{.j}^2}{n} - C$$
$$= \frac{(24.0)^2 + (37.2)^2 + (35.9)^2 + (45.2)^2}{8} - 632.79$$
$$= 661.46 - 632.79$$
$$= 28.67$$

The $SS_{S/A}$ can be obtained as the residual variability,

$$SS_{S/A} = SS_{\text{tot}} - SS_A$$
$$= 76.44 - 28.67$$
$$= 47.77$$

and as a check, by pooling the sums of squares for each group,

$$SS_{S/A} = \sum_{j}\left[\sum_{i} Y^2 - \frac{T_{.j}^2}{n}\right]$$

$$= \left(86.54 - \frac{(24.0)^2}{8}\right) + \cdots + \left(264.24 - \frac{(45.2)^2}{8}\right)$$

$$= 47.77$$

The first procedure corresponds to $(an-1)-(a-1)$ *df*, the second to $(n-1)+ \cdots +(n-1)$. The result in both instances is $a(n-1)$; correspondingly, the two computations on *SS* should yield identical results.

The *MS* and the *F* ratio are now easily computed.

$$MS_A = \frac{SS_A}{a-1} \qquad MS_{S/A} = \frac{SS_{S/A}}{a(n-1)}$$

$$= \frac{28.67}{3} \qquad\qquad = \frac{47.77}{28}$$

$$= 9.56 \qquad\qquad = 1.71$$

$$F = \frac{MS_A}{MS_{S/A}}$$

$$= \frac{9.56}{1.71}$$

$$= 5.59$$

The results of our analysis are summarized in Table 4–3. Turning to Table A–5, we find that for 3 and 28 *df*, an *F* of 4.57 is required for significance at the .01 level. Since our computed *F* exceeds this critical value, we feel justified in concluding that the treatment population means differ. The statement "$p < .01$" indicates that if H_0 is true, the observed *F* of 5.59 will occur in less than 1 percent of the replications of the experiment.

Table 4–3   Analysis of variance for data from a completely randomized one-factor experiment

SV	df	SS	MS	F
Total	31	76.44		
A	3	28.67	9.56	5.59[a]
S/A	28	47.77	1.71	

[a] $p < .01$

In handling actual experimental data, one would not be finished with the analysis at this point. Many interesting questions are still unanswered. Do all four means differ significantly from one another? Or are the 16 percent and 32 percent treatments essentially equivalent, as a quick look at the data suggests? If the treatment means are plotted as a function of concentration, what type of equation best describes the relation? To deal with such questions, numerous additional conceptual and computational factors must be considered. (See Chapters 11 and 17.)

Our presentation so far has been restricted to the case where n is equal for all groups, since derivations of expectations and computations in the analysis of variance are simpler under this condition. This restriction will now be removed in order to present computations for the unequal n case, in which n_j, the number of subjects in the jth treatment group, varies over levels of j. Before considering the computations, note that the analysis of variance model is fundamentally the same as previously; the main difference in the theoretical development is that now

(**4.38**)
$$E(MS_A) = \sigma_e^2 + \frac{\sum_j n_j (\mu_j - \mu)^2}{a - 1}$$

When n_j is constant over j, we have the previously presented EMS. In both cases, the F ratio involves the MS_A and the $MS_{S/A}$.

Table 4–4 contains data for the runway experiment previously analyzed, with some scores randomly discarded from the data matrix of Table 4–2.

Table 4–4   Data for the runway experiment with unequal n

	PERCENTAGE OF SUCROSE IN WATER		
8	16	32	64
1.4	3.2	6.2	5.8
1.4	5.0	3.2	6.6
5.0	2.5	4.5	6.5
2.0	6.4	4.4	5.9
2.3	4.8	4.1	5.9
	4.2		3.0
			5.9
			5.6
$T_{\cdot j} = 12.1$	26.1	22.4	45.2
$n_j = 5$	6	5	8
$\bar{Y}_j = 2.42$	4.35	4.48	5.65
$\sum_i Y_{ij}^2 = 38.21$	123.13	105.10	264.24
$\hat{\sigma}_j^2 = 2.23$	1.92	1.19	1.27

By Equation (4.37), we compute the correction term,

$$C = \frac{(12.1+26.1+22.4+45.2)^2}{(5+6+5+8)}$$

$$= \frac{(105.8)^2}{24} = 466.40$$

The SS_{tot} are now computed as previously:

$$SS_{tot} = 38.21 + 123.13 + 105.10 + 264.24 - C$$

$$= 64.28$$

The revised equation for the SS_A for the unequal n case is

$$SS_A = \sum_j \frac{T_j^2}{n_j} - C$$

Therefore we have

$$SS_A = \frac{(13.1)^2}{5} + \frac{(26.1)^2}{6} + \frac{(22.4)^2}{5} + \frac{(45.2)^2}{8} - C$$

$$= 32.15$$

The $SS_{S/A}$ can again be computed as a residual.

$$SS_{S/A} = SS_{tot} - SS_A = 32.13$$

The mean squares are then computed as before, and F is again the ratio of mean squares.

$$MS_A = \frac{32.15}{3} = 10.72$$

$$MS_{S/A} = \frac{32.13}{(4+5+4+7)} = 1.61$$

$$F = 6.66$$

Entering Table A–5 with 3 and 20 df, we find that the critical value at the 1 percent level, 4.94, is again exceeded by the obtained F. The analysis is summarized in Table 4–5.

Table 4–5   Analysis of variance for data from completely randomized one-factor experiment with unequal n

SV	df	SS	MS	F
Total	23	64.28		
A	3	32.15	10.72	6.66[a]
S/A	20	32.13	1.61	

[a] $p < .01$

4.7 ESTIMATING COMPONENTS OF VARIANCE

Thus far the discussion of the analysis of variance has been concerned solely with tests of significance—with the question, Do the different treatment levels (or treatment combinations) have different effects? There is an equally important question: How greatly do the effects of different treatment levels (or treatment combinations) differ? Indeed, it may be argued that the first question is the less important, that with enough data, the variance of any set of treatment effects is significant. Furthermore, since the psychologist invariably deals with data that are quantitative, it is reasonable to assume that he will profit by examining the magnitude of the quantity. Finally, it is only by estimating population effects that one can assess relative influences of several variables. Since it is desirable to examine the absolute and relative magnitudes of effects, we now turn to estimating components of variance that make up our expected mean squares. We shall first deal with point estimation, and then turn to developing measures of the relative contribution of variables to the total variance.

4.7.1 ABSOLUTE MAGNITUDE OF EFFECTS.   Let us assume that we are interested in estimating θ_A^2, the variance of the means of the four treatment populations defined by the levels of A. From our knowledge of EMS, we know that MS_A estimates $\sigma_e^2 + n\theta_A^2$ and $MS_{S/A}$ estimates σ_e^2. This, together with some simple algebra, yields

$$\frac{MS_A - MS_{S/A}}{n} = \hat{\theta}_A^2$$

where the diacritic "^" denotes "estimate of." In general, *for any design*, to estimate a component of variance, we should:

1. Calculate the difference between the *MS* that would be in the numerator of an *F* test of that component and its error term, or denominator.
2. Divide this difference by the one or more coefficients by which the component of interest is multiplied.

Applying the above procedure to Table 4–3, we conclude that

$$\hat{\theta}_A^2 = \frac{9.56 - 1.71}{8} = .98$$

If H_0 is true, $MS_{S/A}$ may, by chance, be greater than MS_A and the estimate of θ_A^2 will be negative. Such a result is meaningless since variances cannot be negative. We conclude that our best estimate is zero; the negative estimate is assumed to be a chance result.

What do we gain from such estimation? We get a direct measure of how important our independent variable is. Neither the *F* ratio nor its level of significance provide this, since both these quantities are influenced by *n* and error variance. This point should become clear if we suppose that the results presented in Table 4–3 are based on 40 subjects in each group instead of eight, and that

MS_A is now 10.51. The revised F is now 6.15 but the estimate of θ_A^2 is .22, one-fourth its previous value. The larger F is testimony to the greater amount of data collected rather than to increased effectiveness of our experimental manipulation. The F ratio does not tell us whether an effect is large or small but only (with some probability) whether it is zero or not. Nor can F ratios be compared to provide knowledge of the relative effectiveness of two variables—or the same variable in two different experiments—unless the *df* are identical for the two variables and they are tested against the same error variance. Estimates of the magnitudes of effects are prerequisite, moreover, to establishing quantitative behavioral laws.

4.7.2 RELATIVE MAGNITUDE OF EFFECTS.

While the absolute value of the variance due to an independent variable is informative, its size relative to the size of the variances that other sources sponsor is even more so. The values of the F ratio and of $\hat\theta_A^2$ may show that A is an important source of variance but the estimated variance due to A relative to the variance estimated for the population as a whole will reveal to us whether there are other important sources of variation. We use ω^2 to designate this relative magnitude of the effect of a variable. The computation of ω^2 depends on the design, the assumed structural equation that relates data to population parameters, and whether effects are random or fixed.

Assuming Equation (4.1), we have

$$Y_{ij} - \mu = \alpha_j + \varepsilon_{ij}$$

Squaring both sides and taking expectations over the set of treatment populations, we have

$$E(Y_{ij} - \mu)^2 = \frac{\sum \alpha_j^2}{a} + E(\varepsilon_{ij})^2$$

Note that the cross-product term vanishes because of the presumed independence of α and ε. The last equation can be rewritten:

(4.39) $$\sigma_Y^2 = \left(\frac{a-1}{a}\right)\theta_A^2 + \sigma_e^2 = \delta_A^2 + \sigma_e^2$$

Note that whereas $\theta_A^2 [= (\sum \alpha_j^2)/(a-1)]$ is not truly the variance of the α_j, $\delta_A^2 [= (\sum \alpha_j^2)/a]$ is.

We define

(4.40) $$\omega_A^2 = \frac{\delta_A^2}{\sigma_Y^2}$$

the proportion of the total population variance attributable to the effects of A.

From the *EMS* for this design, point estimates are readily obtained:

$$\hat\delta_A^2 = \left(\frac{a-1}{a}\right)\hat\theta_A^2$$

(4.41) $$= \left(\frac{a-1}{a}\right)\left(\frac{MS_A - MS_{S/A}}{n}\right)$$

$$\hat\sigma_e^2 = MS_{S/A}$$

Substituting Equation (4.41) into Equation (4.40), and replacing population parameters by estimates, we get:[3]

(4.42)
$$\hat{\omega}_A^2 = \frac{[(a-1)/a](1/n)(MS_A - MS_{S/A})}{[(a-1)/a](1/n)(MS_A - MS_{S/A}) + MS_{S/A}}$$

Multiplying numerator and denominator by an and dividing by $MS_{S/A}$ yields

(4.43)
$$\hat{\omega}_A^2 = \frac{(a-1)(F_A - 1)}{(a-1)(F_A - 1) + na}$$

where

$$F_A = \frac{MS_A}{MS_{S/A}}$$

For the data of Tables 4–2 and 4–3, we have

$$\hat{\omega}_A^2 = \frac{(3)(4.59)}{(3)(4.59) + 32} = .30$$

We estimate that A accounts for 30 percent of the variance in the set of treatment populations. While this is not trivial, 70 percent of the variance is due to factors for which we have not accounted.

4.8   POWER OF THE F TEST

The power of the F test is usually plotted against an alternative measured by ϕ, where ϕ^2 may be thought of as a ratio of mean squares based on the treatment populations; that is, it is n times the variance among the treatment population means divided by the population error variance. Thus,

(4.44)
$$\phi^2 = \frac{n \sum_j (\mu_j - \mu)^2 / a}{\sigma_e^2}$$

Turning to Table A–8 in the Appendix, we see that power is graphed as a function of ϕ for several combinations of df and α levels. Note that power increases with increasing values of ϕ, df_1, df_2, and α. (These relations have been previously discussed in Chapter 3.) Note also that for any constant variance among the treatment population means, power will increase as error variance decreases, for ϕ will vary inversely with σ; this is apparent from Equation (4.44).

The power functions of Table A–8 provide a basis for deciding how many subjects to include in the experiment. For appropriate use of the table, we require that the following be selected or estimated:

1. *The α level.* This reflects our willingness to risk Type I errors.

---

[3] If we hold ω_A^2 to be $E(\hat{\omega}^2)$, Equation (4.42) provides an approximation since $E(\hat{\delta}_A^2)/E(\hat{\sigma}_e^2) \neq E(\hat{\omega}^2)$; that is, the ratio of expected values does not generally equal the expected value of a ratio. The approximation is reasonably accurate, however, and much simpler than the correct expression.

2. *The level of power.* This is the probability of rejecting a false null hypothesis. This value is not selected in the same way that α is, for power varies as a function of ϕ and the exact value of this latter quantity is unknown. One can select, however, a probability of rejecting H_0, given that ϕ is equal to or greater than some critical value ϕ'.

3. *The error variance,* σ_e^2. Previous experimentation or a pilot study with the dependent variable of interest will provide an estimate of σ_e^2. If several estimates are available, the largest should be chosen. The larger the value of σ_e^2, the larger the value of n required to provide a given level of power.

4. *The critical variance among treatment population means.* This is the minimum value of $\sum(\mu_j - \mu)^2/a$ that it is important to detect. In applied research problems, there may be some magnitude of variance of treatment effects worth considering; below this level any differences in mean performance may be unimportant relative to differences in cost or efficiency in applying the treatments. In other research problems, we may try to find out whether treatment effects estimated from a previous pilot study are significant. Or in the course of developing a theory, we may want at first to model only those independent variables whose effects reach a certain level. The choice of a critical value varies, then, with the purpose of the experiment and our knowledge of the research area.

The following example shows how the above factors combine to permit a selection of n. We assume

$$a = 4 \qquad\qquad \sigma_e^2 = 200$$

$$\alpha = .01 \qquad\qquad \frac{\sum(\mu_j - \mu)^2}{4} = 125$$

$$\text{power} = .80$$

The last figure was arrived at by assuming that it was important to reject H_0 if the successive μ_j were 10 or more units apart. Then,

$$\mu_1 = \mu - 15 \qquad\qquad \mu_3 = \mu + 5$$
$$\mu_2 = \mu - 5 \qquad\qquad \mu_4 = \mu + 15$$

and

$$\frac{\sum(\mu_j - \mu)^2}{4} = \left(\frac{1}{4}\right)[(-15)^2 + (-5)^2 + (5)^2 + (15)^2] = 125$$

We can now calculate $\phi' = \sqrt{n(125)/200} = .79\sqrt{n}$. We require a value of n such that the quantities $\phi' = .79\sqrt{n}$, $\alpha = .01$, $df_1 = 3$, and $df_2 = 4(n-1)$ result in power $= .80$. Turning to the chart for $df_1 = 3$ and $\alpha = .01$, we consider various values of n. Suppose that we first try $n = 5$. Then df_2 is 16 and ϕ' is 1.77. Using the .01 scale, we project upward from a point on the abscissa corresponding to 1.77 until we intersect with a point slightly above the $df_2 = 15$ curve. A horizontal line drawn from that point would intersect the ordinate somewhere between .4 and .5, clearly less than the required power of .8. We must try a larger value of n.

A trial value of 9 results in $df_2 = 32$ and $\phi' = 2.37$; the resulting power value on the ordinate is now slightly less than .9 but clearly above .8. We can run groups smaller than size 9. An n of 8 does quite nicely; the error df is now 28, $\phi' = 2.2$, and the resulting power is just about .80. Therefore we decide on 8 subjects for each treatment group.

4.9   CONCLUDING REMARKS

The one-dimensional design is not often directly applied in psychological research. Many studies involve repeated measurements on subjects, calling for parceling out subject effects in addition to treatment effects. Even studies in which there are no repeated measurements usually involve more than one treatment variable. Nevertheless, notation, derivations, computations, null hypothesis testing, and the model underlying the use of the F test are all extremely important, and will be involved in our further work. If you master them, you will more easily understand the complicated designs and analyses of subsequent chapters.

EXERCISES

4.1   In deriving *EMS*, we assume that $\sum_j^a \alpha_j = 0$. We do not assume, however, that $\sum_i^n \sum_j^a \varepsilon_{ij} = 0$, but rather that

$$E\left(\sum_i^n \sum_j^a \varepsilon_{ij}\right) = 0$$

Why does this distinction between α_j and ε_{ij} exist? How would the situation change if the a levels of the variable A were randomly selected from a population of levels?

4.2   Analyze variance for the following sets of data:

(a)

A_1	A_2	A_3
28	38	64
23	39	73
21	57	61
38	36	48
38	38	72
49	48	52
28	52	54
33	40	60
34	39	54
29	45	60

(b)

A_1	A_2	A_3	A_4	A_5
24	48	42	96	73
07	91	82	67	81
46	63	75	88	33
45	69	76	24	44
97	26		92	94
	22		83	77
	45			60
				89
				25

4.3 Analyze variance for the following: for (a) $n_1 = n_2 = n_3 \doteq 10$; and (b) $n_1 = 6$, $n_2 = 8$, $n_3 = 10$.

Group	A_1	A_2	A_3
$T_{.j}$	30	45	70
$\hat{\sigma}_j^2$	3.2	4.1	5.7

4.4 One of our clinical faculty—call him Norm—wants to compare four methods of therapy. He runs 10 patients in each group and has progress in therapy rated blind by a group of colleagues. The results are:

Source	df	MS
A	3	35
S/A	36	20

which are not significant at any usually accepted level. Along comes Hal, appalled at his colleague's laziness. He redoes the study with 100 patients in each treatment group. The results are:

Source	df	MS
A	3	170
S/A	396	20

The results are now clearly statistically significant. Discuss the discrepancy between Norm's and Hal's findings. Focus on the question whether there is any practical significance in Hal's results. Relate your discussion to the concept of *EMS*.

4.5 One way that has been proposed to measure the relative importance of a variable A is

$$\hat{\omega}_A^2 = \frac{\hat{\delta}_A^2}{\sigma_e^2 + \hat{\delta}_A^2}$$

where $\hat{\delta}_A^2 = [(a-1)/a]\hat{\theta}_A^2$. This is the estimated population variance due to A divided by the estimated total population variance.

(a) (i) Find $\hat{\omega}_A^2$ for this 3-group experiment ($n = 10$):

SV	df	MS
A	2	80
S/A	27	5

(ii) Suppose $n = 5$ and we had the following analysis of variance:

SV	df	MS
A	2	42.5
S/A	12	5

Now find $\hat{\omega}_A^2$.

(b) Comparing your answers to (i) and (ii), what do you conclude about the effect of n on $\hat{\omega}_A^2$?

(c) If $F = 1$, what is $\hat{\omega}_A^2$? Briefly explain.

(d) An alternative measure of the relative importance of a variable is

$$\hat{\eta}_A^2 = \frac{SS_A}{SS_{S/A} + SS_A}$$

   (i) What does SS_A estimate? (In other words, find $E(SS_A)$.)

   (ii) What does $SS_{S/A}$ estimate? (Careful! These are SS, not MS, quantities.)

   (iii) What does η_A^2 estimate?

(e) Compute the value of $\hat{\eta}_A^2$ for:

   (i) The data in (a, (i)).

   (ii) The data in (a, (ii)).

   What do you conclude about the effect of n on $\hat{\eta}_A^2$?

(f) Suppose that $F = 1$. What is $\hat{\eta}_A^2$ for a groups of n subjects?

(g) Which measure ($\hat{\omega}_A^2$ or $\hat{\eta}_A^2$) is preferable, do you think, and why?

4.6 For each of the following cases, state whether you agree or disagree with the conclusion, and *briefly* state why.

(a) An experiment is run with six-year-old and ten-year-old children. There are 4 levels of A, 10 subjects per group. We have

	SV	df	MS	F
6-year-olds	A	3	130	13.0
	S/A	36	10	
10-year-olds	A	3	105	21.0
	S/A	36	5	

We conclude that the variable A has a greater effect with ten-year-olds because the F is larger.

(b) The same experiment is run with 3 groups of 15 six-year-old children and with 3 groups of 10 ten-year-old children.

	SV	df	MS	F
6-year-olds	A	2	160	16.0
	S/A	42	10	
10-year-olds	A	2	130	13.0
	S/A	27	10	

We conclude that A is more effective with six-year-old children.

(c) Suppose the data on the six-year-olds are the same as in (b). However, the results for the ten-year-olds are now

SV	df	MS	F
A	2	95	19.0
S/A	27	5	

We now conclude that A is more effective with ten-year-old children.

(d) Is a comparison of F ratios (or of significance levels) an appropriate way to compare the importance of variables? *Briefly*, why or why not?

4.7 To determine what effect time in therapy has on schizophrenic patients in a state hospital, the clinical staff agree on the following experiment. A group of n subjects will be individually treated on a daily basis for one year, n subjects will be treated for six months, and n others will receive no special therapy. As a measure of the success of therapy, the clinicians will use a 100-point scale that differentially weighs several aspects of personality, which has proved sensitive to experimental manipulations in other studies. They agree that each six months of the individual therapy should produce a mean gain of at least ten points for the technique to be worth the staff's time. They set $\alpha = .05$ and $\beta = .10$, and on the basis of previous research, estimate σ_e^2 at 225. How many patients are needed for the study?

4.8 A previous experiment with a new drug (developed to facilitate the mathematical aptitude factor) has given these results in comparing it with a placebo.

SV	df	MS	F
A	1	40	4.0
S/A	18	10	

The result is not significant at the .01 level.
(a) With $\alpha = .01$, did the experimenter have good power to reject H_0 if the true value of $\sum_i (\mu_i - \mu)^2/2 = 1.5$? Explain briefly showing relevant work.
(b) Why did I choose the value 1.5 as the critical population variance to detect?
(c) If the power in (a) was too low, how many subjects should be run in a repeat of the experiment?

REFERENCES

Anderson, N. H. Scales and statistics: Parametric and nonparametric. *Psychological Bulletin* 58:305–16 (1961).

Bartlett, M. S. Properties of sufficiency and statistical tests. *Proceedings of the Royal Society of London* A 160:238 (1937).

Bartlett, M. S. The use of transformations. *Biometrics* 3:39–52 (1947).

Bevan, M. F., Denton, J. Q., and Myers, J. L. The robustness of the *F* test to violations of continuity and form of treatment population. *British Journal of Mathematical and Statistical Psychology* 27:199–204 (1974).

Boneau, C. A. The effects of violations of assumptions underlying the *t* test. *Psychological Bulletin* 57:49–64 (1960).

Boneau, C. A. A comparison of the power of the *U* and *t* tests. *Psychological Review* 69:246–56 (1962).

Box, G. E. P. Some theories on quadratic forms applied in the study of analysis of variance problems: I. Effect of inequality of variance in the one-way classification. *Annals of Mathematical Statistics* 25:290–302 (1954).

Bradley, J. V. *Studies in Research Methodology VI. The Central Limit Effect for a Variety of Populations and the Robustness of Z, T, and F.* AMRL Technical Documentary Report 64–123, 6570th Aerospace Medical Research Laboratories, Wright-Patterson Air Force Base, Ohio, December, 1964.

Cochran, W. G. The distribution of the largest set of estimated variances as a fraction of their total. *Annals of Eugenics* 11:47–52 (1941).

Cochran, W. G. Some consequences when the assumptions for the analysis of variance are not satisfied. *Biometrics* 3:22–38 (1947).

Donaldson, T. S. Robustness of the *F*-test to errors of both kinds and the correlation between the numerator and denominator of the *F*-ratio. *Journal of American Statistical Association* 322:660–76 (1963).

Fisher, R. A., and Yates, F. *Statistical Tables for Biological, Agricultural and Medical Research.* Edinburgh: Oliver and Boyd, 1955.

Gaito, J. Nonparametric methods in psychological research. *Psychological Review* 5:115–25 (1959).

Games, P. A., and Lucas, P. A. Power of the analysis of variance of independent groups on nonnormal and normally transformed data. *Educational and Psychological Measurement* 26:311–27 (1966).

Hartley, H. O. The maximum *F*-ratio as a short-cut test for heterogeneity of variance. *Biometrika* 37:308–12 (1950).

Hsu, T. C., and Feldt, L. S. The effect of limitations on the number of criterion score values on the significance level of the *F*-test. *American Educational Research Journal* 6:515–27 (1969).

Layard, W. J. Robust large-sample tests for homogeneity of variances. *Journal of the American Statistical Association* 68:195–98 (1973).

Levene, H. Robust tests for equality of variances. Edited by I. Olkins. In *Contributions to Probability and Statistics.* Stanford: Stanford University Press, 1960.

Lindquist, E. F. *Design and Analysis of Experiments in Psychology and Education.* Boston: Houghton Mifflin, 1953, pp. 78–90.

Lunney, G. H. Using analysis of variance with a dichotomous variable: An empirical study. *Journal of Educational Measurement* 7:263–69 (1970).

Scheffé, H. *The Analysis of Variance.* New York: John Wiley & Sons, 1959.

Siegel, S. *Nonparametric Statistics for the Behavioral Sciences.* New York: McGraw-Hill, 1956.

Smith, J. E. K. Data transformations in analysis of variance. *Journal of Verbal Learning and Verbal Behavior* 15:339–46 (1976).

Snedecor, G. W. and Cochran, W. G. *Statistical Methods*, 6th ed. Ames: Iowa State University Press, 1967.

Srivastava, A. B. L. Effect of nonnormality on the power of the analysis of variance test. *Biometrika* 46:114–22 (1959).

Welch, B. L. The significance of the difference between two population means when the population variances are unequal. *Biometrika* 29:350–62 (1937).

SUPPLEMENTARY READING

An excellent review of the status of analysis of variance relative to violations of assumptions is presented in

Glass, G. V., Peckham, P. D., and Sanders, J. R. Consequences of failure to meet assumptions underlying the analysis of variance and covariance. *Review of Educational Research* 42:237–88 (1972).

5 COMPLETELY RANDOMIZED MULTIFACTOR DESIGNS

5.1 INTRODUCTION

The modern theory of experimental design permits the experimenter to study several independent variables, or factors, within the same experiment. This has two big advantages. It is efficient, saving the time and effort of the experimenter, and, equally important, it permits investigating the joint effects of variables. In completely randomized designs involving more than a single factor, subjects are randomly assigned to combinations of treatment levels. The advantages and disadvantages of the completely randomized one-factor design have already been cited (Chapter 4); they hold also for designs involving more than one factor. We again assume that the levels of each variable have been arbitrarily and not randomly selected.

The following experiment is an example of this type of design. We are interested in comparing the effects of 4 different amounts of reward, of 3 different delays of reward, and of the 12 combinations of amount and delay on discrimination learning in rats. Each of 120 rats is randomly assigned to one of the 12 combinations of amount and delay of reward, with the restriction that there be exactly 10 rats exposed to each combination. Ten rats might receive one food pellet 1 sec after a correct response, 10 rats might receive two food pellets 3 sec after a correct response, and so on. In general, such a design involves the random assignment of n subjects to each of ab combinations of treatment levels. In our example, n is 10, and ab has the value $4 \times 3 = 12$. The layout of such a design is presented in Table 5–1. The first subscript, represented by i in the general case, indexes the subjects within each treatment cell and varies from 1 to n. The second subscript j indexes the levels of the independent variable A and varies from 1 to a. The third subscript k indexes the levels of the variable B and varies from 1 to b. Thus, there are a levels of A; b levels of B; and ab cells corresponding to the ab treatment combinations, with abn scores, n in each cell. We shall consider the analysis of variance for this design and for similar designs that differ only in number of independent variables.

Table 5–1   Data matrix for a two-factor design

		B_1	B_2	$\cdots$	B_k	$\cdots$	B_b
A_1		Y_{111}	Y_{112}		Y_{11k}		Y_{11b}
		Y_{211}	Y_{212}		Y_{21k}		Y_{21b}
		$\vdots$	$\vdots$		$\vdots$		$\vdots$
		Y_{i11}	Y_{i12}		Y_{i1k}		Y_{i1b}
		$\vdots$	$\vdots$		$\vdots$		$\vdots$
		Y_{n11}	Y_{n12}		Y_{n1k}		Y_{n1b}
A_2		Y_{121}	Y_{122}		Y_{12k}		Y_{12b}
		Y_{221}	Y_{222}		Y_{22k}		Y_{22b}
		$\vdots$	$\vdots$		$\vdots$		$\vdots$
		Y_{i21}	Y_{i22}		Y_{i2k}		Y_{i2b}
		$\vdots$	$\vdots$		$\vdots$		$\vdots$
$\vdots$							
A_j		Y_{1j1}	Y_{1j2}		Y_{1jk}		Y_{1jb}
		Y_{2j1}	Y_{2j2}		Y_{2jk}		Y_{2jb}
		$\vdots$	$\vdots$		$\vdots$		$\vdots$
		Y_{ij1}	Y_{ij2}		Y_{ijk}		Y_{ijb}
		$\vdots$	$\vdots$		$\vdots$		$\vdots$
$\vdots$							
A_a		Y_{1a1}	Y_{1a2}		Y_{1ak}		Y_{1ab}
		Y_{2a1}	Y_{2a2}		Y_{2ak}		Y_{2ab}
		$\vdots$	$\vdots$		$\vdots$		$\vdots$
		Y_{ia1}	Y_{ia2}		Y_{iak}		Y_{iab}
		$\vdots$	$\vdots$		$\vdots$		$\vdots$

5.2   MODEL FOR THE COMPLETELY RANDOMIZED TWO-FACTOR DESIGN

The *ab* experimental groups of *n* subjects can be considered *ab* random samples, one from each of *ab* treatment populations. These treatment populations, it is assumed, have been drawn from the same infinitely large parent population; thus any differences among the *ab* distributions can be attributed solely to differences among the treatment effects. The basic population parameters are the μ_{jk}, the mean of the treatment population defined by the *j*th level of A and the *k*th level of B; μ_j, the mean of all scores obtained under treatment A_j; μ_k, the mean of all scores obtained under B_k; and μ, the expected value over all *ab* treatment populations, or equivalently, the mean of the original parent population.

The observed data are related to the population parameters by the following equation:

(5.1)
$$Y_{ijk} = \mu + \alpha_j + \beta_k + (\alpha\beta)_{jk} + \varepsilon_{ijk}$$

where

Y_{ijk} = score of ith subject in jth treatment level of A
        and kth treatment level of B

$\alpha_j = \mu_j - \mu$, *main effect* of treatment A_j

$\beta_k = \mu_k - \mu$, *main effect* of treatment B_k

$(\alpha\beta)_{jk} = \mu_{jk} - \mu_j - \mu_k + \mu$, *interaction effect* of treatments A_j and B_k

$\varepsilon_{ijk} = Y_{ijk} - [\mu + \alpha_j + \beta_k + (\alpha\beta)_{jk}]$

$\qquad = Y_{ijk} - \mu_{jk}$, error component

The quantities involved in Equation (5.1) are like those met in Chapter 4, except for the interaction, which therefore merits further definition. Note that

(5.2)
$$(\mu_{jk} - \mu_j - \mu_k + \mu) = (\mu_{jk} - \mu) - (\mu_j - \mu) - (\mu_k - \mu)$$

In words, the interaction effect $(\alpha\beta)_{jk}$ is the combined effect of the jth level of A and the kth level of B, which cannot be accounted for by α_j and β_k, the independent effects of A_j and B_k. The effects α_j and β_k are usually referred to as *main effects*, in contrast with the *interaction effect* $(\alpha\beta)_{jk}$. (The interaction effect and its relation to main effects will be considered further in Section 5.3.2.)

It follows from Equation (5.1) that the variability in our observed data matrix (Table 5–1) has several sources. These are:

1. *The effect of treatment* A_j, namely α_j, is a constant component of all scores obtained under A_j but may vary over levels of j. The variable A is assumed to be a fixed-effect variable; that is, the a levels, it is assumed, have been arbitrarily selected and the population of levels exhausted. Therefore, $\sum_j \alpha_j = 0$. One null hypothesis that will be tested is

$$\alpha_1 = \alpha_2 = \cdots = \alpha_j = \cdots = \alpha_a = 0$$

2. *The effect of treatment* B_k, namely β_k, is a constant component of all scores obtained under B_k but may vary over levels of k. We assume that B, like A, is a fixed-effect variable, and therefore, $\sum_k \beta_k = 0$. We shall test the null hypothesis that

$$\beta_1 = \beta_2 = \cdots = \beta_k = \cdots = \beta_b = 0$$

3. *The interaction effect of* A_j *and* B_k, namely $(\alpha\beta)_{jk}$, is a constant component of all scores obtained under A_j and B_k but may vary over the levels j and k. Since A and B are both fixed-effect variables, $\sum_k (\alpha\beta)_{jk} = \sum_j (\alpha\beta)_{jk} = 0$. The relevant null hypothesis is that

$$(\alpha\beta)_{12} = (\alpha\beta)_{13} = \cdots = (\alpha\beta)_{jk} = \cdots = (\alpha\beta)_{ab} = 0$$

4. *The error component* ε_{ijk} is not accounted for by the systematic manipulation of the variables A and B. This is the only source of variance within the cells that corresponds to the treatment combinations. We assume that the ε_{ijk} are independently and normally distributed with mean zero and variance σ_e^2 within each treatment population defined by a combination of levels of A and B.

From the model presented, we can derive parameter estimates, expected mean squares, and F tests of the null hypotheses stated above. The development is like that of Section 4.2.1 for the one-factor design. The relevant parameter estimates are

$$\hat{\mu} = \frac{T_{...}}{abn}$$
$$= \bar{Y}_{...}$$
$$\hat{\mu}_j = \frac{T_{.j.}}{bn}$$
$$= \bar{Y}_{.j.}$$

(5.3)

$$\hat{\mu}_k = \frac{T_{..k}}{an}$$
$$= \bar{Y}_{..k}$$
$$\hat{\mu}_{jk} = \frac{T_{.jk}}{n}$$
$$= \bar{Y}_{.jk}$$

The expected mean squares and F tests will be considered as part of the general discussion of the analysis of variance (see next section).

5.3   THE ANALYSIS OF VARIANCE FOR THE COMPLETELY RANDOMIZED TWO-FACTOR DESIGN

The sources of variability in our data are specified by Equation (5.1). A general way of proceeding to obtain the appropriate sums of squares is to substitute parameter estimates for parameters in the basic equation. Thus, beginning with Equation (5.1),

$$Y_{ijk} = \mu + \alpha_j + \beta_k + (\alpha\beta)_{jk} + \varepsilon_{ijk}$$

and substituting from Equation (5.3), we arrive at the following identity:

(5.4)
$$Y_{ijk} = \bar{Y}_{...} + (\bar{Y}_{.j.} - \bar{Y}_{...}) + (\bar{Y}_{..k} - \bar{Y}_{...})$$
$$+ (\bar{Y}_{.jk} - \bar{Y}_{.j.} - \bar{Y}_{..k} + \bar{Y}_{...}) + (Y_{ijk} - \bar{Y}_{.jk})$$

Subtracting $\bar{Y}_{...}$ from both sides of Equation (5.4), squaring both sides, and summing over i, j, and k yield

(5.5)
$$\sum_i^n \sum_j^a \sum_k^b (Y_{ijk} - \bar{Y}_{...})^2 = bn \sum_j^a (\bar{Y}_{.j.} - \bar{Y}_{...})^2$$

(sum of squares total, SS_{tot})   (sum of squares for A, SS_A)

$$+ an \sum_k^b (\bar{Y}_{..k} - \bar{Y}_{...})^2$$

(sum of squares for B, SS_B)

$$+ n \sum_j^a \sum_k^b (\bar{Y}_{.jk} - \bar{Y}_{.j.} - \bar{Y}_{..k} + \bar{Y}_{...})^2$$

(sum of squares for interaction, SS_{AB})

$$+ \sum_i^n \sum_j^a \sum_k^b (Y_{ijk} - \bar{Y}_{.jk})^2$$

(sum of squares within groups, $SS_{S/AB}$)

Note that cross-product terms like $2\sum_i \sum_j \sum_k (\bar{Y}_{.j.} - \bar{Y}_{...})(\bar{Y}_{..k} - \bar{Y}_{...})$ equal zero if we have the same number of scores in each cell. This follows if we rewrite the term as $[2\sum_i \sum_j (\bar{Y}_{.j.} - \bar{Y}_{...})]\sum_k (\bar{Y}_{..k} - \bar{Y}_{...})$ and note that $\sum_k (\bar{Y}_{..k} - \bar{Y}_{...}) = 0$.

Equation (5.5) states that the total variability is to be partitioned into four parts: variability due to A, variability due to B, variability due to the AB interaction, and variability due to error. This analysis is not the only one possible, but is the one appropriate for our particular model (Equation (5.1)). If we were to assume, for example, the absence of interaction, our model would assert that

(5.6)
$$Y_{ijk} = \mu + \alpha_j + \beta_k + \varepsilon_{ijk}$$

and the identity that forms the basis for our analysis would be

(5.7)
$$(Y_{ijk} - \bar{Y}_{...}) = (\bar{Y}_{.j.} - \bar{Y}_{...}) + (\bar{Y}_{..k} - \bar{Y}_{...})$$
$$+ (Y_{ijk} - \bar{Y}_{.j.} - \bar{Y}_{..k} + \bar{Y}_{...})$$

In this case, the total variability would be partitioned into three parts from the A effect, the B effect, and error variability. The point of these comments is that the analysis of variance is not an arbitrary set of computations, but a logical consequence of the experimenter's assumptions about the relations between the data and the parameters of the population from which samples have been taken.

5.3.1   SUMMARIZING THE ANALYSIS OF VARIANCE.   Table 5–2 summarizes the analysis of variance for the completely randomized two-factor design. The source of variance (SV) column reflects the partitioning described above and is a direct consequence of Equation (5.1). For the A main effect, there are $(a-1)$ df because of the requirement that the sum of the deviations of the a treatment means about the grand mean must be zero. For a similar reason, df for the B main effect is $b-1$. In the case of the AB interaction, we begin with $(ab-1)$ df, since the sum of the deviations of the ab cell means about the grand mean must

Table 5–2   Analysis of variance for a two-factor design

SV	df	SS	MS	EMS	F
Total	$abn-1$	$\sum_i^n \sum_j^a \sum_k^b Y^2 - C^a$			
A	$a-1$	$\dfrac{\sum_j^a T_{.j.}^2}{nb} - C$	$\dfrac{SS_A}{a-1}$	$\sigma_e^2 + nb\theta_A^2$	$\dfrac{MS_A}{MS_{S/AB}}$
B	$b-1$	$\dfrac{\sum_k^b T_{..k}^2}{na} - C$	$\dfrac{SS_B}{b-1}$	$\sigma_e^2 + na\theta_B^2$	$\dfrac{MS_B}{MS_{S/AB}}$
AB	$(a-1)(b-1)$	$\dfrac{\sum_j^a \sum_k^b T_{.jk}^2}{n} - C - SS_A - SS_B$	$\dfrac{SS_{AB}}{(a-1)(b-1)}$	$\sigma_e^2 + n\theta_{AB}^2$	$\dfrac{MS_{AB}}{MS_{S/AB}}$
S/AB	$ab(n-1)$	$SS_{tot} - SS_A - SS_B - SS_{AB}$	$\dfrac{SS_{S/AB}}{ab(n-1)}$	σ_e^2	

[a] $C = \dfrac{T_{...}^2}{nab}$

be zero. Removing the variability among the cell means because of the A and B effects causes the loss of $(a-1)$ df and $(b-1)$ df. Because of the removal of this variability, the row and column means of Table 5–1 are no longer free to vary, and the corresponding df are lost. Thus, restrictions placed on the variability of the cell means by the grand mean and by the removal of the A and B effects yield

(5.8) $$df_{AB} = (ab-1)-(a-1)-(b-1)$$
$$= (a-1)(b-1)$$

The df associated with S/AB can be viewed as the result of taking the deviations of n scores about their cell mean, which yields $(n-1)$ df, then summing this result over the ab cells, which yields $ab(n-1)$ df. Alternatively, noting that the variability due to S/AB is a residual from the total after the variability due to A, B, and AB was removed, we have

(5.9) $$df_{S/AB} = (abn-1)-(a-1)-(b-1)-(a-1)(b-1)$$
$$= ab(n-1)$$

As a third alternative, we note, the total variability among scores must come either from variability among cell means or variability among scores in the same cell. Therefore, the variability within cells equals the total minus the between-cells variability, or

(5.10) $$df_{S/AB} = (abn-1)-(ab-1)$$
$$= ab(n-1)$$

The computational formulas in the SS column can be derived from the components of Equation (5.5) (see Section 2.2). Each term is expanded, the summations are carried out, raw score formulas are substituted for the means, and the expressions are simplified as far as possible. This lengthy procedure can be bypassed, however, by using the isomorphism of *df* and *SS* (Section 4.6.2). The $a-1$ in the *df* column suggests the sum of a squared quantities minus the correction term, or $\sum_j^a (\quad)^2 - C$. The parentheses can be replaced by $T_{\cdot j \cdot}$, yielding $\sum_j^a T_{\cdot j \cdot}^2 - C$. Finally, we divide the squared total by the number of values on which it is based. Since we have summed over i and k (as indicated by the dot subscripts), we divide by bn, and get

$$SS_A = \frac{\sum_j T_{\cdot j \cdot}^2}{bn} - C$$

Other *SS* formulas can be obtained in the same way.

The *MS* column is again the result of dividing the entries in the *SS* column by the corresponding entries in the *df* column. The entries under *EMS* consist of σ_e^2, plus a term θ^2, representing the corresponding *SV*. The θ^2 term is multiplied by the number of levels of all variables that are not represented in the subscript. The θ^2 terms are defined as follows:

(5.11)
$$\theta_A^2 = \frac{\sum_j^a (\mu_j - \mu)^2}{a-1}$$

(5.12)
$$\theta_B^2 = \frac{\sum_k^b (\mu_k - \mu)^2}{b-1}$$

(5.13)
$$\theta_{AB}^2 = \frac{\sum_j^a \sum_k^b (\mu_{jk} - \mu_j - \mu_k + \mu)^2}{(a-1)(b-1)}$$

The notation θ^2 indicates that the subscripted variable is fixed. This follows from the definitions, Equations (5.11) to (5.13), in which the squared deviations are pooled only over the levels appearing in the experiment. In contrast, σ_e^2 represents the variance of effects for a population of levels of which those in the experiment are only a random sample.

The appropriate *F* tests are also listed in Table 5–2. The logic is exactly that discussed in Chapter 4; we need an error term that consists of all the components of the numerator except the null hypothesis term. Then, if H_0 is true, the ratio of expected mean squares will equal 1. The null hypothesis that the μ_j are all equal (that is, $\theta_A^2 = 0$) is tested by an *F* distributed on $(a-1)$ *df* and $ab(n-1)$ *df*. A significant value of *F* leads to a relatively restricted conclusion. It does not indicate that the levels of *A* differ in their effects at any given level of *B* or at any other level of *B* besides those used in this particular experiment; it does indicate that *averaged over the levels of B used in the present study*, there is an effect from *A*. If the experimenter is interested in comparing the effects of *A* at a particular level of *B*, then a sum of squares of *A* should be computed only for the data at that level of *B*. This sum of squares should then be divided by $a-1$ and tested against the usual error term $MS_{S/AB}$. Such a test is a test of a *simple effect* of *A*, contrasted with the overall test of the main effect of *A*. The null hypothesis that θ_B^2 equals zero is tested by an *F* distributed on $b-1$ and $ab(n-1)$. Rejection of the null hypothesis shows that the μ_k differ, where μ_k is the mean of the kth treatment population of *B*, obtained by averaging over all treatment populations of A. If we want to compare the means of the B treatment populations at one particular level of A, we are again interested in testing simple effects, and the procedure previously described can be applied. The null hypothesis that θ^2_{AB} equals zero is tested by an F distributed on $(a-1)(b-1)$ and $ab(n-1)$ df. Its rejection indicates that variability remains among the μ_{jk} after variability from effects of A and B has been removed.

5.3.2 TWO-FACTOR INTERACTION.   The two-factor interaction merits a detailed interpretation. We begin with Equation (5.14), which follows from Equation (5.1); summing over i and dividing by n yield

(5.14)
$$\bar{Y}_{.jk} = \mu + \alpha_j + \beta_k + (\alpha\beta)_{jk} + \bar{\varepsilon}_{.jk}$$

Substituting sample estimates of population parameters and transposing terms, we obtain

(5.15)
$$\widehat{(\alpha\beta)}_{jk} = (\bar{Y}_{.jk} - \bar{Y}_{...}) - (\bar{Y}_{.j.} - \bar{Y}_{...}) - (\bar{Y}_{..k} - \bar{Y}_{...})$$

the estimate of the interaction effect of the jkth treatment combination (note that the estimate of $\sum_i \varepsilon_{ijk}/n$, $\sum_i (Y_{ijk} - \bar{Y}_{.jk})/n, = 0$). In an example, we next apply Equation (5.15) to the set of values of $\bar{Y}_{.jk} - \bar{Y}_{...}$ in Table 5–3. If in accord with Equation (5.15), we remove the estimate of α_j from each cell mean, the result—Table 5–4—consists of the quantities $(\bar{Y}_{.jk} - \bar{Y}_{...}) - (\bar{Y}_{.j.} - \bar{Y}_{...})$. Note that the two df for A are literally lost; the row means are no longer free to vary but instead stand equal to each other and to the grand mean. Note that further, the column means are unchanged by this manipulation of the data, demonstrating the independence of row and column effects.

Next, remove $(\bar{Y}_{..k} - \bar{Y}_{...})$, the estimate of β_k, from each entry in Table 5–4. The result is Table 5–5, in which there are no df for rows or columns. The entries in this matrix are the estimates of $(\alpha\beta)_{jk}$. Note that they are not all zero, that there is variability in the data not due to A and B effects. If this variability is significantly greater than variability due to individual differences (S/AB), interaction effects, it is said, are present in the data.

Tables 5–4 and 5–5 clarify one interpretation of interaction; namely, it is the variability among cell means that still remains when variability due to the main effects is removed. Ordinarily, however, we want to make some statement about interaction effects in terms of the tabulation or plot of the original cell means. If

Table 5–3   A 3×3 table of values of $\bar{Y}_{.jk} - \bar{Y}_{...}$

	B_1	B_2	B_3	$\bar{Y}_{.j.} - \bar{Y}_{...}$
A_1	−3.5	−3.0	−2.5	−3.0
A_2	−1.0	+1.0	+3.0	+1.0
A_3	−1.5	+2.0	+5.5	+2.0
$\bar{Y}_{..k} - \bar{Y}_{...}$	−2.0	0.0	+2.0	

Table 5–4   Table 5–3 after removal of variability due to A

	B_1	B_2	B_3	$\bar{Y}_{.j.} - \bar{Y}_{...}$
A_1	−0.5	0.0	+0.5	0.0
A_2	−2.0	0.0	+2.0	0.0
A_3	−3.5	0.0	+3.5	0.0
$\bar{Y}_{..k} - \bar{Y}_{...}$	−2.0	0.0	+2.0	

we assume that the data are error-free (or equivalently, that the cell means are means of populations instead of samples), what do the unadjusted means reveal about the presence or absence, as well as the direction, of interaction? To answer this question, consider a data set in which there are no interaction effects; with error components ignored, the appropriate structural equation is

$$(5.14')\qquad\qquad \bar{Y}_{.jk} = \mu + \alpha_j + \beta_k$$

If as in the previous example, the $\alpha_j = -3, 1, 2$ and the $\beta_k = -2, 0, 2$, and if μ is taken as say 6.0, the resulting matrix of cell means would be

$$\begin{bmatrix} 6+(-3)+(-2) & 6+(-3)+(0) & 6+(-3)+(2) \\ 6+(1)+(-2) & 6+(1)+(0) & 6+(1)+(2) \\ 6+(2)+(-2) & 6+(2)+(0) & 6+(2)+(2) \end{bmatrix} = \begin{bmatrix} 1 & 3 & 5 \\ 5 & 7 & 9 \\ 6 & 8 & 10 \end{bmatrix}$$

Plotting these values, as we have done in Figure 5–1, is instructive.

The notable aspect of the two plots is that the curves are parallel. That this must be true in the absence of interaction can be proved for any set of values. Consider four cell means: $\bar{Y}_{.jk}$, $\bar{Y}_{.j'k}$, $\bar{Y}_{.jk'}$, and $\bar{Y}_{.j'k'}$, where $j \neq j'$ and $k \neq k'$. From Equation (5.14'), it follows that $\bar{Y}_{.jk} - \bar{Y}_{.jk'} = \beta_k - \beta_{k'}$. However, $\bar{Y}_{.j'k} - \bar{Y}_{.j'k'}$ yields the same result. Thus, if there are *no* interaction effects, the simple effect of B is constant over levels of A. This is a verbal statement of the parallelism depicted in the right-hand panel of Figure 5–1. Similarly, if interaction is absent, the difference in the effects of A_j and $A_{j'}$ will be $\alpha_j - \alpha_{j'}$ at all levels of B; this is the situation depicted in the left-hand panel of Figure 5–1. Thus, *interaction is a significant departure from parallelism*. The spread among the means for the levels of one variable changes as a function of the level of the second variable. Alternatively, interaction can be viewed as significant variability among simple effects. The simple effects of one variable are not constant at all levels of the second and are therefore not all equal to the main effect.

Table 5–5   Table 5–3 after removal of variability due to A and B

	B_1	B_2	B_3	$\bar{Y}_{.j.} - \bar{Y}_{...}$
A_1	1.5	0	−1.5	0
A_2	0	0	0	0
A_3	−1.5	0	1.5	0
$\bar{Y}_{..k} - \bar{Y}_{...}$	0	0	0	

Figure 5–1   A plot of cell means with no interaction present.

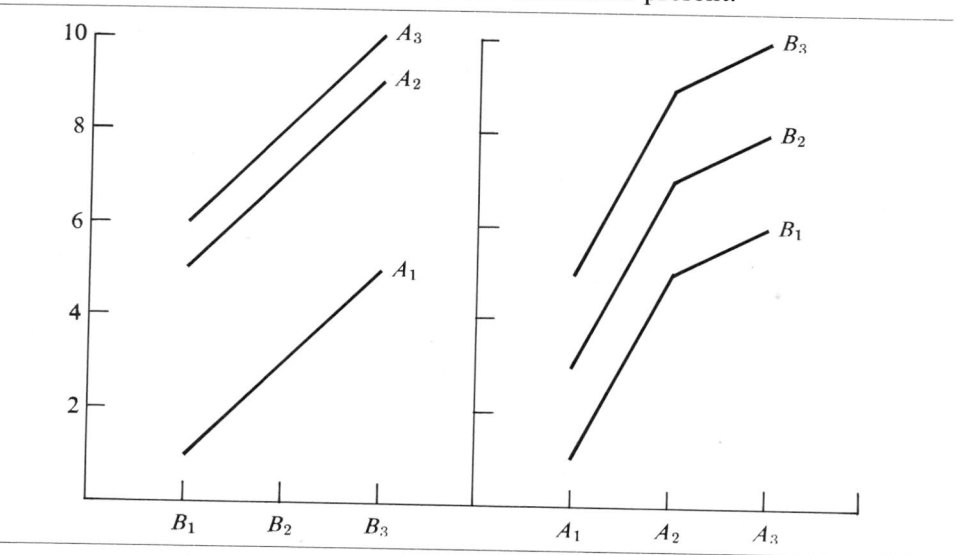

Although the main and interaction effects are independent, adequately interpreting the data requires joint consideration of all three. Consider the two examples in Figure 5–2. For simplicity, again assume errorless data; in other words, assume that the cell means are population means and any nonzero effects are therefore significant. In the left side of Figure 5–2 there are no A or AB effects, but the B effect is significant. Compare this with the right side, in which there is again no A effect and there is again a B effect; but this time there is also an AB interaction effect. The statement that there is no A effect has the same formal meaning for both data plots; if we average over both levels of B, the means for the three levels of A do not differ. In the data on the left, however,

Figure 5–2   Plots of two sets of means.

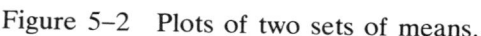
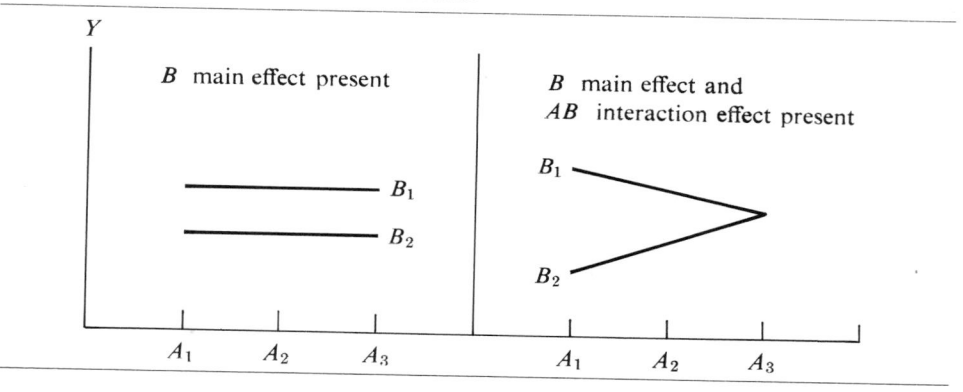

there is no A effect at any level of B, while on the right there are simple A effects at both B_1 and B_2. These effects are in opposite directions and tend to cancel each other. Clearly, if we are to get an accurate picture of the relation among treatment population means, we must consider the total set of effects. Thus, knowing that B is a significant source of variance is not enough; it is only when the condition of the A and AB effects is noted that intelligent discussion of the data can begin.

As a general rule, a significant interaction indicates that main effects are of less interest; tests of simple effects now take on more importance. From the data plotted in the right-hand panel of Figure 5–2, for example, tests of the effects of A at B_1 and at B_2 are appropriate, despite the absence of an overall A main effect. If the cell variances do not differ significantly, the usual overall error term $MS_{S/AB}$ can be used to test simple effects. Otherwise, we must calculate a new within-cell average variance based only on the cells involved in the analysis.

5.3.3 BEHAVIORAL AND STATISTICAL HYPOTHESES.   Understanding of statistical effects is not complete without some feeling for their relation to behavioral hypotheses. Although there are no rules for translating hypotheses about behavior into hypotheses about statistical effects, we shall try to show how to do it.

Consider the following experiment. On each of 100 trials, each subject has a choice between gambling and not gambling. There are two independent variables: A, the amount that can be won or lost on a gamble, and B, the consequences of not gambling. The levels of A are:

$A_1 = 5\cancel{c}$ is won or lost on a gamble
$A_2 = 15\cancel{c}$ is won or lost on a gamble
$A_3 = 25\cancel{c}$ is won or lost on a gamble

On each gamble the subject has an equal chance of winning or losing. The levels of B are:

$B_1 =$ the subject always pays out $1\cancel{c}$ on trials in which he does not gamble
$B_2 =$ the subject always receives $1\cancel{c}$ on trials in which he does not gamble

One theory of choice behavior would predict the following:

1. The percentage of gambles P will increase as the amount risked (the levels of A) increases when the consequence of not gambling is a sure gain (B_2).
2. P will decrease as the amount risked increases when the consequence of not gambling is a sure loss (B_1).
3. P will increase (when the alternative is B_2) over the levels of A at about the same rate that P will decrease (when the alternative is B_1) over the levels of A.
4. P will be greater when the alternative to gambling is a sure loss (B_1) than when it is a sure gain (B_2), regardless of the level of A.

These hypotheses about the performance of subjects would be confirmed if the plot of the means for the AB combinations looked like the right-hand side of

Figure 5–2. Thus, the translation into statistical hypotheses would be as follows:

1'. There will be a significant *AB* interaction (from hypotheses 1 and 2).
2'. The *A* main effect will not be significant (from 3).
3'. The *B* main effect will be significant (from 4).

If these three hypotheses are verified and also the effects are in the predicted direction, our behavioral hypotheses are supported.

Let us consider a second example of the relation between behavioral and statistical hypotheses. Schizophrenics and normal subjects, equated on an intelligence test, are required to perform on a concept formation task. Two sets of stimuli are used; both involve the same concepts. One set contains pictures that express social approval (a woman patting a boy on the head, for instance); the second set contains pictures expressing social disapproval (a woman shaking her finger at a boy, for example). Certain theories of personality and intellectual performance might lead to the following hypotheses:

1. Normal subjects will take fewer trials to reach the relevant concept than schizophrenics will, whichever stimulus set is the basis for comparison.
2. The performances of normal subjects will not be influenced by the type of stimulus set (approving or disapproving pictures).
3. Schizophrenics will perform less well on the disapproving set of pictures than on the approving.

If all three hypotheses were correct, the data might look something like the plot in Figure 5–3. The following results are hypothesized for the statistical analysis:

1'. There will be a significant main effect due to the personality variable (from No. 1).
2'. There will be a significant main effect due to stimulus sets (from Nos. 2 and 3).
3'. There will be a significant interaction of personality and stimulus (from Nos. 2 and 3).

It is a good idea to go through the process just discussed before collecting data. The experimenter should try to visualize the sorts of results that might be obtained and should consider which sources of variance in the analysis will reflect these results. This kind of activity will help to ensure that the experimental design makes possible adequate tests of the behavioral hypotheses or provides answers to the questions posed. In addition, this kind of thinking promotes a feeling for the relation of data to psychological processes. Remember, the end goal of experimentation is not a statistical statement (like "The stimulus set variable is a significant source of variance"), but a statement about behavior (for example, "Stimuli that express social disapproval are less readily categorized than stimuli expressing approval by a population consisting of schizophrenic and normal subjects").

It is also important to plot the data once they are obtained and to study carefully the plot of data. Plot the same set of cell means several ways, as in

Figure 5–3 Performance of schizophrenic and normal subjects on a concept formation task.

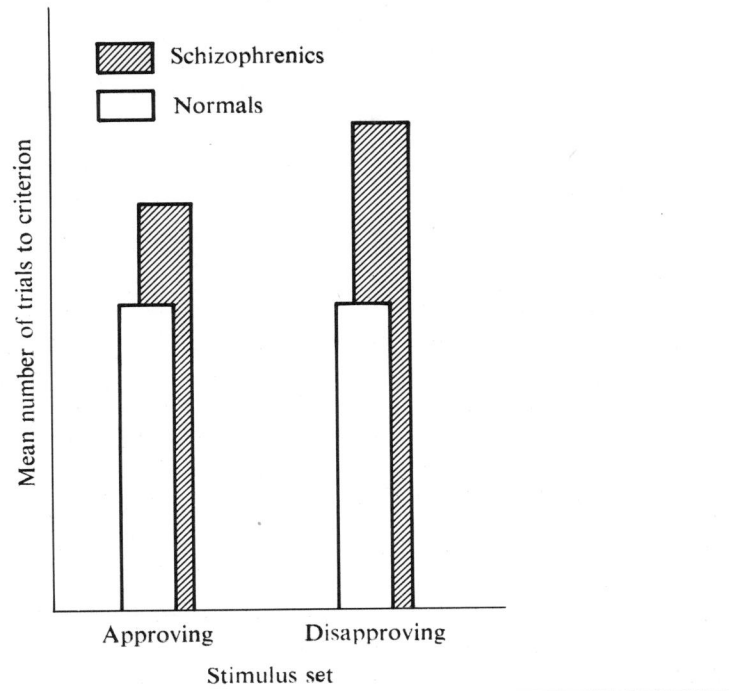

Figure 5–1. Plot the means at the levels of *B*, averaging over the levels of *A*, to get a picture of the *B* main effect. Look at the *A* main effect by averaging over levels of *B*. Keep these plots at hand while reading the entries in the analysis of variance table.

5.3.4 A NUMERICAL EXAMPLE FOR THE TWO-FACTOR DESIGN, EQUAL *n*. Table 5–6 presents data from an experiment involving two independent variables, *A* and *B*. Obtain, as a first step in the analysis, the sum of scores and the sum of squared scores for each cell (combination of *j* and *k*).

The formulas for the sums of squares of Table 5–2 now can be applied. The correction term is again the sum of all scores squared, divided by the total number of scores; thus,

$$C = \frac{(730)^2}{48} = 11{,}102.080$$

The SS_{tot} is

$$SS_{tot} = 3{,}577 + \cdots + 1{,}088 - C$$
$$= 16{,}090 - C$$
$$= 4{,}987.920$$

Table 5-6   Data from a completely randomized two-factor experiment

	A_1B_1	A_1B_2	A_1B_3	
	7	6	9	
	33	11	12	
	26	11	6	
	27	18	24	
	21	14	7	
	6	18	10	
	14	19	1	
	19	14	10	
$T_{.1k} = 153$		111	79	$T_{.1.} = 343$
$\sum_i Y^2_{i1k} = 3{,}577$		1,679	1,087	
	A_2B_1	A_2B_2	A_2B_3	
	42	28	13	
	25	6	18	
	8	1	23	
	28	15	1	
	30	9	3	
	22	15	4	
	17	2	6	
	32	37	2	
$T_{.2k} = 204$		113	70	$T_{.2.} = 387$
$\sum Y^2_{i2k} = 5{,}934$		2,725	1,088	
$T_{..k} = 357$		224	149	$T_{...} = 730$

Next,

$$SS_A = \frac{(343)^2 + (387)^2}{24} - C$$
$$= 11{,}142.417 - C$$
$$= 40.337$$

Similarly,

$$SS_B = \frac{(357)^2 + (224)^2 + (149)^2}{16} - C$$
$$= 12{,}489.125 - C$$
$$= 1{,}387.045$$

The calculation of the total sum of squares is the same as it was for the one-factor design (Chapter 4). In fact, for any design, SS_{tot} is the sum of all squared scores minus the correction term. The sum of squares for the A

treatment variable is also computed as it was in Chapter 4. In general, the design is collapsed so that it consists of a groups of scores; then the SS_A is computed as it is for a one-factor design. In our example, we ignore the B variable and view the design as consisting of two groups of 24 (in general, bn) scores each. To compute SS_B, we ignore the A variable and view the design as a one-factor design consisting of three groups of 16 scores (in general, b groups of an scores). This is exactly what has been done in the immediately preceding computations.

To calculate the SS_{AB}, one might first calculate the sum of squares for the cell means, which will be denoted $SS_{\overline{AB}}$. Essentially, the design is again viewed as a one-factor design, this time consisting of six (ab) groups of eight (n) scores each. Consequently,

$$SS_{\overline{AB}} = \frac{(153)^2 + (111)^2 + \cdots + (70)^2}{8} - C$$
$$= 12{,}657.000 - C$$
$$= 1{,}554.920$$

Next, remove that portion of the variability among cells due to A and B effects. We get

$$SS_{AB} = 1{,}554.920 - SS_A - SS_B$$
$$= 127.538$$

Now the sum of squares for the error term, S/AB, is required. This may be computed as a residual from the total variability; that is,

$$SS_{S/AB} = SS_{tot} - SS_A - SS_B - SS_{AB}$$
$$= 3{,}433.000$$

Alternatively, the error sum of squares could be computed as the difference between the total and between-cells variability, as suggested by Equation (5.10):

$$SS_{S/AB} = SS_{tot} - SS_{\overline{AB}}$$
$$= 4{,}987.920 - 1{,}554.920$$
$$= 3{,}433.000$$

A third approach to calculating $SS_{S/AB}$ is possible because the $MS_{S/AB}$ equals the average cell variance. Therefore,

(5.16)
$$MS_{S/AB} = \sum_j \sum_k \hat{\sigma}_{jk}^2 / ab$$

and by substituting and transposing terms, we get

$$SS_{S/AB} = (n-1) \sum_j \sum_k \hat{\sigma}_{jk}^2$$

(5.17)
$$= 7(490.428)$$
$$= 3{,}432.996$$

which is correct within a slight rounding error. These alternative methods provide a *partial* check on the calculations. Thus, an error in calculating SS_A might affect the residual method of obtaining the error sum of squares and not affect the third method used. On the other hand, an error in squaring the individual scores would lead to the same result with all approaches.

Knowing about these alternative calculations is also worth while because there are situations in which the within-cell variability must be obtained before treatment effects are calculated. An instance of this will be noted in the discussion of calculations for unequal *n*. An additional reason for demonstrating these alternative calculations is to emphasize the underlying identity of several views of within-cell variability—it is residual when all treatment effects have been accounted for; it is the difference between total variability and variability among cell means; and it is the average within-cell variability.

The mean squares and the *F* ratios are now easily computed. Dividing sums of squares by the appropriate *df*, we obtain the entries in the *MS* column of Table 5–7. The *A*, *B*, and *AB* mean squares are each tested against the same error term, $MS_{S/AB}$, and we get the ratios reported in the *F* column. Only the *B* main effect is significant.

Turn now to Figure 5–4, which shows a plot of the group means. The source of the *B* main effect is clear; $\bar{Y}_{..k}$ consistently declines as *k* increases. The figure also suggests that scores are higher under A_2 than under A_1 and that there is an *AB* interaction, the decline over levels of *B* appearing to be more rapid under A_2 than under A_1. This sort of result more than any other suggests how important statistical analysis is. Although inspection of the figure suggests *A* and *AB* effects, the statistical analysis reveals that the *A* and *AB* variabilities are well within the range that might be expected on the basis of individual differences.

In connection with Figure 5–4, note that this method of graphing the data is appropriate only if *B* is a quantitative variable. If there is no rationale for ordering the levels of *B*, or if there is no rationale for spacing the levels along the abscissa, then a histogram and not a line graph should be presented. The line graph suggests a function about which one can ask questions of shape and curvature, questions appropriate only when the independent variable is a quantitative scaled variable.

Table 5–7   Analysis of variance for data from a two-factor experiment

SV	df	SS	MS	F
Total	47	4,987.92		
A	1	40.34	40.34	.49
B	2	1,387.05	693.52	8.48[a]
AB	2	127.54	63.77	.78
S/AB	42	3,433.00	81.74	

[a] $p < .01$

Figure 5–4   Mean scores plotted as a function of B for both levels of A.

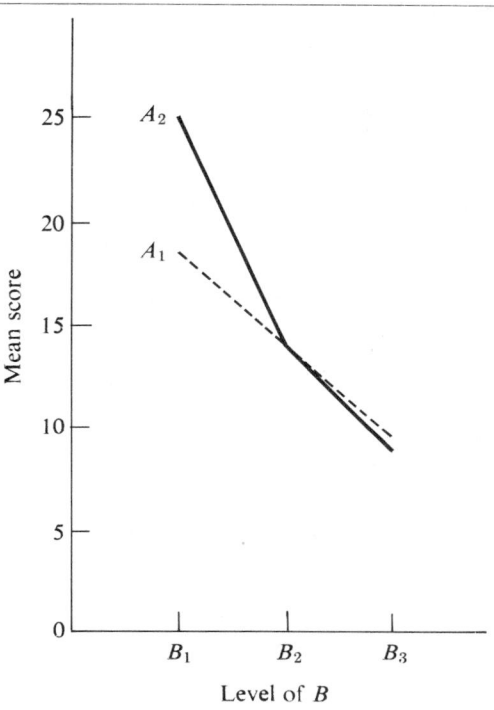

5.4   UNEQUAL CELL FREQUENCIES

We have proceeded as though equal-size experimental groups have been sampled from equal-size treatment populations. When these conditions are not met, problems may arise. We have noted (Chapter 4) that unequal ns exaggerate the consequences of violating the analysis of variance model. Furthermore, the situation of unequal ns calls for somewhat different analyses from those considered thus far. The choice of analysis is not always simple, the interpretation is occasionally complicated, and the actual computational labor is always more than in the equal n case.

The least troublesome departure from equality of cell frequencies is the unequal but proportional case. Proportionality is illustrated by the following set of cell frequencies:

$n_{11} = 8$	$n_{12} = 6$	$n_{13} = 6$	$n_{1.} = 20$
$n_{21} = 4$	$n_{22} = 3$	$n_{23} = 3$	$n_{2.} = 10$
$n_{.1} = 12$	$n_{.2} = 9$	$n_{.3} = 9$	

Table 5–8   Example involving disproportionate cell frequencies

		B_1	B_2	TOTALS	$\bar{Y}_{.j.}$
A_1	$T_{.1k}$	40	200	240	
	n_{1k}	2	8	10	
	$\bar{Y}_{.1k}$	20	25		24
A_2	$T_{.2k}$	40	20	60	
	n_{2k}	8	2	10	
	$\bar{Y}_{.2k}$	5	10		6
	$T_{..k}$	80	220		
	$n_{.k}$	10	10	$\bar{Y}_{...} = 15$	
	$\bar{Y}_{..k}$	8	22		

The critical feature of these n_{jk}s is that the ratio of row frequencies is constant over columns, 2:1 in this example. Equivalently, the ratio of column frequencies is constant over rows, 4:3:3 in the example. (We shall consider the data analysis for this case in Section 5.4.1.)

Disproportionate ns are more serious. Something of the nature of the problem is shown in Table 5–8. Looking only at the cell means, we find no evidence of an AB interaction; both $\bar{Y}_{.11} - \bar{Y}_{.21}$ and $\bar{Y}_{.12} - \bar{Y}_{.22}$ equal 15. According to the discussion of Section 5.3.2, in the absence of interaction we should expect main effects to equal simple effects; $\bar{Y}_{.1.} - \bar{Y}_{.2.}$ should also equal 15. The obtained difference, however, is 18. We obtain a result that appears even stranger if we do an analysis of variance. We find that $SS_A = 1620$, $SS_B = 980$, and $SS_{AB} = (6200 - 4500) - 1620 - 980 = -900$. The minus sign is neither a computational nor a typographical error. Nevertheless, from the development of analysis of variance presented thus far, the result is utter nonsense. From the cell means, the value of SS_{AB} should be zero; a negative value bears no interpretation within our framework.

The reason for these strange computational results may become clearer if we consider an extreme case of disproportionate cell frequency. Suppose the ns were

$$
\begin{array}{c}
\quad\quad B_1 \quad B_2 \\
\begin{array}{c} A_1 \\ A_2 \end{array}
\begin{bmatrix} 0 & 8 \\ 8 & 0 \end{bmatrix}
\end{array}
$$

Now, SS_A and SS_B are identical; both are based solely on the difference between the A_2B_1 and A_1B_2 means and thus the A and B main effects are perfectly correlated. If we now enter two scores in the other two cells, as in Table 5–8, the correlation of effects is no longer perfect, although it will still be large. The magnitude of both SS_A and SS_B will depend primarily (though no longer entirely) on the difference between the A_2B_1 and A_1B_2 means.

Another way to picture the problem is to consider the partitioning of $SS_{\overline{AB}}$, the variability among the cell means. We have

$$\sum_j \sum_k n_{jk}(\bar{Y}_{.jk} - \bar{Y}_{...})^2 = \sum_j \sum_k n_{jk}(\bar{Y}_{.j.} - \bar{Y}_{...})^2 + \sum_j \sum_k n_{jk}(\bar{Y}_{..k} - \bar{Y}_{...})^2$$

$$+ \sum_j \sum_k n_{jk}(\bar{Y}_{.jk} - \bar{Y}_{.j.} - \bar{Y}_{..k} + \bar{Y}_{...})^2$$

(5.18)
$$+ 2\sum_j \sum_k n_{jk}(\bar{Y}_{.j.} - \bar{Y}_{...})(\bar{Y}_{..k} - \bar{Y}_{...})$$

$$+ 2\sum_j \sum_k n_{jk}(\bar{Y}_{.j.} - \bar{Y}_{...})(\bar{Y}_{.jk} - \bar{Y}_{.j.} - \bar{Y}_{..k} + \bar{Y}_{...})$$

$$+ 2\sum_j \sum_k n_{jk}(\bar{Y}_{..k} - \bar{Y}_{...})(\bar{Y}_{.jk} - \bar{Y}_{.j.} - \bar{Y}_{..k} + \bar{Y}_{...})$$

If the n_{jk} are proportional or equal, the last three terms are zero. Otherwise, they can be either positive or negative. The reason $SS_{\overline{AB}} - SS_A - SS_B$ is negative in our example is that it includes the three cross-product terms, which have a relatively large negative total in this data set. The important point is that the cross-product terms are numerators of correlation coefficients, between the A and B effects, the A and AB effects, and the B and AB effects, respectively. If these correlations are not zero, then the three sum-of-squares terms are not independently distributed and the usual F test computations cannot independently assess the significance of main and interaction effects.

In what follows, we consider three different conditions in which cell frequencies are not equal. First, we shall discuss the analysis of data from multifactor designs for which proportionality can reasonably be assumed to hold in the population. We shall then deal with experiments in which although treatment populations can reasonably be presumed to be equal in size, observations are by chance missing from one or more cells. Finally, we shall briefly take up disproportionality in the treatment populations; where there is disproportionality, effects are correlated not merely in the experimental sample but in the parent populations. (See Chapter 15 for detail on the issues and possible solutions for this particular case, which follow development of a general regression framework.)

5.4.1   PROPORTIONAL FREQUENCIES IN THE TREATMENT POPULATIONS.   Assume that we have designed a study of how political party preference and previous political participation affect attitudes toward current political issues. Further assume that large-scale sampling studies have reliably established that Democrats, Republicans, and Independents exist in the population in the ratio $4:3:3$ and that two-thirds of the individuals in each of these categories voted in the previous political election. Consequently, we might have six groups in our study with the ns that were presented at the beginning of this section to demonstrate proportionality. Calculations in this proportional but unequal n case are straightforward. The analysis of variance is carried out as usual, always dividing each squared total by the number of observations composing the total. The difference is that in the equal n case, for example,

$$SS_A = \frac{\sum_j T_{.j.}^2}{bn} - C$$

and in the proportional n case, we have

$$SS_A = \sum_j \frac{T_{.j.}^2}{n_{j.}} - C$$

If the n_{jk} are all equal,

$$n_{1.} = \cdots = n_{a.} = bn$$

Table 5–9   Two-dimensional data matrix with disproportionate cell frequencies

A_1B_1	A_1B_2	A_1B_3		
7	6	9		
33	11	12		
26	18	6		
27	14	24		
21	19	1		
6	14	10		
14				
19				
$T_{.1k} = 153$	82	62	$T_{.1.} = 297$	
$\bar{Y}_{1k} = 19.125$	13.667	10.333	$\sum_k \bar{Y}_{1k} = 43.125$	
$\sum_i Y_{i1k}^2 = 3{,}577$	1,234	938		
A_2B_1	A_2B_2	A_2B_3		
42	28	13		
8	6	10		
28	1	1		
30	2	6		
22	37	10		
17				
32				
$T_{.2k} = 179$	74	40	$T_{.2.} = 293$	
$\bar{Y}_{2k} = 25.571$	14.800	8.000	$\sum_k \bar{Y}_{2k} = 48.371$	
$\sum_i Y_{i2k}^2 = 5{,}309$	2,194	406		
$\sum_j \bar{Y}_{.jk} = 44.696$	28.467	18.333	$\sum_j \sum_k \bar{Y}_{.jk} = 91.496$	
$T_{..k} = 332$	156	109		

and the equations for the equal and proportional cases are identical. The specific formulas for the analysis of variance of the proportional n two-factor design are:

(5.19)
$$SS_{tot} = \sum_i \sum_j \sum_k Y_{ijk}^2 - C$$

(5.20)
$$SS_A = \sum_j \frac{T_{.j.}^2}{n_{j.}} - C$$

(5.21)
$$SS_B = \sum_k \frac{T_{..k}^2}{n_{.k}} - C$$

(5.22)
$$SS_{AB} = \sum_j \sum_k \frac{T_{.jk}^2}{n_{jk}} - C - SS_A - SS_B$$

(5.23)
$$SS_{S/AB} = SS_{tot} - SS_A - SS_B - SS_{AB}$$

(5.24)
$$C = \frac{T_{...}^2}{n_{..}}$$

In the example of the two-factor study of political attitudes described above, we assumed that sampling was stratified, that relative frequencies in the study matched relative frequencies in the populations. Suppose we had instead sampled 37 subjects at random from the voting population at large and that these subjects fell as indicated in Table 5–9. These cell frequencies are not proportional. If, however, proportionality in the treatment populations can reasonably be assumed, and if the obtained cell frequencies do not depart from proportionality so notably as to render the assumption suspect, the *method of expected cell frequencies* is appropriate.

Method of expected cell frequencies. The method of expected cell frequencies, appropriate when the observed cell frequencies are almost proportional and there is good reason to assume proportionality in the populations, can be applied to the data of Table 5–9, a subset of the data of Table 5–6.

STEP 1.   Compute $SS_{S/AB}$.

$$SS_{S/AB} = SS_{tot} - SS_{\overline{AB}}$$
$$= \sum_k^b \sum_j^a \sum_i^{n_{jk}} Y_{ijk}^2 - \sum_k^b \sum_j^a \frac{T_{.jk}^2}{njk}$$
$$= (3{,}577 + \cdots + 406) - \left[\frac{(153)^2}{8} + \cdots + \frac{(40)^2}{5} \right]$$
$$= 13{,}658 - 10{,}679.915$$
$$= 2{,}978.085$$

STEP 2.   Compute the expected cell frequencies. If $n_{j.}/n_{..}$ is the probability of sampling an individual from population A_j, and if $n_{.k}/n_{..}$ is the probability of sampling an individual from population B_k, then

(5.25)
$$E(n_{jk}) = \left(\frac{n_{j.}}{n_{..}} \right) \left(\frac{n_{.k}}{n_{..}} \right) (n_{..}) = \frac{n_{j.} n_{.k}}{n_{..}}$$

Table 5–10 Expected cell frequencies and sums for data of Table 5–9

		B_1	B_2	B_3	
A_1	$E(n_{1k})$	$\dfrac{(20)(15)}{37}=8.11$	$\dfrac{(20)(11)}{37}=5.95$	$\dfrac{(20)(11)}{37}=5.95$	$n_{1.}=20$
	$E(T_{.1k})$	$(19.13)(8.11)=155.14$	$(13.67)(5.95)=81.34$	$(10.33)(5.95)=61.40$	$E(T_{.1.})=297.88$
A_2	$E(n_{2k})$	$\dfrac{(17)(15)}{37}=6.89$	$\dfrac{(17)(11)}{37}=5.05$	$\dfrac{(17)(11)}{37}=5.05$	$n_{2.}=17$
	$E(T_{.2k})$	$(25.57)(6.89)=176.18$	$(14.80)(5.05)=74.74$	$(8.00)(5.05)=40.40$	$E(T_{.2.})=291.32$
	$n_{.k}$	15	11	11	$n_{..}=37$
	$E(T_{..k})$	331.32	156.08	101.80	$E(T_{...})=589.20$

where n_j. and $n_{.k}$ have been previously defined, and $n_{..}$ is the total number of observations in the experiment. These frequencies are tabulated in Table 5–10.

STEP 3.   Compute the expected cell totals. Multiply the obtained cell mean (Table 5–9) by its expected denominator, obtained in the previous step. The result is an estimate of what the sum of scores should be for each cell, if we use the denominator calculated by Equation (5.24). These expected totals are also tabulated in Table 5.10.

STEP 4.   Apply Equations (5.19), (5.20), and (5.21), using the frequencies and sums of Table 5–10. Thus,

$$SS_A = \frac{(297.88)^2}{20} + \frac{(291.32)^2}{17} - \frac{(589.20)^2}{37}$$

$$= 46.216$$

$$SS_B = \frac{(331.32)^2}{15} + \frac{(156.08)^2}{11} + \frac{(101.80)^2}{11} - \frac{(589.20)^2}{37}$$

$$= 1{,}092.330$$

$$SS_{AB} = \frac{(155.14)^2}{8.11} + \cdots + \frac{(40.40)^2}{5.05} - \frac{(589.20)^2}{37} - 46.216 - 1{,}092.330$$

$$= 126.504$$

The mean squares and F ratios are presented in Table 5–11 and are consistent with those of Table 5–7.

5.4.2   EQUAL FREQUENCIES IN THE TREATMENT POPULATIONS.   It is often reasonable to assume that the treatment populations are of equal size but ill luck caused disproportionate ns to appear. The air conditioning may fail and rats, ever contrary, die in unequal numbers over treatment groups. Or the apparatus is disbanded, and too late we note that several subjects failed to appear in some groups, again not with equal frequencies. If one or two scores are missing, they may be estimated by the means of their cells; dfs are lost from the error df for each score so estimated. Occasionally, discarding one or two scores at random from one or more cells will achieve equal ns; although this may seem to be a

Table 5–11   Analysis of variance for the data of Table 5–9 by method of expected cell frequencies

SV	df	SS	MS	F
A	1	46.216	46.216	.48
B	2	1,092.330	546.165	5.69[a]
AB	2	126.504	63.252	.58
S/AB	31	2,978.085	96.067	

[a] $p < .01$

sinful waste of data, if we start with a large number of observations, the skies will not fall nor—more to the point—will loss of power be noticeable. If neither of the two approaches just sketched is appropriate, and if the ratio of largest to smallest cell frequency is no more than $2:1$, the *method of unweighted means* provides a reasonably straightforward method of analysis. Before the computational steps are detailed, a reminder is in order. Regardless of cell frequencies, none of the methods just cited is appropriate if the experimenter has reason to believe that the populations are not of equal size, that the *ns* are systematically related to the treatment level. Unless the pattern of cell frequencies is truly a chance departure from equality or proportionality, any method that treats the frequencies as equal or proportional misrepresents the treatment populations and may lead to erroneous inferences.

Method of unweighted means. This method, appropriate when treatment populations can be assumed to be equal in size, will also be illustrated by application to the data of Table 5–9. We weight the treatment combinations equally by computing the main and interaction sums of squares as though each cell contained exactly one score, its mean. We then adjust the error variance appropriately.

STEP 1.   Calculate SS_A, SS_B, and SS_{AB} for the cell means rather than for the original data.

$$SS_A = \frac{\sum_j (\sum_k \bar{Y}_{.jk})^2}{b} - \frac{(\sum_j \sum_k \bar{Y}_{.jk})^2}{ab}$$

$$= \frac{(43.125)^2 + (48.371)^2}{3} - \frac{(91.496)^2}{6}$$

$$= 1{,}399.840 - 1{,}395.263$$

$$= 4.577$$

$$SS_B = \frac{\sum_k (\sum_j \bar{Y}_{.jk})^2}{a} - \frac{(\sum_j \sum_k \bar{Y}_{.jk})^2}{ab}$$

$$= \frac{(44.696)^2 + (28.467)^2 + (18.333)^2}{2} - 1{,}395.263$$

$$= 176.838$$

$$SS_{AB} = \sum_j \sum_k \bar{Y}_{.jk}^2 - \frac{(\sum_j \sum_k \bar{Y}_{.jk})^2}{ab} - SS_A - SS_B$$

$$= (19.125)^2 + \cdots + (8.000)^2 - 1{,}395.263 - 4.577 - 176.838$$

$$= 19.561$$

Compute mean squares as in the proportional case, dividing each sum of squares by its *df*. The error mean square is ordinarily the average within-cell variance. In the present case, each cell contains only one "score," the mean of the original entries. Therefore, we require the average variance of the cell means. Since the variance of a mean is known to be the variance of the scores divided by the cell frequency—that is, $\sigma_{\bar{Y}}^2 = \sigma^2/n$—we compute $MS'_{S/AB}$, the average variance of the cell means, as

(5.26)
$$MS'_{S/AB} = \frac{1}{ab} \sum_j \sum_k \frac{\hat{\sigma}_{jk}^2}{n_{jk}}$$

With homogeneity of variance assumed, the best estimate of σ_{jk}^2, for all j and k, is $MS_{S/AB}$, the average within-cell variance for the original data.

STEP 2.   Compute $SS_{S/AB}$ from the data of Table 5–9.

$$SS_{S/AB} = SS_{tot} - SS_{\overline{AB}} = 2{,}978.085$$

STEP 3.   Compute the reciprocal of the harmonic mean of the cell entries.

$$\frac{1}{\tilde{n}_h} = \left(\frac{1}{ab}\right)\left(\sum_j \sum_k \frac{1}{n_{jk}}\right)$$
$$= \frac{1}{6}\left(\frac{1}{8} + \frac{1}{6} + \cdots + \frac{1}{5}\right)$$
$$= .163$$

STEP 4.   It follows from Equation (5.26) and the use of $MS_{S/AB}$ as the estimate of σ_{jk}^2 that

$$MS'_{S/AB} = \frac{1}{\tilde{n}_h} MS_{S/AB}$$
$$= (.163)\left(\frac{2{,}978.085}{31.0}\right)$$
$$= 15.65$$

Note that the error df are

$$df_{S/AB} = df_{total} - df_{betw.\ cells}$$
$$= (37 - 1) - (6 - 1)$$
$$= 31$$

The F ratios are formed in the usual way. The results of the analysis are summarized in Table 5–12 and lead to the same conclusions as the analyses reported in Tables 5–7 and 5–11.

Table 5–12   Analysis of variance for data of Table 5–9 by method of unweighted means

SV	df	SS	MS	F
A	1	4.58	4.58	.29
B	2	176.84	88.42	5.64[a]
AB	2	19.56	9.78	.62
S/AB(adj)	31	485.43	15.65	

[a] $p < .01$

5.4.3 DISPROPORTIONATE FREQUENCIES IN THE TREATMENT POPULATIONS. Frequently, the most plausible, perhaps the only defensible, assumption is that unequal n typifies the treatment populations. This is often true when levels of the independent variable are not imposed by the experimenter but are attributes of the subjects—for example, age, attitude, socioeconomic level, personality classification. In a study of attitudes toward certain current problems, we might classify subjects by political party and by whether they consider themselves liberal or conservative. It is doubtful that the four treatment populations defined by these two variables are proportional.

A better understanding of the issues posed by such disproportionality, or *nonorthogonality* of independent variables, can be derived by considering Figure 5–5. The circle in each panel represents the total variability about the grand mean μ. The left-hand panel represents the situation in which treatment populations are equal or proportional in size; the two shaded areas, a and b, represent the proportions of variability due to the two independent variables A and B. (For simplicity, we have ignored the contribution of the interaction.) As the lack of overlap in the shaded areas indicates, each independent variable makes an independent, and thus separately estimable, contribution to the total variability.

In the right-hand panel, the shaded areas again represent the contributions of the variables A and B to the total variability. In this case, however, the independent variables are not orthogonal; the overlap in the shaded area c represents the joint contribution of A and B.[1] This failure to have orthogonal variables confronts us with a difficult problem in data analysis and interpretation. What do we mean by the effects of A and B? If we compute the SS_A in the usual way, we are essentially treating the areas labeled a and c as representing variation

Figure 5–5   Partitioning of variability in the dependent variable when the independent variables are orthogonal (left panel) and when they are not (right panel).

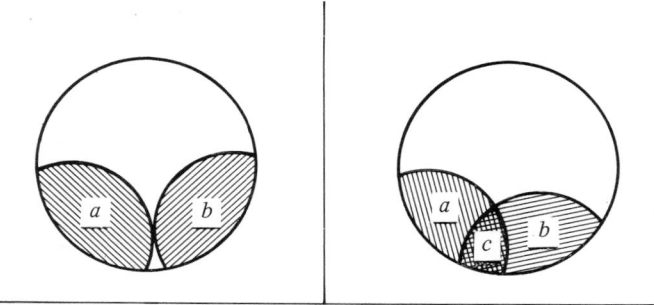

---

[1] In the interest of simplicity, Figure 5–5 somewhat misrepresents the situation. First, we have ignored the contribution of the AB interaction. Second, the area labeled c actually represents the covariation due to A and B, a quantity that can take on positive or negative values. As a consequence, the variability due to A after the joint contribution of B (area a) has been adjusted for may be smaller (as the figure implies) or larger than the unadjusted variability due to A (areas a and c combined).

due to A; the consequence of this is that we treat the area labeled b as representing the variation due to B. That is, B now is actually "B adjusted for the contribution of A." By contrast, we could first compute the usual SS_B, then compute an SS_A adjusted for the effects of B; then b plus c would represent the variation due to B, and a represent the variation due to A. We could analyze the data so that A was adjusted for B effects and B for A effects; then the A and B SV would be represented by areas a and b. There are other choices also. No one choice is always correct; the choice depends on the nature of the independent variables and the purpose of the experiment. (See Chapter 15 for more detail. We also consider there appropriate computational procedures and we provide references to several published articles in which the issues raised about Figure 5–5 have been discussed.)

The computations for the disproportionate n case are tedious to do without a computer. Fortunately, appropriate programs are available at most computing centers. Nevertheless, investigators eager to avoid this complication, as well as the interpretative problems just noted, have frequently forced orthogonality on their data by sampling equal numbers of subjects from each treatment combination. This is unwise because parameters so estimated may bear little relation to true parameter values. For example, the mean at A_j will be an average over all levels of B; in the experiment, the data at each level of B will carry equal weight in calculations for $\bar{Y}_{.j.}$, whereas in the population, the larger B populations (for that level of A) will make a greater contribution to the value of μ_j than the smaller ones will.

Snedecor and Cochran (1967, pp. 475–77) suggest that even with very disproportionate ns, the method of unweighted means (see Section 5.4.2) provides a reasonable test of interaction. If the F proves significant, evaluation of main effects is of little interest. The appropriate procedure would be to test simple effects. For these tests, unequal ns pose no problem because we are now essentially dealing with one-factor designs. If the test of interaction does not yield a significant result, regression analysis (Chapter 15) would be the appropriate next step.

5.5 MODEL FOR THE COMPLETELY RANDOMIZED THREE-FACTOR DESIGN

The approach we have developed for the two-factor design can be extended to designs involving more than two factors. The layout of a three-factor design is given in Table 5–13. The independent variables are A, B, and C, and the relevant indices are

$$i = 1, 2, \ldots, n \qquad k = 1, 2, \ldots, b$$
$$j = 1, 2, \ldots, a \qquad m = 1, 2, \ldots, c$$

The abc experimental groups of n subjects can be viewed as abc random samples, one from each of abc populations, which systematically differ among themselves

Table 5–13   Data matrix for three-dimensional design

			B_1	$\cdots$	B_k	$\cdots$	B_b
			Y_{1111}		Y_{11k1}		Y_{11b1}
			$\vdots$		$\vdots$		$\vdots$
		A_1	Y_{i111}		Y_{i1k1}		Y_{i1b1}
			$\vdots$		$\vdots$		$\vdots$
			Y_{n111}		Y_{n1k1}		Y_{n1b1}
	$\vdots$						
			$\vdots$		$\vdots$		$\vdots$
C_1		A_j	Y_{1j11}		Y_{ijk1}		Y_{ijb1}
	$\vdots$						
		A_a	Y_{ia11}		Y_{iak1}		Y_{iab1}
			$\vdots$		$\vdots$		$\vdots$
	$\vdots$						
			$\vdots$		$\vdots$		$\vdots$
C_m		A_j	Y_{ij1m}		Y_{ijkm}		Y_{ijbm}
			$\vdots$		$\vdots$		$\vdots$
	$\vdots$						
			$\vdots$		$\vdots$		$\vdots$
C_c		A_j	Y_{ij1c}		Y_{ijkc}		Y_{ijbc}
			$\vdots$		$\vdots$		$\vdots$

only in the treatment combination applied. The mean of the treatment population defined by the jth level of A, the kth level of B, and the mth level of C is μ_{jkm}. The mean of the population of all scores obtained under the jth level of A is μ_j, the expected value over the bc populations to which treatment A_j has been applied. Similarly, μ_k is the expected value computed over the ac treatment populations to which B_k has been applied; and μ_m is the expected value over the ab treatment populations to which C_m has been applied. Also, we can conceptualize the expected value of all scores under the treatment combinations A_jB_k, A_jC_m, and B_kC_m; such means will be denoted by μ_{jk}, μ_{jm}, and μ_{km}, respectively. Finally, the expected value over all abc treatment populations is denoted by μ.

The observed data are related to the population parameters by the equation

(5.27) $Y_{ijkm} = \mu + \alpha_j + \beta_k + \gamma_m + (\alpha\beta)_{jk} + (\alpha\gamma)_{jm} + (\beta\gamma)_{km} + (\alpha\beta\gamma)_{jkm} + \varepsilon_{ijkm}$

where

$\alpha_j = \mu_j - \mu$, main effect of treatment A_j

$\beta_k = \mu_k - \mu$, main effect of treatment B_k

$\gamma_m = \mu_m - \mu$, main effect of treatment C_m

$(\alpha\beta)_{jk} = \mu_{jk} - \mu_j - \mu_k + \mu$, interaction effect of A_j and B_k

$(\alpha\gamma)_{jm} = \mu_{jm} - \mu_j - \mu_m + \mu$, interaction effect of A_j and C_m

$(\beta\gamma)_{km} = \mu_{km} - \mu_k - \mu_m + \mu$, interaction effect of B_k and C_m

$(\alpha\beta\gamma)_{jkm} = \mu_{jkm} - \mu_{jk} - \mu_{jm} - \mu_{km} + \mu_j + \mu_k + \mu_m - \mu$, interaction effect of A_j, B_k, and C_m

$\varepsilon_{ijkm} = Y_{ijkm} - [\mu + \alpha_j + \beta_k + \gamma_m + (\alpha\beta)_{jk} + (\alpha\gamma)_{jm} + (\beta\gamma)_{km} + (\alpha\beta\gamma)_{jkm}]$
$= Y_{ijkm} - \mu_{jkm}$, error component

If the levels of all three independent variables are arbitrarily chosen, then the sum of *any* effect over *any* of its indices is zero; for example,

$$\sum_j (\alpha\beta\gamma)_{jkm} = \sum_k (\alpha\beta\gamma)_{jkm} = \sum_m (\alpha\beta\gamma)_{jkm} = 0$$

The same is true for all main and first-order (two-factor) interaction effects. The error component ε_{ijkm} is assumed to be independently and normally distributed, with zero mean and variance σ_e^2 within each of the *abc* treatment populations defined by the selected treatment combinations. As in the simpler designs already considered, the error component is that component of each score not accounted for by the contribution of the grand mean μ or by the systematic manipulation of the independent variables.

   With one exception, the quantities defined above have all been encountered in the preceding sections of this chapter. The exception is $(\alpha\beta\gamma)_{jkm}$, which is designated the *second-order interaction effect* to distinguish it from the first-order interactions, which involve only two independent variables. Note that

(5.28) $(\alpha\beta\gamma)_{jkm} = (\mu_{jkm} - \mu) - [\alpha_j + \beta_k + \gamma_m + (\alpha\beta)_{jk} + (\alpha\gamma)_{jm} + (\beta\gamma)_{km}]$

The interaction effect $(\alpha\beta\gamma)_{jkm}$ is the combined effect of the jth level of A, the kth level of B, and the mth level of C adjusted for the independent contributions of the main and interaction effects that have been removed in Equation (5.28).

5.6   THE ANALYSIS OF VARIANCE FOR THE COMPLETELY RANDOMIZED THREE-FACTOR DESIGN

The relation between any single score Y_{ijkm} and the single and joint effects of the variables A, B, and C is described in the identity of Equation (5.29).

$$
\begin{aligned}
Y_{ijkm} - \bar{Y}_{\ldots} = &(Y_{ijkm} - \bar{Y}_{.jkm}) + (\bar{Y}_{.j..} - \bar{Y}_{\ldots}) \\
&+ (\bar{Y}_{..k.} - \bar{Y}_{\ldots}) + (\bar{Y}_{...m} - \bar{Y}_{\ldots}) \\
&+ (\bar{Y}_{.jk.} - \bar{Y}_{.j..} - \bar{Y}_{..k.} + \bar{Y}_{\ldots}) \\
&+ (\bar{Y}_{.j.m} - \bar{Y}_{.j..} - \bar{Y}_{...m} + \bar{Y}_{\ldots}) \\
&+ (\bar{Y}_{..km} - \bar{Y}_{..k.} - \bar{Y}_{...m} + \bar{Y}_{\ldots}) \\
&+ (\bar{Y}_{.jkm} + \bar{Y}_{.j..} + \bar{Y}_{..k.} + \bar{Y}_{...m} \\
&\quad - \bar{Y}_{.jk.} - \bar{Y}_{.j.m} - \bar{Y}_{..km} - \bar{Y}_{\ldots})
\end{aligned}
$$

(5.29)

Only one of the terms in the right-hand side of Equation (5.29) is new. The expression $(\bar{Y}_{.jkm} + \bar{Y}_{.j..} + \bar{Y}_{..k.} + \bar{Y}_{...m} - \bar{Y}_{.jk.} - \bar{Y}_{.j.m} - \bar{Y}_{..km} - \bar{Y}_{....})$ is the least-squares estimate of $(\alpha\beta\gamma)_{jkm}$, the joint effect of the jth, kth, and mth levels of the variables A, B, and C adjusted for the main effects of these variables and for all first-order interaction effects of these variables. Equation (5.30) expresses this interpretation of the second-order, or three-variable, interaction effect more explicitly.

$$
\begin{aligned}
(\bar{Y}_{.jkm} &+ \bar{Y}_{.j..} + \bar{Y}_{..k.} + \bar{Y}_{...m} - \bar{Y}_{.jk.} - \bar{Y}_{.j.m} - \bar{Y}_{..km} - \bar{Y}_{....}) \\
&= (\bar{Y}_{.jkm} - \bar{Y}_{....}) - (\bar{Y}_{.j..} - \bar{Y}_{....}) - (\bar{Y}_{..k.} - \bar{Y}_{....}) - (\bar{Y}_{...m} - \bar{Y}_{....}) \\
&\quad - (\bar{Y}_{.jk.} - \bar{Y}_{.j..} - \bar{Y}_{..k.} + \bar{Y}_{....}) - (\bar{Y}_{.j.m} - \bar{Y}_{.j..} - \bar{Y}_{...m} + \bar{Y}_{....}) \\
&\quad - (\bar{Y}_{..km} - \bar{Y}_{..k.} - \bar{Y}_{...m} + \bar{Y}_{....})
\end{aligned}
$$

(5.30)

Squaring both sides of Equation (5.29) and summing over all indices, as done in the early part of this chapter and in Chapter 4, results in Equation (5.31) (note that the cross-product terms have again vanished).

(5.31)

$$
\sum_i^n \sum_j^a \sum_k^b \sum_m^c (Y_{ijkm} - \bar{Y}_{....})^2 = nbc \sum_j^a (\bar{Y}_{.j..} - \bar{Y}_{....})^2
$$
$$
(SS_{tot}) \qquad\qquad (SS_A)
$$
$$
+ nac \sum_k^b (\bar{Y}_{..k.} - \bar{Y}_{....})^2
$$
$$
(SS_B)
$$
$$
+ nab \sum_m^c (\bar{Y}_{...m} - \bar{Y}_{....})^2
$$
$$
(SS_C)
$$
$$
+ nc \sum_j^a \sum_k^b (\bar{Y}_{.jk.} - \bar{Y}_{.j..} - \bar{Y}_{..k.} + \bar{Y}_{....})^2
$$
$$
(SS_{AB})
$$
$$
+ nb \sum_j^a \sum_m^c (\bar{Y}_{.j.m} - \bar{Y}_{.j..} - \bar{Y}_{...m} + \bar{Y}_{....})^2
$$
$$
(SS_{AC})
$$
$$
+ na \sum_k^b \sum_m^c (\bar{Y}_{..km} - \bar{Y}_{..k.} - \bar{Y}_{...m} + \bar{Y}_{....})^2
$$
$$
(SS_{BC})
$$
$$
+ n \sum_j^a \sum_k^b \sum_m^c (\bar{Y}_{.jkm} + \bar{Y}_{.j..} + \bar{Y}_{..k.} + \bar{Y}_{...m} - \bar{Y}_{.jk.} - \bar{Y}_{.j.m} - \bar{Y}_{..km} - \bar{Y}_{....})^2
$$
$$
(SS_{ABC})
$$
$$
+ \sum_i^n \sum_j^a \sum_k^b \sum_m^c (Y_{ijkm} - \bar{Y}_{.jkm})^2
$$
$$
(SS_{S/ABC})
$$

We turn next to Table 5–14, which summarizes the analysis of variance for the three-dimensional case. The sources of variance follow immediately from Equation (5.31). The *df* are derived as before; for example, to obtain df_{ABC} we refer to Equation (5.30), which states that the second-order interaction effect is the cell effect (deviation of the cell mean from the grand mean) adjusted for first-order interaction effects and for main effects. This suggests

(**5.32**)
$$df_{ABC} = (abc - 1) - (a-1)(b-1) - (a-1)(c-1)$$
$$- (b-1)(c-1) - (a-1) - (b-1) - (c-1)$$
$$= (a-1)(b-1)(c-1)$$

The *SS* formulas are simple extensions of those developed earlier. Again note that one *df* is gained, or lost, for each squared quantity added, or subtracted.

The mean square column has been omitted since it has been obtained, as before, by dividing the entries in the *SS* column by the corresponding *df*. The entries in the *EMS* column have a similar form to those previously presented. With the exception of the entry for *S/ABC*, they consist of two components, σ_e^2 and some term involving a θ^2 quantity. The subscript of the θ^2 term is always the entry in the *SV* column for that line of the analysis of variance table. The multiplier of the θ^2 term consists of the product of all dimensions of the design not included in the subscript. Thus, θ_{BC}^2 does not include *S* or *A* in the subscripts and is therefore multiplied by the levels of these, *n* and *a*. The definitions of the θ^2 term are like those encountered previously:

(**5.33**)
$$\theta_A^2 = \frac{\sum_j (\mu_j - \mu)^2}{a - 1}$$

$$\theta_B^2 = \frac{\sum_k (\mu_k - \mu)^2}{b - 1}$$

$$\theta_C^2 = \frac{\sum_m (\mu_m - \mu)^2}{c - 1}$$

$$\theta_{AB}^2 = \frac{\sum_j \sum_k (\mu_{jk} - \mu_j - \mu_k + \mu)^2}{(a-1)(b-1)}$$

$$\theta_{AC}^2 = \frac{\sum_j \sum_m (\mu_{jm} - \mu_j - \mu_m + \mu)^2}{(a-1)(c-1)}$$

$$\theta_{BC}^2 = \frac{\sum_k \sum_m (\mu_{km} - \mu_k - \mu_m + \mu)^2}{(b-1)(c-1)}$$

$$\theta_{ABC}^2 = \frac{\sum_j \sum_k \sum_m (\mu_{jkm} + \mu_j + \mu_k + \mu_m - \mu_{jk} - \mu_{jm} - \mu_{km} - \mu)^2}{(a-1)(b-1)(c-1)}$$

Tests of null hypotheses about the various θ^2s are readily made by dividing the corresponding mean squares by $MS_{S/ABC}$. Since the *EMS* for this term differs from the expectations for all other terms only by the null hypothesis component, $MS_{S/ABC}$ is clearly the appropriate error term. We again note that this situation holds only when the levels of the independent variables have been arbitrarily

Table 5–14 Analysis of variance for three-factor design

SV	df	SS	EMS	F
A	$a-1$	$\dfrac{\sum_i^a T_{.j..}^2}{nbc} - C$	$\sigma_e^2 + nbc\theta_A^2$	$\dfrac{MS_A}{MS_{S/ABC}}$
B	$b-1$	$\dfrac{\sum_k^b T_{..k.}^2}{nac} - C$	$\sigma_e^2 + nac\theta_B^2$	$\dfrac{MS_B}{MS_{S/ABC}}$
C	$c-1$	$\dfrac{\sum_m^c T_{...m}^2}{nab} - C$	$\sigma_e^2 + nab\theta_C^2$	$\dfrac{MS_C}{MS_{S/ABC}}$
AB	$(a-1)(b-1)$	$\dfrac{\sum_i^a \sum_k^b T_{.jk.}^2}{nc} - C - SS_A - SS_B$	$\sigma_e^2 + nc\theta_{AB}^2$	$\dfrac{MS_{AB}}{MS_{S/ABC}}$
AC	$(a-1)(c-1)$	$\dfrac{\sum_i^a \sum_m^c T_{.j.m}^2}{nb} - C - SS_A - SS_C$	$\sigma_e^2 + nb\theta_{AC}^2$	$\dfrac{MS_{AC}}{MS_{S/ABC}}$
BC	$(b-1)(c-1)$	$\dfrac{\sum_k^b \sum_m^c T_{..km}^2}{na} - C - SS_B - SS_C$	$\sigma_e^2 + na\theta_{BC}^2$	$\dfrac{MS_{BC}}{MS_{S/ABC}}$
ABC	$(a-1)(b-1)(c-1)$	$\dfrac{\sum_i^a \sum_k^b \sum_m^c T_{.jkm}^2}{n} - C - SS_A - SS_B - SS_C - SS_{BC} - SS_{AB} - SS_{AC}$	$\sigma_e^2 + n\theta_{ABC}^2$	$\dfrac{MS_{ABC}}{MS_{S/ABC}}$
S/ABC	$abc(n-1)$	$\displaystyle\sum_i^n \sum_j^a \sum_k^b \sum_m^c Y^2 - \dfrac{\sum_i^a \sum_k^b \sum_m^c T_{.jkm}^2}{n}$	σ_e^2	

selected; when one or more of the independent variables have random effects (in the sense defined in Chapter 1), different expectations will be encountered, and consequently, other error terms besides the within-cell mean square may be required.

5.7   INTERACTION EFFECTS IN THE THREE-FACTOR DESIGN

The interpretation of first-order interactions is like that made for the two-factor designs. A significant AB interaction again indicates that the differences among the means of the populations defined by the levels of A change as a function of the level of B (or equivalently, that the differences among the means for the levels of B change as a function of the level of A). Assume, for example, that the entries in the $2 \times 2 \times 2$ (or 2^3) design of Table 5–15 are treatment population means. To investigate the AB interaction, we average over the levels of C, obtaining

$$\begin{array}{cc} & B_1 \quad B_2 \\ A_1 & \begin{bmatrix} 12 & 6 \\ 8 & 18 \end{bmatrix} \\ A_2 & \end{array}$$

An AB interaction is clearly present, since

$$(\bar{Y}_{.11.} - \bar{Y}_{.21.}) - (\bar{Y}_{.12.} - \bar{Y}_{.22.}) = (12-8) - (6-18) = 16$$

and not zero, which would be the case if all AB interaction effects were absent. The AC and BC interactions also contribute to the variability among cell means; the relevant matrices are

$$\begin{array}{cc} & C_1 \quad C_2 \\ A_1 & \begin{bmatrix} 11 & 7 \\ 9 & 17 \end{bmatrix} \\ A_2 & \end{array} \quad \text{and} \quad \begin{array}{cc} & C_1 \quad C_2 \\ B_1 & \begin{bmatrix} 10 & 10 \\ 10 & 14 \end{bmatrix} \\ B_2 & \end{array}$$

A first-order interaction is significant when the differences among simple effects of one variable change significantly over the levels of the second variable. Similarly, a second-order interaction effect is significant when the simple interaction effects of two variables change as a function of the level of the third variable. This is easily investigated for the 2^3 design. For the data of Table 5–15, the measure of the simple AB interaction at C_1 is $(15-5) - (7-13) = 16$ and at C_2, $(9-11) - (5-23) = 16$. According to the definition just given, the ABC interaction does not contribute to the variability among cell means, since the measure of

Table 5–15   Set of treatment population means

		C_1		C_2
	B_1	B_2	B_1	B_2
A_1	15	7	9	5
A_2	5	13	11	23

the *AB* interaction is the same at both levels of the third variable *C*. It is important to realize that the same conclusion follows if the *AC* interaction is considered at each level of *B*, or the *BC* interaction at each level of *A*. These are all equivalent approaches, yielding identical information.

The approach just exemplified can be extended to any three-factor design regardless of how many treatment levels are involved. Assuming that no error variance exists, if there is no second-order interaction,

(**5.34**) $[(\bar{Y}_{.jkm} - \bar{Y}_{.j'km}) - (\bar{Y}_{.jk'm} - \bar{Y}_{.j'k'm})]$
$$-[(\bar{Y}_{.jkm'} - \bar{Y}_{.j'km'}) - (\bar{Y}_{.jk'm'} - \bar{Y}_{.j'k'm'})] = 0$$

for all values of j, j', k, k', m, and *m'.* Suppose a third level of *A* is added to the data of Table 5–15, yielding

$$
\begin{array}{cc}
 & C_1 & & C_2 \\
 & B_1 \quad B_2 & & B_1 \quad B_2 \\
A_1 & \begin{bmatrix} 15 & 7 \\ 5 & 13 \\ 8 & 3 \end{bmatrix} & & \begin{bmatrix} 9 & 5 \\ 11 & 23 \\ 10 & 9 \end{bmatrix} \\
A_2 & & & \\
A_3 & & &
\end{array}
$$

Applying Equation (5.46), when $j = 1$, $j' = 2$, $k = 1$, $k' = 2$, $m = 1$, and $m' = 2$, gives

$$[(15-5)-(7-13)]-[(9-11)-(5-23)] = 16-16 = 0$$

and when $j = 1$, $j' = 3$, $k = 1$, $k' = 2$, $m = 1$, and $m' = 2$,

$$[(15-8)-(7-3)]-[(9-10)-(5-9)] = 3-3 = 0$$

and when $j = 2$, $j' = 3$, $k = 1$, $k' = 2$, $m = 1$, and $m' = 2$,

$$[(5-8)-(13-3)]-[(11-10)-(23-9)] = -13-(-13) = 0$$

One can again conclude that there is no *ABC* interaction.

An alternative method of investigating the *ABC* interaction is suggested by Tables 5–3, 5–4, and 5–5, which are the result of directly applying the analysis of variance model to first-order interaction effects. For second-order effects, Equation (5.28) could be applied, removing all main and first-order interaction effects from the three-dimensional matrix of means. If the *ABC* interaction is not significant, the variance among the adjusted cell means would be insignificant.

To promote further an understanding of interaction effects in three-factor designs, we next consider graphs of several sets of means for various combinations of effects. Assuming that no error variance is present, we find that the data of Figure 5–6 clearly reflect the presence of *A*, *B*, and *C* main effects. Averaging over the combinations of *B* and *C*, we have

$$
\begin{array}{ccc}
A_1 & A_2 & A_3 \\
[25 & 20 & 18]
\end{array}
$$

Averaging over the combinations of *A* and *C*, we have

$$
\begin{array}{ccc}
B_1 & B_2 & B_3 \\
[27 & 17 & 19]
\end{array}
$$

Figure 5–6   Data from a 3^3 design with only main effects present.

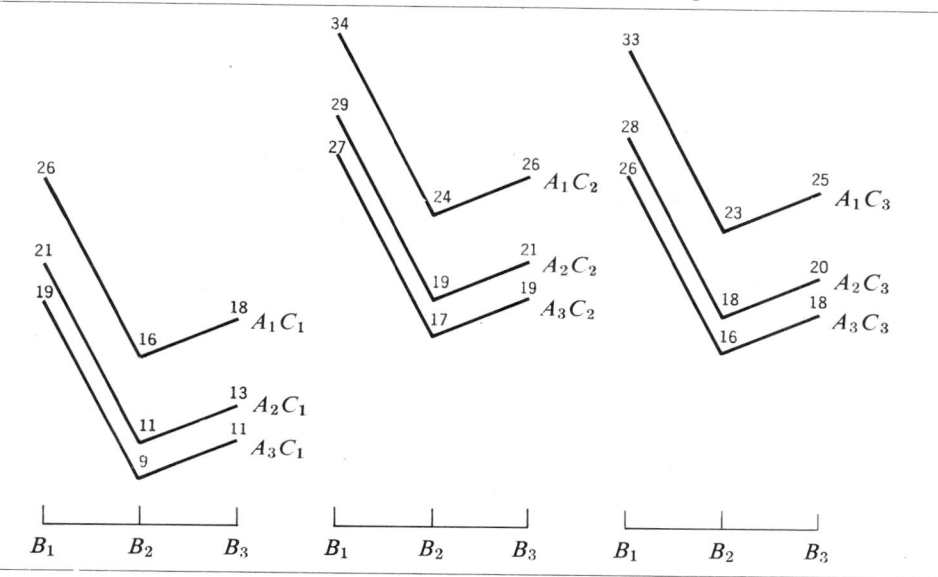

And averaging over the combinations of A and B, we have

$$\begin{array}{ccc} C_1 & C_2 & C_3 \\ [16 & 24 & 23] \end{array}$$

We next note that at each level of C, the three curves are parallel; there are no simple AB interaction effects at any of the levels of C. This absence of simple interaction effects is a *sufficient* condition for the absence of an overall AB interaction, since the overall effect is merely an average of the simple effects at the different C levels and the average of a set of zeros is just zero. The absence of simple effects is also a sufficient condition to conclude that there is no ABC interaction. The existence of the ABC interaction requires that the simple interaction effects of two variables change as a function of the level of the third variable. This condition is not met, since the AB interaction effects are zero at all levels of C. Using other sets of data, we shall soon show that the absence of simple interaction effects is not a *necessary* condition for overall zero AB or ABC interactions.

The status of the AC and BC interactions is not so clear as the status of the AB. When the curves in Figure 5–6 are shifted so that the three curves in each set are similar for level of A, it becomes clearer that no BC interaction exists. Alternatively, we can average over the levels of A, to get the matrix

$$\begin{array}{c} \\ B_1 \\ B_2 \\ B_3 \end{array} \begin{array}{ccc} C_1 & C_2 & C_3 \\ \begin{bmatrix} 22 & 30 & 29 \\ 12 & 20 & 19 \\ 14 & 22 & 21 \end{bmatrix} \end{array}$$

A plot of this set of means will result in three parallel lines. This is also true of

$$
\begin{array}{c}
\begin{array}{ccc} A_1 & A_2 & A_3 \end{array} \\
\begin{array}{c} C_1 \\ C_2 \\ C_3 \end{array}
\begin{bmatrix}
20 & 15 & 13 \\
28 & 23 & 21 \\
27 & 22 & 20
\end{bmatrix}
\end{array}
$$

the matrix of means for the AC combinations.

Two plots of another data set are given in the upper and lower halves of Figure 5–7. It is again assumed that the means are populations means, and that there is no error variance in the data set. Considering first the upper half of the figure, we find that an A main effect exists (the means are lower under the A_3 conditions than under the A_1 and A_2 conditions at all combinations of levels of B and C); that a B main effect exists (the B_1 means are highest, B_2 next, and B_3 lowest at combinations of A and C levels); and that a C main effect exists (the means generally increase as the level of C does). It is more difficult to determine the status of interaction effects than it was the data of Figure 5–6. The added complication is the presence of simple interaction effects; there is definitely an AB interaction at each level of C. The replot of the data in the bottom half of Figure 5–7 provides some illumination in this case. Shifting curves so that each set of three is at a single level of A, we find that there are no simple BC interaction effects. At each level of A, the spread among the curves does not change as a function of the level of B. As with the data of Figure 5–6, we can draw two conclusions from the absence of BC effects at the levels of A: (1) there is no overall BC interaction, and (2) there is no overall ABC interaction. A plot of means for the BC combinations, averaging over levels of A, will further verify (1), and the truth of (2) can be tested by investigating the validity of Equation (5.34) for the data of Figure 5–7. The appropriate data plots will also verify that AC and AB interactions do exist.

It has been seen that the absence of simple first-order interaction effects means that the overall first-order interaction, as well as the second-order interaction, cannot be a source of variance. The next question is whether it is possible to have simple interaction effects but no overall interaction, and what the implications of such results are for the second-order interaction. The relevant data are plotted in Figure 5–8. At each level of C, AB interaction effects are present. Alternative plots of the data will verify that simple AC and BC interaction effects also exist. When we table the means for the AB combinations, first averaging over levels of C, however, we obtain

$$
\begin{array}{c}
\begin{array}{ccc} B_1 & B_2 & B_3 \end{array} \\
\begin{array}{c} A_1 \\ A_2 \\ A_3 \end{array}
\begin{bmatrix}
18 & 15 & 12 \\
18 & 15 & 12 \\
24 & 21 & 18
\end{bmatrix}
\end{array}
$$

Figure 5–7   Two plots of a set of data with *A*, *B*, *C*, *AB*, and *AC* effects present.

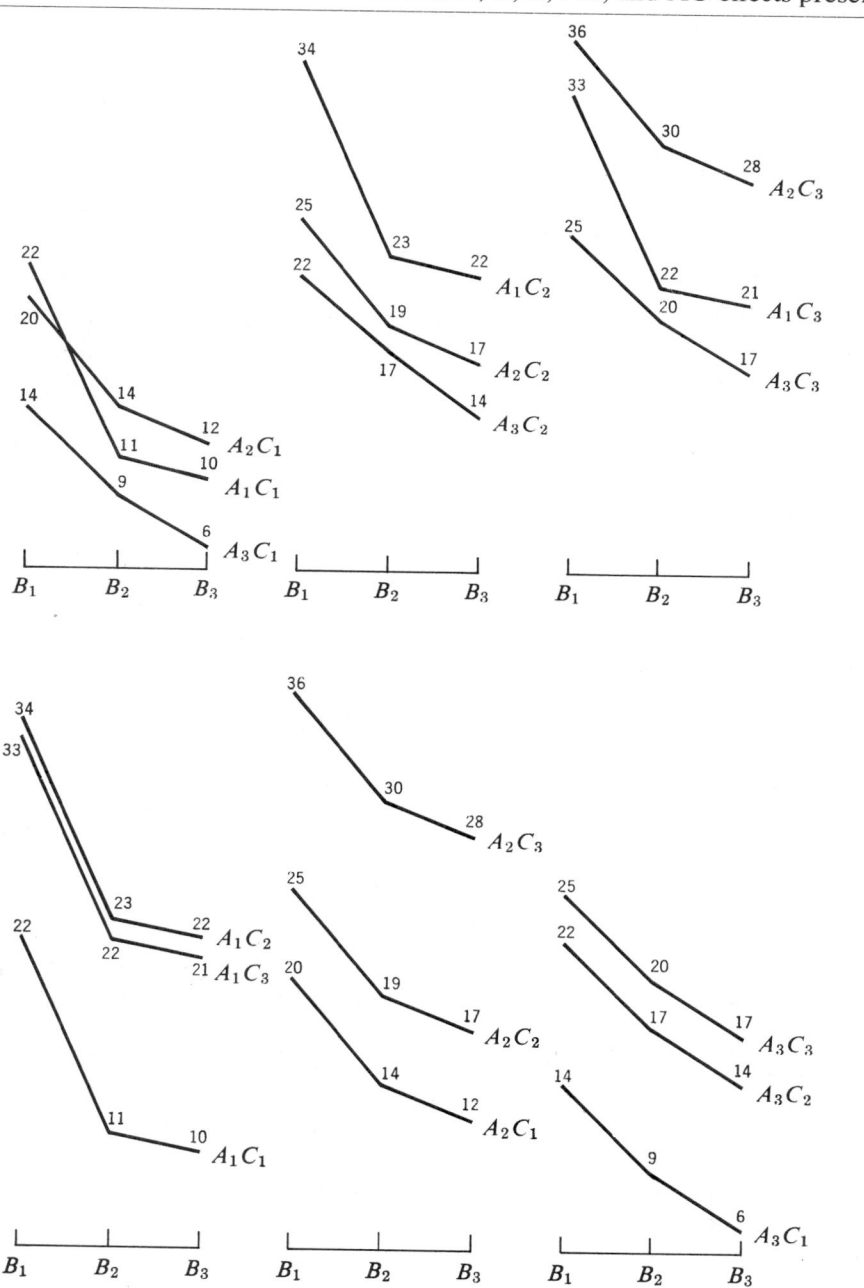

Figure 5–8   Data from a 3^3 design with A, B, and ABC effects present.

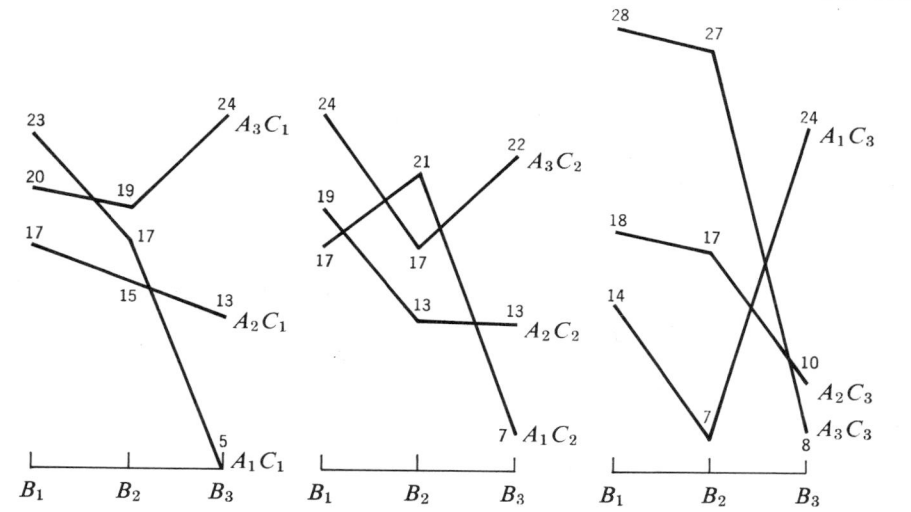

which in a plot gives a set of parallel curves. Similarly, we get

$$
\begin{array}{c@{\quad}ccc}
 & C_1 & C_2 & C_3 \\
A_1 & 15 & 15 & 15 \\
A_2 & 15 & 15 & 15 \\
A_3 & 21 & 21 & 21
\end{array}
$$

which shows that there is no AC interaction. A third tabulation shows that no BC interaction exists:

$$
\begin{array}{c@{\quad}ccc}
 & B_1 & B_2 & B_3 \\
C_1 & 20 & 17 & 14 \\
C_2 & 20 & 17 & 14 \\
C_3 & 20 & 17 & 14
\end{array}
$$

Thus it is possible that simple two-variable interaction effects may exist at each level of the third variable but so vary over levels of the third variable as to cancel each other. The result is that there is no overall first-order interaction. An additional consequence of this situation is that a second-order, or three-variable, interaction must exist, since by definition, if the interaction effects of two variables change over the levels of the third variable, there is a second-order interaction. The hypothesis of an ABC interaction can be independently verified by applying Equation (5.34) to the data of Figure 5–8. For example, let

$$
\begin{array}{ll}
j = 1 & j' = 3 \\
k = 1 & k' = 2 \\
m = 1 & m' = 3
\end{array}
$$

Then,

$$[(\bar{Y}_{jkm}-\bar{Y}_{j'km})-(\bar{Y}_{jk'm}-\bar{Y}_{j'k'm})]-[(\bar{Y}_{jkm'}-\bar{Y}_{j'km'})-(\bar{Y}_{jk'm'}-\bar{Y}_{j'k'm'})]$$
$$=[(23-20)-(17-19)]-[(14-28)-(7-27)]$$
$$=5-6=-1$$

Since the above result does not equal zero, the *ABC* interaction contributes to the variability among cell means.

Still assuming errorless data (or equivalently, that we are dealing with population means), we can summarize the preceding results very generally. Let X be any source of variance (say A, AB, ABC) and Y be some variable not involved in the effect. Then it is possible to speak of the simple X effects at the levels of Y, the overall X effects obtained by averaging over the levels of Y, and the interaction of X and Y:

1. If the X effects are zero at all levels of Y, the overall X effects will be zero. Thus, if there is no effect due to A at any level of B, there will be no A main effect; if there are no AB interactions at any level of C, there will be no overall AB effects.
2. If the X effects at all levels of Y are zero, the overall interaction of X and Y will be zero. Thus, if the simple effects due to A are zero at all levels of B, the overall AB effects will be zero; if the effects due to AB are zero at all levels of C, the overall effects due to ABC will be zero.
3. If the overall effects due to X are zero but some of the simple effects of X at the levels of Y are not zero, effects due to the interaction XY will not be zero. Thus, if the main effects due to A are zero but simple effects at various levels of B are not, the AB interaction effects will not be zero; if the overall AB interaction effects are zero and if the simple AB effects at levels of C are not, the ABC interaction effects will be nonzero.
4. If the interaction effects of X and Y are zero and if the overall effects of X are zero, then the simple effects of X at each level of Y will be zero. Thus, if the overall A main effects and the overall AB interaction effects are zero, the simple effects of A at each level of B will be zero. If the overall AB and ABC effects are zero, the AB simple interaction effects at each level of C will all be zero.

The preceding discussion has dealt solely with relations among population means. With actual data, apparent violations of our conclusions will occasionally occur. For example, overall tests of A and AB effects may yield nonsignificant results, but tests of simple effects of A may prove significant at one or more levels of B. This set of statistical outcomes is inconsistent with statement 4, which is a true statement about population means. Such inconsistencies between the results of the analysis and our conclusions about relations in the population are a warning to view our inferences with more than the usual skepticism. In the present example, either a Type II error has occurred when the overall effects were tested or a Type I error has occurred when the simple effects were tested.

5.8 NUMERICAL EXAMPLE FOR THE THREE-FACTOR DESIGN

Assume that each of 120 subjects is given 45 trials on a discrimination problem. The subjects are equally divided in age (A: 6, 8, 10 years), amount of reward (R: 1¢, 5¢), and delay of reward (D: 0 seconds, 10 seconds). The data have been summarized in Table 5–16 in the most convenient way for analyzing variance. The indices of notation are to be interpreted as follows:

i indexes the scores in each cell: $i = 1, 2, \ldots, 10$
j indexes the age levels: $j = 1$ (6 yr), 2 (8 yr), 3 (10 yr)
k indexes the reward levels: $k = 1$ (1¢), 2 (5¢)
m indexes the delay levels: $m = 1$ (0 sec), 2 (10 sec).

Table 5–16   Data from three-factor experiment

		6 Yr	8 Yr	10 Yr	
1¢	$T_{.jkm} =$	258	329	368	$T_{..km} = 955$
0 sec	$\sum_i Y^2 =$	6,864	11,209	13,704	
5¢	$T_{.jkm} =$	262	351	383	$T_{..km} = 996$
	$\sum_i Y^2 =$	7,306	12,585	14,789	
	$T_{.j.m} =$	520	680	751	$T_{...m} = 1,951$
	$\sum_k \sum_i Y^2 =$	14,170	23,794	28,493	
1¢	$T_{.jkm} =$	217	238	308	$T_{..km} = 763$
10 sec	$\sum_i Y^2 =$	4,887	5,928	9,696	
5¢	$T_{.jkm} =$	220	263	381	$T_{..km} = 864$
	$\sum_i Y^2 =$	5,220	7,221	14,625	
	$T_{.j.m} =$	437	501	689	$T_{...m} = 1,627$
	$\sum_k \sum_i Y^2 =$	10,107	13,149	24,321	
	$T_{.j..} =$	957	1,181	1,440	$T_{....} = 3,578$
1¢	$T_{.jk.} =$	475	567	676	$T_{..k.} = 1,718$
5¢	$T_{.jk.} =$	482	614	764	$T_{..k.} = 1,860$
	$T_{.j..} =$	957	1,181	1,440	$T_{....} = 3,578$
				$\sum_m \sum_k \sum_j \sum_i Y^2 = 114,030$	

Totals of scores and of squared scores have been calculated for each cell of the design and for combinations of cells. With these quantities calculated, it is necessary only to substitute in the formulas of Table 5–14 to complete the analysis of variance. These totals will also be found useful in subsequent tests (Chapters 13 and 17) and in tests of simple effects.

The correction term C is again the sum of all scores, squared and divided by the total number of scores.

$$C = \frac{T^2_{....}}{120} = \frac{(3,578)^2}{120} = 106,684.03$$

The SS_{tot} is again

$$SS_{tot} = \sum_i \sum_j \sum_k \sum_m Y^2 - C$$
$$= 114,030 - 106,684.03$$
$$= 7,345.97$$

To obtain the sum of squares for *any* main effect:

1. Square the sum for each level of the variable.
2. Sum the squared quantities.
3. Divide by the number of scores at each level.
4. Subtract C.

Thus,

$$SS_A = \frac{\sum_j T^2_{.j..}}{40} - C$$
$$= \frac{(957)^2 + (1,181)^2 + (1,440)^2}{40} - \frac{(3,578)^2}{120}$$
$$= 109,605.25 - 106,684.03$$
$$= 2,921.22$$

$$SS_R = \frac{\sum_k T^2_{..k.}}{60} - C$$
$$= \frac{(1,718)^2 + (1,860)^2}{60} - \frac{(3,578)^2}{120}$$
$$= 106,852.06 - 106,684.03$$
$$= 168.03$$

$$SS_D = \frac{\sum_m T^2_{...m}}{60} - C$$
$$= \frac{(1,951)^2 + (1,627)^2}{60} - \frac{(3,578)^2}{120}$$
$$= 107,558.83 - 106,684.03$$
$$= 874.80$$

To obtain the sums of squares for *any* interaction:

1. Square the sum for each combination of levels of the variables involved in the interaction.
2. Sum the squared quantities.
3. Divide by the number of scores in each combination of levels.
4. Subtract C.
5. Subtract the sums of squares for all effects contributing to the variability among the means for the treatment combinations involved in the interaction. To obtain the *ARD* interaction sum of squares, for example, we remove the sums of squares for A, R, D, AR, AD, and DR.

Thus,

$$SS_{AR} = \frac{\sum_j \sum_k T^2_{.jk.}}{20} - C - SS_A - SS_R$$

$$= \frac{(475)^2 + \cdots + (764)^2}{20} - \frac{(3,578)^2}{120} - 2,921.22 - 168.03$$

$$= 109,855.30 - 106,684.03 - 2,921.22 - 168.03$$

$$= 82.02$$

$$SS_{AD} = \frac{\sum_j \sum_m T^2_{.j.m}}{20} - C - SS_A - SS_D$$

$$= \frac{(520)^2 + \cdots + (689)^2}{20} - \frac{(3,578)^2}{120} - 2,921.22 - 874.80$$

$$= 110,674.60 - 106,684.03 - 2,921.22 - 874.80$$

$$= 194.55$$

$$SS_{RD} = \frac{\sum_k \sum_m T^2_{..km}}{30} - C - SS_R - SS_D$$

$$= \frac{(955)^2 + \cdots + (864)^2}{30} - \frac{(3,578)^2}{120} - 168.03 - 874.80$$

$$= 107,756.86 - 106,684.03 - 168.03 - 874.80$$

$$= 30.00$$

$$SS_{ARD} = \frac{\sum_j \sum_k \sum_m T^2_{.jkm}}{10} - C - SS_A - SS_R - SS_D - SS_{AR} - SS_{AD} - SS_{RD}$$

$$= \frac{(258)^2 + \cdots + (381)^2}{10} - \frac{(3,578)^2}{120} - 2,921.22 - 168.03$$

$$- 874.80 - 82.02 - 194.55 - 30.00$$

$$= 111,009.00 - 106,684.03 - 2,921.22 - 168.03 - 874.80 - 82.02$$

$$- 194.55 - 30.00$$

$$= 54.35$$

The sum of squares for *subjects-within-ARD treatment combinations* can be most simply computed as

$$SS_{S/ARD} = SS_{tot} - SS_{\overline{ARD}}$$
$$= 7{,}345.97 - (111{,}009.00 - 106{,}684.03)$$
$$= 3{,}021.00$$

The results of the complete analysis of variance are summarized in Table 5–17. It is clear that performance improves with age and deteriorates when reinforcement is delayed. The main effect of reward is less clearly established, but the F ratio is almost significant at the .01 level, and the effects are in the direction that theory and experimental data would suggest—namely, performance is better under the higher reward. It appears reasonable to conclude that the levels of reward used in this study do differ in their effects. The failure to obtain any significant interactions suggests that the effects of any one variable do not change markedly over levels of the other two variables.

5.9   MORE THAN THREE INDEPENDENT VARIABLES

The analysis and interpretation of data for completely randomized designs involving four or more independent variables are in all respects straightforward generalizations of the material presented for the simpler cases. Each variable and each possible combination of variables is a potential contributor to the total variability, and so is the variability among subjects within each combination of variables.

The df for main effects are again the number of levels of the variable minus 1; those for an interaction are the product of the df for the variables entering into the interaction. Assume, for example, a design involving six independent variables labeled $A, B, \ldots, F$. Then the df for the $ABDE$ interaction are $(a-1)(b-1) \times (d-1)(e-1)$, where a, b, d, and e are the numbers of levels of A, B, D, and E, following our previous notational usage.

Table 5–17   Analysis of variance for data from three-factor experiment

SV	df	SS	MS	F
A	2	2,921.22	1,460.61	55.22[a]
R	1	168.03	168.03	6.01
D	1	874.80	874.80	31.27[a]
AR	2	82.02	41.01	1.47
AD	2	194.55	97.28	3.48
RD	1	30.00	30.00	1.07
ARD	2	54.35	27.18	.97
S/ARD	108	3,020.76	27.97	

[a] $p < .01$

The rules for computing *SS* for main and for interaction effects (Section 5.8) apply in all designs. Thus, to obtain the *SS* for the *ABDE* interaction:

1. Square the sum of scores for each of the *abde* combinations of levels of the four variables.
2. Sum the *abde* squared quantities.
3. Divide by the number of scores obtained in each of the *abde* combinations.
4. Subtract *C*.
5. Subtract the sums of squares for all main and interaction effects embedded in the *ABDE* combination.

The above steps can be summarized by

(5.35)
$$SS_{ABDE} = \frac{\sum^a \sum^b \sum^d \sum^e (\sum^n \sum^c \sum^f Y)^2}{ncf} - C - SS_A - SS_B$$
$$- SS_D - SS_E - SS_{AB} - SS_{AD} - SS_{AE} - SS_{BD}$$
$$- SS_{BE} - SS_{DE} - SS_{ABD} - SS_{ADE} - SS_{BDE} - SS_{ABE}$$

Note that if each term in Equation (5.35) is replaced by the corresponding *df*, the result is

$$(abde - 1) - (a - 1) - (b - 1) - \cdots - (a - 1)(b - 1)(e - 1) = (a - 1)(b - 1)(d - 1)(e - 1)$$

The computations above are applicable when all cells contain equal numbers of subjects. Alternative analyses have been previously presented in the two-factor case for unequal but proportionate *n*s and for disproportionate *n*s. These analyses can be extended to data for designs involving more than two factors.

In all designs, regardless of the number of variables, mean squares are ratios of sums of squares to degrees of freedom. Still assuming completely randomized designs in which the levels of all variables have been arbitrarily chosen, we find that the *EMS* will consist of σ_e^2 plus some quantity θ^2 with a subscript. This subscript denotes the source of variance being referred to and is multiplied by *n* (where *n* is the number of subjects in each cell) and by the numbers of levels of those variables in the design which are not in the subscript. For example, the *EMS* for the *ABDE* interaction in the six-variable experiment referred to before is $\sigma_e^2 + cfn\theta_{ABDE}^2$, where *n* is the number of subjects in each of the *abcdef* cells. It follows from this discussion that the mean square for subjects within cells will be the error term (the denominator, that is) for all *F* tests to be made in completely randomized designs. The assumptions underlying the *F* test for multifactor designs parallel the earlier assumptions for the two- and three-factor designs. Specifically, we assume that the error components of the scores are independently and normally distributed, with mean zero and variance σ_e^2 within each population defined by the total number of treatment combinations.

5.10   POOLING IN MULTIFACTOR DESIGNS

Experiments frequently involve sources of variance that although necessary to the design, are really of no interest to the investigator. An experiment might be designed, for example, to study the effects of several drugs D and several drug concentrations C on maze learning performance in rats. Position P of the food reinforcement is carefully balanced, the reward appearing in the right arm for half of the subjects in each CD combination and in the left arm for the other half of the subjects. Because running the experiment is time-consuming, two experimenters E share the work; each one runs exactly half of the subjects undergoing each CDP combination. We now have a four-factor design; two of the factors, C and D, are of interest while P has been introduced as a control for position preferences and E has been introduced for practical reasons. We have three choices in analyzing this design. One possibility is to perform a complete analysis, testing separately each of the four main effects and all possible interactions. Excluding the within-cell error term, there are 15 such terms. There are three objections to this procedure: (1) it involves computational labor, which our interests do not require; (2) we are performing 12 F tests that are of no interest (assuming that only the C, D, and CD effects are of interest) and are greatly increasing the probability of obtaining significance by chance, with consequent difficulty in interpreting the three F tests that are of interest; and (3) we are cluttering the analysis of variance results with 12 added terms, making it more difficult to pick out the terms of interest. A second possible analysis, which avoids such difficulties, is to compute SS_C, SS_D, SS_{CD}, and a residual error sum of squares; the residual equals the difference between SS_{total} and the first three terms. This is frequently done but it is a dangerous procedure. If the null hypothesis is false for any of the 12 terms whose sums of squares are added to $SS_{S/CDPE}$, then the error mean square is an estimate of both σ_e^2 *and variability due to treatment effects.* In testing C, D, or CD, the ratio of expected mean squares will be less than 1 if H_0 is true; that is, the F test will be negatively biased.

There is a third procedure that avoids the pitfalls of the first two. In the present example, it would entail calculating sums of squares for C, D, CD, $S/CDPE$, and one for the pool of all the 12 terms involving P and E. This combined, or pooled, sum of squares can be readily obtained by subtraction and so can its associated df. Thus

$$df_{P,E \text{ pool}} = (cdpen - 1) - [(c-1) + (d-1) + (c-1)(d-1) \\ + cdpe(n-1)] = cd(pe-1)$$

This result suggests a computational check. Since calculations of sums of squares follow from knowledge of the df, then $SS_{P,E \text{ pool}}$ can be calculated directly as

$$\frac{\sum^c \sum^d \sum^p \sum^e (\sum^n Y)^2}{n} - \frac{\sum^c \sum^d (\sum^p \sum^e \sum^n Y)^2}{pen}$$

The approach taken here is flexible. We might be interested in the P and E main effects but not in their interactions. These would then be calculated separately and the pool based on only 10 terms. Note that the pool is of no interest in itself but provides a way of obtaining an appropriate error term without calculating a large number of uninteresting effects.

The issue of pooling will arise again in Chapter 9, where we find designs in which error df tend to be small. It might happen that other mean squares would have the same expectation as the error term, say if we could reasonably assume some interaction variance to be negligible. If they did, they could be pooled with the usual error term to provide a new error mean square with more df and consequently more power. Pooling, we should note now, is a weighted average of several mean squares. In pooling two mean squares, we get

$$MS_{pool} = \frac{SS_1 + SS_2}{df_1 + df_2} = \left(\frac{df_1}{df_1 + df_2}\right)MS_1 + \left(\frac{df_2}{df_1 + df_2}\right)MS_2$$

It may be of interest to realize that the denominator of the familiar t statistic for two independent groups is merely a function of the pool of two variances under the assumption that they both estimate the same population variance; that is,

$$\hat{\sigma}_{Y_1 - Y_2} = \sqrt{\left(\frac{1}{n_1} + \frac{1}{n_2}\right)\left[\left(\frac{n_1 - 1}{n_1 + n_2 - 2}\right)\hat{\sigma}_1^2 + \left(\frac{n_2 - 1}{n_1 + n_2 - 2}\right)\hat{\sigma}_2^2\right]}$$

5.11   COMPUTATIONS FOR SINGLE DEGREE-OF-FREEDOM EFFECTS

Consider an experimental design in which there are two levels of the variables A and B. We then have four treatment combinations: A_1B_1, A_1B_2, A_2B_1, and A_2B_2. Designate the sum of n scores in each cell T_{11}, T_{12}, T_{21}, and T_{22}, respectively. The usual computational formula for the SS_A is then

(5.36) $$SS_A = \frac{(T_{11} + T_{12})^2}{2n} + \frac{(T_{21} + T_{22})^2}{2n} - \frac{(T_{11} + T_{12} + T_{21} + T_{22})^2}{4n}$$

Let

(5.37) $$T_{1.} = T_{11} + T_{12} = \sum_{i=1}^{n} \sum_{k=1}^{2} Y_{i1k}$$

$$T_{2.} = T_{21} + T_{22} = \sum_{i=1}^{n} \sum_{k=1}^{2} Y_{i2k}$$

Substituting Equation (5.37) into Equation (5.36) and expanding gives

$$SS_A = \frac{T_{1.}^2 + T_{2.}^2}{2n} - \frac{T_{1.}^2 + T_{2.}^2 + 2T_{1.}T_{2.}}{4n}$$

(5.38) $$= \frac{T_{1.}^2 + T_{2.}^2 - 2T_{1.}T_{2.}}{4n}$$

$$= \frac{(T_{1.} - T_{2.})^2}{4n}$$

Equation (5.38) provides a much faster way of calculating SS_A than Equation (5.36) does. The reduction in computational effort is even more marked when we consider the shortcut formula for SS_{AB}:

(5.39)
$$SS_{AB} = \frac{(T_{11} + T_{22} - T_{12} - T_{21})^2}{4n}$$

Note that there is no need to subtract sums of squares for main effects or to remove a correction term. Equation (5.39) can be derived from the general interaction formula presented earlier; this was done for Equation (5.37). A quick and very general method for arriving at the sum of squares formulas for any single *df* term is available if we redesignate the treatment combinations:

$$A_1 B_1 = ab$$
$$A_1 B_2 = a$$
$$A_2 B_1 = b$$
$$A_2 B_2 = (1)$$

The subscript 1 indicates the presence of the lowercase letter, and the subscript 2 indicates the absence of the letter. If all subscripts are 2s, the new designation is (1). This notation can be extended to any number of variables, all of which are tested at two levels. Thus, the cell $A_1 B_2 C_2 D_1$ in a 2^4 design is relabeled *ad*. Given these new labels, we may readily state the appropriate shortcut formula. The following steps are involved:

1. List the new designations for all cells. For a 2^3 design, these would be

$$abc(=A_1 B_1 C_1) \qquad a(=A_1 B_2 C_2)$$
$$ab(=A_1 B_1 C_2) \qquad b(=A_2 B_1 C_2)$$
$$ac(=A_1 B_2 C_1) \qquad c(=A_2 B_2 C_1)$$
$$bc(=A_2 B_1 C_1) \qquad (1)(=A_2 B_2 C_2)$$

2. Divide the cells into two classes, those that have an even number of letters in common with the effect and those that have an odd number of letters in common with the effect. For the *AC* interaction, the two classes are

1	2
abc	*a*
ac	*c*
b	*ab*
(1)	*bc*

Each of the designations in class 1 contains either two or none of the letters *a* and *c*. All designations in class 2 contain one of the letters *a* and *c*.

3. Obtain the sum of scores for classes 1 and 2 above and then square the difference in the sums. Divide this quantity by the total number of scores.

In the preceding 2^3 example, we might represent the operation of subtracting class 2 from class 1 totals by

$$[abc + ac + b + (1)] - [a + c + ab + bc]$$

which equals

$$[(abc + ac) - (ab + a)] - [(bc + c) - (b + (1))]$$

This last quantity is the difference between C_1 and C_2 totals at A_1 minus the difference between C_1 and C_2 totals at A_2, which we had previously defined (Section 5.3.2) as a measure of first-order interaction. To show further the development of single *df* formulas and their relation to the meaning of interaction, we next consider the *ABC* interaction for a 2^3 design. According to the development of Section 5.7 (in particular, see Equation (5.34)), the *ABC* interaction is a measure of the variation in the interaction of two variables that occurs over the levels of the third variable. In our new notation, the appropriate contrast is

$$[(abc - ac) - (bc - c)] - \{(ab - a) - [b - (1)]\}$$

which equals

$$[abc + a + b + c] - [ab + ac + bc + (1)]$$

exactly the contrast that the odd-even rule, rule 2 above, demands. All designations to the left of the minus sign have one or three letters in common with *ABC*. To obtain SS_{ABC} we therefore insert the appropriate cell totals in the preceding formula for the contrast, square the result, and divide by $8n$, the total number of measurements.

Consider, as an alternative to the odd-even rule, this algebraic technique for generating contrasts. If the SS_{AC} is required, expand the quantity $(a - 1)(b + 1) \times (c - 1)$. We then have

$$(abc + ac + b + 1) - (a + c + ab + bc)$$

the contrast previously arrived at by the odd-even rule. For the SS_A, expand $(a - 1)(b + 1)(c + 1)$ and obtain

$$(a + ab + ac + abc) - (b + c + bc + 1)$$

The approach is simply to insert a minus sign only within those parentheses containing the letters appearing in the designation of the effect of interest.

In Section 5.8, sums of squares for a $2 \times 2 \times 3$ design were computed. The *D*, *R*, and *DR* effects are all on one *df*. Applying our single *df* approach to the cell totals of Table 5–16, we obtain

$$SS_D = \frac{(1{,}951 - 1{,}627)^2}{120}$$

$$= 874.80$$

$$SS_R = \frac{(1{,}718 - 1{,}860)^2}{120}$$

$$= 168.03$$

$$SS_{DR} = \frac{(955 + 864 - 996 - 763)^2}{120}$$

$$= 30.00$$

These are exactly the results obtained in Section 5.8.

5.12 CONCLUDING REMARKS

The completely randomized designs have several advantages. The analysis of the data is simpler than for most other designs. For any given number of measurements, the error *df* will be larger for these designs than for comparable designs. The requirements of the underlying model are most easily met by completely randomized designs, and violations of the assumptions embedded in the model are least likely to affect inferences derived from the *F* ratio. These designs share one main deficit. Since the within-cell variability that forms the error term is a function of individual differences, the efficiency of this design is relatively low. Other designs, which allow the experiment to remove from the error term the variability due to individual differences, will generally yield a more precise estimate of population effects. A completely randomized approach should be considered whenever subjects are reasonably homogeneous for the variable being measured; whenever a large *n* is available, compensating to some extent for the variability of measurements; or whenever the available *n* is so small that the loss in error degrees of freedom that always accompanies more efficient designs yields a considerable loss in power. There are many experimental situations in which it is impossible to do anything but assign different subjects to different levels of the variables. This is self-evident when the independent variable is personality type or training technique. It may also be true when much time is needed to obtain a measure from the subject and it is therefore preferable to obtain only one measure from each subject.

EXERCISES

5.1 Plot *all* main and interaction effects. Assuming errorless data, which effects are significant?

	A_1			A_2			A_3		
	B_1	B_2	B_3	B_1	B_2	B_3	B_1	B_2	B_3
C_1	22	12	14	19	6	8	16	5	6
C_2	18	8	10	21	8	10	18	7	8
C_3	14	4	6	20	7	9	23	12	13

5.2 Assume a two-factor completely randomized design with n subjects in each of ab cells. Derive formulas for $\hat{\omega}_A^2$, $\hat{\omega}_B^2$, and $\hat{\omega}_{AB}^2$ (see Section 4.7.2).

5.3 We have the following set of cell means and variances, based on 10 scores per cell:

	Means			Variances		
	B_1	B_2	B_3	B_1	B_2	B_3
A_1	7	9	10	16	20	16
A_2	3	6	4	18	17	21

(a) Test *B* for significance.
(b) We are particularly interested in whether A_1 and A_2 differ at B_3. Do the significance test. Briefly justify your choice of error term.

(c)   Compute:

$$\sum_k \frac{(T_{.1k} - T_{.2k})^2}{20} - SS_A$$

(d)   How many df are associated with the term in (c)? Suppose I had an $a \times b$ design. I compute the SS_{A/B_1} (SS for A at B_1), SS_{A/B_2}, and so on. Then I subtract SS_A from $\sum_k SS_{A/B_k}$. This is what I did in (c). What are the df in this general case?

(e)   There is a point to this. What is the formula in (c) computing? Provide a one or two-sentence intuitive argument supporting your claim. (Do not make your case by recalculating the term for the data, using another formulation. Also, merely mumbling something about df is not sufficient, though it is necessary.)

5.4   Patients in a mental hospital are divided into experimental groups on the basis of their socioeconomic level (3 levels) and the kind of treatment they receive (2 levels, psychotherapy and behavior therapy). The investigator predicts that (1) psychotherapy will be less effective than behavior therapy, and (2) psychotherapy will be more effective the higher the socioeconomic level of the patient, but this will not be true for behavior therapy. In fact, no main effect of socioeconomic level is predicted.

State which SVs (sources of variance) should be significant. Then set up a matrix of cell means consistent with our hypotheses, assuming errorless data.

5.5   Consider a study in which each S is presented with a set of digits. He is then presented with a probe digit and must respond "yes" or "no"; the probe was or wasn't in the list. We hypothesize that observed reaction time is $RT = t_e + t_c + t_r$, where t_e = time to encode (read) the probe digit, t_c = time to compare it with all the members of the memorized list, and t_r is time to say "yes" or "no." Note that only t_c should change with the number of digits in memory.

Suppose we have a two-factor design. One variable is L, list length, the number of digits memorized. The other is Q, quality of the probe. It may be clear and easy to read or fuzzy and tough to read. We think that this should only influence the time it takes to encode the stimulus. Let $t_r = 300$. Make up values of t_e and t_c for high- and low-quality stimulus and $L = 2$, 3, and 4. The numbers should be consistent with the theory. Then plot your "observed" RT. If you analyze the variance, what terms should be significant? Which not?

5.6   Consider each of the following sets of hypotheses. Which SV should be significant? Plot a data set consistent with the theory.

(a)   In a bar press experiment, we believe that

$$Y = K \times D \times P$$

where Y = bar pressing rate (our dependent variable)
D = hours of deprivation
K = constant
P = number of practice trials

(b)   In impression formation studies, we give subjects some information on the attractiveness A and intelligence I of an individual and then ask them to rate the individual. We believe that R (rating) = $(A + I)/2$.

(c)   We have depressives and nondepressives D, who are given either a success or a failure experience (outcome O) and are led to believe the task involves luck or effort (attribution A). We hypothesize that:

(i) Expectancies about success in future tasks are always higher for non-depressives than for depressives.

(ii) Under effort instructions, success leads to higher expectancies than failure does; this difference is larger for depressives than for nondepressives.

(iii) Under luck instructions, success raises expectancies by the same amount for both depressives and nondepressives. Furthermore, the average difference in expectancies between success and failure is lower than it was under effort instructions.

5.7 Ninety-six children are subjects in a study of perceptual discrimination. Half of the children are six years old, half are nine years old. Half of the subjects are tested with two-dimensional objects, half with three-dimensional objects. Half are required to discriminate for shape, half for color. Thus, there are eight groups differing in age (A: 6 and 9), dimensions (D: 2 or 3), and relevant cue (C: shape and color). We have the following hypotheses:

1. Nine-year-olds will make fewer errors than six-year-olds.
2. On the average, it will be easier to discriminate three-dimensional objects than two-dimensional objects.
3. The difference between two and three-dimensional objects will be more marked for six-year-olds than for nine-year-olds.
4. The difference will hold for shape but not color.
   What effects should be significant?

5.8 Consider a two-factor design with n subjects in each cell. One possible model is

$$Y_{ijk} = \mu + \alpha_j + \beta_k + (\alpha\beta)_{jk} + \varepsilon_{ijk}$$

where

$$\alpha_j = \mu_j - \mu$$

$$\beta_k = \mu_k - \mu$$

$$(\alpha\beta)_{jk} = \mu_{jk} - \mu_j - \mu_k + \mu$$

$$\varepsilon_{ijk} = Y_{ijk} - \mu_{jk}$$

Then the analysis of variance table is

SV	df	EMS
A	$a-1$	$\sigma_e^2 + nb\theta_a^2$
B	$b-1$	$\sigma_e^2 + na\theta_b^2$
AB	$(a-1)(b-1)$	$\sigma_e^2 + n\theta_{ab}^2$
S/AB	$ab(n-1)$	σ_e^2

Note:

$$\theta_{AB}^2 = \frac{\sum_j \sum_k (\alpha\beta)_{jk}^2}{(a-1)(b-1)}$$

An alternative model is

$$Y_{ijk} = \mu + \alpha_j + \beta_k + \varepsilon_{ijk}$$

where $\varepsilon_{ijk} = Y_{ijk} - \mu_j - \mu_k + \mu$. Assuming the second model to be true, and that

$\hat{\varepsilon}_{ijk} = Y_{ijk} - \bar{Y}_j - \bar{Y}_k + \bar{Y}_{...}$, prove that

$$E\left(\sum_i \sum_j \sum_k \hat{\varepsilon}_{ijk}^2\right) = (abn - a - b + 1)\sigma_e^2$$

If the second model is valid, what might be the advantage in applying it? (*Hint:* How does the analysis of variance table change?) What problem arises if the second model is assumed but is not valid (that is, $\theta_{AB}^2 \neq 0$)?

5.9   Children are divided into *abce* groups of n each according to age A, level of B, level of C, and experimenter E. Only the main and interaction effects of B and C are of interest. To save computational time, it is decided to pull out one SV containing all terms that include A, E, or both. The SVs are therefore

B

C

BC

Pooled A, E effects

Within-cell error

(a) Give the SS and df formulas for the pooled term.
(b) Suppose we analyzed the data as having the following SVs:

B

C

BC

Residual

What are the residual df? Why might the first analysis be preferred?

5.10   We have the following cell totals with eight subjects in a cell.

		D_1		D_2	
		C_1	C_2	C_1	C_2
	B_1	14	22	31	18
A_1	B_2	12	34	33	21
	B_3	26	24	43	19
		52	80	107	58
	B_1	42	46	20	41
A_2	B_2	15	18	25	30
	B_3	27	17	15	44
		84	81	60	115

Set up the single df computations for (a) SS_D; (b) SS_{AC}; (c) SS_{ACD}.

5.11   Estimate the proportion ω^2 that each SV contributes to the total population variance for the following table (see Exercise 5.2).

SV	df	MS
A	2	512
B	4	512
AB	8	152
S/AB	75	62

5.12 (a) Estimate main and interaction effects for the following table of means, assuming equal cell frequencies.

$$\bar{Y}_{.jk} = \begin{array}{c} A_1 \\ A_2 \\ A_3 \end{array} \begin{array}{ccc} B_1 & B_2 & B_3 \\ \begin{bmatrix} 12 & 8 & 22 \\ 8 & 6 & 13 \\ 1 & 4 & 16 \end{bmatrix} \end{array}$$

(b) Subtract the calculated value of α_j from each cell mean. How do the means of the columns for this new data matrix compare with those of part a?

(c) Suppose we had the following cell frequencies for the table of cell means in (a).

$$n_{jk} = \begin{array}{c} A_1 \\ A_2 \\ A_3 \end{array} \begin{array}{ccc} B_1 & B_2 & B_3 \\ \begin{bmatrix} 4 & 8 & 12 \\ 3 & 6 & 9 \\ 1 & 2 & 3 \end{bmatrix} \end{array}$$

Estimate row and column main effects. Again, subtract the α_j from the cell means. What effect does this have on the column means?

(d) Now assume the following cell frequencies for the table of means.

$$n_{jk} = \begin{array}{c} A_1 \\ A_2 \\ A_3 \end{array} \begin{array}{ccc} B_1 & B_2 & B_3 \\ \begin{bmatrix} 5 & 10 & 5 \\ 5 & 5 & 10 \\ 10 & 5 & 5 \end{bmatrix} \end{array}$$

Again, subtract the α_j from the cell means and note the effect on the column means.

(e) As you review your answers to (b), (c), and (d), what conclusions do you draw?

REFERENCE

Snedecor, G. W., and Cochran, W. G. *Statistical Methods*, 6th ed. Ames: Iowa State University, 1967.

6 DESIGNS USING A CONCOMITANT VARIABLE

6.1 TREATMENTS × BLOCKS: INTRODUCTION

The main disadvantage of completely randomized designs is their relative inefficiency. The variability among subjects within groups, the error term against which the variability among treatment means is tested, is generally large. Much of this error variance can be attributed to individual differences in factors that contribute to performance. Even people treated alike will differ in their scores because of differences in such factors as attitude, previous experience, and intelligence. If the contribution of such individual-difference variables could somehow be removed from the data matrix, the error variance would be reduced and it would be easier to detect the effects of the independent variable. A procedure for doing this, for removing some of the error variance attributable to individual differences, has been developed.

In the design under consideration, subjects are divided into b blocks according to their scores on a concomitant variable, a measure thought to be highly correlated with the dependent variable. In a study in which some measure of paired-associate learning is of interest, the available population of subjects might be divided into blocks by intelligence test scores or even by scores in another paired-associate task besides the one to be used in the experiment. The simplest way to distribute subjects among blocks is to rank them on the basis of the concomitant score. Assume that we have abn subjects. Then the highest-scoring an subjects will be assigned to one block (say B_1), the next an subjects will be assigned to B_2, and so on. The an subjects within each block are then randomly assigned to the levels of A, with the sole restriction that there be an equal number at each level. The result of the procedure that we have described is a two-factor (A, B) design with n subjects in each of the ab cells. The an subjects in each block are considered a random sample from an infinitely large population defined by the range of concomitant scores for that block. The members of this population have been randomly assigned to the a levels of A.

The treatments × blocks design has several advantages over the completely randomized, one-factor design. First, treatment groups are roughly matched for at least one measure that should affect performance. Second, since the design is essentially a two-factor design, the treatments×blocks interaction effects can be investigated. This means that we can consider such a question as, Are differences in the effects of different training methods greater at one level of intelligence than at another? Third, and most important, the treatments×blocks design will generally be much more efficient than a one-factor design involving the same total number of dependent measures at each treatment level. To see why this is so, we next analyze in detail the efficiency of the treatments×blocks design as it compares with the completely randomized design.

6.2 RELATIVE EFFICIENCY

We shall prove the contention that the treatments×blocks design is usually more efficient than the completely randomized design. As a by-product of the proof, an estimate of the ratios of error variances will be obtained for the two designs. If it is known how much more efficient the treatments×blocks design has been than the completely randomized design would have been, there is a basis for judging whether it would be worth while to establish blocks in subsequent related experiments. Finally, although a specific derivation will be given for the relative efficiency of the treatments×blocks and the completely randomized one-factor designs, the procedure to be presented can be readily generalized for investigating relative efficiencies of various other designs.

Table 6–1 contains the information necessary for a statement of the relative efficiency of the completely randomized and treatments×blocks designs. Instead of the usual σ_e^2, the symbols $\sigma_{e_{cr}}^2$ and $\sigma_{e_{t\times b}}^2$ are used to distinguish between the error variances. Sums of squares formulas have been omitted, since those for the one-factor case were previously presented in Chapter 4 and those for the treatments×blocks design are the same as those for the two-factor design of Chapter 5. For the completely randomized design, bn scores are assumed at each level of A so that both designs are based on a total of abn scores.

The expected total sum of squares for the completely randomized design is

(6.1)
$$ESS_{tot_{cr}} = (a-1)(\sigma_{e_{cr}}^2 + bn\theta_A^2) + a(bn-1)\sigma_{e_{cr}}^2$$

Table 6–1   Expectations for two designs

SV	df	EMS
Completely Randomized		
A	$a-1$	$\sigma_{e_{cr}}^2 + bn\theta_A^2$
S/A	$a(bn-1)$	$\sigma_{e_{cr}}^2$
Treatments×Blocks		
A	$a-1$	$\sigma_{e_{t\times b}}^2 + bn\theta_A^2$
B	$b-1$	$\sigma_{e_{t\times b}}^2 + an\theta_B^2$
AB	$(a-1)(b-1)$	$\sigma_{e_{t\times b}}^2 + n\theta_{AB}^2$
S/AB	$ab(n-1)$	$\sigma_{e_{t\times b}}^2$

and the expected total sum of squares for the treatments×blocks design is

(**6.2**)
$$ESS_{tot_{t \times b}} = (a-1)(\sigma^2_{e_{t \times b}} + bn\theta^2_A) + (b-1)(\sigma^2_{e_{t \times b}} + an\theta^2_B)$$
$$+ (a-1)(b-1)(\sigma^2_{e_{t \times b}} + n\theta^2_{AB}) + ab(n-1)\sigma^2_{e_{t \times b}}$$

Since both designs involve a groups of bn subjects, it is reasonable to assume that $ESS_{tot_{cr}} = ESS_{tot_{t \times b}}$. Then, setting the right-hand side of Equation (6.1) equal to the right-hand side of Equation (6.2), canceling $bn(a-1)\theta^2_A$, and simplifying, we have

(**6.3**)
$$(abn-1)\sigma^2_{e_{cr}} = (abn - ab + a - 1)\sigma^2_{e_{t \times b}} + (b-1)(\sigma^2_{e_{t \times b}} + an\theta^2_B)$$
$$+ (a-1)(b-1)(\sigma^2_{e_{t \times b}} + n\theta^2_{AB})$$
$$= (abn - ab + a - 1)\sigma^2_{e_{t \times b}} + ESS_B + ESS_{AB}$$

and

(**6.4**)
$$\sigma^2_{e_{cr}} = \sigma^2_{e_{t \times b}} + \frac{1}{abn-1}[(ESS_B + ESS_{AB}) - a(b-1)\sigma^2_{e_{t \times b}}]$$

We now have the population error variance for the completely randomized design as a function of population parameters estimated from data obtained with the treatments×blocks design. But relative efficiency involves comparing mean squares rather than expected mean squares. Therefore, Equation (6.4) is replaced by a similar statement in which sample statistics are substituted for population parameters:

(**6.5**)
$$MS_{S/A} = MS_{S/AB} + \frac{1}{abn-1}[(SS_B + SS_{AB}) - a(b-1)MS_{S/AB}]$$

By Equation (6.5), the data from a treatments×blocks design can be used to estimate what the error variance would have been if the completely randomized design had been used. Such information should aid in selecting designs for future experiments in an area.

The efficiency of the treatments×blocks design relative to that of the completely randomized design is defined as

$$RE = \frac{MS_{S/A}}{MS_{S/AB}}$$

The item $MS_{S/A}$ is therefore replaced by the right side of Equation (6.5), resulting in

(**6.6**)
$$RE = 1 - \frac{a(b-1)}{abn-1} + \frac{SS_B + SS_{AB}}{(abn-1)MS_{S/AB}}$$

From Equation (6.6) it is seen that RE will be greater than 100 percent (the treatments×blocks design will be more efficient than the completely randomized) whenever

$$\frac{SS_B + SS_{AB}}{(abn-1)MS_{S/AB}} > \frac{a(b-1)}{abn-1}$$

or if both sides are multiplied by

$$\frac{(abn-1)}{a(b-1)}$$

whenever

$$\frac{(SS_B + SS_{AB})/a(b-1)}{MS_{S/AB}} > 1$$

The quantity $SS_B + SS_{AB}$ is distributed on

$$(b-1)+(a-1)(b-1) = a(b-1)$$

df. Therefore

$$\frac{SS_B + SS_{AB}}{a(b-1)}$$

is a mean square and

$$\frac{(SS_B + SS_{AB})/a(b-1)}{MS_{S/AB}}$$

is essentially an F ratio. Thus, RE is likely to be greater than 1 whenever an F test of the combined B and AB effects is greater than 1 (not necessarily significant, merely greater than 1) and will increase as the variability due to either B or AB increases. The condition that this F be greater than 1 will have high probability whenever the concomitant and dependent variables are correlated in the population. If intelligence and paired-associate scores are correlated, for example, highly intelligent subjects should have higher paired-associate scores than subjects of low intelligence. This implies differences among the blocks based on level of intelligence and will generally be reflected in large values of SS_B and will result in high relative efficiency.

While $MS_{S/A}$ will generally be larger than $MS_{S/AB}$, the former is distributed on $abn - a$ df whereas the latter is distributed on $abn - ab$ df (assuming abn measures for both designs). Thus, the completely randomized design is less efficient than the treatments × blocks, but its greater number of error df may result in a more powerful F ratio. Fisher (1952) has proposed an adjustment to account for the discrepancies in df. He suggests that relative efficiency be defined as

(6.7)
$$RE = \frac{(df_{S/AB}+1)(df_{S/A}+3)}{(df_{S/AB}+3)(df_{S/A}+1)} \cdot \frac{MS_{S/A}}{MS_{S/AB}}$$

For example, if $a = 2$ and N (the total number of observations) $= 8$, for the completely randomized design $df_{S/A} = 2(4-1) = 6$, and for the treatments × blocks design with $b = 2$, $df_{S/AB} = 2[2(2-1)] = 4$. Then the adjusted relative efficiency is

$$RE = \left(\frac{5}{7}\right)\left(\frac{9}{7}\right)\frac{MS_{S/A}}{MS_{S/AB}}$$

$$= (.92)\frac{MS_{S/A}}{MS_{S/AB}}$$

6.3   SELECTING THE OPTIMAL NUMBER OF BLOCKS

The number of blocks b influences experimental results in two ways. On the one hand, as b increases, the number of error df decreases (assuming that the total number of observations is held constant), resulting in reduced power of the F test. On the other hand, increasing the number of blocks reduces error variance. The variability within a block $\times$ treatment combination should be smaller when there are three blocks (for example, high, medium, and low intelligence) than when there are two blocks (high and low intelligence). Because of these opposed effects of increased block number, an optimal block number exists. There is some level of b such that lower and higher values result in less precise tests of treatment effects. This optimal value of b changes as a function of a (number of treatment levels), n, and ρ (the correlation coefficient for the population of concomitant and dependent measures). We intend to recommend optimal values of b for various combinations of a, n, and ρ. To do this, we need to develop some measure of error variance, which we intend to minimize by our choice of b (Section 6.3.1). We must also specify the way in which blocks will be established because the optimal value of b may depend on this (Section 6.3.2). We shall present a set of recommendations, and also a discussion of the functional relation between optimal b and design parameters (a, n, and ρ) (Section 6.3.3).

6.3.1   DESIGN IMPRECISION.   Some measure of the adequacy of the design is needed so that we can decide on the value of b for various experiments. Such a measure should also clearly reflect the influences of a, n, and ρ. Since the influence of ρ is not clear in the RE formula, and since RE must be computed relative to another design, we look for some other index of error variability for the present purpose. The most generally accepted measure is I_a, the *apparent inprecision* of the design, the formula for which is

(6.8)
$$I_a = \frac{df_{S/AB} + 3}{df_{S/AB} + 1} I_t$$

$$= \left[\frac{ab(n-1) + 3}{ab(n-1) + 1} \right] \left\{ \frac{1 - \rho^2[1 - (\bar{\sigma}_x^2 / \sigma_X^2)]}{1 - \rho^2} \right\}$$

where σ_X^2 is the variance of concomitant measures for the jth treatment population, and $\bar{\sigma}_x^2$ is computed by obtaining the variance of concomitant measures for each block in the jth treatment population and then averaging over the b blocks; ρ is the correlation between X and Y measures in the jth treatment population. The value I_t is the true imprecision. The value I_a is the preferred measure, since Fisher's correction, $(df_{S/AB} + 3)/(df_{S/AB} + 1)$, adjusts for the loss of df (and therefore power) due to estimation of B and AB effects. To provide some idea of how we arrive at Equation (6.8), and also to give a better idea of what imprecision means, we consider the sample-to-sample fluctuation in the difference between two treatment group means. We denote this variance of the difference between the means as $\sigma^2_{\bar{Y}_j - \bar{Y}'_j}$. An expression for *aver* $\sigma^2_{\bar{Y}_j - \bar{Y}'_j}$ (*aver* refers to the average variance over the b blocks) can be derived if it is assumed that:

1. Within the population from which each block is sampled, Y (the dependent measure) is a linear function of X (the concomitant measure) and the slope of this function is the same for all b populations.
2. The population variability of the Ys about the best-fitting straight line is the same for all values of X. This is often referred to as *homoscedasticity*.

The appropriate expression is

(6.9)
$$ aver\; \sigma^2_{\bar{Y}_j - \bar{Y}'_j} = \frac{2\sigma^2_Y}{n}\left[1 - \rho^2\left(1 - \frac{\bar{\sigma}^2_x}{\sigma^2_X}\right)\right] $$

where σ^2_Y is the variance of dependent measures for the jth treatment population; σ^2_X, $\bar{\sigma}^2_x$, and ρ have been defined above. It is assumed that these population values are the same for all a treatment populations.

It is evident from Equation (6.9) that as the number of observations n is increased, the average sampling error decreases. The same relation holds for ρ and sampling error. Note that when ρ is zero,

$$ aver\; \sigma^2_{\bar{Y}_j - \bar{Y}'_j} = \frac{2\sigma^2_Y}{n} $$

When ρ is 1,

$$ aver\; \sigma^2_{\bar{Y}_j - \bar{Y}'_j} = \left(\frac{2\sigma^2_Y}{n}\right)\left(\frac{\bar{\sigma}^2_x}{\sigma^2_X}\right) $$

Since the average variability over the blocks in a population will be less than the total variability in the population, $(\bar{\sigma}^2_x/\sigma^2_X) < 1$, and the sampling error when ρ is 1 will be some fraction of the sampling error when ρ is zero.

While the number of blocks b does not explicitly appear in Equation (6.9), its influence can be understood by considering $\bar{\sigma}^2_x$. If there is only one block, we actually have a completely randomized design, $\bar{\sigma}^2_x$ is identical to σ^2_X, and

$$ aver\; \sigma^2_{\bar{Y}_j - \bar{Y}'_j} = \frac{2\sigma^2_Y}{n} $$

As the number of blocks increases, the range of concomitant scores within each block necessarily decreases, and $\bar{\sigma}^2_x$ becomes progressively smaller. Therefore, when the number of blocks is very large, $\bar{\sigma}^2_x$ is near zero and the average error is approximately

(6.10)
$$ min\; \sigma^2_{\bar{Y}_j - \bar{Y}_{j'}} = \frac{2\sigma^2_Y}{n}(1 - \rho^2) $$

Cox has proposed that the ratio of the average to the theoretically minimum sampling error (that is, $min\; \sigma^2_{\bar{Y}_j - \bar{Y}_{j'}}$) be used as an index of the precision of the design. Specifically, for the treatments×blocks design, we have the *true impre-cision* I_t, where

(6.11)

$$I_t = \frac{aver\ \sigma^2_{\bar{Y}_j - \bar{Y}_{j'}}}{min\ \sigma^2_{\bar{Y}_j - \bar{Y}_{j'}}}$$

$$= \frac{1 - \rho^2[1 - (\bar{\sigma}^2_x/\sigma^2_X)]}{1 - \rho^2}$$

Multiplying by Fisher's adjustment, we have Equation (6.8), which defines I_a.

6.3.2 THREE APPROACHES TO BLOCKING.   Thus far, we have used the term *blocking* without specifying how subjects will be assigned to blocks. Several different methods are conceivable and the choice may have implications for both the validity of the assumptions underlying the F test and for the specific value of optimal b, the value of b that minimizes I_a. There are at least three possibilities:

1. *Equal proportions.* The range of X (the concomitant variable) is divided into b segments so that equal proportions of the population fall within each seg-ment. Under this procedure, the ranges of X defining the different segments generally will not be equal. From the individuals whose scores fall within a particular segment, an are randomly sampled and distributed among the a treatment conditions with n subjects in each cell.
2. *Equal segments.* The range of X is divided into b segments so that each encompasses an equal range of scores. In this case, the population propor-tions represented by each segment would not be equal unless X were uniformly distributed. Again, an subjects are randomly sampled from each of the b subpopulations and randomly assigned to the a treatment conditions.
3. *Blocking by ranks.* The an subjects who scored highest on X are designated as Block 1, the next-highest-scoring an subjects become Block 2, and so on. As in methods 1 and 2, assignment to treatments is random within each block.

Investigators of the consequences of using these three methods have made several assumptions:

1. Both X and Y are independently and normally distributed.
2. Variances for Y are the same for all a treatment populations.
3. Assumptions 1 and 2, used in Section 6.3.2 to derive an expression for I_a, hold. Briefly, the same linear relation holds between X and Y in all population blocks, and homoscedasticity prevails.

Making the above assumptions, Feldt (1958) has investigated methods 1 and 2. He has found that there is some small degree of heterogeneity of variance over blocks within a treatment population. Although the heterogeneity increases with ρ, it is never large enough to be important, in view of our discussion in Chapter 4. Hornbeck and Alf (1972) have reached a similar conclusion for method 3. Although heterogeneity of variance over blocks is somewhat greater with this method, only at values of ρ of .8 or greater is the violation of the homogeneity of variance assumption important. Relations of this size are fairly infrequent in the behavioral and social sciences.

Assumptions 1 through 3 have also been used in deriving optimal values of b. We shall now review Feldt's recommendations under these assumptions and under the additional constraint that subjects have been assigned to blocks by method 1, equal proportions.

6.3.3 OPTIMAL VALUES OF b.

Feldt has used I_a as the basis for determining the optimal number of blocks to be included in an experiment. That number of blocks resulting in a lower value of I_a compared with any other number of blocks is considered the optimal number. Table 6-2 has been reproduced from Feldt's article. Data available from previous research or from pilot studies can be used to estimate ρ. With this estimate and a knowledge of a and N (the total number of scores), the experimenter has a sound basis for deciding what number of blocks to include in the experiment.

The relations among the optimal value of b and the values of a, ρ, and N can be summarized as follows:

1. As ρ increases, the optimal value of b also increases. The increased loss in error df due to increase in b is more than compensated for by the improvement in precision due to increase in ρ.
2. As N increases, the optimal value of b increases. The increased loss in error df due to increases in b has less effect on precision as the total number of observations increases.

Table 6–2 Optimal values of b for selected experimental conditions, levels assumed defined by equal proportions of the population

ρ	a	\multicolumn{6}{c}{N}
		20	30	50	70	100	150
.2	2	2	3	4	5	7	9
	5	1	2	2	3	4	6
.4	2	3	4	6	9	13	17
	5	2	3	4	5	7	10
.6	2	4	6	9	13	17	25[b]
	5	2	3	5	7	9	14
.8	2	5[a]	7[a]	12[a]	17	23	25[b]
	5	2[a]	3[a]	5[a]	7[a]	10[a]	15[a]

SOURCE: L. S. Feldt, "A Comparison of the Precision of Three Experimental Designs Employing a Concomitant Variable," *Psychometrika* 23:335–54 (1958). By permission of the author and the editor.
NOTE: The numerical entries have been rounded to the nearest integer.
[a] Limit imposed by the requirement $N \geq ab$.
[b] Slight improvement possible with more than 25 levels.

3. As *a* decreases, the optimal value of *b* increases. The increased loss in error *df* due to increases in *b* is compensated for by the reduced loss due to df_{AB} as *a* decreases.

The imprecision for a completely randomized design is easily obtained if we realize that this is actually a treatments×blocks design with only one block. Consequently $\bar{\sigma}_x^2 = \sigma_X^2$, and substituting in Equation (6.8) yields

(6.12)
$$I_a = \left[\frac{df_{S/A}+3}{df_{S/A}+1}\right]\frac{1}{1-\rho^2}$$

Table 6–3 was derived from Equations (6.8) and (6.12); it presents the ratio of values of I_a for the completely randomized design to values of I_a for the treatments×blocks design (assuming that the optimal value of *b* is used). The tabled values are consistently greater than 1, indicating that the completely randomized design is more imprecise, or less precise. The advantage of the treatments×blocks design increases as ρ and *N* increase and as *a* decreases. To put it another way, the effort involved in assigning subjects to blocks is most worth while when the optimal number of blocks is large.

In using Table 6–2 to design experiments, it is helpful to note that the experimenter can readily use other values of ρ, *N*, and *a* besides those in the table. Suppose, for example, that $\rho = .6$, $N = 30$, and $a = 4$. The value of optimal *b* for $a = 4$ is not tabled, but can be approximated by linear interpolation. When $\rho = .6$, $N = 30$, and $a = 5$, the value of *b* is 3. When $a = 2$, then $b = 6$. Since $a = 4$ is one-third the distance between $a = 5$ and $a = 2$, we take as the required value of *b* the number falling one-third the distance between 3 and 6. The optimal value chosen for the experiment is therefore 4. We can interpolate among various values of *N* in exactly the same way.

Table 6–3   Ratio I_a for completely randomized design to I_a for treatment×blocks design

		\multicolumn{6}{c}{*N*}				
ρ	*a*	20	30	50	70	100	150
.2	2	1.015	1.023	1.029	1.034	1.036	1.038
	5	1.000	1.009	1.021	1.027	1.031	1.035
.4	2	1.117	1.141	1.161	1.170	1.176	1.181
	5	1.060	1.097	1.136	1.153	1.165	1.174
.6	2	1.388	1.443	1.494	1.516	1.531	1.541
	5	1.235	1.340	1.431	1.470	1.500	1.523
.8	2	2.196	2.378	2.543	2.615	2.666	2.699
	5	1.608	1.943	2.276·	2.427	2.543	2.629

SOURCE: L. S. Feldt, "A Comparison of the Precision of Three Experimental Designs Employing a Concomitant Variable," *Psychometrika* 23:335–354 (1958). By permission of the author and the editor.

If the value of ρ estimated for an experiment does not appear in Table 6–2, we can linearly interpolate between values of ρ^2. Suppose, for example, that $N = 150$, $\rho = .3$, and $a = 2$. The square of ρ is .09, which falls five-twelfths of the distance between $.2^2$ ($= .04$) and $.4^2$ ($= .16$). When $N = 150$, $\rho = .2$, and $a = 2$, the optimal value of b is 9; when $\rho = .4$, then $b = 17$. The level of b for our example should therefore fall at the value five-twelfths of the distance between 9 and 17. To the nearest integer, the result is 12.

In many circumstances, the values of b presented in Table 6–2 might best be viewed as rough guidelines. If method 3, blocking by ranks, is used, slightly different values are optimal for some of the conditions we have tabled (Hornbeck and Alf, 1972). Nor it is clear how much the optimality of our tabled values of b depends on the validity of assumptions made in deriving them. Nevertheless, the tabled values will probably be closer to optimal b than any uninformed guess. Furthermore, relations between optimal b and a, n, and ρ are likely to hold under departures from our assumed conditions; knowing such relations should generally guide decisions about the value of b. Finally, *any* blocking that produces clear B or AB effects or both will improve precision and should be done whenever it is practical.

6.4   BLOCKING AFTER DATA COLLECTION

Sometimes, having run a completely randomized design, investigators have tried to salvage a noisy data set by blocking after the fact. Given five groups of ten scores each, for example, we might order subjects within each group by their scores. We could then designate the highest-ranking pair in each group B_1, the next-ranking pair B_2, and so on; we should then carry out a two-factor analysis of variance, and a very efficient one at that. Alternatively, the individuals in each treatment group could be blocked on the basis of some concomitant measure.

The above approach is ingenious; unfortunately, it is also dishonest. To understand why, first consider the *EMS* for the A source of variance. Reordering scores within each group will not change the group mean; thus the MS_A will be exactly what it would have been under the usual one-factor analysis, and its expectation will be

$$E(MS_A) = \sigma^2_{e_{cr}} + bn\theta^2_A$$

as usual. The error term has, however, been changed by our post hoc partitioning into blocks. The variability within each of our newly created cells should be considerably smaller than for an entire treatment group undifferentiated into blocks. Its expectation will be

$$E(MS_{error}) = \sigma^2_{e_{t \times b}}$$

Under the null hypothesis, the ratio of *EMS* will be $\sigma^2_{e_{cr}}/\sigma^2_{e_{t \times b}}$, a value generally much greater than 1. A significant F may merely reflect the reduction in error variance due to blocking rather than any variability due to treatments.

We have in the past used *EMS* to justify error terms, and to derive expressions for relative efficiency and estimates of population variances and of population variance proportions. What we have just shown is still another application, the evaluation of possible biases inherent in some *F* tests.

6.5   EXTENSIONS OF THE TREATMENTS × BLOCKS DESIGN

We have been limited so far to the case in which only one treatment variable is under investigation. Extension to experiments in which several treatment variables are of interest is straightforward. If one wants to test hypotheses about the single and joint effects of two variables *A* and *C*, for example, the model and the analysis is the same as for the three-factor design (Chapter 5), blocks *B* being the third factor. Subjects are divided into blocks on the basis of the control variable; within each block, there is random assignment of subjects to the *ac* treatment combinations. The derivation of relative efficiency follows directly from that of Section 6.2. The value I_t, the true imprecision, will not be influenced by the number of treatment variables. Apparent imprecision will be increased, however, if additional treatments are investigated and *N* is held constant. To see why this is so, note that the adjustment factor that transforms I_t into I_a is $(df_{error} + 3)/(df_{error} + 1)$. If, for example, $N = 80$, $a = 2$, and $b = 4$,

$$df_{error} = 80 - 1 - (2-1) - (4-1) - (2-1)(4-1)$$
$$= 72$$

and

$$I_a = \frac{75}{73} I_t = 1.03 I_t$$

If five levels of a variable *C* are added to the design, we should then have

$$df_{error} = 72 - (5-1) - (2-1)(5-1) - (4-1)(5-1)$$
$$-(2-1)(4-1)(5-1)$$
$$= 40$$

In this case,

$$I_a = \frac{43}{41} I_t = 1.09 I_t$$

When the error *df* are considered, the design involving more treatment variables is slightly less precise.

Tables like Table 6–2 have not been designed for the multivariable cases. Because of the reduction in precision just noted, however, values of *b* smaller than those presented in Table 6–2 should be used whenever there is more than one treatment variable. As N becomes larger, the loss of error df, occasioned by the inclusion of additional variables, has a correspondingly smaller effect on I_a. Therefore, the tabled values become better approximations to optimal b.

6.6   THE EXTREME-GROUP DESIGN

A design related to the treatments × blocks design is one in which only two extreme blocks of subjects are used. The eyelid conditioning rates of subjects who score high on the Taylor Manifest Anxiety Scale are compared with the tests for low-scoring subjects, for example, or the gambling behavior of subjects scoring high on the MMPI Psychopathic Deviant Scale is compared with the gambling behavior of subjects who score low on the scale. At the outset, the distinction between this design and the treatments × blocks should be made clear. In the treatments × blocks design, subjects are divided into blocks on the assumption that the block, or concomitant, variable is correlated with the dependent variable. The purpose of the design is to improve the precision of the test of some treatment effect. In the extreme-group design, the block variable is the variable whose effect is of interest. Rather than assume a correlation between concomitant and dependent variable, the experimenter investigates whether such a correlation exists. Do high-anxiety subjects give more conditioned responses than low-anxiety subjects? Do those subjects who score high on a psychopathic deviant scale take more risks in a laboratory gambling task than those who score low on the scale?

In what follows, we consider three strategies that can be used in evaluating the relation between X and Y. Our premise is that the X variable (say anxiety or deviance scores) can be measured inexpensively, perhaps by administering a test to a large group of individuals in a single session. The Y variable, however, is more expensive to measure, and therefore fewer observations are available. If N X and Y scores can be obtained, the remainder of this section is not of interest; in such cases, the most powerful approach is the usual significance test of the null hypothesis that ρ, the population correlation coefficient, is zero. That test is

(6.13)
$$t = \frac{r\sqrt{n-2}}{\sqrt{1-r^2}}$$

where t is distributed on $n-2$ df. The strategies to be considered are of interest, however, when only some proportion p of the N subjects who provided X scores are to be tested in the second stage. We shall compare strategies in their relative powers against alternatives to the null hypothesis. Conclusions are derived assuming that X and Y are linearly related and have a bivariate normal distribution.

Strategy 1 involves obtaining X scores from a sample of N subjects. The Y scores are then obtained only from some subset of subjects who had very high or very low X scores. The proportions of high-scoring and low-scoring subjects need not be the same. After collecting the Y data, we estimate the *full-range correlation* (the correlation that would have resulted if all N Y scores had been available), using a formula that Alf and Abrahams (1975) provided:

(6.14)
$$\hat{R} = \frac{r'(S_x/s_x)}{\sqrt{1-(r')^2+(r')^2(S_x/s_x)}}$$

where

> r' = correlation between X and Y for the combined high and low groups
> S_x = standard deviation of the full set of n values of X
> s_x = standard deviation of X scores for the combined high and low groups

Having calculated $\hat{R}$, we can test the null hypothesis that ρ equals zero by applying Equation (6.13), with r replaced by $\hat{R}$ and n replaced by the number of individuals in the combined high and low groups. Alf and Abrahams (1975) have demonstrated that the power of this test increases as a function of p and of n.

In *Strategy* 2, extreme groups again are made up of a subset of the subjects who provided an X score in the first stage of the research. In this approach, however, the relation between X and Y is tested by the usual t test of the $\bar{Y}_{.j}$; strictly speaking, the null hypothesis is that $\mu_{\text{high}} - \mu_{\text{low}} = 0$. Feldt (1961) has considered the optimal size of the extreme groups—the criterion for optimality being the power of the t test of the means. He has shown that when ρ is .10, the power of the t test is greatest if each extreme group consists of 27 percent of the population tested on the concomitant measure. As the correlation between the concomitant and dependent measures increases, the optimal percentage decreases, but not markedly. When ρ is .80, power is greatest if each extreme group contains 23.3 percent of the population. Obviously, we never know ρ; the purpose of the experiment is to determine whether ρ is different from zero. It seems reasonable to conclude, however, that if we use the top and bottom fourths of concomitant measures, we shall not depart far from optimality, regardless of the true value of ρ. The Taylor Manifest Anxiety Scale, for example, might be administered to a class of 100 elementary psychology students. Fifty of these, the 25 scoring highest and the 25 scoring lowest, would then be asked to be subjects in an eyelid-conditioning experiment. Mean number of conditioned responses would be compared for the two extreme groups.

Strategy 3 is not really an extreme-group strategy but has been considered by both Feldt (1961) and Alf and Abrahams (1975) as an alternative to the extreme-group strategies. In this method, some proportion of the original N subjects are randomly sampled and tested in the second stage. A correlation coefficient is computed for this subset of individuals and its significance is tested by Equation (6.13); again, n should be replaced by the number of individuals for whom there are both X and Y scores.

Alf and Abrahams have shown that strategy 1, estimating and testing the full-range correlation, is uniformly more powerful than both strategies 2 and 3. Since the power of strategy 1 increases with n, the authors recommend using strategy 1 with as large a sample as is practical in stage 2. This demonstrated superiority of strategy 1 over strategies 2 and 3 should be carefully noted, since

Figure 6–1   Example of a nonlinear relation between Y and X.

the application of the two weaker strategies seems to be more frequent in the research literature.

The results just summarized rest on the assumptions that the distribution of the concomitant and dependent measures is bivariate normal and that the relation between the two variables is linear. If the latter assumption is incorrect, the extreme-group approach is at worst misleading, and at best relatively uninformative. Consider Figure 6–1 to understand why the extreme-group approach might be misleading. If the extreme groups are tested, we shall surely conclude that they do not differ relative to Y, the dependent measure. This is true, but completely misses the important fact that there is a relation between X and Y, that the extreme groups differ from a middle group. Even if the relation between X and Y is monotonic (that is, if Y consistently increases or consistently decreases as X increases), if it is not linear, the extreme-group approach results in a loss of information. Whenever there is reason to suspect a nonlinear relation, it is better to sample the available proportion randomly, regardless of its size, from the total population from which the concomitant measure was taken. In this way, information about the shape of the function relating X and Y can be obtained.

EXERCISES

6.1   Assume that we are interested in the main and joint effects of two variables, A and B. Derive the efficiency of an $A \times B \times$ blocks design relative to the efficiency of a two-factor (A, B) completely randomized design.

6.2 Analyze the following set of data obtained with an $A \times$ blocks design:

	A_1	A_2	A_3
B_1	1	16	3
	2	10	1
B_2	4	13	3
	3	12	3
B_3	6	7	9
	8	10	5
B_4	10	7	5
	8	7	12

Estimate the efficiency relative to a completely randomized design. What is the primary reason that the $A \times$ blocks design is more efficient?

6.3 Several previous studies have yielded estimates of the correlation between an intelligence test and paired-associate learning ranging from .25 to .35. For four levels of A and a total of 80 available subjects, what value of b would yield optimal precision?

6.4 Assume that X and Y are correlated so that a reliable estimate of ρ is .6. Assume that 36 Ss are available for an experiment and that there are three levels of A. We divide the subjects into 6 blocks on the basis of the X score. The anova table is:

SV	df	MS
A	2	100
B	5	160
AB	10	22
S/AB	18	16

If we wanted to reject H_0 at the 1 percent level when $\mu_3 = \mu_2 + 4$ and $\mu_2 = \mu_1 + 4$, or when the spread is greater:

(a) What estimated power do we have? (*Note:* For the T×B design, Equation (4.44) should read

$$\phi^2 = \frac{bn \sum_j (\mu_j - \mu)^2/a}{\sigma_e^2}$$

(b) What power should we expect to have if we had used a completely randomized design instead?

6.5 Suppose there is a **U**-shaped relation between performance Y and some concomitant variable X.

(a) In investigating effects of A on Y, shall we find a gain due to blocking on X? Explain in a sentence or two.

(b) If we did block, would Feldt's table for optimal blocks b be useful? Again, explain in one or two sentences.

REFERENCES

Alf, E. F., and Abrahams, N. M. The use of extreme groups in assessing relationships. *Psychometrika* 40:563–72 (1975).

Cox, D. R. The use of a concomitant variable in selecting an experimental design. *Biometrika* 44:150–58 (1957).

Feldt, L. S. A comparison of the precision of three experimental designs employing a concomitant variable. *Psychometrika* 23:335–54 (1958).

Feldt, L. S. The use of extreme groups to test for the presence of a relationship. *Psychometrika* 26:307–16 (1961).

Fisher, R. A. *Statistical Methods for Research Workers,* 12th ed. London: Oliver & Boyd, 1952.

Hornbeck, F. W., and Alf, E. F. Precision of the treatments-by-blocks analysis of variance, paper presented at Fifth Annual Meeting in Mathematical Psychology, La Jolla, California, August 10, 1972.

7 REPEATED-MEASUREMENT DESIGNS

7.1 INTRODUCTION

We have now learned how to separate error variance from the variance among treatment effects. Within the completely randomized designs, error variance is a large complex to which individual differences and errors of measurement contribute. As one way of reducing the error variance, some of the variability due to individual differences can be removed if subjects are divided into blocks on the basis of a concomitant variable correlated with the dependent variable. We can take this further, using designs in which error variability due to individual differences is further reduced, in fact completely removed.

The one-factor repeated-measurement design entails a test of each subject under each of the a levels of the treatment variable A. The order of presentation of the A_j is randomized independently for each subject. The design can be laid out as a two-factor design in which subjects constitute the second factor. This has been done in Table 7–1. The important point is that subjects can be treated like any main effect in a two-factor design, the variability associated with the effect can be isolated, and as a consequence the error variance in this design is not inflated by variability due to individual differences. The design will generally be more efficient than even the treatments × blocks design, since variability due to individual differences will be more effectively separated from error variance.

Increased precision is not the sole reason for using the repeated-measurement design. Both the completely randomized and the treatments × blocks designs require more subjects than the repeated-measurement design does to achieve the same power of the F test. Therefore, using the repeated-measurement design should be considered whenever a limited number of subjects are available for long periods. This will frequently be the case in small clinics, military research installations, and industrial settings. It will often be a factor in choosing a design in large universities.

The repeated-measurement design is the natural one to select when we are concerned with performance trends over time. If we want to measure dark adaptation over time, for example, the most efficient use of subjects requires that each subject be tested at all points of time that are of interest. In this instance, time is the treatment variable A.

Table 7-1   Data matrix for one-factor repeated-measurement design

	A_1	$\cdots$	A_2	$\cdots$	A_j	$\cdots$	A_a
S_1	Y_{11}		Y_{12}		Y_{1j}		Y_{1a}
S_2	Y_{21}		Y_{22}		Y_{2j}		Y_{2a}
$\vdots$							
S_i	Y_{i1}		Y_{i2}		Y_{ij}		Y_{ia}
$\vdots$							
S_n	Y_{n1}		Y_{n2}		Y_{nj}		Y_{na}

Although the repeated-measurement design does not involve any new computational problems, two other issues arise that have not been considered with previous designs. For the first time, the levels of one of our variables, S (subjects), can be considered a random sample from a population of levels. In Sections 7.2–7.4, some implications of random, as opposed to fixed, effects will be considered. There is also the strong possibility that scores for two treatments will be correlated, since both sets of scores are obtained from the same individuals. (The implication of this violation of the usual independence assumption will also be dealt with in Sections 7.2–7.4.)

7.2   ONE-FACTOR CASE: ASSUMPTIONS AND DEFINITIONS

7.2.1   TERMS IN THE MODEL.   Consider a group of n subjects, each of whom is tested once under each level of A, the treatment variable. The order of exposing the subject to the treatment levels is random, and the randomization is carried out independently for each subject. It is assumed that the n subjects are a random sample from an infinite population of subjects. We can picture a "true" score μ_{ij} for subject i under treatment A_j; that is,

(7.1) $$Y_{ij} = \mu_{ij} + \varepsilon_{ij}$$

where ε_{ij} is the error of measurement, distributed independently and normally, and

(7.2) $$E(\varepsilon_{ij}) = 0$$

(7.3) $$E(\varepsilon_{ij}^2) = \sigma_e^2$$

Note that the independence assumption implies

(7.4) $$E(\varepsilon_{ij}\varepsilon_{i'j'}) = 0$$

for $i \neq i'$ or $j \neq j'$. This last condition will be important in deriving *EMS*.

We define the average true score under A_j for the population of subjects as

(7.5) $$\mu_j = E(\mu_{ij})$$

and the treatment effect as

(7.6)
$$\alpha_j = \mu_j - \mu$$

Ordinarily, the experimenter arbitrarily selects the a levels of the treatment variable. It is as though there were only a levels in the population and they were all represented in the study. If this is the case,

(7.7)
$$\sum_{j=1}^{a} \alpha_j = 0$$

because the sum of all deviations of treatment means about their mean μ must be zero. This fact is used in deriving the expected mean squares (cf. Chapter 4). A second implication of the fact that the effects of A are fixed (that is, its levels are arbitrarily chosen) is that inferences about treatment effects should be limited to those treatment levels included in the experiment. To put it another way, $\theta_A^2 [= \sum_j (\mu_j - \mu)^2 / (a-1)]$ is a measure of the variability of only those a treatment effects included in the population, and strictly speaking, our null hypothesis is that θ_A^2 equals zero. Of course, we expect experimenters to use their brains as well as their F ratios to draw inferences. An investigator working with a set of a drugs may well extend conclusions to other drugs of related chemical composition. Such conclusions may be as correct as those based on the statistical analysis and are desirable whenever the investigator's substantive knowledge permits extrapolation beyond the treatment levels used. Nevertheless, it is wise to recognize that conclusions about the effects of the treatments arbitrarily included in the experiment are based on different grounds from those underlying the effects of related treatments not included in the experiment.

The average true score for subject i is

(7.8)
$$\mu_i = \sum_{j=1}^{a} \frac{\mu_{ij}}{a}$$

and the subject effect is

(7.9)
$$\eta_i = \mu_i - \mu$$

The situation is different for subject effects from what it is for treatment effects. The most reasonable assumption about the subjects is that they are a random sample from an infinite population. Admittedly, it would be hard to defend the usual procedures of picking subjects as truly random, and it is sometimes difficult to characterize precisely the population sampled. It is clear, however, that the n subjects have not been arbitrarily chosen because of a desire to compare n qualitatively or quantitatively differing characteristics. The random sampling assumption best seems to describe the true state of affairs. In view of this, $\sum_{i=1}^{n} \eta_i \neq 0$, but

(7.10)
$$E(\eta_i) = 0$$

and we define the variance of the population of subject effects as

(7.11)
$$\sigma_S^2 = E(\eta_i^2)$$

The interaction effect associated with the ijth cell is

(7.12)
$$(\eta\alpha)_{ij} = \mu_{ij} - \mu_i - \mu_j + \mu$$

Because A is a fixed-effect variable,

(7.13)
$$\sum_{j=1}^{a} (\eta\alpha)_{ij} = 0$$

However, $\sum_{i=1}^{n} (\eta\alpha)_{ij} \neq 0$. This is because we are summing only a subset of the population of interaction effects at A_j. (This point is discussed further in Section 7.3.1.) Although the sum of any n interaction effects is not zero, the expectation of the population of effects is zero. That is, at any level A_j, taking the expectation over the population of subjects, we have

(7.14)
$$E(\eta\alpha)_{ij} = 0$$

Furthermore, we define the variance of interaction effects as

(7.15)
$$\sigma_{SA}^2 = \sum_{j=1}^{a} \frac{E(\eta\alpha)_{ij}^2}{a-1}$$

We assume that the η_i and the $(\eta\alpha)_{ij}$ have a jointly normal distribution.

Adding to μ the effects defined by Equations (7.6), (7.9), and (7.12), as well as the error component, we arrive at the following structural equation, algebraically equivalent to Equation (7.1):

(7.16)
$$Y_{ij} = \mu + \eta_i + \alpha_j + (\eta\alpha)_{ij} + \varepsilon_{ij}$$

7.2.2   THE VARIANCE-COVARIANCE MATRIX.   When several observations are obtained from the same individual, there is the possibility, even the likelihood, that they will not be independent. We can represent the lack of independence between Y_{ij} and $Y_{ij'}$ by the μ_{ij} and the $\mu_{ij'}$. Considering treatments A_j and $A_{j'}$, we define the covariance term,

(7.17)
$$\mathrm{cov}(\mu_{ij}, \mu_{ij'}) = E(\mu_{ij} - \mu_j)(\mu_{ij'} - \mu_{j'})$$

where, as usual, the expectation is taken over the population of individuals. The expression in Equation (7.17) equals zero only if observations obtained under A_j and $A_{j'}$ are independent of each other. It may help the reader to imagine the covariance by thinking of it as the numerator of a population correlation of the scores obtained under two treatments A_j and $A_{j'}$. If the correlation is $\rho_{jj'}$,

$$\rho_{jj'} = \frac{\mathrm{cov}(\mu_j, \mu_{j'})}{\sigma_j \sigma_{j'}}$$

It follows that

(7.18)
$$\mathrm{cov}(\mu_{ij}, \mu_{ij'}) = \sigma_j \sigma_{j'} \rho_{jj'}$$

where σ_j is the standard deviation of true scores under treatment A_j, that is,

(7.19)
$$\sigma_j^2 = E(\mu_{ij} - \mu_j)^2$$

and $\rho_{jj'}$ is the correlation of the μ_{ij} and $\mu_{ij'}$ for the population of subjects. We can form an $a \times a$ matrix, the variance-covariance matrix:

$$(7.20) \qquad \Sigma = \begin{bmatrix} \sigma_1^2 & \sigma_1\sigma_2\rho_{12} & \cdots & \sigma_1\sigma_{j'}\rho_{1j'} & \cdots & \sigma_1\sigma_a\rho_{1a} \\ \sigma_2\sigma_1\rho_{12} & \sigma_2^2 & \cdots & \sigma_2\sigma_{j'}\rho_{2j'} & \cdots & \sigma_2\sigma_a\rho_{2a} \\ \vdots & \vdots & & \vdots & & \vdots \\ \sigma_j\sigma_1\rho_{1j} & \sigma_j\sigma_2\rho_{2j} & \cdots & \sigma_j\sigma_{j'}\rho_{jj'} & \cdots & \sigma_j\sigma_a\rho_{ja} \\ \vdots & \vdots & & \vdots & & \vdots \\ \sigma_a\sigma_1\rho_{1a} & \sigma_a\sigma_2\rho_{2a} & \cdots & \sigma_a\sigma_{j'}\rho_{j'a} & \cdots & \sigma_a^2 \end{bmatrix}$$

Note that the matrix is symmetric; the entry in the jth row and the j'th column is the same as the one in the j'th row and jth column.

Since we are soon going to take up the implications of nonindependence for EMS and F tests (Sections 7.3 and 7.4), it is helpful now to note the relation between variance components defined earlier (Section 7.2.1) and the terms of Equation (7.20). From Equations (7.9) and (7.11),

$$\sigma_S^2 = E(\mu_i - \mu)^2$$

From Equation (7.8) and the definition of μ,

$$\sigma_S^2 = E\left(\frac{\sum_j \mu_{ij}}{a} - \frac{\sum_j \mu_j}{a}\right)^2$$

$$= \left(\frac{1}{a}\right)^2 E\left[\sum_j (\mu_{ij} - \mu_j)^2 + \sum_j \sum_{\substack{j' \\ j \neq j'}} (\mu_{ij} - \mu_j)(\mu_{ij'} - \mu_{j'})\right]$$

We can reverse the order of E and Σ; in that case, it should be clear that σ_S^2 is a function of variances $E(\mu_{ij} - \mu_j)^2$, and covariances $E(\mu_{ij} - \mu_j)(\mu_{ij'} - \mu_{j'})$. In fact, we have

$$(7.21) \qquad \sigma_S^2 = \left(\frac{1}{a}\right)^2 \left[\sum_j \sigma_j^2 + \sum_j \sum_{\substack{j' \\ j \neq j'}} \sigma_j\sigma_{j'}\rho_{jj'}\right] = \Sigma_{..}$$

where $\Sigma_{..}$ is the average of the a^2 entries in the variance-covariance matrix defined by Equation (7.20).

We can likewise show that

$$(7.22) \qquad \sigma_{SA}^2 = \frac{\sum_j (\sigma_j^2 - \Sigma_{..})}{a - 1}$$

We have defined several terms to which we shall subsequently refer and have established relations among them. For ease of reference, these terms have been collected in Table 7–2.

Table 7–2 Parameters defined for general-model, one-factor repeated-measurement design

	PARAMETER	DEFINITION	COMMENT
Means	μ_{ij}	"True" score	
	μ_i	$\dfrac{\sum_j \mu_{ij}}{a}$	
	μ_j	$E(\mu_{ij})$	
	μ	$\dfrac{\sum_j \mu_j}{a}$	
Effects	ε_{ij}	$Y_{ij} - \mu_{ij}$	$E(\varepsilon_{ij}) = 0$
	α_j	$\mu_j - \mu$	Fixed, $\sum_j \alpha_j = 0$
	η_i	$\mu_i - \mu$	Random, $E(\eta_i) = 0$
	$(\eta\alpha)_{ij}$	$\mu_{ij} - \mu_i - \mu_j + \mu$ or $(\mu_{ij} - \mu) - \alpha_j - \eta_i$	$\sum_j (\eta\alpha)_{ij} = 0, \quad E(\eta\alpha)_{ij} = 0$
Components of variance	σ_e^2	$E(\varepsilon_{ij}^2)$	
	θ_A^2	$\dfrac{\sum_j \alpha_j^2}{a-1}$	
	σ_S^2	$E(\eta_i^2)$	$= \sum_{\cdot\cdot}$ (see Eq. (7.21))
	σ_{SA}^2	$\dfrac{\sum_j E(\eta\alpha)_{ij}^2}{a-1}$	$= \dfrac{\sum_j (\sigma_j^2 - \sum_{\cdot\cdot})}{a-1}$
Variance-covariance terms	$\boldsymbol{\Sigma}$	$a \times a$ variance-covariance matrix	
	$\mathrm{cov}(\mu_{ij}, \mu_{ij'})$	$E(\mu_{ij} - \mu_j)(\mu_{ij'} - \mu_{j'})$	Off-diagonal element of $\boldsymbol{\Sigma}$, $= \sigma_j \sigma_{j'} \rho_{jj'}$
	σ_j^2	$E(\mu_{ij} - \mu_j)^2$	Diagonal element of $\boldsymbol{\Sigma}$
	ρ_{ij}	Correlation of μ_{ij} and $\mu_{ij'}$	
	$\Sigma_{\cdot\cdot}$	Mean of a^2 elements of $\boldsymbol{\Sigma}$	

7.3   ONE-FACTOR CASE: EXPECTED MEAN SQUARES

7.3.1   COMPONENTS OF VARIANCE.   Since *EMS* is so important in justifying the F ratio, estimating variance components and ω^2, and calculating relative efficiency, we wish to consider *EMS* for the repeated-measurement design. The definitions, assumptions, and conditions set forth in Section 7.2 are basic to what follows, and accordingly, you may find it useful now and then to refer to Table 7–2.

We first consider $E(MS_A)$. From Equation (7.16),

(7.23) $$\bar{Y}_{\cdot j} = \mu + \alpha_j + \bar{\eta}_\cdot + \overline{(\eta\alpha)}_{\cdot j} + \bar{\varepsilon}_{\cdot j}$$

Summing over levels of j and dividing by a, and noting that $\sum_j \alpha_j = 0$ and $\sum_j (\eta\alpha)_{ij} = 0$, we have

(7.24) $$\bar{Y}_{\cdot\cdot} = \mu + \bar{\eta}_\cdot + \bar{\varepsilon}_{\cdot\cdot}$$

where

$$\bar{\eta}_{.} = \sum_{i=1}^{n} \frac{\eta_i}{n} \quad (\overline{\eta\alpha})_{.j} = \sum_{i=1}^{n} \frac{(\eta\alpha)_{ij}}{n}$$

$$\bar{\varepsilon}_{.j} = \sum_{i=1}^{n} \frac{\varepsilon_{ij}}{n} \quad \bar{\varepsilon}_{..} = \sum_{j=1}^{a} \sum_{i=1}^{a} \frac{\varepsilon_{ij}}{an}$$

From Equations (7.23) and (7.24), and the definition of SS_A, we have

(7.25)
$$E(SS_A) = n \sum_j E[\alpha_j + (\overline{\eta\alpha})_{.j'} + (\bar{\varepsilon}_{.j} - \bar{\varepsilon}_{..})]^2$$

Because the α_j, $(\eta\alpha)_{ij}$, and ε_{ij} are distributed independently of each other, expectations of their cross-products equal zero. Therefore, when we expand the right side of Equation (7.25), we obtain

(7.26)
$$E(SS_A) = n \sum_j [E(\alpha_j^2) + E(\overline{\eta\alpha})_{.j}^2 + E(\bar{\varepsilon}_{.j} - \bar{\varepsilon}_{..})^2]$$

Since α_j is a fixed effect, it will not vary over replications of the experiment; therefore, $E(\alpha_j^2) = \alpha_j^2$. Other terms can be simplified on the basis of the developments of Section 3.5. We showed there that $\sigma_{\bar{Y}}^2 = \sigma^2/n$ provided that the observations are independently distributed. In this case, instead of the mean of a sample of n Ys, we have the mean of a sample of n $(\overline{\eta\alpha})_{ij}$s. Its sampling variance $E(\overline{\eta\alpha})_{.j}^2$ equals the variance of the interaction effects in the jth population divided by n, $E(\eta\alpha)_{ij}^2/n$. Similar arguments underlie

$$E(\bar{\varepsilon}_{.j} - \bar{\varepsilon}_{..})^2 = E(\bar{\varepsilon}_{.j}^2 + \bar{\varepsilon}_{..}^2 - 2\bar{\varepsilon}_{.j}\bar{\varepsilon}_{..})$$

$$= \frac{\sigma_e^2}{n} + \frac{\sigma_e^2}{an} - \frac{2\sigma_e^2}{an}$$

We now have

(7.27)
$$E(SS_A) = n \sum_j \left[\alpha_j^2 + \frac{E(\eta\alpha)_{ij}^2}{n} + \left(\frac{1}{n} - \frac{1}{an}\right)\sigma_e^2 \right]$$

Simplifying further, dividing by $a-1$, and noting the definitions of variance components in Table 7–2, we finally reach our goal:

(7.28)
$$E(MS_A) = \sigma_e^2 + \sigma_{SA}^2 + n\theta_A^2$$

The result derived above has one big difference from the result derived earlier for two-factor designs (Chapter 5); in the earlier work, the interaction component σ_{AS}^2 did not contribute to the variability among the $\bar{Y}_{.j}$. Why does it in this case? To answer this, we note that $\sum_i(\eta\alpha)_{ij}$ neither equals zero (as $\sum_j(\eta\alpha)_{ij}/n$ does) nor is constant over levels of j (as $\sum_i \eta_i/n$ is). It does not represent the sum of all population interaction effects obtainable under A_j but rather a random sample of size n from that infinitely large set of values. Furthermore, the composition of the sample of n interaction effects will vary over treatment levels (as the subscript j suggests). The consequence of this argument is that the SS_A will reflect variability among interaction effects (except where $\rho = 1$ and no interactions are present), as the $E(MS_A)$ in Table 7–3 indicates.

Table 7-3 Analysis of variance, one-factor repeated-measurement design

SV	df	SS	EMS[a,b] GENERAL CASE	HOMOGENEITY OF VARIANCES OF DIFFERENCES	HOMOGENEOUS VARIANCES AND $\rho = 1$
S	$n-1$	$\dfrac{\sum_i T_{i.}^2}{a} - C$	$\sigma_e^2 + a\sigma_s^2$ $\;\cdots\;$ $\sigma_e^2 + a\bar\sigma_{..}$	$\sigma_e^2 + a\sigma_s^2$ $\;\cdots\;$ $\sigma_e^2 + \left(\dfrac{1}{a}\right)\left[\sum_i \sigma_i^2 - (a-1)\lambda\right]$	$\sigma_e^2 + a\sigma_s^2$ $\;\cdots\;$ $\sigma_e^2 + a\sigma^2$
A	$a-1$	$\dfrac{\sum_j T_{.j}^2}{n} - C$	$\sigma_e^2 + \sigma_{SA}^2 + n\theta_A^2$ $\;\cdots\;$ $\sigma_e^2 + \dfrac{\sum_i(\sigma_i^2 - \bar\sigma_{..})}{a-1} + n\theta_A^2$	$\sigma_e^2 + \sigma_{SA}^2 + n\theta_A^2$ $\;\cdots\;$ $\sigma_e^2 + \lambda + n\theta_A^2$	$\sigma_e^2 + n\theta_A^2$ $\;\cdots\;$ $\sigma_e^2 + n\theta_A^2$
SA	$(n-1)(a-1)$	$\sum\sum Y^2 - C - SS_S - SS_A$	$\sigma_e^2 + \sigma_{SA}^2$ $\;\cdots\;$ $\sigma_e^2 + \dfrac{\sum_i(\sigma_i^2 - \bar\sigma_{..})}{a-1}$	$\sigma_e^2 + \sigma_{SA}^2$ $\;\cdots\;$ $\sigma_e^2 + \lambda$	σ_e^2 $\;\cdots\;$ σ_e^2

[a] In each pair of EMS, the expression above the dashed line is in terms of the components of variance defined in Table 7-2. The expression below the line is the equivalent expression in terms of the elements of the variance-covariance matrix (Equation 7.20).

[b] As noted in the text, the two leftmost cases, in which $\sigma_{SA}^2 \neq 0$, are often referred to as *nonadditive*; when variances are homogeneous and all $\rho = 1$, we have the *additive* case.

Table 7–3 also presents *EMS* for subjects *S* and for the interaction *SA*. Note that σ^2_{SA} does not contribute to $E(MS_S)$. To see why this is so, consider $\overline{Y}_{i.}$, the mean of the *a* scores obtained from the *i*th subject. Summing over *j* and dividing by *a* in Equation (7.3), we have

$$\textbf{(7.29)} \qquad \overline{Y}_{i.} = \mu + \frac{\sum_j \eta_i}{a} + \frac{\sum_j \alpha_j}{a} + \frac{\sum_j (\eta\alpha)_{ij}}{a} + \frac{\sum_j \varepsilon_{ij}}{a}$$

Since *A* is a fixed-effect variable, $\sum_j \alpha_j = 0$. Furthermore, $\sum_j (\eta\alpha)_{ij}$ represents the sum of *all* interaction effects for the *i*th subject and therefore will also be zero. Thus, Equation (7.4) can be replaced by the simpler result:

$$\textbf{(7.29')} \qquad \overline{Y}_{i.} = \mu + \eta_i + \frac{\sum_j \varepsilon_{ij}}{a}$$

The variability among the means of the *n* subjects in the experiment does not involve variability among interaction effects, since such effects do not contribute to any subject's mean performance. This is reflected in $E(MS_S)$ in Table 7–3.

A numerical example may help to further clarify why interaction effects do not contribute to variability among subject means but do contribute to variability among treatment level means. Table 7–4 contains data for a population consisting of four subjects. Assume that two subjects S_1 and S_2 are sampled for an "experiment." We "test" both subjects at all three levels of *A*. Thus, "subjects" is a random-effect variable, and *A* is a fixed-effect variable. Check the calculations for $S(\eta_i)$, $A(\alpha_j)$, and $SA[(\eta\alpha)_{ij}]$ effects. The mean interaction effect is zero for each subject, and therefore interaction variability does not contribute to the variability among subject means. Calculate also the mean interaction effect for each level of *A*, limiting the calculations to the two subjects included in the experiment. We have

$$(\eta\alpha)_{.1} = \tfrac{1}{2}(-1+3) = 1$$
$$(\eta\alpha)_{.2} = \tfrac{1}{2}(1+2) = 1.5$$
$$(\eta\alpha)_{.3} = \tfrac{1}{2}(0-5) = -2.5$$

Clearly, the variability among treatment means is based partly on variability among interaction effects.

Table 7–4   Data for population of four subjects

	A_1		A_2		A_3			
	Y_{i1}	$(\eta\alpha)_{i1}$	Y_{i2}	$(\eta\alpha)_{i2}$	Y_{i3}	$(\eta\alpha)_{i3}$	μ_i	η_i
S_1	2	−1	6	1	10	0	6	−.5
S_2	4	3	5	2	3	−5	4	−2.5
S_3	5	−1	8	0	14	1	9	2.5
S_4	3	−1	3	−3	15	4	7	.5
μ_j	3.5		5.5		10.5		$\mu = 6.5$	
α_j	−3		−1		4			

7.3.2 *EMS* IN TERMS OF THE VARIANCE-COVARIANCE MATRIX. Referring to σ_S^2 and σ_{SA}^2 in Table 7–2, we can rewrite the *EMS* in terms of the elements of Σ. We distinguish three sets of assumptions about Σ:

1. *General case.* The elements are essentially unrestricted. To obtain the equivalent *EMS* in terms of the elements of Σ, we merely substitute the expressions in Table 7–2 for σ_S^2 and σ_{SA}^2.

2. *Homogeneous variances of difference scores.* Consider treatment levels A_j and $A_{j'}$. For each subject in the population, we can compute a difference score, $d_{jj'} = Y_j - Y_{j'}$. It is assumed that $\sigma_{d_{jj'}}^2$ is homogeneous for all pairs of treatments A_j and $A_{j'}$. Huynh and Feldt (1970) have pointed out that this condition is equivalent to assuming that

$$\text{cov}\,(\mu_{ij}, \mu_{ij'}) = \frac{\sigma_j^2 + \sigma_{j'}^2}{2} - \lambda$$

where λ is a value independent of j and j'. With this restriction, Equation (7.21) becomes

(7.30)
$$\sigma_S^2 = \Sigma_{..} = \left(\frac{1}{a}\right)^2 \left\{ \sum_j \sigma_j^2 + \frac{1}{2}[(\sigma_1^2 + \sigma_2^2) + \cdots \right.$$
$$\left. + (\sigma_a^2 + \sigma_{a-1}^2)] - a(a-1)\lambda \right\} = \frac{1}{a}\left[\sum_j \sigma_j^2 - (a-1)\lambda\right]$$

and Equation (7.22) becomes

(7.31)
$$\sigma_{SA}^2 = \frac{\sum_j (\sigma_j^2 - \Sigma_{..})}{a - 1} = \lambda$$

3. *Homogeneous variances, $\rho = 1$.* The results in Table 7–3 follow immediately on setting all variances and covariances in Equations (7.21) and (7.22) equal to σ^2. This condition is equivalent to assuming that the $(\eta\alpha)_{ij}$ are all zero; in either event, $\sigma_{SA}^2 = 0$.

Conditions 1 and 2 are often lumped together as the *nonadditive* case, and 3 is referred to as the *additive case*. Nevertheless, the distinctions among the three sets of conditions are important in data analyses and in interpreting results of those analyses. As usual, we do not get something for nothing. The first, least restrictive condition poses the greatest problems for investigators. We now look into the implications of the three sets of assumptions about the variance-covariance matrix.

7.4 ONE-FACTOR CASE: IMPLICATIONS OF THE MODELS

7.4.1 VALIDITY OF F TESTS. Three models for the analysis of variance have been proposed, for the repeated-measurement design. Though neither the partitioning of the total variability nor the computations for the various sums of squares

differ under the three models, what model we choose does affect our inferences. The most obvious implication is that nonadditivity permits only a negatively biased test of the subject effect. By negatively biased, we mean that
$E(MS_S)/E(MS_{SA}) < 1$ when H_0 is true. Returning to Table 7–3, we find that

$$\frac{E(MS_S)}{E(MS_{SA})} = \frac{\sigma_e^2 + a\sigma_S^2}{\sigma_e^2 + \sigma_{SA}^2}$$

and if H_0 is true ($\sigma_S^2 = 0$), the ratio is clearly less than 1. This tendency for too
many Type II errors to result is not a particularly important consequence of
nonadditivity, since subject variability is usually large enough to be detected by
even the most conservative test. In fact, if there were not strong a priori evidence
for significant subject variability, using the design would have questionable value
since there would be no basis for expecting precision to be better than with any
other design.

There is a more important reason for distinguishing among the three cases in
Table 7–3. Although the observations for any individual are not independently
distributed in any of the three cases (unless $\rho = 0$ for all j and j'), MS_A/MS_{AS} is
distributed exactly like F if variances of difference scores are the same for all
treatment pairs A_j and $A_{j'}$.[1] In the general case of Table 7–3, however, the F test
is only approximate. Several questions follow from these observations.

First, how can we determine whether or not subjects and treatments
interact? There is no obvious error term against which to test MS_{SA} in Table 7–3.
Nevertheless, additivity is of interest because it provides a sufficient condition for
an exact F test of treatments. Second, if the data suggest that effects are not
additive, is there any way of transforming the data to additivity? Third, if it is
assumed that effects are nonadditive and no appropriate transformation is available, what is the extent and direction of bias in the F test? Note that even with
nonadditive data, the F test will still be exact if we have homogeneous variances
of difference scores. Fourth, if we assume neither additivity nor homogeneity of
variance for the difference scores, and an appropriate transformation to additivity
cannot be found, how can we improve on the approximate F test? We shall
consider the first two questions—how to test the hypothesis of additivity, and how
to transform to additivity—in Section 7.5. In the remainder of Section 7.4.1, we
shall discuss the nature of the bias in the general case and some ways of dealing
with it.

If the variances of difference scores are heterogeneous, the bias is generally
positive; too many Type I errors are made in testing treatment effects against
MS_{SA}. We could make the test of treatment effects contingent on the outcome of
a preliminary test of the assumption that the variances of difference scores were
homogeneous (Huynh and Feldt, 1970). A simpler procedure stems from a paper

---

[1] Most writers have noted Box's (1954) proof that the ratio is distributed like F under conditions of
homogeneous variances ($\sigma_1^2 = \cdots = \sigma_a^2 = \sigma^2$) and covariances ($\rho_{12}\sigma_1\sigma_2 = \cdots = \rho\sigma^2$). This "compound symmetry" implies homogeneity of variances of differences but the converse is not true.
Therefore, the weakest condition for MS_A/MS_{AS} to be distributed like F is homogeneity of variances
of differences. Compound symmetry requires the additional constraint that all $\lambda = (1-\rho)\sigma^2$ and
additivity implies that $\rho = 1$.

by Box (1954), in which he showed that the statistic MS_A/MS_{AS} is distributed as F but on $(a-1)\varepsilon$ and $(a-1)(n-1)\varepsilon$ df. As Huynh and Feldt have pointed out, homogeneity of variances of difference scores implies $\varepsilon = 1$, and therefore the usual F test. Under conditions in which there is extreme heterogeneity of variance of difference scores, ε approaches a lower bound, $1/(a-1)$. Therefore, for an extreme violation of the homogeneity assumptions, F would be distributed on 1 and $(n-1)$ df. In view of these comments about the F distribution, it is suggested that the F statistics first be assessed against the F required for significance on 1 and $(n-1)$ df. In other words, first assume the worst possible degree of heterogeneity of variances of differences and consequently set ε equal to $1/(a-1)$. If ε is actually larger than this value, that is, if the variances are more homogeneous than has been assumed, we have thrown away df and are using a negatively biased F test. Thus, if the obtained F is significant with this generally conservative approach, we can have reasonable faith in rejecting the null hypothesis.

What if the conservative F test just outlined is not significant? It may mean that the null hypothesis is true. On the other hand, nonsignificance may merely indicate that our homogeneity assumptions are not as badly violated as the conservative test implies. To check this last possibility, now assess the obtained F against the F required for significance on $a-1$ and $(n-1)(a-1)$ df. In other words, assume homogeneity of variances of differences and consequently set ε equal to 1. If this positively biased F test is not significant, we can be reasonably sure that the null hypothesis should be accepted.

It is possible for the two F tests to yield contradictory results. The F assessed on 1 and $(n-1)$ df might not be significant, whereas the F assessed on $a-1$ and $(n-1)(a-1)$ df might be significant. Since the first F may be negatively biased and the second F may be positively biased, it is difficult to make a clear-cut decision about the null hypothesis. A statement about the approximate level of significance can generally be made, however. The negatively biased F may not, for example, be significant at the 5 percent level but may be significant at the 10 percent level, while the positively biased F is significant at the 5 percent level, leading to the conclusion that the true significance level is somewhere between 5 percent and 10 percent. If this sort of approximation is not sufficient, an exact test can be computed (Rao, 1952). (For this, see Chapter 18.) A slightly simpler procedure, which yields a reasonable approximation to the nominal α level (Collier, Baker, Mandeville, and Hayes, 1967; Stoloff, 1967), involves computing an estimate of ε, the adjustment for df.

The calculation of $\hat{\varepsilon}$ requires values of the variances for each level of A and covariances for each pair of levels of A. We first define the obtained covariances,

$$S_{jj'} = \frac{\sum_{i=1}^{n} (Y_{ij} - \overline{Y}_{.j})(Y_{ij'} - \overline{Y}_{.j'})}{n-1}, \qquad j \neq j'$$

and variances,

$$S_{jj} = \frac{\sum_{i=1}^{n} (Y_{ij} - \overline{Y}_{.j})^2}{n-1}$$

Furthermore, $\bar{S}_j$ is an average of the variance and $a - 1$ covariances associated with treatment A_j; that is,

$$\bar{S}_j = \frac{1}{a}(S_{j1} + S_{j2} + \cdots + S_{jj} + \cdots + S_{ja})$$

Finally, $\bar{S}_{..}$ is the average of the a^2 variances and covariances, or equivalently,

$$\bar{S}_{..} = \frac{1}{a}\sum_{j=1}^{a}\bar{S}_j$$

By these definitions of terms, the appropriate equation for adjusting df for the variance-covariance matrix is

(7.32)
$$\hat{\varepsilon} = \frac{a^2(\bar{S}_{jj} - \bar{S}_{..})^2}{(a-1)(\sum_{j=1}^{a}\sum_{j'=1}^{a} S_{jj'}^2 - 2a\sum_{j=1}^{a}\bar{S}_j^2 + a^2\bar{S}_{..}^2)}$$

To illustrate the computations, we use a data set from an article by Greenhouse and Geisser (1959). The variance-covariance matrix is presented in Table 7–5, together with summary statistics; note that the matrix must be symmetric with $S_{jj'} = S_{j'j}$ for all j and j'. Assume $n = 10$. Rounding the adjusted df to the nearest integer, we use 4 df and 37 df in testing the A effects in Table 7–5.

7.4.2   EFFICIENCY OF THE DATA ANALYSIS.   Even if the variances of difference scores are homogeneous, efficiency may be a problem. If the data are best described by a nonadditive model, efficiency will be less than in the additive case ($\rho = 1$), and the resulting F test will be less powerful. In other words, if the null

Table 7–5   Variance-covariance matrix

	A_1	A_2	A_3	A_4	A_5	A_6	$\sum_{j'} S_{jj'}$	$\bar{S}_j$
A_1	3.100	.101	−.279	−.083	−.009	1.557	4.387	.731
A_2	.101	5.780	1.013	−.114	−1.014	.039	5.805	.968
A_3	−.279	1.013	5.560	1.039	1.366	−.169	8.530	1.422
A_4	−.083	−.114	1.039	5.600	3.080	.258	9.780	1.630
A_5	−.009	−1.014	1.366	3.080	6.820	.222	10.465	1.744
A_6	1.557	.039	−.169	.258	.222	5.170	7.077	1.179
								7.674

$\bar{S}_{jj}$ is the average of the variances, the diagonal entries $= 5.34$

$\bar{S}$ is the average of all 36 entries $= 1.28$

$(\bar{S}_{jj} - \bar{S}_{..})^2 = 16.48$

$\sum_j \sum_{j'} S_{jj'}^2 = 212.86$

$2a \sum_j \bar{S}_j^2 = 126.98$

$a^2 \bar{S}_{..}^2 = 58.91$

$$\hat{\varepsilon} = \frac{(6^2)(16.48)}{(6-1)(212.86 - 126.98 + 58.91)} = .82$$

hypothesis is false, $E(MS_A)/E(MS_{SA})$, where

$$\frac{E(MS_A)}{E(MS_{SA})} = \frac{\sigma_e^2 + n\theta_A^2}{\sigma_e^2}$$

will be greater than $E(MS_A)/E(MS_{SA})$ under the nonadditive model where

$$\frac{E(MS_A)}{E(MS_{SA})} = \frac{\sigma_e^2 + \sigma_{SA}^2 + n\theta_A^2}{\sigma_e^2 + \sigma_{SA}^2}$$

To see why this is so, consider the following illustrative ratios:

$$\frac{15}{5}, \quad \frac{20}{10}, \quad \frac{50}{40}, \quad \frac{100}{90}$$

As we add the same constant to the numerator and denominator of any one ratio, we obtain a smaller ratio. Thus,

$$\frac{15+5}{5+5} < \frac{15}{5} \quad \text{and} \quad \frac{50+50}{40+50} < \frac{50}{40}$$

Values of $E(MS_A)/E(MS_{SA})$ for the two models are like any two of the illustrative ratios. If the nonadditive model is correct, the constant σ_{SA}^2 has essentially been added to both numerator and denominator of the value of $E(MS_A)/E(MS_{SA})$ for additive data; thus the size of the expected F ratio has been decreased.

A numerical example should further make our point about efficiency and simultaneously show the calculations involved in analyzing data from the one-factor repeated-measurement design. Table 7–6 contains response time scores to which we shall apply the analysis of variance. For subjects,

$$SS_S = \frac{(5.6)^2 + (14.6)^2 + (24.5)^2}{3} - \frac{(44.7)^2}{9} = 59.58$$

For the treatment effects, we have

$$SS_A = \frac{(12.7)^2 + (13.8)^2 + (18.2)^2}{3} - \frac{(44.7)^2}{9} = 5.65$$

For the error variability, we have

$$SS_{SA} = 291.17 - \frac{(44.7)^2}{9} - 59.58 - 5.65 = 3.93$$

Table 7–6   Response time data

	A_1	A_2	A_3	$T_{i.}$
S_1	1.7	1.9	2.0	5.6
S_2	4.4	4.5	5.7	14.6
S_3	6.6	7.4	10.5	24.5
$T_{.j} = 12.7$		13.8	18.2	$T_{..} = 44.7$

Table 7–7   Response speed data obtained by taking reciprocals of the scores in Table 7–6

	A_1	A_2	A_3	$T_{i.}$
S_1	.589	.526	.500	1.615
S_2	.227	.222	.175	.624
S_3	.152	.135	.095	.382
$T_{.j} = .968$		.883	.770	$T.. = 2.621$

To test the treatment effects, we compute

$$F = \frac{5.65/2}{3.93/4} = 2.87$$

which is not significant at the 5 percent level.

In many experiments, response speed, the reciprocal of response time, is a perfectly meaningful measure. Instead of the time per response, the number of responses per unit of time is noted. Speed measurements have been made by taking the reciprocals of data in Table 7–6; the results are presented in Table 7–7. Variance is analyzed as before, but to help prevent misplaced decimal points, the precaution has been taken of multiplying all the entries of Table 7–7 by 1,000. (This will not affect the F ratios.) Then,

$$SS_S = \frac{(1,615)^2 + (624)^2 + (382)^2}{3} - \frac{(2,621)^2}{9} = 284,548.3$$

and

$$SS_A = \frac{(968)^2 + (883)^2 + (770)^2}{3} - \frac{(2,621)^2}{9} = 6,577.6$$

and

$$SS_{SA} = 1,055,389 - \frac{(2,621)^2}{9} - 284,548.3 - 6,577.6 = 969.7$$

The test of treatment effects yields

$$F = \frac{6,577.6/2}{969.7/4} = 13.56$$

a result significant at the .01 level.

Why do such diverse inferences result from the two sets of data that are only transforms of each other? The answer lies in looking more closely at the data in the two tables. Returning to Table 7–6, we note that the spread among scores is much greater at A_3 than that at A_1. In Table 7–7, the variability among subjects seems less affected by the level of A. The point is more clearly made in Figure 7–1, in which both data sets are plotted, that of Table 7–6 against the left-hand axis and that of Table 7–7 against the right-hand axis. The response time data

Figure 7–1   A plot of the data in Tables 7–6 and 7–7.

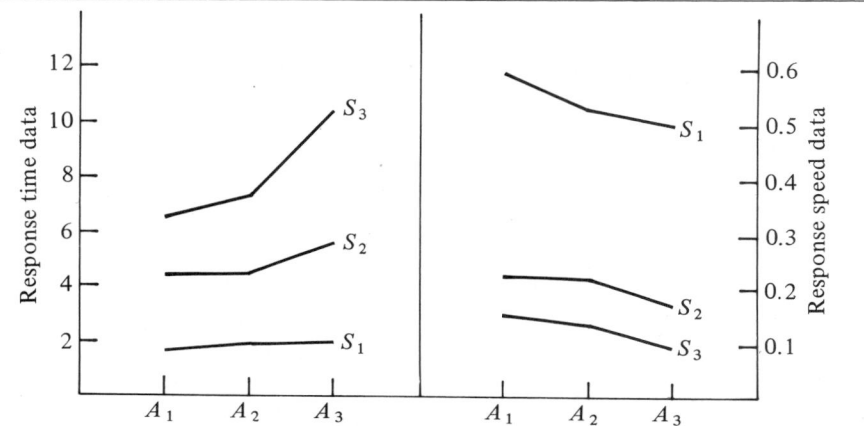

more clearly contain interaction effects. As a result of this nonadditivity, efficiency is lower and the F test is less powerful.[2]

7.4.3   MISSING DATA.

Let us assume that the additive model adequately describes our data. Further suppose that one score is missing; subject i has failed to appear to be tested under A_j, or the recording apparatus was not working in one session. Call the missing score X_{ij}. Then,

$$E(X_{ij}) = \mu + \eta_i + \alpha_j$$

Our best estimate of μ is

$$\hat{\mu} = \frac{\text{grand total}}{an}$$

In this case, the grand total is $T_{..} + X_{ij}$, where $T_{..}$ is the sum of the $an-1$ scores that were obtained. Similarly, we estimate

$$\hat{\eta}_i = \hat{\mu}_i - \hat{\mu}$$

where

$$\hat{\mu}_i = \frac{T_i + X_{ij}}{a}, \qquad T_i = \text{obtained total for } S_i$$

and

$$\hat{\alpha}_j = \hat{\mu}_j - \hat{\mu}$$

[2] The transformations to additivity will not always result in larger F ratios. If there is extreme heterogeneity of covariance in the original data set, the transformation will cause a reduction in positive bias (variances and covariances are now roughly homogeneous on the transformed, additive, scale). If the null hypothesis is true, the result of the transformation could be a reduced F ratio.

where

$$\hat{\mu}_j = \frac{T_j + X_{ij}}{n}, \qquad T_j = \text{obtained total for } A_j$$

then

$$\hat{X}_{ij} = \frac{T_{..} + \hat{X}_{ij}}{an} + \left(\frac{T_i + \hat{X}_{ij}}{a} - \frac{T_{..} + \hat{X}_{ij}}{an}\right) + \left(\frac{T_j + \hat{X}_{ij}}{n} - \frac{T_{..} - \hat{X}_{ij}}{an}\right)$$

from which we obtain our estimate,

(7.33)
$$\hat{X}_{ij} = \frac{nT_i + aT_j - T_{..}}{(n-1)(a-1)}$$

Suppose we have several missing values—$X_{ij}^{(1)}$, $X_{ij}^{(2)}$, and so on. Guess at all, except $X_{ij}^{(1)}$, which is approximated by Equation (7.2), using guesses of $X_{ij}^{(2)}$, $X_{ij}^{(3)}, \ldots$, as though they were part of the data. With our approximation of $X_{ij}^{(1)}$ and guesses of $X_{ij}^{(3)}$, $X_{ij}^{(4)}, \ldots$, we can now apply Equation (7.2) to approximate $X_{ij}^{(2)}$. We continue this process, repeating the cycle with first-round approximations of X_{ij}, until two successive cycles show little change. One final point: a *df* is lost from the error term *df* for each value estimated.

We originally assumed that the additive model described our data. If $(\eta\alpha)_{ij}$ is not zero for all *i* and *j*, we have

$$\hat{X}_{ij} = \hat{\mu} + \hat{\eta}_i + \hat{\alpha}_j + (\eta\hat{\alpha})_{ij}$$

which, when we substitute for the effects in terms of $\hat{\mu}_{ij}$, $\hat{\mu}_i$, and $\hat{\mu}_j$ (see Table 7–2), yields

$$\hat{X}_{ij} = \hat{\mu}_{ij}$$

Without other measurements in the *ij*th cell, no estimate is possible.

7.4.4 ESTIMATING ω^2. Assuming the additive case, we have the following estimates of components of variance. (You may want to review the developments in Section 4.7.)

$$\hat{\sigma}_S^2 = \frac{MS_S - MS_{SA}}{a}$$

$$\hat{\delta}_A = \left(\frac{a-1}{a}\right)\left(\frac{MS_A - MS_{SA}}{n}\right)$$

$$\hat{\sigma}_e^2 = MS_{SA}$$

Then,

(7.34)
$$\hat{\omega}_S^2 = \frac{\hat{\sigma}_S^2}{\hat{\sigma}_e^2 + \hat{\sigma}_S^2 + \hat{\delta}_A^2}$$

$$= \frac{n(F_S - 1)}{an + n(F_S - 1) + (a-1)(F_A - 1)}$$

and

(7.35)

$$\hat{\omega}_A^2 = \frac{\hat{\delta}_A^2}{\hat{\sigma}_e^2 + \hat{\sigma}_S^2 + \hat{\delta}_A^2}$$

$$= \frac{(a-1)(F_A - 1)}{an + n(F_S - 1) + (a-1)(F_A - 1)}$$

where $F_A = MS_A/MS_{SA}$ and $F_S = MS_S/MS_{SA}$.

If treatments and subjects interact ($\sigma_{SA}^2 \neq 0$), the *EMS* provide no estimate of σ_S^2; as a result, we cannot calculate $\hat{\omega}_A^2$ or $\hat{\omega}_S^2$. At best, we can get crude bounds between which the variance proportion must fall. We can get an upper bound on $\hat{\omega}_A^2$ ($\omega_{A,u}^2$) by assuming $\sigma_S^2 = 0$.

(7.36)

$$\hat{\omega}_{A,u}^2 = \frac{\hat{\delta}_A^2}{\hat{\sigma}_e^2 + \hat{\sigma}_{SA}^2 + \hat{\delta}_A^2}$$

$$= \frac{(a-1)(F_A - 1)}{an + (a-1)(F_A - 1)}$$

A lower bound ($\hat{\omega}_{A,l}^2$) follows by noting that

$$\hat{\sigma}_S^2 = \frac{MS_S - \hat{\sigma}_e^2}{a}.$$

Under the nonadditive model, we lack an estimate of σ_e^2 and therefore of σ_S^2. However, if we use the approximation

(7.37)

$$\hat{\sigma}_S^2 \approx \frac{MS_S}{a}$$

we have an overestimate of σ_S^2 and accordingly, an underestimate of ω_A^2. Then,

(7.38)

$$\hat{\omega}_{A,l}^2 = \frac{(a-1)(F_A - 1)}{an + (a-1)(F_A - 1) + nF_S}$$

We can also get boundaries on $\hat{\omega}_S^2$. With $\hat{\sigma}_S^2 = 0$,

(7.39)

$$\hat{\omega}_{S,l}^2 = 0$$

and by Equation (7.37),

(7.40)

$$\hat{\omega}_{S,u}^2 = \frac{nF_S}{an + (a-1)(F_A - 1) + nF_S}$$

7.4.5   RELIABILITY AND ANALYSIS OF VARIANCE.   An important application of estimating the components of variance is estimating reliability. Suppose that we have several judges rate several individuals on some personality trait. Perhaps we need such ratings because we want to correlate the personality measure with some performance measure, or perhaps we want to pick subjects having various degrees of the trait for a subsequent experiment. If the judges agree closely in their ratings of each subject, we feel reasonably confident that we have a good estimate of where each subject stands in the trait being measured. If the judges vary greatly in their ratings of each subject, we are less confident in our assessment of each individual for the trait in question. True differences among individuals may be obscured by the variability in our measuring instrument, the set of judges.

The same problem of how consistent internally a measuring instrument is arises in many other circumstances. We may be concerned with developing a test that discriminates among the arithmetic abilities of individuals; we want the items to correlate well with each other. We want to choose a set of pictures that represent different points along a scale of attractiveness; they will be used later in an experiment on aesthetic judgment of different personality types. Here, we are again interested in the degree of consistency among the judges rating the pictures.

The problem, a common one, is assessing the consistency of measurements (items, ratings) of individual subjects or stimuli—this with the hope of establishing an internally consistent measuring instrument and thus clearly discriminating among the individuals or objects to be measured. One way of doing this is to compute correlation coefficients for all pairs of items and examine the average intercorrelation. This is a laborious procedure. The analysis of variance can provide equivalent information with much less effort.

We assume n subjects (or in general, objects to be measured) drawn at random from an infinite population of individuals and a set of a items (or judges or other measuring instruments) drawn at random from an infinite population of items. Then the score for subject i on item j can be represented as

$$(7.41) \qquad Y_{ij} = \mu_i + \varepsilon_{ij}$$

We may think of μ_i as the "true" score for the individual, his expected score over the population of items (judges, and so on) from which we have sampled. The value ε_{ij} is the error of measurement associated with the ijth observation. The error of measurement is due to chance variability in performance. The subject guesses correctly on one item and incorrectly on another; he is more alert or better motivated at one moment than at another. Thus, one subject scores higher on one member of a pair of items and another subject scores higher on the other member of the pair; the variability does not reflect true differences in ability of the subjects nor true differences in the difficulty of the items. Indeed, the model presented does not envisage any systematic differences among items; there is no α_j component associated with item j that raises or lowers all scores on that item by a constant amount. Thus, this model is invalid if, for example, some items are intrinsically more difficult than others or if some judges consistently rate higher than others.

We can subtract μ, the grand mean of the subject-item population, from both sides of Equation (7.41). Defining, as usual, $\eta_i = \mu_i - \mu$, we have

$$(7.41') \qquad (Y_{ij} - \mu) = \eta_i + \varepsilon_{ij}$$

We assume that the ε_{ij} are independently and normally distributed with variance σ_e^2 and that the η_i are independently and normally distributed with variance σ_S^2.

Table 7–8   Partitioning of variance for a measure of R_{11}

SV	df	EMS
S	$n-1$	$\sigma_e^2 + a\sigma_s^2$
Within S	$n(a-1)$	σ_e^2

Squaring both sides of Equation (7.41′) and taking expectations over the population of subjects and items, we have

$$(7.42) \qquad \sigma_Y^2 = \sigma_S^2 + \sigma_e^2$$

A reasonable definition of reliability is that it is the proportion of total variance attributable to the individuals or objects being measured. That is,

$$(7.43) \qquad R_{11} = \frac{\sigma_S^2}{\sigma_Y^2}$$

where R_{11} indicates the reliability coefficient. Note that R_{11} is merely a special instance of $\hat{\omega}^2$. Consequently, we derive a computational formula for it in the same way. Our model suggests the partitioning of variance displayed in Table 7.8. Our variance estimates are

$$(7.44) \qquad \hat{\sigma}_S^2 = \frac{MS_{BS} - MS_{WS}}{a}$$

and

$$(7.45) \qquad \hat{\sigma}_e^2 = MS_{WS}$$

On the basis of Equation (7.43), it follows that

$$(7.46) \qquad R_{11} = \frac{\hat{\sigma}_S^2}{\hat{\sigma}_S^2 + \hat{\sigma}_e^2}$$

and substituting from Equations (7.44) and (7.45), we have

$$(7.47) \qquad R_{11} = \frac{(MS_{BS} - MS_{WS})/a}{(MS_{BS} - MS_{WS})/a + MS_{WS}}$$
$$= \frac{F-1}{F-1+a}$$

The reliability coefficient defined by Equation (7.47) can be thought of as a measure of the degree to which the a items are measuring the same trait. More specifically, R_{11} can be shown to be an average of the $[a(a-1)]/2$ intercorrelations which can be computed among items when the average within-item variance is used as the denominator for all correlations.

We noted earlier that R_{11} was a special case of $\hat{\omega}_S^2$. The connection is evident if we reconsider Equation (7.34), which provides a formula for $\hat{\omega}_S^2$ in the subjects $\times$ treatments design. Under the assumptions of the present section, $\theta_A^2 = 0$. Therefore, let F_A in Equation (7.34) equal 1; the result is Equation (7.47).

Suppose that we want to find out how well another set of a items, drawn from the same item population, will correlate with the present set. How reliable is the subject's average score $\bar{Y}_{i.}$?

Returning to Equation (7.41), we sum over j and divide by a to obtain

$$\bar{Y}_{i.} = \mu_i + \bar{\varepsilon}_{i.}$$

Since μ_i is assumed to be independent of ε_{ij},

(7.48)
$$\sigma_{\bar{Y}_{i.}}^2 = \sigma_S^2 + \sigma_{\bar{\varepsilon}_{i.}}^2$$
$$= \sigma_S^2 + \frac{\sigma_e^2}{a}$$

Then the reliability of the mean R_{aa} is

(7.49)
$$R_{aa} = \frac{\hat{\sigma}_S^2}{\hat{\sigma}_S^2 + \hat{\sigma}_e^2/a}$$

Substituting from the *EMS* column of Table 7–8, we have

(7.50)
$$R_{aa} = \frac{(MS_{BS} - MS_{WS})/a}{(MS_{BS} - MS_{WS})/a + MS_{WS}/a}$$
$$= 1 - \frac{1}{F}$$

We can interpret R_{aa} as a measure of the correlation between the mean scores for two sets of a randomly sampled items administered at different times to the same sample of subjects. If R_{aa} were high, we could expect a second sample of a items to provide a similar inference about the relative performance of our subjects.

Equation (7.41) implies that any differences among items are chance happenings. We have pooled the A and SA sources in the usual repeated-measurement design to give a single estimate of error. Sometimes there is reason to believe that systematic differences among items exist. This might be so if the items had been intentionally selected at different levels of difficulty. As a second example, subjects might be rated by a raters, each of whom may be suspected to have a different set of standards. In any event, we could use the additive model

$$Y_{ij} = \mu + \eta_i + \alpha_j + \varepsilon_{ij}$$

Table 7–3 ($\rho = 1$) now provides the appropriate analysis of variance and Equation (7.34) provides the appropriate formula for R_{11}. If the additivity assumption is suspect, Equations (7.39) provides a lower bound on R_{11} and (7.40) an upper bound.

The approach just sketched can be extended to a variety of problems. We can administer parallel forms of a test at different times, that is, two sets of items randomly drawn from the same item population. If our concern is whether or not the two forms are consistently measuring the same attributes, we can assess the proportion of total variance associated with the *form* main effect. Alternatively, we may be interested in test-retest reliability: Does a measuring instrument give consistent results over time? Again, the analysis of variance provides an answer. In fact, a well-designed psychometric study could let us separate the relative contributions of error variance due to inconsistency among items from inconsistency due to variability over time or forms of a test. We require a statement of the model. Then our general approach for estimating ω (Section 7.4.4) is applied.

7.5 ADDITIVITY

Several implications of the distinction between the additive and nonadditive models for repeated-measurement designs have been noted. First, if subjects and treatments interact, the usual F test of treatments may be positively biased, necessitating either an approximate test or calculation of covariance terms. Second, some sources of variance (say the subjects) are not testable unless additivity is assumed. Third, the presence of subjects × treatments interactions will tend to reduce design efficiency. Fourth, procedures for estimating missing scores are based on assuming additivity. Fifth, exact estimates of all variance components, and of ω^2, require the additivity assumption. Finally, there is another, nonstatistical consideration. Additive data may be more parsimoniously described than nonadditive data. Take the data of Figure 7–1. The experimenter who presents a mathematical model of response time as a function of level of A must estimate a different regression coefficient for each subject; each curve has a different slope. On the other hand, by the response speed data, only the slope-intercepts differ among subjects. One regression coefficient describes the inclination of all three curves.

In view of the potential advantages of working with additive data, two questions arise. How can we detect the presence of subjects × treatments interactions? If we assume that nonadditivity is present, how can we transform the data to achieve additivity? The answers to the question generally involve computational labor beyond the usual analysis of variance and are not always satisfactory. Nevertheless, they merit consideration.

7.5.1 TUKEY'S TEST OF ADDITIVITY. The choice between data models reduces to a test of the null hypothesis that σ_{SA}^2 is zero. If the null hypothesis is rejected, the nonadditive model is appropriate; otherwise, the additive model applies. Tukey (1949) has proposed a single df test that involves further analyzing the interaction sum of squares into two components. One of these represents nonadditivity and is distributed on one df. The remaining interaction variability provides the error term for the nonadditivity test and is distributed on $(n-1)(a-1)-1\ df$.

To exemplify the calculations, we shall do the Tukey test on the response time data originally presented in Table 7–6, and reproduced in Table 7–9. Note that the deviations of each column and row mean from the grand mean have also been tabled. This is the first step in calculating the F statistic for additivity. Next

Table 7–9   Data for Tukey's single *df* test

	A_1	A_2	$A_{3.}$	$\bar{Y}_{i.}$	$\bar{Y}_{i.} - \bar{Y}_{..}$	$\sum_j Y_{ij}(\bar{Y}_{.j} - \bar{Y}_{..})$
S_1	1.7	1.9	2.0	1.9	−3.1	.25
S_2	4.4	4.5	5.7	4.9	−.1	1.39
S_3	6.6	7.4	10.5	8.2	3.2	3.97
$\bar{Y}_{.j}$	4.3	4.6	6.1	$\bar{Y}_{..} = 5.0$		
$\bar{Y}_{.j} - \bar{Y}_{..}$	−.7	−.4	1.1			

obtain the cross-product terms, $\sum_j Y_{ij}(\bar{Y}_{.j} - \bar{Y}_{..})$. These are

$$(-.7)(1.7) + (-.4)(1.9) + (1.1)(2.0) = .25$$
$$(-.7)(4.4) + (-.4)(4.5) + (1.1)(5.7) = 1.39$$
$$(-.7)(6.6) + (-.4)(7.4) + (1.1)(10.5) = 3.97$$

Multiplying by the row deviations, we obtain the numerator N for a sum of squares to test additivity:

(7.51)
$$N = \sum_i \sum_j Y_{ij}(\bar{Y}_{.j} - \bar{Y}_{..})(\bar{Y}_{i.} - \bar{Y}_{..})$$
$$= (.25)(-3.1) + (1.39)(-.1) + (3.97)(3.2)$$
$$= 11.8$$

The denominator is

(7.52)
$$D = \sum_j (\bar{Y}_{.j} - \bar{Y}_{..})^2 \sum_i (\bar{Y}_{i.} - \bar{Y}_{..})^2$$
$$= (1.8)(19.9) = 35.82$$

and Tukey's single *df* sum of squares due to nonadditivity is

(7.53)
$$SS_{nonadd} = \frac{N^2}{D} = 3.89$$

Subtracting the above from the SS_{SA} computed previously for these data, we have

$$SS_{bal} = 3.93 - 3.89 = .04$$

The *F* ratio is

$$F = \frac{MS_{nonadd}}{MS_{bal}} = \frac{3.89}{.04/3} = 292$$

a result that is clearly significant, even on one and three *df*.

What do the SS_{nonadd} represent? You can get an idea by considering an experiment in which each of *n* subjects is tested on each of *a* successive days. We should expect an individual subject's scores to be correlated with the daily

Table 7–10   An additive data set

	A_1	A_2	A_3	$\bar{Y}_{i.}$	$\bar{Y}_{i.} - \bar{Y}_{..}$	$\sum_j Y_{ij}(\bar{Y}_{.j} - \bar{Y}_{..})$
S_1	3	5	1	3	-2	8
S_2	4	6	2	4	-1	8
S_3	8	10	6	8	3	8
$\bar{Y}_{.j}$	5	7	3	$\bar{Y}_{..} = 5$		
$\bar{Y}_{.j} - \bar{Y}_{..}$	0	2	-2			

averages, the $\bar{Y}_{.j}$; if the group means show improvement over days, for example, then for any individual subject, we expect the values of Y_{ij} to increase over days. We should have an SA interaction if the rate at which Y_{ij} changed as a function of $\bar{Y}_{.j}$ varied for individual subjects, for then the curves for subjects would not be parallel. Further suppose that an individual's rate of change over values of $\bar{Y}_{.j}$ was correlated to his average performance. The subject with the highest average performance over days, for example, might show little change over days because he was already performing well on day 1, while other subjects with lower values of $\bar{Y}_{i.}$ might show more rapid improvement relative to the changes in $\bar{Y}_{.j}$. It is this kind of interaction, this correlation between a subject's average performance and the rate at which his performance changes relative to the changes in the group performance, to which Tukey's test is sensitive.

You can further appreciate how the test works by looking at the cross-products column in Table 7–9. Note that the cross-products (.25, 1.39, 3.97) increase as a function of the row means (1.9, 4.9, 8.2). The nonadditivity sum of squares depends on the slope of the function relating the cross-products and the means. Consider, by contrast, the perfectly additive data set of Table 7–10. Note that there is no change in the cross-products as a function of changes in the row means. The SS_{nonadd} will be zero for this data set.

Tukey's test will not be sensitive to all interactions. Clearly, there are interaction effects in Table 7–11, but the SS_{nonadd} equals zero since $[(1)(4) + (0)(0) + (-1)(4)]^2$ equals zero. Why doesn't the Tukey test work here? The answer lies in the cross-products column. Note that the entries decrease and then increase as the row means increase. In other words, the slope of the straight line that best relates

Table 7–11   Nonadditive data not sensitive to Tukey's F test

	A_1	A_2	A_3	$\bar{Y}_{i.}$	$\bar{Y}_{i.} - \bar{Y}_{..}$	$\sum_j Y_{ij}(\bar{Y}_{.j} - \bar{Y}_{..})$
S_1	4	7	4	5	1	4
S_2	4	4	4	4	0	0
S_3	2	5	2	3	-1	4
$\bar{Y}_{.j}$	10/3	16/3	10/3	$\bar{Y}_{..} = 4$		
$\bar{Y}_{.j} - \bar{Y}_{..}$	$-2/3$	4/3	$-2/3$			

Figure 7-2   A plot of cross-products against row means for three data sets.

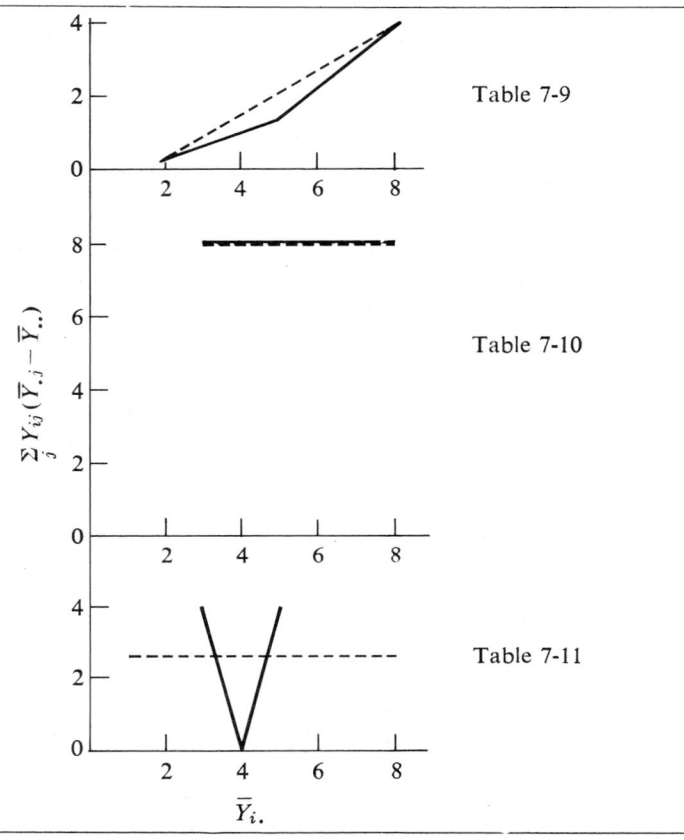

cross-products to row means is zero. Figure 7–2 contains the plot of cross-products against row means for the data sets of Tables 7–9 to 7–11. It is the slope of the dashed line, the best-fitting straight line, which is of interest. If Tukey's test is to be sensitive to interaction effects, the slope of the dashed line must deviate from a slope of zero; the greater the departure from a horizontal line, the greater the SS_{nonadd}. Fortunately, the nature of most interaction effects is such that the cross-product row-mean relation will have a linear component and Tukey's test will therefore be serviceable.

7.5.2 TRANSFORMATIONS OF THE DATA.   We turn next to how to deal with data when the evidence suggests that the nonadditive model is appropriate. The example of Figure 7–1 suggests one solution. It was possible to transform the original data to a data set that appears to be well described by an additive model. One approach to selecting a data transformation stems from an intermediate result obtained in calculating SS_{nonadd}. The slope of regression lines like those in

Figure 7–2 is given by N/D (see Equations (7.51) and (7.52)). Anscombe and Tukey (1963) have proposed the transformation $Y' = Y^p$, where

(7.54) $$p = 1 - \frac{N}{D}\,\bar{Y}_{..}$$

For the data of Table 7–9, $p = -.65$. Applying this transformation, and calculating the nonadditivity test for the transformed data, we obtain an F less than 1, strong evidence that the transformation has been successful in producing additivity.[3] Corroborative evidence, and a clear indication of the sense in which the transformation has achieved additivity, is provided by the value of N/D. For the transformed data, this is $-.0002$; the function corresponding to those in Figure 7–2 is quite flat.

In considering transformations to additivity, we should remember several points. First, it is possible that no appropriate transformation exists. An obvious case is that in which curves cross in the original plot of scores for subject-treatment combinations. Second, Equation (7.54) may not always provide the best transformation. Anscombe and Tukey make this point and suggest that investigators try several transformations, relying on their sense of the data to aid in their selection. Third, several transformations may achieve similar results. The reciprocal transformation ($p = -1.$), which we originally used to transform the data of Table 7–6 into those of Table 7–7 (refer to Figure 7–1), yields an F of only 1.33 for nonadditivity; $N/D = .0012$, further indicating that additivity has been achieved. In many cases, this transformation would be preferred; for example, most investigators would find that the reciprocal transformation of time to rate seemed better intuitively than the transformation $p = -.65$. Finally, theory or practical requirements may dictate a particular data scale. (We have discussed this point in Chapter 4 in considering transformations to achieve homogeneous variances.)

7.6 MULTIFACTOR REPEATED-MEASUREMENT DESIGN

Although the discussion has thus far been limited to two-dimensional designs involving subjects and one other variable, there is no reason why several other variables cannot be investigated. As before, n subjects would be randomly sampled, and each one would be tested under all combinations of all other variables. Thus, if there were the variables A and B, each subject would be tested ab times, once under each treatment combination, with the order of the ab presentations independently randomized for each subject. If A and B both have fixed effects, the three-factor $S \times A \times B$ design is a simple extension of the $S \times A$ design we have been considering thus far in this chapter. If the additional variable

---

[3] Such transformations can be readily carried out on a high-speed computer and even on many electronic pocket calculators. If you have no other aid but a table of logarithms, the following approach will work. If $Y' = Y^p$, $\log Y' = p \log Y$. Therefore, the antilog of $p \log Y$ gives the desired value of Y'.

B is assumed to have random effects, significance tests and estimation of variance components are more complicated.

7.6.1   THE $S \times A \times B$ DESIGN, A AND B FIXED.   Determining SV and df, as well as calculating SS, is straightforward. We proceed exactly as with the three-factor designs (Chapter 5), with the qualification that there is only one score in each cell. The result is displayed in Table 7–12.

We have assumed a nonadditive model in deriving the EMS. With A and B both having fixed effects, the expectations for S, A, and AB are the same as earlier (Table 7–3, nonadditive case). Furthermore, expectations for B and SB parallel the expectations for A and SA. Finally, note that σ^2_{ABS} contributes to the AB mean square. The reason is that a different set of randomly sampled values of $(\eta\alpha\beta)_{ijk}$, the second-order interaction effect, contributes to each AB cell mean.

As usual, F tests are based on the rule that the ratio of EMS should be 1, assuming H_0 to be true. Therefore, A, B, and AB are tested against AS, BS, and ABS, respectively. Under the nonadditive model, no appropriate error term exists against which to test the other terms in Table 7–12.

7.6.2   THE $S \times A \times B$ DESIGN, A FIXED AND B RANDOM.   Suppose that each of n subjects is required to read aloud each of b words under each of a treatments, these might be a different print sizes or types of print, or perhaps a different trials. It is desirable to select the words so that they can be reasonably viewed as a random sample from some large population of words. Otherwise, we have no statistical grounds for generalizing our conclusions about the effects of A on reading time beyond the particular set of b words chosen for the experiment. Of course, there are many other types of experiments in which we should like to be able to generalize beyond our particular set of stimuli. The discussion that follows can be applied to many situations—for example, ratings of pictures along some dimension, trials to solve each of several problems, time to comprehend sentences, galvanic skin response to each of a series of words. What reasonably can be viewed as a random set of b stimuli in such experiments is not a simple issue. We postpone discussion of it until Section 7.6.5. Until then, we shall take up the implications for data analysis, assuming that the effects of B are a random sample from some very large population of such effects.

Comparing the two sets of EMS in Table 7–12, it is apparent that this assumption results in the addition of components of variance to most expectations. There are differences between the two sets of expectations because when B has random effects, the sum of interaction effects involving B is no longer zero. Take the specific example of $E(MS_A)$. When B is a fixed-effect variable, $\sum_{k=1}^{b}(\alpha\beta)_{jk} = 0$ at each A_j; when B is a random-effect variable, this is no longer true. Variability among the $\overline{Y}_{.j.}$ will be partly due to variability among the quantities $\sum_{k=1}^{b}(\alpha\beta)_{jk}$. Therefore, σ^2_{AB} contributes to $E(MS_A)$ in this case. Similar arguments hold for other differences between the two sets of EMS in Table 7–12. These matters lead to a useful rule for generating expectations for the designs under consideration. A term (θ^2 or σ^2) will contribute to the EMS for some

Table 7-12 Analysis of variance for two-factor repeated-measurement design

SV	df	SS	EMS A, B FIXED	EMS A FIXED, B RANDOM
Total	$abn-1$	$\sum_i\sum_j\sum_k Y^2_{ijk} - C$		
A	$a-1$	$\dfrac{\sum_i T^2_{.i.}}{nb} - C$	$\sigma_e^2 + b\sigma_{AS}^2 + bn\theta_A^2$	$\sigma_e^2 + b\sigma_{AS}^2 + n\sigma_{AB}^2 + \sigma_{ABS}^2 + bn\theta_A^2$
B	$b-1$	$\dfrac{\sum_k T^2_{..k}}{na} - C$	$\sigma_e^2 + a\sigma_{BS}^2 + an\theta_B^2$	$\sigma_e^2 + a\sigma_{BS}^2 + an\sigma_B^2$
S	$n-1$	$\dfrac{\sum_i T^2_{i..}}{ab} - C$	$\sigma_e^2 + ab\sigma_S^2$	$\sigma_e^2 + a\sigma_{BS}^2 + ab\sigma_S^2$
AB	$(a-1)(b-1)$	$\dfrac{\sum_i\sum_k T^2_{.ik}}{n} - C - SS_A - SS_B$	$\sigma_e^2 + \sigma_{ABS}^2 + n\theta_{AB}^2$	$\sigma_e^2 + \sigma_{ABS}^2 + n\sigma_{AB}^2$
AS	$(a-1)(n-1)$	$\dfrac{\sum_i\sum_i T^2_{ii.}}{b} - C - SS_A - SS_S$	$\sigma_e^2 + b\sigma_{AS}^2$	$\sigma_e^2 + \sigma_{ABS}^2 + b\sigma_{AS}^2$
BS	$(b-1)(n-1)$	$\dfrac{\sum_k\sum_i T^2_{i.k}}{a} - C - SS_B - SS_S$	$\sigma_e^2 + a\sigma_{BS}^2$	$\sigma_e^2 + a\sigma_{BS}^2$
ABS	$(a-1)(b-1)(n-1)$	$SS_{tot} - SS_A - \cdots - SS_{BS}$	$\sigma_e^2 + \sigma_{ABS}^2$	$\sigma_e^2 + \sigma_{ABS}^2$

source if (1) its subscripts include all those that designate the *SV* in question, and (2) none of the remaining subscripts represent a fixed-effect variable. For example, σ^2_{ABS} contributes to MS_{AB}; *ABS* contains *AB* and *S* represents a random-effect variable. However, σ^2_{ABS} does not contribute to MS_B; although *ABS* contains *B*, one of the remaining subscripts, *A*, represents a fixed-effect variable.

The immediate result of all this is that there is no obvious error term against which to test *A*, the variable of primary interest. We require a mean square whose expectation is $\sigma^2_e + b\sigma^2_{AS} + n\sigma^2_{AB} + \sigma^2_{ABS}$. There is no single line with this expectation in Table 7–12. We now turn to some ways of dealing with this awkward problem.

7.6.3 CONSEQUENCES OF ADDITIVITY FOR *F* TESTS.

Let us focus our attention on the *EMS* of Table 7–12 for the case in which *B* is assumed to have random effects. It should be immediately apparent that there is no problem in testing any main effect if there is complete additivity, that is, if it is assumed that there is no variance due to interactions in the population. In that case, the first three mean squares in the table are estimates of σ^2_e plus a null hypothesis term, and the last four terms estimate only σ^2_e. With several terms that have precisely the same expectation, it makes sense to average them to achieve a single, more reliable estimate on more *df* and consequently greater power than any one of them alone would afford. We have already introduced the method of averaging called *pooling* (Section 5.10). In the present example, we obtain the sum of the four *SS* for interaction and divide by the total of the associated *df*.

A complete absence of interaction variability would be unlikely. A more promising possibility is that either σ^2_{AB} or σ^2_{AS} is zero. Suppose that *A* and *B* do not interact. Then we can delete σ^2_{AB} from all *EMS* in Table 7–12, and now MS_{AS} provides an appropriate error term against which to test *A*.

In the developments thus far, we have been able to provide a test of *A* by assuming that one or more interaction components is zero. Unfortunately, wishing will not make it so, and there are dangers inherent in tests of *A* based on the false assumption that certain variables do not interact. For example, if we assume $\sigma^2_{AB} = 0$, and if that assumption is incorrect, the ratio of *EMS* for the test of *A* against *AS* will be

$$\frac{\sigma^2_e + \sigma^2_{ABS} + b\sigma^2_{AS} + n\sigma^2_{AB} + bn\theta^2_A}{\sigma^2_e + \sigma^2_{ABS} + b\sigma^2_{AS}}$$

A significant *F* ratio may reflect the *AB* variance rather than a nonzero value of θ^2_A.

This raises the following issue. On what basis is it reasonable to conclude that one or more components of variance is zero when such a conclusion, if erroneous, may result in a biased *F* test of other null hypotheses? Statisticians do not completely agree on the answer to this question; we propose the following guidelines, however, based on studies by Bozivich, Bancroft, and Hartley (1956), Forster and Dickinson (1976), and Srivastava and Bozivich (1961). We propose that before a variance component is deleted from the *EMS*, there should be prior grounds for believing that the interaction does not contribute to variability, *and* a preliminary F test should fail to be significant at the .25 level. Prior belief in additivity may derive from knowing the experimental situation or from analyzing results of preliminary tests in related experiments. The preliminary test is a test of the interaction null hypothesis. Before deleting σ^2_{AB} from our set of expectations, for example, we should want some a priori reason for believing it to be zero and should want the ratio MS_{AB}/MS_{ABS} to be nonsignificant at the .25 level.

7.6.4 QUASI F RATIOS.

There rarely will be prior grounds for simplifying the design model. The most reasonable expectation usually will be that the effects of treatment variables A will vary over subjects S and stimuli B. Thus, the issue remains: How do we test the hypothesis that $\theta^2_A = 0$ when we have the EMS of Table 7–12 (B random)? The approach generally taken is to calculate a quasi $F(F')$ ratio—a ratio of combinations of mean squares whose expectations are equal under the null hypothesis. One possibility is

(7.55)
$$F'_1 = \frac{MS_A}{MS_{AB} + MS_{AS} - MS_{ABS}}$$

To understand the reasoning, note that the ratio of EMS is

$$\frac{\sigma^2_e + \sigma^2_{ABS} + b\sigma^2_{AS} + n\sigma^2_{AB} \ominus bn\theta^2_A}{\sigma^2_e + \sigma^2_{ABS} + b\sigma^2_{AS} + n\sigma^2_{AB}}$$

which equals one if H_0 ($\theta^2_A = 0$) is true.

Note that this is not a sufficient condition for the ratio to be a proper F statistic. Both numerator and denominator must be at least approximately distributed as χ^2/df. Satterthwaite (1946) has shown that under the usual assumptions of analysis of variance, a linear combination of mean squares has approximately this distribution; the appropriate df for the denominator are

(7.56)
$$df = \frac{(MS_{AB} + MS_{AS} - MS_{ABS})^2}{MS^2_{AB}/df_{AB} + MS^2_{AS}/df_{AS} + MS^2_{ABS}/df_{ABS}}$$

which should be rounded to the nearest integer. Other quasi Fs can be computed. One that has been considered by several statisticians is

(7.57)
$$F'_2 = \frac{MS_A + MS_{ABS}}{MS_{AB} + MS_{AS}}$$

As with F'_1, the ratio of EMS is again 1 under the null hypothesis. The numerator df are

(7.58)
$$df_{\text{num}} = \frac{(MS_A + MS_{ABS})^2}{MS^2_A/df_a + MS^2_{ABS}/df_{ABS}}$$

In general, given a linear *combination of mean squares*, $CMS = MS_1 \pm MS_2 \pm \cdots \pm MS_K$, we get

(7.59)
$$df_{CMS} = \frac{(MS_1 \pm MS_2 \pm \cdots \pm MS_K)^2}{MS^2_1/df_1 + MS^2_2/df_2 + \cdots + MS^2_K/df_K}$$

Which quasi F ratio should be computed? There is not any clear answer to this question yet. The value F_1' has the disadvantage that negative values will occur whenever $MS_{AB} + MS_{AS} < MS_{ABS}$. In addition, Davenport and Webster (1973) found that the true probability of Type I error was closer to the nominal level when F_1' was used and df were small ($a = b = n = 3$) or σ_{AS}^2, σ_{AB}^2, and σ_{ABS}^2 were all small (.01). Both approximations were very poor in the case of small variance. For more df and larger variance components, Davenport and Webster found virtually no difference in either Type I or Type II error rates between the two procedures. Hudson and Krutchkoff (1968) have investigated somewhat different df and σ^2 conditions. They got slightly different results; F_1' generally yielded a result a fraction of a percentage closer to the nominal α level but F_2' had slightly more power to detect false null hypotheses. The power advantage was generally less than 1 percent and was never over 4 percent. A case can be made for F_1' on intuitive grounds. Only one value of df has to be approximated and one could expect the ratio based on a chi square variable and an approximate chi square variable to provide a better approximation to the F ratio than a ratio based on two approximate chi squares. Our own recommendation would be to use F_1' except where it will be negative or where df are all three or less.

How good is the approximation? All the available evidence suggests that it is quite good except when the variance components are very small. In that case, a preliminary test might permit using MS_{AS} or MS_{AB} as the error term (as we suggested in the preceding section). We should qualify this rosy picture. First, "all the available evidence" does not constitute a vast body of research. There are the two papers cited above (Davenport and Webster, Hudson and Krutchkoff) and one other by Forster and Dickinson (1976). Altogether, they have considered relatively few combinations of df and values of interaction variances. Our second reservation arises because the three investigations cited all assume independent observations. We have discussed this assumption in detail earlier and shall only note here that the consequence of violating the independence assumption when using combinations of mean squares is not evident.

One last comment is in order. When some variable besides subjects is to be viewed as a random-effect variable, it is important that we include as many levels of it as of subjects. Interaction of A with both subjects and B contribute to the error variance against which treatment effects are to be evaluated. It is important to have sufficient df associated with both variables to ensure good power. Furthermore, the limited evidence that we have suggests that F' more closely approximates F as a, b, and n increase.

7.6.5   FIXED OR RANDOM EFFECTS?

It should be clear from the presentation to this point that designating variables fixed or random has important implications for both significance testing and parameter estimation, as well as for the degree to which we can generalize our results. Therefore let us look further into classifying effects.

Despite the importance of the decision how effects should be classified, it is not always simple to make. At one extreme, we have variables that should be clearly viewed as fixed in effect. The levels of the variable have been arbitrarily selected for inclusion in the experiment, and because of the way they were selected, there is no basis for viewing them as a random sample from a population of levels. This class includes most treatment variables such as type of drug or amount of reward, and also such characteristics of individuals as clinical category.

At the other extreme, we have random sampling from some large well-defined population. This is rarely realized in practice, and it is therefore difficult to determine to what population of subjects it is appropriate to generalize the conclusions from our current experiment. Can we reasonably view our subjects as a random sample of adults? College students? College sophomores? College sophomores currently enrolled in Introductory Psychology? College sophomores currently enrolled in Introductory Psychology at the university in which the experiment was run? The answer largely depends on the research. In studies of a sensory process, like visual acuity, we might generalize to the population of adults having normal vision. In studies of human learning, we might define our population more narrowly, reserving judgment on whether our conclusions will hold for populations having a markedly different average level of ability from that characterizing those within the institution at which the current experiment was run. When in doubt, we prefer restricting generalizations to the more narrowly defined population.

Even though the population is rarely as well defined as we should like, it should be clear from the preceding comments that we do view subjects as a random-effect variable. Our justification is that subjects are not arbitrarily selected. Other individuals are provided an equal opportunity to participate and might well serve if replications of the experiment were run. Furthermore, those who are subjects are not necessarily more representative of the potential subject pool; there is some likelihood, although it is small, that extreme behaviors will be sampled. This is important because we assume such sampling variability in applying our statistical procedures.

Classifying stimuli such as words or pictures provides greater difficulty. For many experiments, we can argue, on much the same grounds that we presented in discussing subjects, that such stimuli are random samples from a population (perhaps ill-defined) of stimuli. That is, the stimuli are not arbitrarily selected and there are many other stimuli that have an equal opportunity for being chosen under the procedure used. In many other experiments, however, the choice of stimulus is so constrained that it is difficult to imagine a population from which this set of materials is one relatively small sample. In studies involving word pairs, for example, restrictions are often placed on the grammatical class, length in both syllables and letter, familiarity, and number of associates, of each word, and also on the rated similarity, associative strength, and other forms of possible relation between the two words. The experimenter may have great difficulty in meeting those restrictions. Under such conditions, it is not clear that stimuli should be treated as having random effects. Two rough guidelines may prove useful. First, under the constraints extant, could independent investigators produce other samples of stimuli? Second, if we assume that the answer to the first question is positive—there is a reasonably large population available—was there an equal likelihood that all members of that population could be included in the experiment? If this answer also proves positive, it seems reasonable to treat the stimuli as having random effects with all that this implies for our data analyses and the scope of our conclusions.

7.7   CONCLUDING REMARKS

The psychological literature abounds with examples of designs involving repeated measurements—both the relatively simple designs we have described and more complex variations. The reasons for this are not difficult to discern. This design has potentially far greater precision than the designs considered before. Furthermore, the design is natural when the supply of subjects is limited relative to the number of treatment combinations to be studied, or when the experimenter's goal is to collect data on some performance measure plotted as a function of time. In using these designs, however, the experimenter must realize the potential problems. We have been especially concerned with the possible consequences of population interactions between subjects and treatments. We have suggested using alternative F tests when heterogeneity of variances of differences is suspected, an F test to detect nonadditivity, approximate F tests when several variables have random effects, and possible ways of transforming the data to an additive scale. We have also indicated that these techniques will not always suffice; adjusting *df* when heterogeneity is suspected may still not permit a clear inference about the null hypothesis, Tukey's test will not detect all departures from additivity, the properties of quasi F ratios are not completely known, and appropriate transformations will not always be found. These techniques nevertheless provide a starting point for coping with some of the problems that may result from using the repeated-measurement design. More important, awareness of the potential problems is the first requirement for deciding whether to use the design and for intelligently evaluating summary statistics like the F ratio.

When the independent variable is something besides time or trial number, it is important that great care be taken to randomize the order of presentation of treatments independently for each subject. This is done partly to guard against confounding time and treatments. What inference can be drawn in the extreme case in which A_1 is always presented first, A_2 always second, and so on? Proper randomization of the order of presentation is also partly important for minimizing heterogeneity of covariance. Scores for treatments close together in time should be more highly correlated than scores for treatments further apart. By randomizing the treatment presentations independently for each subject, each pair of treatments is given an equal opportunity to appear any given length of time apart.

It is also helpful to provide sufficient time between presentations of treatments to minimize "carry-over" effects. If rats, for example, are being tested in a Skinner box under each of several drugs, time between testings should be long enough to allow the effects of the last drug to wear off. Even if the different orders of presentation balance so that treatments and trials are not confounded, carry-over effects, if present, will result in increased variability among orders of presentation and thus reduce the efficiency of the design.

The design we have examined is the simplest possible design involving repeated measurements. Presently we shall look into complications such as having subjects randomly distributed over levels of a second treatment variable and systematically counterbalancing the orders of presentation of treatments. We now have a relatively simple context in which to consider aspects of many repeated-measurement designs—*EMS* under nonadditivity, implications of nonadditivity, the use of transformations, the implications of heterogeneity of variances of differences, and quasi *F* ratios. With some understanding of these aspects of repeated-measurements design, we can later concentrate more on the actual data analysis.

EXERCISES

7.1 The repeated-measurement design is really a special case of the randomized block design. Suppose that in the design of Chapter 6, the blocks were chosen at random; they might be social groups or litters of rats. On the basis of the discussion of the present chapter, how would the analysis of the treatments × blocks experiment change?

7.2 Apply Tukey's single *df* test to the following data set. Calculate the *F*, then find a transformation to additivity. Calculate Tukey's test for the transformed data.

$$
\begin{array}{c}
\quad\quad A_1 \quad A_2 \quad A_3 \\
\begin{array}{c} S_1 \\ S_2 \\ S_3 \\ S_4 \end{array}
\left[\begin{array}{ccc}
44 & 62 & 72 \\
25 & 40 & 48 \\
22 & 37 & 44 \\
10 & 20 & 25
\end{array}\right]
\end{array}
$$

7.3 Consider the following summary statistics for three sets of data. Which data sets should be sensitive to the Tukey test?

		$\bar{Y}_{i.} - \bar{Y}_{..}$	$\sum_j Y_{ij}(\bar{Y}_{.j} - \bar{Y}_{.1})$
Set 1	S_1	-3	2
	S_2	-1	4
	S_3	1	4
	S_4	3	2
Set 2	S_1	-3	2
	S_2	-1	6
	S_3	1	2
	S_4	3	6
Set 3	S_1	-3	2
	S_2	-1	5
	S_3	1	4
	S_4	3	7

7.4   Prove that $\sum_i(\eta\alpha)_{ij} = 0$ where α is a fixed-effect variable and η is a random-effect variable.

7.5   Compute the *df* adjustment for heterogeneity of covariance for the following data set.

$$
\begin{array}{cccc}
 & A_1 & A_2 & A_3 \\
S_1 & 12 & 14 & 20 \\
S_2 & 9 & 6 & 18 \\
S_3 & 10 & 11 & 23 \\
S_4 & 8 & 10 & 20
\end{array}
$$

7.6   Consider a design in which each of n subjects reads aloud each of b words under each of a experimental conditions. In such cases, the experimenter often calculates the average of the b scores for each combination of subject and experimental condition, then performs an analysis of variance on the cell means, as though there were *an* scores in a subjects × treatments design. What are the consequences of this procedure (a) if words are viewed as having fixed effects? (b) if words are viewed as having random effects?

7.7   Assume that the μ_{ij} and $\mu_{ij'}$ are independently distributed and that $\sigma_1^2 = \cdots = \sigma_j^2 = \cdots = \sigma_a^2 = \sigma^2$ in the subjects × treatments design. Express σ_{SA}^2 and σ_S^2 as functions of σ^2.

7.8   Huynh and Feldt (1970) present the following variance-covariance matrix in their paper. Does it satisfy the condition of homogeneous variances of differences? of homogeneity of variances and covariances (compound symmetry)? Justify your answer. Also calculate the *df* adjustment factor $\hat{\varepsilon}$.

$$
\Sigma = \begin{bmatrix}
1.0 & .5 & 1.5 \\
.5 & 3.0 & 2.5 \\
1.5 & 2.5 & 5.0
\end{bmatrix}
$$

REFERENCES

Anscombe, F. J., and Tukey, J. W. The examination and analysis of residuals. *Technometrics* 5:141–60 (1963).

Box, G. E. P. Problems in the analysis of growth and wear curves. *Biometrics* 6: 362–89 (1950).

Box, G. E. P. Some theories on quadratic forms applied in the study of analysis of variance problems: II. Effects of inequality of variance and covariance between errors in the two-way classification. *Annals of Mathematical Statistics* 35: 484–98 (1954).

Bozivich, H., Bancroft, T. A., and Hartley, H. O. Power of analysis of variance test procedures for certain incompletely specified models. *Annals of Mathematical Statistics* 27: 1017–43 (1956).

Collier, R. O., Baker, F. D., Mandeville, G. K., and Hayes, T. F. Estimates of test size for several test procedures based on conventional variance ratios in the repeated measure design. *Psychometrika* 32: 339–53 (1967).

Davenport, J. M., and Webster, J. T. A comparison of some approximate *F* tests. *Technometrics* 15: 779–89 (1973).

Forster, K. I., and Dickinson, R. G. More on the language as fixed-effect fallacy: Monte Carlo estimates of error rates for F_1, F_2, F', and min *F*. *Journal of Verbal Learning and Verbal Behavior* 15: 135–42 (1976).

Greenhouse, S. W., and Geisser, S. On methods in the analysis of profile data. *Psychometrika* 24: 95–112 (1959).

Hudson, J. D., and Krutchkoff, R. C. A Monte Carlo investigation of the size and power of tests employing Satterthwaite's synthetic mean squares. *Biometrika* 55: 431–33 (1968).

Huynh, H., and Feldt, L. S. Conditions under which mean square ratios in repeated measurements designs have exact *F*-distributions. *Journal of the American Statistical Association* 65: 1582–89 (1970).

Rao, C. R. *Advanced Statistical Methods in Biometric Research.* Canada: John Wiley & Sons, 1952.

Satterthwaite, F. E. An approximate distribution of estimates of variance components. *Biometrics Bulletin* 2: 110–14 (1946).

Srivastava, S. R., and Bozivich, H. Power of certain analysis of variance test procedures involving preliminary tests. *Bulletin de l'Institut international statistique*, 33rd Session (1961).

Stoloff, P. H. An empirical evaluation of the effects of violating the assumption of homogeneity of covariance for the repeated measures design of the analysis of variance. University of Maryland, Technical Report TR–66–28, NSG–398, May 1966.

Tukey, J. W. One degree of freedom for nonadditivity. *Biometrics*, 5: 232–42 (1949).

SUPPLEMENTARY READING

The primary source for our derivation of expected mean squares is Chapter 8 of

Scheffé, H. *The Analysis of Variance.* New York: John Wiley & Sons, 1959.

Personality researchers have been concerned with the relative contributions of individuals and situations to behavioral variability and have frequently used ω^2 in investigating this issue. The following article provides a criticism, and proposal of an alternative measure, also based on variance components.

Golding, S. L. Flies in the ointment: Methodological problems in the analysis of the percentage of variance due to persons and situations. *Psychological Bulletin* 82: 278–88 (1975).

Several investigators have been interested in comparing the effects of a variable when subjects are exposed to all levels (this chapter) with the effects when subjects are tested at only one level of the variable (Chapter 4). A clever solution to the data analysis problem, using quasi *F* ratios, has been given in

Erlebacher, A. Design and analysis of experiments contrasting the within- and between-subjects manipulation of the independent variable. *Psychological Bulletin* 84: 212–19 (1977).

As we have seen, the question when to treat a variable as a random-effect variable is not easily answered. Clark has argued that stimuli in many cognitive psychology experiments have been erroneously treated as fixed, and has presented what he believes to be more appropriate analyses, including quasi *F* tests. Wike and Church wrote a detailed response; and Clark and others wrote comments on that paper, which were presented in a single article. The relevant references are:

Clark, H. H. The language-as-fixed-effect fallacy: A critique of language statistics in psychological research. *Journal of Verbal Learning and Verbal Behavior* 12: 335–39 (1973).

Clark, H. H., Cohen, J., Keppel, G., and Smith, J. F. K. Discussion of Wike and Church's comments. *Journal of Verbal Learning and Verbal Behavior* 15: 257–66 (1976).

Wike, E., and Church, J. Comments on Clark's "The language-as-fixed-effect fallacy." *Journal of Verbal Learning and Verbal Behavior* 15: 249–55 (1976).

8 MIXED DESIGN: BETWEEN-SUBJECTS AND WITHIN-SUBJECT VARIABILITY

8.1 INTRODUCTION

Thus far, we have examined two kinds of designs, those in which different treatments involve different subjects and those in which all subjects are tested under each treatment. The most prevalent design in the psychological literature is a combination of these two approaches. For example, n subjects might be tested at A_1, n other subjects tested at A_2, and so on, until an subjects have been accounted for. The subjects will have been randomly assigned to the a treatments, and thus the design will seem to be a completely randomized one-factor design. Each of the an subjects, however, is also tested at each of the b levels of the independent variable B, and the order of presentation of the b treatments randomized independently for each subject, as in the repeated-measurement design. We shall generally call A a *between-subjects* variable and B a *within-subject* variable. The data matrix for this mixed design is shown in Table 8–1.

It is not difficult to determine why mixed designs are so frequently used in psychological research. One reason is that psychologists are often interested in comparing group performances over time or trials. In a free operant situation, for example, different subjects may be tested at different percentages of reinforcement for several successive minutes. Percentage is the A variable and blocks of time the B variable. In another example, one could measure the signal brightness subjects would need for detecting the signal at various times after entering a darkened room, different groups of subjects having been exposed to different illuminations before the dark-adaptation test. In experiments like these, the interest lies in comparing the average performance over time for the various levels of A (the A main effect); in determining whether, averaging over all subjects, performance is modified over time (the B main effect); and in determining whether the shapes and slopes of the performance curves are similar for the a groups (the AB interaction effect).

A second frequent use of the mixed designs is in psychometric research. Arts, science, engineering, and education majors could be tested on each of several measures in some standard battery of tests. One usually wants to know whether the average performance over the tests is the same for all groups and

Table 8–1   Data matrix for mixed design, one between-subjects and one within-subject variable

		B_1	B_2	$\cdots$	B_k	$\cdots$	B_b
	S_{11}	Y_{111}	Y_{112}		Y_{11k}		Y_{11b}
	S_{21}	Y_{211}	Y_{212}		Y_{21k}		Y_{21b}
A_1	$\vdots$						
	S_{i1}	Y_{i11}	Y_{i12}		Y_{i1k}		Y_{i1b}
	$\vdots$						
	S_{n1}	Y_{n11}	Y_{n12}		Y_{n1k}		Y_{n1b}
$\vdots$							
	S_{1j}	Y_{1j1}	Y_{1j2}		Y_{1jk}		Y_{1jb}
	S_{2j}	Y_{2j2}	Y_{2j2}		Y_{2jk}		Y_{2jb}
A_j	$\vdots$						
	S_{ij}	Y_{ij1}	Y_{ij2}		Y_{ijk}		Y_{ijb}
	$\vdots$						
	S_{nj}	Y_{nj1}	Y_{nj2}		Y_{njk}		Y_{njb}
$\vdots$							
	S_{1a}	Y_{1a1}	Y_{1a2}		Y_{1ak}		Y_{1ab}
	S_{2a}	Y_{2a1}	Y_{2a2}		Y_{2ak}		Y_{2ab}
A_a	$\vdots$						
	S_{ia}	Y_{ia1}	Y_{ia2}		Y_{iak}		Y_{iab}
	$\vdots$						
	S_{na}	Y_{na1}	Y_{na2}		Y_{nak}		Y_{nab}

whether the group profiles are of the same shape; profiles are the bar graphs showing group performance on each measure. Thus, if arts and science majors perform better on a verbal aptitude measure and engineers perform better on a quantitative aptitude measure, the profiles will not be parallel. This variability in the simple effects for each measure will be reflected in the magnitude of the AB interaction. Unless the b measures are on the same scale, the profile analysis (test of interaction) is difficult to interpret. Generally, people working with these measures have used z or T scores; data from different levels of B are then comparable in the sense that the b populations can be considered to have been drawn from the same parent population.

    The mixed designs are also appropriate whenever B and AB effects are of greater interest than A effects, since the B and AB effects will usually be tested against a smaller error term with more df than the A effects will. There will also be instances in which the mixed design is required by the nature of the independent variables; one variable is clearly a between-subjects variable while the second seems to meet the requirements of the repeated-measurement model. Variables that will generally be between-subjects variables are those that will entail carry-over effects (say method of training) and individual characteristics (say age or personality characteristic as evidenced by position on some scale). On the other hand, stage of practice is naturally a within-subject variable.

8.2 ONE BETWEEN-SUBJECTS AND ONE WITHIN-SUBJECT VARIABLE

8.2.1 THE ANALYSIS OF VARIANCE MODEL. Hitherto, we have begun with a detailed statement of the model, relating individual scores to parameters of the population from which they have been sampled. We now have a more complicated design. Consequently, it will be easier to begin by partitioning the variance of the scores in Table 8–1 in a way that appears reasonable in light of our previous experience with designs (Chapters 4 through 7).

One approach is to construct the design so that it looks more familiar. We could ignore the variable A, for example, and then we should have the simple repeated-measurement design of Chapter 7, with an subjects and b levels of the treatment variable. By analogy to Table 7–2, we should have an arrangement like the accompanying chart.

SV	df
S	$an-1$
B	$b-1$
$S \times B$	$(an-1)(b-1)$

This yields the required total of $abn-1$ df. A second look at the data matrix of Table 8–1 suggests that the above analysis is unrealistic; it clearly neglects the variable A, which is presumably of more than passing interest. Since the three terms above account for all the possible df, the A source of variance cannot be an additional term but must be a component of one of the above terms. Table 8–1 suggests the answer; the an subjects may differ because they are at a different levels of A. Thus the SS_S should be partitioned into two components, an A and an S/A term on $a-1$ and $a(n-1)$ df respectively, just as in our treatment of the completely randomized design (Chapter 4). We now have an arrangement as shown.

SV	df
S	$an-1$
A	$a-1$
S/A	$a(n-1)$
B	$b-1$
$S \times B$	$(an-1)(b-1)$

Returning to Table 8–1, we see that the design could be viewed as a two-factor (A and B) design, with n scores in each cell; our analysis thus far, however, has failed to take into account any possible AB interaction. The necessary $(a-1)(b-1)$ df can be extracted only from the SB term. We note that

$$(an-1)(b-1) = (a-1)(b-1) + a(n-1)(b-1)$$

This suggests that

$$SS_{SB} = SS_{AB} + SS_{SB/A}$$

Table 8–2 contains the complete analysis. The terms A, S/A, B, AB, and SB/A are the ones of interest; the Between S and Within S terms represent intermediate steps designed to ease the calculations.

The foregoing discussion of the partitioning of the total variability should help us to understand the model now presented. We assume that the a groups of n subjects are random samples from a corresponding treatment populations consisting of infinite numbers of subjects. This is the exact approach of Chapters 4 and 5. We assume (as in Chapter 7) that Y_{ijk}, the observed score for the ijth subject (or the ith subject at the jth level of A) under treatment B_k is randomly sampled from an infinite population of independent measurements taken on that subject under B_k; the expected value of this hypothetical population is μ_{ijk}. Other relevant parameters are μ_{ij}, the expected value of the total population of scores for the ijth subject; μ_j, the expected value of the total population of scores under treatment A_j; μ_k, the expected value for the total population of scores under B_k; and μ_{jk}, the expected value for the total population of scores under A_j and B_k. The structural model is

(8.1)
$$Y_{ijk} = \mu + \alpha_j + \eta_{i/j} + \beta_k + (\alpha\beta)_{jk} + (\eta\beta)_{ik/j} + \varepsilon_{ijk}$$

where α_j, β_k, and $(\alpha\beta)_{jk}$ are defined as previously, and

$\eta_{i/j} = \mu_{ij} - \mu_j$, a measure of the unique contribution of the ijth subject

$$\begin{aligned}(\eta\beta)_{ik/j} &= (\mu_{ijk} - \mu) - (\mu_k - \mu) - (\mu_{ij} - \mu) - (\alpha\beta)_{jk}\\ &= \mu_{ijk} - \mu_{ij} - \mu_{jk} + \mu_j,\end{aligned}$$ a measure of the interaction effect associated with the ijth subject and kth level of B, and adjusted for the interaction of A_j and B_k

$\varepsilon_{ijk} = Y_{ijk} - \mu_{ijk}$, the error of measurement.

Several statements about the population parameters follow, in which it is assumed that A and B are variables having fixed effects and that subjects are randomly sampled from a large population of subjects. With regard to fixed effects, $\sum \alpha_j = 0$, $\sum \beta_k = 0$, and $\sum (\alpha\beta)_{jk} = 0$. It is further assumed that the $\eta_{i/j}$, $(\eta\beta)_{ik/j}$, and ε_{ijk} are all randomly sampled from normally distributed populations with mean zero and variances $\sigma^2_{S/A}$, $\sigma^2_{SB/A}$, and σ^2_e, respectively.

8.2.2   THE ANALYSIS OF VARIANCE.   Table 8–2 presents the SV, df, SS, EMS, and F (MS have been omitted; as usual, they are simply ratios of SS to df). The SV are consistent with Equation (8.1), with one source for each term in the equation except ε_{ijk}. Omitting an independent error term is consistent with the analyses of Chapter 7. Since there is only a single score in each combination of S, A, and B, no within-cell variability exists, and the sources of Table 8–2 therefore account for the total variability in the data matrix. For convenience, we have grouped the sources into two sets, those that account for the variability between subjects and those that account for the variability within a subject. Within the first set, we have A and S/A, corresponding to α_j and $\eta_{i/j}$ effects. Within the second set of terms, we have B, AB, and SB/A, corresponding to β_k, $(\alpha\beta)_{jk}$, and $(\eta\beta)_{ik/j}$.

The nature of the correspondence will become clearer shortly when we turn to the *EMS*.

Scanning the *SV*, you may wonder why there is no *SA* term present. The answer lies in the distinction between *crossing* and *nesting*. When data are obtained for all combinations of two variables, the variables, it is said, cross. In this case, an interaction sum of squares can be computed for the two variables, since the question, Is the difference in the effects of *A* a function of the level of *B*?, has meaning. Subjects and *B*, and *A* and *B*, cross in the design under discussion.

Scores cannot be obtained for all combinations of subjects and levels of *A*, since any given subject appears in combination with only one level of *A*. The question of interaction is meaningless. Consider asking whether the difference between Subject 1 and Subject 2 is greater at A_1 than at A_2. In our design, both subjects appear only at A_1, or both appear only at A_2, or one appears only at A_1 while the other appears only at A_2. In any of these cases, the question posed above has no answer. What we have instead of crossing is the nesting of subjects within levels of *A*—there are *n* subjects at A_1, *n* others at A_2, and so on.

How do we interpret the nested terms? It is as though the analysis of variance were carried out separately at each level of *A* and the resulting sums of squares were then pooled. Thus, $SS_{S/A}$ could be obtained by computing the variability for each set of *n* subjects' means about their mean and summing the *a* terms so obtained. Equivalently, it is that variability among subjects that remains after we remove the variability due to the fact that subjects are at different levels of *A*. Similarly, we could obtain $SS_{SB/A}$ by computing an *SB* interaction within each level of *A* and then summing the resulting *a* terms. Or we can view $SS_{SB/A}$ as the interaction variability for subjects and *B* that remains after we remove the contribution due to the interaction variability of *A* and *B*.

Next turn to the *df* column of Table 8–2. No discussion of the entries for *A*, *B*, or *AB* seems necessary; their rationale has been previously considered. The *df* for the *between-subjects* source reflect the variability of *an* means about the grand mean. The $df_{S/A}$ can be calculated as a residual:

$$df_{S/A} = df_{B.S} - df_A$$
$$a(n-1) = (an-1) - (a-1)$$

or we can note that at each level of *A*, the variability of subject means about their mean is based on $n-1$ *df*. Pooling over levels of *A* gives the appropriate result.

The *within-subject* variability is obtained by computing for each subject the variability of his *b* measures about their mean and then pooling over subjects. Therefore, we have $b-1$ *df* for each subject, and since there are *an* subjects, pooling results in $an(b-1)$ *df*. Alternatively,

$$df_{W.S} = df_{tot} - df_{B.S}$$
$$an(b-1) = (abn-1) - (an-1)$$

The $df_{SB/A}$ can be computed in several ways. It is a residual from the

Table 8–2 Analysis of variance for mixed design, one between-subjects and one within-subject variable

SV	df	SS	EMS	F
Total	$abn-1$	$\sum_i^n \sum_j^a \sum_k^b Y_{ijk}^2 - C$		
Between S	$an-1$	$\dfrac{\sum_i^n \sum_j^a T_{ij.}^2}{b} - C$		
A	$a-1$	$\dfrac{\sum_j^a T_{.j.}^2}{nb} - C$	$\sigma_e^2 + b\sigma_{S/A}^2 + nb\theta_A^2$	$\dfrac{MS_A}{MS_{S/A}}$
S/A	$a(n-1)$	$SS_{B.S} - SS_A$	$\sigma_e^2 + b\sigma_{S/A}^2$	
Within S	$an(b-1)$	$SS_{tot} - SS_{B.S}$		
B	$b-1$	$\dfrac{\sum_k^b T_{..k}^2}{na} - C$	$\sigma_e^2 + \sigma_{SB/A}^2 + na\theta_B^2$	$\dfrac{MS_B}{MS_{SB/A}}$
AB	$(a-1)(b-1)$	$\dfrac{\sum_j^a \sum_k^b T_{.jk}^2}{n} - C - SS_A - SS_B$	$\sigma_e^2 + \sigma_{SB/A}^2 + n\theta_{AB}^2$	$\dfrac{MS_{AB}}{MS_{SB/A}}$
SB/A	$a(n-1)(b-1)$	$SS_{W.S} - SS_B - SS_{AB}$	$\sigma_e^2 + \sigma_{SB/A}^2$	

within-subject variability:

$$df_{SB/A} = df_{W.S} - df_B - df_{AB}$$
$$= an(b-1) - (b-1) - (a-1)(b-1)$$
$$= a(n-1)(b-1)$$

The $df_{SB/A}$ reflects the difference between the overall interaction of subjects and levels of B (disregarding the presence of the variable A) and the AB interaction:

$$df_{SB/A} = df_{SB} - df_{AB}$$
$$= (an-1)(b-1) - (a-1)(b-1)$$
$$= a(n-1)(b-1)$$

The $df_{SB/A}$ reflects the SB variability at each level of A, pooled over levels of A:

$$df_{SB/A} = (n-1)(b-1) + \cdots + (n-1)(b-1)$$
$$= a(n-1)(b-1)$$

The computational formulas for sums of squares present no difficulties. They follow the same logic just developed for *df*. In those cases for which several approaches to the calculation of *df* (say *SB/A*) have been presented, the simplest one has been used as a basis for computing sums of squares. An alternative calculation for $SS_{SB/A}$ will be suggested to prepare the reader for subsequent designs in which similar nested terms cannot always be calculated as simple residual quantities. The alternative approach rests on the expansion of *df* proposed earlier. We begin with

$$df_{SB/A} = a(n-1)(b-1)$$
$$= anb - an - ab + a$$

We require a squared quantity for each *df*; therefore we have

$$\sum_i^n \sum_j^a \sum_k^b (\quad)^2 - \sum_i^n \sum_j^a (\quad)^2 - \sum_j^a \sum_k^b (\quad)^2 + \sum_j^a (\quad)^2$$

Since all indices of summation must be represented, we now have

$$\sum_i^n \sum_j^a \sum_k^b (Y_{ijk})^2 - \sum_i^n \sum_j^a (T_{ij.})^2 - \sum_j^a \sum_k^b (T_{.jk})^2 + \sum_j^a (T_{.j.})^2$$

Since each squared quantity must be divided by the number of scores summed before squaring, the final result is

(8.2)
$$S_{SB/A} = \sum_i^n \sum_j^a \sum_k^b Y_{ijk}^2 - \frac{\sum_i^n \sum_j^a T_{ij.}^2}{b} - \frac{\sum_j^a \sum_k^b T_{.jk}^2}{n} + \frac{\sum_j^a T_{.j.}^2}{nb}$$

Note that each of the four component terms has been computed previously as part of some other sum-of-squares quantity. For example, $\sum_i^n \sum_j^a T_{ij.}^2/b$ is calculated in obtaining $SS_{B.S}$. Where do the other three terms appear among other sum-of-squares entries?

The rationale for the *EMS* is essentially that of Chapter 7 for the nonadditive repeated-measurement model. The designs are now becoming so complex, however, that some rules for generating *EMS* would seem helpful. The following will apply to designs involving fixed effects, random effects, or both. The chief stipulation in using these rules is that an infinite number of levels are assumed in populations from which a random sample is obtained. This assumption will be approximately correct in most psychological research.

RULES OF THUMB FOR GENERATING *EMS*

RULE 1.   *Decide for each independent variable (including subjects) whether it is fixed or random. Assign a letter to designate each variable. Assign another letter to be used as a coefficient that represents the number of levels of each variable. In the example of Table 8–2, the variables are designated A, B, and S; the coefficients are a, b, and n; A and B are fixed-effect variables while S is random.*

RULE 2.   *List σ_e^2 as part of each EMS.*

RULE 3.   *For each EMS, list the null hypothesis component, that is, the component corresponding directly to the SV under consideration. Thus we add $nb\theta_A^2$ to the EMS for the A line, $b\sigma_{S/A}^2$ to the EMS for the S/A line. Note that a component consists of three parts:*

1. *A coefficient representing the number of scores at each level of the effect (for example, nb scores at each level of A, or b scores for each subject).*

2. *A* σ^2 *or* θ^2, *depending on whether the effect is assumed to be random or fixed* [σ^2 *is the variance of the population of effects; for example,* $\sigma^2_{S/A} = E(\eta^2_{i/j})$, $\theta^2_A = \sum_j \alpha^2_j/(a-1)$].

3. *As subscripts, those letters that designate the effect under consideration.*

RULE 4.   *Now add to each EMS all components whose subscripts contain all the letters designating the SV in question. Since the subscript SB/A contains the letters S and A, for example, add* $\sigma^2_{SB/A}$ *to the EMS for the S/A line (this is later deleted according to Rule 6).*

RULE 5.   *Next, examine the components for each SV. If a slash appears in the subscript, define only the letters to the left of the slash as "essential." If there are several slashes (as in the next chapter), only the letters preceding the leftmost slash are essential. If there is no slash in the subscript, all letters are considered essential.*

RULE 6.   *Among the essential letters, ignore any that are necessary to designate the SV. If the source is A, in considering* $n\theta^2_{AB}$, *for example, ignore the A. If the source is S/A, in considering the* $\sigma^2_{SB/A}$ *component, S and B are essential subscripts and S is to be ignored. If any of the remaining essential letters designate fixed variables, delete the entire component from the EMS. Thus, in the preceding examples, since B represents a fixed variable,* $n\theta^2_{AB}$ *does not contribute to the EMS for A and* $\sigma^2_{SB/A}$ *does not contribute to the EMS for S/A.*

In Table 8–3 are listed the *EMS* as they would appear after Rules 1 to 4 had been carried out. The underlined components are those deleted on the basis of Rule 6. If the underlined components are erased, we have the results in the *EMS* column of Table 8–2.

With the *EMS* available, the *F* tests readily follow. As always, error terms are required so that the ratio of *EMS* equals 1 when the null hypothesis is true. Such error terms are possible for tests of *A*, *B*, and *AB* effects. One word of caution about the tests of *B* and *AB* effects; heterogeneity of covariance can positively bias the test, as in the simpler repeated-measurement designs (Chapter 7). We shall consider this problem in more detail (Section 8.4) after several more mixed designs.

8.2.3   A NUMERICAL EXAMPLE.   An illustrative set of data is given in Table 8–4. Running only two subjects at each level of *A* is not generally recommended; we hope you will use this example as a model for computation rather than for experimentation.

Before the actual analysis of variance, subtotals are obtained for each subject, for each level of *A*, for each level of *B*, and for each *AB* combination. Since all these quantities are eventually used in the analysis, it is wise to have them available from the start.

We first calculate the correction term. As usual, this is the squared total of all scores divided by the total number of observations. In this case,

$$C = \frac{(99)^2}{12} = 816.75$$

Table 8–3 *EMS* for mixed design, one between-subjects and one within-subject variable

SV	EMS		
A	$\sigma_e^2 + nb\theta_A^2$	$+ \underline{n\theta_{AB}^2 + b\sigma_{S/A}^2} + \sigma_{SB/A}^2$	
S/A	σ_e^2	$+ b\sigma_{S/A}^2 + \sigma_{SB/A}^2$	
B	σ_e^2	$+ na\theta_B^2 + \underline{n\theta_{AB}^2}$	$+ \sigma_{SB/A}^2$
AB	σ_e^2	$+ n\theta_{AB}^2$	$+ \sigma_{SB/A}^2$
SB/A	σ_e^2		$+ \sigma_{SB/A}^2$

Note: Underscored terms are not part of the final *EMS*.

The SS_{tot} is obtained by squaring each score, summing the squared scores and subtracting C. Thus, we have

$$SS_{tot} = \sum_i^n \sum_j^a \sum_k^b Y_{ijk}^2 - C$$
$$= (7)^2 + (1)^2 + \cdots + (14)^2 + (9)^2 - C$$
$$= 1,023.00 - 816.75$$
$$= 206.25$$

We next turn to the first principal component of the SS_{tot}, the $SS_{B.S}$:

$$SS_{B.S} = \frac{\sum_i^n \sum_j^a T_{ij.}^2}{b} - C$$
$$= \frac{(15)^2 + (21)^2 + (24)^2 + (39)^2}{3} - C$$
$$= \frac{2,763}{3} - 816.75$$
$$= 104.25$$

Table 8–4 Data for a numerical example, one between-subjects and one within-subject variable

		B_1	B_2	B_3	$T_{ij.}$
A_1	S_{11}	7	1	7	15
	S_{21}	9	2	10	21
	$T_{.1k} = 16$		3	17	$T_{.1.} = 36$
A_2	S_{12}	11	6	7	24
	S_{22}	16	14	9	39
	$T_{.2k} = 27$		20	16	$T_{.2.} = 63$
	$T_{..k} = 43$		23	33	$T_{...} = 99$

The $SS_{B.S}$ are then partitioned:

$$SS_A = \frac{\sum_j^a T_{.j.}^2}{nb} - C$$

$$= \frac{(36)^2 + (63)^2}{6} - C$$

$$= \frac{5,265}{6} - 816.75$$

$$= 60.75$$

and

$$SS_{S/A} = SS_{B.S} - SS_A$$

$$= 104.25 - 60.75$$

$$= 43.50$$

The second principal component of the SS_{tot} is the $SS_{w.s}$. This is simply the difference between the SS_{tot} and $SS_{B.S}$. Therefore,

$$SS_{w.s} = SS_{tot} - SS_{B.S}$$

$$= 206.25 - 104.25$$

$$= 102.00$$

The within-subject variability is now analyzed into its components:

$$SS_B = \frac{\sum_k^b T_{..k}^2}{na} - C$$

$$= \frac{(43)^2 + (23)^2 + (33)^2}{4} - C$$

$$= \frac{3,467}{4} - 816.75$$

$$= 50.00$$

and

$$SS_{AB} = \frac{\sum_j^a \sum_k^b T_{.jk}^2}{n} - C - SS_A - SS_B$$

$$= \frac{(16)^2 + (3)^2 + \cdots + (16)^2}{2} - C - SS_A - SS_B$$

$$= \frac{1,939}{2} - 816.75 - 60.75 - 50.00$$

$$= 42.00$$

and

$$SS_{SB/A} = SS_{w.s} - SS_B - SS_{AB}$$

$$= 102.00 - 50.00 - 42.00$$

$$= 10.00$$

Table 8–5 Analysis of variance for data of Table 8–4

SV	df	SS	MS	F
Total	11	206.25		
Between S	3	104.25		
A	1	60.75	60.75	2.79
S/A	2	43.50	21.75	
Within S	8	102.00		
B	2	50.00	25.00	10.00[a]
AB	2	42.00	21.00	8.40[a]
SB/A	4	10.00	2.50	

[a] $p < .05$

Table 8–5 summarizes the analysis. Assuming that our α level is .05, both the B and AB effects are significant. To instill some meaning into this last statement, we turn to Figure 8–1, which contains a plot of the six cell means. The dashed line represents the main effect of B. The source of the significant B effect is due to the roughly **V**-shaped function obtained when the average performance is plotted against the levels of B. If high scores are desirable, it seems that B_1 is the preferred treatment. It is tempting to go somewhat further and more precisely conclude that B_1 is better than both B_2 and B_3, and that B_3 is better than B_2; our F test only permits the inference that at least one of these three contrasts involves

Figure 8–1 A plot of cell means of Table 8–4

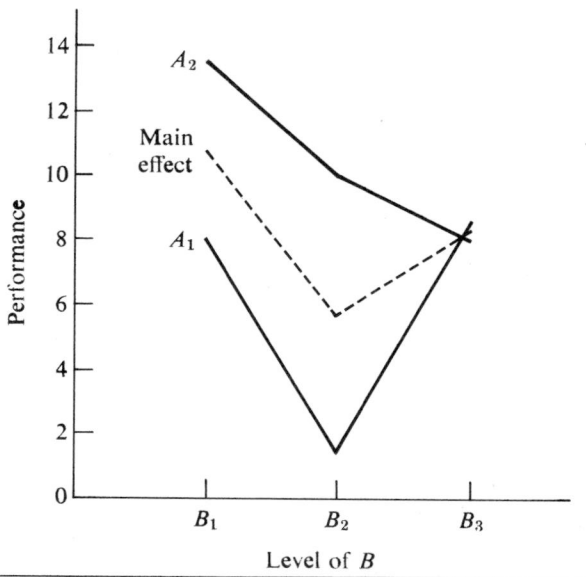

a difference in the population, however. Presumably it is safe to assume that B_1 and B_2 differ in their effects, since the difference in their means is the greatest. (In Chapter 11 the set of three contrasts will be evaluated as well as others that are not quite so obvious.)

It has just been noted that inferences about the population effects of B are limited until further comparisons can be made within pairs of means. Our conclusions about B are now further qualified by noting that the AB interaction is significant. The simple effects of B at each level of A are not identical to the main effect. It is not difficult to detect the source of the AB interaction in Figure 8–1. The B means, when plotted at A_1, form a symmetrical **V**-shaped function; when the means are plotted for the A_2 data, the function decreases monotonically as the level of B increases.

Testing simple effects is straightforward. If we want to test the effects of A at any single level of B, the appropriate error term is based on the scores at that level of B. In short, we calculate our mean squares as though we had a one-factor (A) completely randomized design, using only the data for the level of B in which we are interested. To test the effects of B at any level of A, we could consider only the data for that group; our error term would be MS_{SB/A_j}, subjects $\times B$ at A_j. Somewhat more power would be achieved by using SB/A, the error term for the overall B main effect, as the error term against which to evaluate the simple effect of B at A_j; this term is on $a(b-1)(n-1)$ df rather than $(b-1)(n-1)$ df. This approach should be used only if the SB interaction variance is stable over the levels of A.

8.3   ADDITIONAL MIXED DESIGNS

Countless variations of the design in the preceding section have appeared in the experimental journals. They should present no insurmountable difficulties. The developments follow those of Section 8.2; to investigate this, let us look into two additional mixed designs.

8.3.1   TWO BETWEEN-SUBJECTS AND ONE WITHIN-SUBJECT VARIABLE.   Table 8–6 presents the data matrix for the design to be considered next. A random sample from a large population of subjects has been randomly distributed among the ab combinations of levels of the variables A and B, with the restriction that there be n subjects in each combination. Each subject is tested at c levels of the variable C, and the order of presentation of the c treatments is independently randomized for each subject. The notation used in Table 8–6, and throughout this section, will be as follows:

$$i = 1, 2, \ldots, n$$
$$j = 1, 2, \ldots, a$$
$$k = 1, 2, \ldots, b$$
$$m = 1, 2, \ldots, c$$

Table 8–6 Data matrix for mixed design, two between-subjects variables and one within-subject

			C_1	$\cdots$	C_m	$\cdots$	C_c
		S_{111}	Y_{1111}		Y_{111m}		Y_{111c}
	A_1	$\vdots$					
		S_{i11}	Y_{i111}		Y_{i11m}		Y_{i11c}
		$\vdots$					
		S_{n11}	Y_{n111}		Y_{n11m}		Y_{n11c}
B_1	A_j	S_{ij1}	Y_{ij11}		Y_{ij1m}		Y_{ij1c}
	A_a	S_{ia1}	Y_{ia11}		Y_{ia1m}		Y_{ia1c}
B_k	A_j	S_{ijk}	Y_{ijk1}		Y_{ijkm}		Y_{ijkc}
B_b	A_j	S_{ijb}	Y_{ijb1}		Y_{ijbm}		Y_{ijbc}

Developing the model. As in Section 8.2.1, we initially view the design as consisting of two factors: abn subjects and c levels of the within-subject variable C. This suggests the accompanying chart.

SV	df
Between S	$abn-1$
C	$c-1$
Between $S \times C$	$(abn-1)(c-1)$

which accounts for the total df, $abcn-1$. We next consider the main and interaction effects involving A and B, the between-subjects variables. These sources are extracted from the between-subjects variability, yielding the arrangement as shown.

SV	df
Between S	$abn-1$
A	$a-1$
B	$b-1$
AB	$(a-1)(b-1)$
S/AB	$ab(n-1)$
C	$c-1$
Between $S \times C$	$(abn-1)(c-1)$

This breakdown is still incomplete; it is clear from Table 8–6 that A and B, as well as subjects, cross with C and that such interaction terms must be taken into account. If we consider the interaction of each of the four between-subjects terms with C, we get four additional terms: AC, BC, ABC, and SC/AB. The corresponding df are

$$(a-1)(c-1) \qquad (b-1)(c-1) \qquad (a-1)(b-1)(c-1)$$

and

$$ab(n-1)(c-1)$$

which together add to $(abn-1)(c-1)$, the df from the Between $S \times C$ term. Replacing that term by its four components, we get the final result, Table 8–7.

A model consistent with the above is provided by

$$(\textbf{8.3}) \quad Y_{ijkm} = \mu + \alpha_j + \beta_k + (\alpha\beta)_{jk} + \eta_{i/jk} + \gamma_m + (\alpha\gamma)_{jm} + (\beta\gamma)_{km}$$
$$+ (\alpha\beta\gamma)_{jkm} + (\eta\gamma)_{im/jk} + \varepsilon_{ijkm}$$

where α_j, β_k, and γ_m are the main effects associated with A_j, B_k, and C_m, their interaction effects are defined as usual, and $\eta_{i/jk}$ is the unique contribution associated with the ith subject in A_jB_k; also $(\eta\gamma)_{im/jk}$ is the nested subject $\times C$ interaction effect, and ε_{ijkm} is the error of measurement associated with sampling Y_{ijkm}. It is assumed that A, B, and C are fixed-effect variables; this, as usual, implies that the sum of effects over the appropriate indices is zero. The ε_{ijkm}, $\eta_{i/jk}$, and $(\eta\gamma)_{im/jk}$ effects are randomly sampled from infinite populations of such effects. The population distributions are normal, with zero expected value and with variances σ_e^2, $\sigma_{S/AB}^2$, and $\sigma_{SC/AB}^2$, respectively.

The analysis of variance. The analysis of variance for the two between-subjects and one within-subject design is summarized in Table 8–6. The SV are consistent with Equation (8.3). The df are derived as in Section 8.2.2. The df_{tot}, as usual, are one less than the total number of scores, which is $abcn$ in the present case. This quantity is partitioned into two parts, the $df_{\text{B.S}}$, which is one less than the total number of subjects, and $df_{\text{W.S}}$, which is the df for each subject $(c-1)$ pooled over the number of subjects. The df for all treatment main and interaction effects are identical to those presented for previous designs, and the df for the two

Table 8–7 Analysis of variance for mixed design, two between-subjects variables and one within-subject

SV	df	SS	EMS	F
Total	$abnc - 1$	$\sum_i^a \sum_j^b \sum_k^c \sum_m^n Y_{ijkm}^2 - C$		
Between S	$abn - 1$	$\dfrac{\sum_i^n \sum_j^a \sum_k^b T_{ijk.}^2}{c} - C$		
A	$a - 1$	$\dfrac{\sum_j^a T_{.j..}^2}{nbc} - C$	$\sigma_e^2 + c\sigma_{S/AB}^2 + nbc\theta_A^2$	$\dfrac{MS_A}{MS_{S/AB}}$
B	$b - 1$	$\dfrac{\sum_k^b T_{..k.}^2}{nac} - C$	$\sigma_e^2 + c\sigma_{S/AB}^2 + nac\theta_B^2$	$\dfrac{MS_B}{MS_{S/AB}}$
AB	$(a-1)(b-1)$	$\dfrac{\sum_j^a \sum_k^b T_{.jk.}^2}{nc} - C - SS_A - SS_B$	$\sigma_e^2 + c\sigma_{S/AB}^2 + nc\theta_{AB}^2$	$\dfrac{MS_{AB}}{MS_{S/AB}}$
S/AB	$ab(n-1)$	$SS_{B.S} - SS_A - SS_B - SS_{AB}$	$\sigma_e^2 + c\sigma_{S/AB}^2$	
Within S	$abn(c-1)$	$SS_{tot} - SS_{B.S}$		
C	$(c-1)$	$\dfrac{\sum_m^c T_{...m}^2}{nab} - C$	$\sigma_e^2 + \sigma_{SC/AB}^2 + nba\theta_C^2$	$\dfrac{MS_C}{MS_{SC/AB}}$
AC	$(a-1)(c-1)$	$\dfrac{\sum_j^a \sum_m^c T_{.j.m}^2}{nb} - C - SS_A - SS_C$	$\sigma_e^2 + \sigma_{SC/AB}^2 + nb\theta_{AC}^2$	$\dfrac{MS_{AC}}{MS_{SC/AB}}$
BC	$(b-1)(c-1)$	$\dfrac{\sum_k^b \sum_m^c T_{..km}^2}{na} - C - SS_B - SS_C$	$\sigma_e^2 + \sigma_{SC/AB}^2 + na\theta_{BC}^2$	$\dfrac{MS_{BC}}{MS_{SC/AB}}$
ABC	$(a-1)(b-1)(c-1)$	$\dfrac{\sum_j^a \sum_k^b \sum_m^c T_{.jkm}^2}{n} - C - SS_A - SS_B - SS_C$ $- SS_{BC} - SS_C - SS_{AC} - SS_{BC} - SS_{ABC}$	$\sigma_e^2 + \sigma_{SC/AB}^2 + n\theta_{ABC}^2$	$\dfrac{MS_{ABC}}{MS_{SC/AB}}$
SC/AB	$ab(n-1)(c-1)$	$SS_{W.S} - SS_C - SS_{AC} - SS_{BC} - SS_{ABC}$	$\sigma_e^2 + \sigma_{SC/AB}^2$	

Table 8–8   Data for numerical example, two between-subjects and one within-subject design

			C_1	C_2	C_3	$T_{ijk.}$
A_1	B_1	S_{111}	4	8	10	22
		S_{211}	6	9	12	27
		$T_{.j1m} = 10$		17	22	$T_{.11.} = 49$
	B_2	S_{112}	3	7	11	21
		S_{212}	5	11	12	28
		$T_{.j2m} = 8$		18	23	$T_{.12.} = 49$
		$T_{.j.m} = 18$		35	45	$T_{.1..} = 98$
A_2	B_1	S_{121}	4	6	9	19
		S_{221}	5	8	8	21
		$T_{.21m} = 9$		14	17	$T_{.21.} = 40$
	B_2	S_{122}	4	3	9	16
		S_{222}	1	6	8	15
		$T_{.22m} = 5$		9	17	$T_{.22.} = 31$
		$T_{.2.m} = 14$		23	34	$T_{.2..} = 71$
		$T_{...m} = 32$		58	79	$T_{....} = 169$

Subtotals for *BC* Cells

	C_1	C_2	C_3	$T_{..k.}$
B_1	19	31	39	89
B_2	13	27	40	80
$T_{...m} = 32$		58	79	$T_{....} = 169$

nested effects can be readily obtained as residuals. Computational formulas for sums of squares all follow the precedents set in the preceding section and in preceding chapters. The results of applying the rules of thumb for generating *EMS* are similar to those achieved in Section 8.2; we again obtain two error terms, one for between-subjects and one for within-subject effects.

A numerical example. An analysis of variance on the data of Table 8–8 will show the computation. We first compute

$$SS_{\text{tot}} = \sum_{i}^{n} \sum_{j}^{a} \sum_{k}^{b} \sum_{m}^{c} Y_{ijkm}^2 - C$$

$$= (4)^2 + (8)^2 + \cdots + (6)^2 + (8)^2 - \frac{(169)^2}{24}$$

$$= 1{,}403.000 - 1{,}190.042$$

$$= 212.958$$

The total variability among subjects is

$$SS_{\text{B.S}} = \frac{\sum_i^n \sum_j^a \sum_k^b T_{ijk.}^2}{c} - C$$

$$= \frac{(22)^2 + (27)^2 + \cdots + (15)^2}{3} - \frac{(169)^2}{24}$$

$$= 3{,}721.000 - 1{,}190.042$$

$$= 50.291$$

This variability can be partitioned into several components:

$$SS_A = \frac{\sum_j^a T_{.j..}^2}{nbc} - C$$

$$= \frac{(98)^2 + (71)^2}{12} - \frac{(169)^2}{24}$$

$$= 1{,}220.417 - 1{,}190.042$$

$$= 30.375$$

$$SS_B = \frac{\sum_k^b T_{..k.}^2}{nac} - C$$

$$= \frac{(89)^2 + (80)^2}{12} - \frac{(169)^2}{24}$$

$$= 1{,}193.417 - 1{,}190.042$$

$$= 3.375$$

$$SS_{AB} = \frac{\sum_j^a \sum_k^b T_{.jk.}^2}{nc} - C - SS_A - SS_B$$

$$= \frac{(49)^2 + (49)^2 + (40)^2 + (31)^2}{6} - \frac{(169)^2}{24} - SS_A - SS_B$$

$$= 1{,}227.167 - 1{,}190.042 - 30.375 - 3.375$$

$$= 3.375$$

$$SS_{S/AB} = SS_{\text{B.S}} - SS_A - SS_B - SS_{AB}$$

$$= 50.291 - 30.375 - 3.375 - 3.375$$

$$= 13.166$$

Note that the *A*, *B*, and *AB* terms can be swiftly computed by the single *df* approach first presented in Section 5.11. Thus,

$$SS_A = \frac{(98-71)^2}{24} = 30.375$$

$$SS_B = \frac{(89-80)^2}{24} = 3.375$$

$$SS_{AB} = \frac{(49-49-40+31)^2}{24} = 3.375$$

The within-subject variability can be obtained as the difference between the total and the between-subjects variability:

$$SS_{W.S} = SS_{tot} - SS_{B.S}$$

$$= 212.958 - 50.291$$

$$= 162.667$$

This term is next partitioned into its components. We have

$$SS_C = \frac{\sum_m^c T^2_{...m}}{nab} - C$$

$$= \frac{(32)^2+(58)^2+(79)^2}{8} - \frac{(169)^2}{24}$$

$$= 1{,}328.625 - 1{,}190.042$$

$$= 138.583$$

$$SS_{AC} = \frac{\sum_j^a \sum_m^c T^2_{.j.m}}{nb} - C - SS_A - SS_C$$

$$= \frac{(18)^2+(35)^2+\cdots+(34)^2}{4} - \frac{(169)^2}{24} - SS_A - SS_C$$

$$= 1{,}363.750 - 1{,}190.042 - 30.375 - 138.583$$

$$= 4.750$$

$$SS_{BC} = \frac{\sum_k^b \sum_m^c T^2_{..km}}{na} - C - SS_B - SS_C$$

$$= \frac{(19)^2+(31)^2+\cdots+(40)^2}{4} - \frac{(169)^2}{24} - SS_B - SS_C$$

$$= 1{,}335.250 - 1{,}190.042 - 3.375 - 138.583$$

$$= 3.250$$

$$SS_{ABC} = \frac{\sum_j^a \sum_k^b \sum_m^c T_{\cdot jkm}^2}{n} - C - SS_{AB} - SS_{AC} - SS_{BC} - SS_A - SS_B - SS_C$$

$$= \frac{(10)^2 + (17)^2 + \cdots + (9)^2 + (17)^2}{2} - \frac{(169)^2}{24} - SS_A - SS_B$$

$$- SS_C - SS_{AB} - SS_{AC} - SS_{BC}$$

$$= 1{,}375.500 - 1{,}190.042 - 30.375 - 3.375 - 138.583 - 3.375$$

$$- 4.750 - 3.250$$

$$= 1.750$$

$$SS_{SC/AB} = SS_{\text{w.s}} - SS_C - SS_{AC} - SS_{BC} - SS_{ABC}$$

$$= 162.667 - 138.583 - 4.750 - 3.250 - 1.750$$

$$= 14.334$$

The final results of the analysis are summarized in Table 8–9. Only the A and C main effects are significant; the population mean is higher for the A_1 treatment than for the A_2 treatment, and the mean performance is also an increasing function of the level of C.

8.3.2 ONE BETWEEN-SUBJECTS AND TWO WITHIN-SUBJECT VARIABLES. This design is interesting because it involves a more complex analysis of within-subject variability than any previously examined. The design is common; indeed, three and even four within-subject variables are frequently manipulated. An example of one experiment with this design is a study of paired-associate learning in which the subject must learn the 16 responses that are correct for the 16 stimuli presented to him. Trials to criterion (perhaps two errorless runs) is the dependent variable. The stimuli and responses are nonsense syllables. Half of the stimuli have high association values, the other half have low association values; similarly, the

Table 8–9 Analysis of variance for the data of Table 8–8

SV	df	SS	MS	F
Total	23	212.958		
Between S	7	50.291		
A	1	30.375	30.375	9.228[a]
B	1	3.375	3.375	1.025
AB	1	3.375	3.375	1.025
S/AB	4	13.166	3.292	
Within S	16	162.667		
C	2	138.583	69.292	38.674[a]
AC	2	4.750	2.375	1.325
BC	2	3.250	1.625	.907
ABC	2	1.750	.875	.488
SC/AB	8	14.334	1.792	

[a] $p < .01$

responses are divided between high and low-association syllables. Thus, four pairs are high-high (B_1C_1), four are low-high (B_2C_1), four are high-low (B_1C_2), and four are low-low (B_2C_2). Since each S must learn all 16 pairs, stimulus association value B and response association value C are both within-subject variables.

Table 8–10 presents a data matrix for this class of design. A random sample from a large population of subjects has been randomly distributed among the a levels of the between-subjects variable A, with the restriction that there are exactly n subjects at each level. The within-subject variables are B and C, and there are bc scores obtained from each subject. The order of presentation of the bc treatment combinations is randomized independently for each subject. The indices i, j, k, and m are used for each of the levels of subject A, B, and C.

Development of the model. As in all our investigating of mixed designs, we use the developments of Chapter 7 as a point of departure. The design may initially be viewed as an $S \times B \times C$ design with an subjects. Following the pattern set by Table 7–13, we develop the accompanying table.

SV	df
Between S	$an - 1$
B	$b - 1$
Between $S \times B$	$(an - 1)(b - 1)$
C	$c - 1$
Between $S \times C$	$(an - 1)(c - 1)$
BC	$(b - 1)(c - 1)$
Between $S \times BC$	$(an - 1)(b - 1)(c - 1)$

The terms account for the total of $abcn - 1$ df. This intermediate analysis clearly neglects possible effects due to A and its interactions with the within-subject variables B and C. The further breakdown of the above table to get the remaining terms is straightforward. As in the past,

$$SS_{B.S} = SS_A + SS_{S/A}$$

Now consider the crossing of A and S/A with B; this yields AB and SB/A on $(a-1)(b-1)$ and $a(n-1)(b-1)$ df, which together account for $(an-1)(b-1)$ df. Therefore, the Between $S \times B$ term above can be replaced by two terms, AB and SB/A. Crossing A and S/A with C yields the two components of Between $S \times C$, AC and SC/A; crossing A and S/A with BC yields the two components of Between $S \times BC$, ABC and SBC/A. The complete set of SV is given in Table 8–11.

Equation (8.4) presents a model that provides for the general set of effects suggested by the above discussion:

(8.4) $Y_{ijkm} = \mu + \alpha_j + \eta_{i/j} + \beta_k + \gamma_m + (\beta\gamma)_{km} + (\alpha\beta)_{jk} + (\alpha\gamma)_{jm}$

$$+ (\alpha\beta\gamma)_{jkm} + (\eta\beta)_{ik/j} + (\eta\gamma)_{im/j} + (\eta\beta\gamma)_{ikm/j} + \varepsilon_{ijkm}$$

As usual, we need specific assumptions about the population parameters to be able to derive *EMS* and so that the ratios of mean squares testing the null

Table 8–10   Data matrix for mixed design, one between-subjects and two within-subject variables

		B_1				$\cdots$	B_k				$\cdots$	B_b			
		C_1	$\cdots$	C_m	C_c	$\cdots$	C_1	C_m	$\cdots$	C_c	$\cdots$	C_1	C_m	$\cdots$	C_c
A_1	S_{11}	Y_{1111}	$\cdots$	Y_{111m}	Y_{111c}	$\cdots$	Y_{11k1}	Y_{11km}	$\cdots$	Y_{11kc}	$\cdots$	Y_{11b1}	Y_{11bm}	$\cdots$	Y_{11bc}
	$\vdots$														
	S_{i1}	Y_{i111}	$\cdots$	Y_{i11m}	Y_{i11c}	$\cdots$	Y_{i1k1}	Y_{i1km}	$\cdots$	Y_{i1kc}	$\cdots$	Y_{i1b1}	Y_{i1bm}	$\cdots$	Y_{i1bc}
	$\vdots$														
	S_{n1}	Y_{n111}	$\cdots$	Y_{n11m}	Y_{n11c}	$\cdots$	Y_{n1k1}	Y_{n1km}	$\cdots$	Y_{n1kc}	$\cdots$	Y_{n1b1}	Y_{n1bm}	$\cdots$	Y_{n1bc}
$\cdots$															
A_j	S_{ij}	Y_{ij11}	$\cdots$	Y_{ij1m}	Y_{ij1c}	$\cdots$	Y_{ijk1}	Y_{ijkm}	$\cdots$	Y_{ijkc}	$\cdots$	Y_{ijb1}	Y_{ijbm}	$\cdots$	Y_{ijbc}
	$\vdots$														
$\cdots$															
A_a	S_{ia}	Y_{ia11}	$\cdots$	Y_{ia1m}	Y_{ia1c}	$\cdots$	Y_{iak1}	Y_{iakm}	$\cdots$	Y_{iakc}	$\cdots$	Y_{iab1}	Y_{iabm}	$\cdots$	Y_{iabc}
	$\vdots$														

hypotheses are distributed as *F*. It is assumed that the effects of the variables *A*, *B*, and *C* are fixed; the levels of these variables have been arbitrarily chosen. Consequently, the fixed-effect population components (for example, θ_A^2) are defined exactly as in Section 8.3.1. It is assumed that subjects have been randomly sampled from a large population of subjects. Consequently, $\eta_{i/i}$, $(\eta\beta)_{ik/i}$, $(\eta\gamma)_{imj}$, and $(\eta\beta\gamma)_{ikm/j}$, as well as ε_{ijkm}, are assumed to be random samples from normally distributed populations with mean zero and variances $\sigma_{S/A}^2$, $\sigma_{SB/A}^2$, $\sigma_{SC/A}^2$, $\sigma_{SBC/A}^2$, and σ_e^2, respectively.

Analysis of variance. The *SV* of Table 8–11 parallel the terms in Equation (8.4), with the qualification that there is no separate error term corresponding to ε_{ijkm}. In the *df* column, only the nested subjects × treatments interactions warrant comment. The previous section on the model provides the clue to these quantities. For example, the *SB/A* effect is the difference between the *Between S × B (SB)* effect and the *AB* effect. It follows that $df_{SB/A} = df_{SB} - df_{AB}$, or

$$a(n-1)(b-1) = (an-1)(b-1) - (a-1)(b-1)$$

An alternative way of arriving at the above result is to view the nested interaction as a pool of *a SB* interactions, each computed at a different level of *A*. Then there are $(n-1)(b-1)$ *df* at each level of *A*, and the pooled result is $a(n-1)(b-1)$ *df*.

The *SS* calculations are also familiar, with the possible exception of the nested interaction terms. Either of the two approaches indicated for *df* can be used to calculate *SS*. We could calculate, for example, an SS_{SB} as we would for a simple repeated-measurement design and then subtract SS_{AB}, thereby obtaining $SS_{SB/A}$:

$$SS_{SB/A} = \left[\frac{\sum_i \sum_j \sum_k T_{ijk.}^2}{c} - C - SS_{BS} - SS_B \right] - SS_{AB}$$

As an alternative, the overall design could be treated as *a* two-factor repeated-measurement designs. An *SB* interaction sum of squares could be computed for each of the *a* sets of data, and these then pooled. In Table 8–11, the calculations are based on the isomorphism of *df* and *SS*. (This approach has been presented several times in the text, most recently in Section 8.2.2.)

The *EMS* follow from the rules of thumb presented earlier. You can profit by checking through each line to be sure that you understand how the rules apply. Once the *EMS* have been generated, the *F* tests present no problem. Each main or interaction effect is tested against the nested effect that follows most closely in the table.

A numerical example. Table 8–12 contains data for an illustrative analysis. Besides the data, subtotals are given that will be useful during the analysis. We begin by computing the total sum of squares:

$$SS_{tot} = \sum_i \sum_j \sum_k \sum_m Y_{ijkm}^2 - C$$

$$= (3)^2 + (5)^2 + \cdots + (11)^2 - \frac{(412)^2}{48}$$

$$= 3{,}912.000 - 3{,}536.333$$

$$= 375.667$$

Table 8–11 Analysis of variance for mixed design, one between-subjects and two within-subject variables

SV	df	SS	EMS	F
Total	$anbc - 1$	$\sum_i^n \sum_j^a \sum_k^b \sum_m^c Y_{ijkm}^2 - C$		
Between S	$an - 1$			
A	$a - 1$	$\dfrac{\sum_j^a T_{.j..}^2}{nbc} - C$	$\sigma_e^2 + bc\sigma_{S/A}^2 + nbc\theta_A^2$	$\dfrac{MS_A}{MS_{S/A}}$
S/A	$a(n-1)$	$SS_{B.S} - SS_A$	$\sigma_e^2 + bc\sigma_{S/A}^2$	
Within S	$an(bc-1)$	$SS_{tot} - SS_{B.S}$		
B	$b - 1$	$\dfrac{\sum_k^b T_{..k.}^2}{nac} - C$	$\sigma_e^2 + c\sigma_{SB/A}^2 + nac\theta_B^2$	$\dfrac{MS_B}{MS_{SB/A}}$
AB	$(a-1)(b-1)$	$\dfrac{\sum_j^a \sum_k^b T_{.jk.}^2}{nc} - C - SS_A - SS_B$	$\sigma_e^2 + c\sigma_{SB/A}^2 + nc\theta_{AB}^2$	$\dfrac{MS_{AB}}{MS_{SB/A}}$
SB/A	$a(n-1)(b-1)$	$\dfrac{\sum_i^n \sum_j^a \sum_k^b T_{ijk.}^2}{c} - \dfrac{\sum_i^n \sum_j^a T_{ij..}^2}{bc} - \dfrac{\sum_j^a \sum_k^b T_{.jk.}^2}{nc} + \dfrac{\sum_j^a T_{.j..}^2}{nbc}$	$\sigma_e^2 + c\sigma_{SB/A}^2$	
C	$c - 1$	$\dfrac{\sum_m^c T_{...m}^2}{nab} - C$	$\sigma_e^2 + b\sigma_{SC/A}^2 + nab\theta_C^2$	$\dfrac{MS_C}{MS_{SC/A}}$
AC	$(a-1)(c-1)$	$\dfrac{\sum_j^a \sum_m^c T_{.j.m}^2}{nb} - C - SS_A - SS_C$	$\sigma_e^2 + b\sigma_{SC/A}^2 + nb\theta_{AC}^2$	$\dfrac{MS_{AC}}{MS_{SC/A}}$
SC/A	$a(n-1)(c-1)$	$\dfrac{\sum_i^n \sum_j^a \sum_m^c T_{ij.m}^2}{b} - \dfrac{\sum_i^n \sum_j^a T_{ij..}^2}{bc} - \dfrac{\sum_j^a \sum_m^c T_{.j.m}^2}{nb} + \dfrac{\sum_j^a T_{.j..}^2}{nbc}$	$\sigma_e^2 + b\sigma_{SC/A}^2$	
BC	$(b-1)(c-1)$	$\dfrac{\sum_k^b \sum_m^c T_{..km}^2}{na} - C - SS_B - SS_C$	$\sigma_e^2 + \sigma_{SBC/A}^2 + na\theta_{BC}^2$	$\dfrac{MS_{BC}}{MS_{SBC/A}}$
ABC	$(a-1)(b-1)(c-1)$	$\dfrac{\sum_j^a \sum_k^b \sum_m^c T_{.jkm}^2}{n} - C - SS_A - SS_B - SS_{AB} - SS_C - SS_{AC} - SS_{BC}$	$\sigma_e^2 + \sigma_{SBC/A}^2 + n\theta_{ABC}^2$	$\dfrac{MS_{ABC}}{MS_{SBC/A}}$
SBC/A	$a(n-1)(b-1)(c-1)$	$SS_{W.S} - SS_B - SS_{AB} - SS_{SB/A} - SS_C - SS_{AC} - SS_{SC/A} - SS_{BC} - SS_{ABC}$	$\sigma_e^2 + \sigma_{SBC/A}^2$	

Table 8-12 Data and subtotals for an analysis of variance, one between-subjects and two within-subject variables

		B_1					B_2					
		C_1	C_2	C_3	C_4	$T_{ij.}$	C_1	C_2	C_3	C_4	$T_{ij2.}$	$T_{ij..}$
A_1	S_{11}	3	5	7	8	23	7	6	9	9	31	54
	S_{21}	2	5	6	8	21	8	5	10	12	35	56
		$T_{.11m}=5$	10	13	16	$T_{.11.}=44$	$T_{.12m}=15$	11	19	21	$T_{.12.}=66$	$T_{.1..}=110$
A_2	S_{12}	8	9	9	11	37	6	7	7	9	29	66
	S_{22}	7	7	9	10	33	7	8	9	11	35	68
		$T_{.21m}=15$	16	18	21	$T_{.21.}=70$	$T_{.22m}=13$	15	16	20	$T_{.22.}=64$	$T_{.2..}=134$
A_3	S_{13}	10	12	13	15	50	5	7	8	10	30	80
	S_{23}	11	11	14	16	52	8	8	9	11	36	88
		$T_{.31m}=21$	23	27	31	$T_{.31.}=102$	$T_{.32m}=13$	15	17	21	$T_{.32.}=66$	$T_{.3..}=168$
		$T_{..1m}=41$	49	58	68	$T_{..1.}=216$	$T_{..2m}=41$	41	52	62	$T_{..2.}=196$	$T_{....}=412$

Table 8-12. (continued)

Subjects × C Subtotals ($T_{ij,m}$)

		C_1	C_2	C_3	C_4	$T_{ij..}$
A_1	S_{11}	10	11	16	17	54
	S_{21}	10	10	16	20	56
	$T_{.1.m} = $	20	21	32	37	$T_{.1..} = 110$
A_2	S_{12}	14	16	16	20	66
	S_{22}	14	15	18	21	68
	$T_{.2.m} = $	28	31	34	41	$T_{.2..} = 134$
A_3	S_{13}	15	19	21	25	80
	S_{23}	19	19	23	27	88
	$T_{.3.m} = $	34	38	44	52	$T_{.3..} = 168$
	$T_{...m} = $	82	90	110	130	$T_{....} = 412$

For the variability among subjects, we have

$$SS_{\text{B.S}} = \frac{\sum_i \sum_j T_{ij..}^2}{bc} - C$$

$$= \frac{(54)^2 + (56)^2 + \cdots + (80)^2 + (88)^2}{8} - \frac{(412)^2}{48}$$

$$= 3{,}647.000 - 3{,}536.333$$

$$= 110.667$$

This term has two components

$$SS_A = \frac{\sum_j T_{.jkm}^2}{nbc} - C$$

$$= \frac{(110)^2 + (134)^2 + (168)^2}{16} - \frac{(412)^2}{48}$$

$$= 3{,}642.500 - 3{,}536.333$$

$$= 106.167$$

and

$$SS_{B/A} = SS_{\text{B.S}} - SS_A$$

$$= 110.667 - 106.167$$

$$= 4.500$$

We turn next to the within-subject variability:

$$SS_{\text{W.S}} = SS_{\text{tot}} - SS_{\text{B.S}}$$

$$= 375.667 - 110.667$$

$$= 265.000$$

For the main effect of B, we can use the single df formula, thus saving some computational effort:

$$SS_B = \frac{(216 - 196)^2}{48} = 8.333$$

For the AB interaction, we have

$$SS_{AB} = \frac{\sum_j \sum_k T_{.jk.}^2}{nc} - C - SS_A - SS_B$$

$$= \frac{(44)^2 + (70)^2 + \cdots + (66)^2}{8} - \frac{(412)^2}{48} - 106.167 - 8.333$$

$$= 3{,}756.000 - 3{,}536.333 - 106.167 - 8.333$$

$$= 105.167$$

The error variability for the previous two terms is

$$SS_{SB/A} = \frac{\sum_i \sum_j \sum_k T_{ijk.}^2}{c} - \frac{\sum_i \sum_j T_{ij..}^2}{bc} - \frac{\sum_j \sum_k T_{.jk.}^2}{nc} + \frac{\sum_j T_{.j..}^2}{nbc}$$

$$= \frac{(23)^2 + (31)^2 + \cdots + (52)^2 + (36)^2}{4} - 3{,}647.000 - 3{,}756.000 + 3{,}642.500$$

$$= 9.500$$

For the C main effect, we have

$$SS_C = \frac{\sum_m T_{...m}^2}{abn} - C$$

$$= \frac{(82)^2 + (90)^2 + (110)^2 + (130)^2}{12} - \frac{(412)^2}{48}$$

$$= 3{,}652.000 - 3{,}536.333$$

$$= 115.667$$

For the AC interaction, we have

$$SS_{AC} = \frac{\sum_j \sum_m T_{.j.m}^2}{nb} - C - SS_A - SS_C$$

$$= \frac{(20)^2 + (21)^2 + \cdots + (44)^2 + (52)^2}{4} - \frac{(412)^2}{48} - 106.167 - 115.667$$

$$= 3{,}764.000 - 3{,}536.333 - 106.167 - 115.667$$

$$= 5.833$$

The error variability for the previous two terms is

$$SS_{SC/A} = \frac{\sum_i \sum_j \sum_m T_{ij.m}^2}{b} - \frac{\sum_i \sum_j T_{ij..}^2}{bc} - \frac{\sum_j \sum_m T_{.j.m}^2}{nb} + \frac{\sum_j T_{.j..}^2}{nbc}$$

$$= \frac{(10)^2 + (11)^2 + \cdots + (23)^2 + (27)^2}{2} - 3{,}647.000 - 3{,}764.000 + 3{,}642.500$$

$$= 5.500$$

The BC interaction sum of squares is obtained by

$$SS_{BC} = \frac{\sum_k \sum_m T_{..km}^2}{na} - C - SS_B - SS_C$$

$$= \frac{(41)^2 + (49)^2 + \cdots + (52)^2 + (62)^2}{6} - \frac{(412)^2}{48} - 8.333 - 115.667$$

$$= 3{,}663.333 - 3{,}536.333 - 8.333 - 115.667$$

$$= 3.000$$

Table 8–13   Analysis of variance for data of Table 8–12

SV	df	SS	MS	F
Total	47	375.667		
Between S	5	110.667		
A	2	106.167	53.083	35.389[a]
S/A	3	4.500	1.500	
Within S	42	265.000		
B	1	8.333	8.333	2.631
AB	2	105.167	52.583	16.603[b]
SB/A	3	9.500	3.167	
C	3	115.667	38.556	63.103[a]
AC	6	5.833	.972	1.591
SC/A	9	5.500	.611	
BC	3	3.000	1.000	2.571
ABC	6	8.500	1.417	3.630
SBC/A	9	3.500	.389	

[a] $p < .025$
[b] $p < .01$

We have next

$$SS_{ABC} = \frac{\sum_j \sum_k \sum_m T_{.jkm}^2}{n} - C - SS_A - SS_B - SS_C - SS_{AB} - SS_{AC} - SS_{BC}$$

$$= \frac{(5)^2 + (10)^2 + \cdots + (17)^2 + (21)^2}{2} - C - SS_A - SS_B - SS_C - SS_{AB}$$

$$- SS_{AC} - SS_{BC}$$

$$= 3{,}889.000 - 3{,}536.333 - 106.167 - 8.333 - 115.667 - 105.167$$

$$- 5.833 - 3.000$$

$$= 8.500$$

The error variability for the two previous terms can be taken as a residual from the $SS_{\text{w.s}}$:

$$SS_{SBC/A} = SS_{\text{w.s}} - SS_B - SS_{AB} - SS_{SB/A} - SS_C - SS_{AC} - SS_{SC/A} - SS_{BC} - SS_{ABC}$$

$$= 265.000 - 8.333 - 105.167 - 9.500 - 115.667 - 5.833 - 5.500$$

$$- 3.333 - 8.167$$

$$= 3.500$$

The final results of the analysis are summarized in Table 8–13.

8.4   CONCLUDING REMARKS

The mixed design is a compromise between the simplicity of the completely randomized design and the high relative efficiency of the simple repeated-measurement design. In testing the null hypothesis that $\theta_A^2 = 0$, where A is a between-subjects variable, precision is usually low; this part of the design is essentially completely randomized. By introducing one or more within-subject variables, fewer subjects are needed than if all variables were between subjects. A potentially precise test of the within-subject effects is provided, since the error terms are not inflated by individual differences. But saving subjects and increasing precision costs something. As in the simpler repeated-measurement designs, the possibility of subjects × treatments interactions is introduced, with the consequent possibilities of lowered efficiency (if the interaction with subjects is large) and heterogeneity of covariance. If the population covariances are not similar for all possible pairs of levels of within-subject variables, only the F tests of within-subject effects will be positively biased. For levels of the between-subjects variables, the expected covariances should all be zero if randomization has been properly carried out.

Tukey's single df test (Chapter 7) can be extended to test for nonadditivity in the mixed design. For example, in a one between-subjects and one within-subject design, a sum of squares for nonadditivity could be computed at each level of A. These a quantities could then be pooled, yielding a sum of squares for nonadditivity on $a\,df$ that would be tested against the balance of the SB/A interaction. Transformations to additivity can also be sought following the guidelines of Chapter 7. If additivity is achieved, the nested interaction effects are dropped from the model and such terms as $\sigma_{SB/A}^2$ disappear from the EMS. The covariances will be homogeneous, and the tests of within-subject effects will generally be efficient and unbiased.

Without strong evidence to the contrary, most investigators assume that the nonadditive model as we have presented it is appropriate. Therefore, it becomes important to be aware of the possibility of heterogeneity of covariance and the consequent positive bias of the tests of within-subject effects. Conservative F tests parallel to those presented in Section 7.4.1 are available. Consider, for example, the one between-subjects and one within-subject design of the preceding section. The statistics MS_B and MS_{AB} are distributed on $(b-1)\varepsilon$ and $(a-1)(b-1)\varepsilon\,df$ and their common error term $MS_{SB/A}$ is distributed on $a(n-1)(b-1)\varepsilon\,df$; the value ε lies between $1/(b-1)$ (maximum heterogeneity of variance and covariance) and 1 (homogeneous variances and covariances). The conservative test of the B effect would therefore use 1 and $a(n-1)\,df$, and a similar test of the AB interaction would use $a-1$ and $a(n-1)\,df$.

If the conservative test is not significant and the usual test on the full df is, the investigator may wish to estimate directly the adjustment ε. Equation (7.32) is still appropriate but the variances and covariances are slightly modified for the mixed design. To obtain ε, we need a $b \times b$ variance-covariance matrix whose elements are obtained by averaging over the levels of the between-subjects variable A. Thus we have

(8.5)
$$S_{kk'} = \frac{\sum_j \sum_i (Y_{ijk} - \bar{Y}_{..k})(Y_{ijk'} - \bar{Y}_{..k'})}{a(n-1)}$$

(For a more detailed discussion of covariance and of nonadditivity in general, see Chapter 7.)

EXERCISES

8.1 Analyze variance appropriately for these data.

		B_1C_1	B_1C_2	B_1C_3	B_2C_1	B_2C_2	B_2C_3
	S_{11}	18	8	7	14	9	4
A_1	S_{21}	20	10	13	11	6	10
	S_{31}	23	12	14	13	12	3
	S_{12}	16	5	7	10	15	5
A_2	S_{22}	13	8	4	16	11	6
	S_{32}	18	6	10	17	11	4
	S_{13}	14	6	8	11	3	1
A_3	S_{23}	20	10	8	7	9	2
	S_{33}	13	7	7	13	5	6

8.2 Consider the following data set:

		B_1	B_2	B_3	$\bar{Y}_{ij.}$
A_1	S_{11}	4	5	6	5
	S_{21}	9	3	6	6
A_2	S_{12}	6	2	4	4
	S_{22}	8	3	7	6

(i) Find SS_A. (Don't do it the hard way!)
(ii) Find $SS_{S/A}$.
(iii) Suppose you had a one-factor completely randomized design with two Ss in each of two groups:

$$\begin{array}{cc} A_1 & A_2 \\ 5 & 4 \\ 6 & 6 \end{array}$$

Find SS_A, $SS_{S/A}$. How do your answers compare with (i), (ii)? How do the F ratio compare? What do you conclude?
(iv) SS_{SB/A_1}, the sum of squares for $S \times B$ operating as though there were no A_2 data, is 13; SS_{SB/A_2} is 1.0. What is $MS_{SB/A}$, based on these two values?

8.3 In the design of Exercise 8.2, suppose B were three randomly selected paragraphs in an experiment on memory for prose materials. Now, give the *EMS*, the F test of A, and the *df* associated with the test. How is your answer changed if there are grounds for assuming $\sigma_{AB}^2 = 0$? Does any other assumption have an equivalent consequence?

8.4 We have a mixed design with A (between Ss) and B (within S). The data are punched on cards and we wish to analyze the whole on the computer. Unfortunately, the only available program is for a 2-factor (crossing) anova. Show how each *SS* in the mixed anova would be obtained if you are allowed to put the data through the computer any number of times. You are allowed to combine (add or subtract) the results of the several computer runs on a desk calculator.

8.5 Consider a design with three groups of 10 subjects, four trials for each. Let us assume extreme heterogeneity of covariance. We calculate $F = 2.20$ for *trials*. What is the approximate probability of an F this large or larger if H_0 is true?

9 HIERARCHICAL DESIGNS

9.1 INTRODUCTION

Now that we have examined the models and analyses that are appropriate when subjects are nested within levels of the variable A and are crossed with the levels of a second variable B, the pattern can be extended in two ways. One design has subjects nested within levels of a variable G, while the levels of G are in turn nested within the levels of A (Sections 9.2–9.4). Another involves nesting several within-subject variables (Section 9.5). We refer to these two kinds of design that involve nesting as "hierarchical," referring to the hierarchy of variables that typifies them. The statistical models and computations are straightforward extensions of earlier developments.

The $S/G/A$ designs have an important place in both psychological and educational research. To cite one example, consider a group dynamics experiment designed for studying the effects of stress on attitude change in the members of four-man conference teams. We might have several conference groups under high stress and an equal number under low stress; the resultant design might be characterized as *subjects within conference groups within levels of stress*. A similar example can be taken from educational psychology. Several first-grade classes are taught reading under one method and an equal number are taught under a second method; all students are tested at the end of the term. The design might be characterized as *subjects within classes within methods*. Still another example might be taken from the animal laboratory. Different methods of rearing rats might be compared, with each method applied to several different litters. We might characterize this design as *subjects within litters within methods of rearing*. In all these examples, there are three potential sources of variability among the means computed for the levels of A. First, there may be a treatment effect in the sampled population. Second, differences in the compositions of the groups nested within the different levels of A may be a factor. Third, variability among treatment-level means will partly reflect that there are different subjects at each level of A.

The primary new aspect of the hierarchical design is the assumption that an individual's score is partly influenced by the social unit of which he is a member.

Even though the same experimental treatment is applied to both of two individuals in different social groups (or school classes or litters), they will differ from each other, not only because they are different individuals, but also because they are subject to interactions with different sets of individuals and occurrences. Once we recognize that some experiments involve such group effects, we have to consider those effects within the statistical model.

Nesting of within-subject variables is also a common occurrence in psychological research. Subjects are frequently tested with several sets of stimuli. Experimenters may measure response time to rare words and to common words (words within frequency level), for example, or may obtain ratings of pictures depicting several forms of social interaction (pictures within social interaction levels). If stimuli, say words or pictures, are viewed as having random effects, and this will usually be appropriate, there will be several potential causes of the variability among the means of the different conditions in which stimuli are nested. First, there may be reliable effects of the conditions, as when word frequency affects response time. Second, means at the different treatment levels may vary due to chance variability among the stimuli selected for the experiment. Third, as in repeated-measurement designs generally, variability due to interactions of subjects with within-subject variables may contribute to variability among the obtained treatment means. Recognizing stimulus variability necessitates finding an error term that includes such variability. In many instances, it will be necessary to calculate quasi F ratios to incorporate both stimuli-within-treatment and subjects × treatments variability in the error term.

The importance of studying hierarchical designs goes beyond application of the design. The presentation of these designs and their analyses should further an understanding of how to establish structural models and then translate these models into analysis of variance tables. Every design that the researcher may encounter cannot possibly be considered here. The material of this chapter should, however, help establish certain fundamental but widely applied principles of data analysis (first introduced in Chapter 8).

9.2   GROUPS WITHIN TREATMENTS

9.2.1   THE ANALYSIS OF VARIANCE MODEL.   We begin by imagining a treatment populations, differing systematically only in level of treatment variable A. From each population, subjects are randomly selected in groups of size n; the sampling ends when g groups of n subjects have been sampled from each population. The resulting experimental layout is presented in Table 9–1. The design involves a total of agn subjects with n subjects in each group, g groups at each level of A, and a levels of A. The notational indices are

$$i = 1, 2, \ldots, n$$
$$j = 1, 2, \ldots, g$$
$$k = 1, 2, \ldots, a$$

Table 9–1   Data matrix for the group-within-treatment design

	A_1			A_k			A_a	
G_{11} $\cdots$	G_{j1} $\cdots$	G_{g1}	G_{1k} $\cdots$	G_{jk} $\cdots$	G_{gk}	G_{1a} $\cdots$	G_{ja} $\cdots$	G_{ga}
Y_{111}	Y_{1j1}	Y_{1g1}	Y_{11k}	Y_{1ja}	Y_{1gk}	Y_{11a}	Y_{1ja}	Y_{1ga}
Y_{i11}	Y_{ij1}	Y_{ig1}	Y_{i1k}	Y_{ijk}	Y_{igk}	Y_{i1a}	Y_{ija}	Y_{iga}
Y_{n11}	Y_{nj1}	Y_{ng1}	Y_{n1k}	Y_{njk}	Y_{ngk}	Y_{n1a}	Y_{nja}	Y_{nga}

Next, an equation is required to relate the representative score Y_{ijk} to the parameters of the population from which the sample is drawn. We begin by ignoring the variable A; the design is viewed as a one-factor design, the "factor" is groups G, and there are ag levels and n subjects at each level. In accord with the one-factor model (Chapter 4), this view suggests

(9.1) $$Y_{ijk} = \mu + \gamma_{jk} + \varepsilon_{ijk}$$

where $\gamma_{jk} = \mu_{jk} - \mu$, the overall effect of the jth group at the kth level of A, and $\varepsilon_{ijk} = Y_{ijk} - \mu_{jk}$, the residual error component. Equation (9.1) disregards the possibility of an effect due to A_k. Presumably, group means differ not only because the groups have different compositions, but also because some groups are at one level of A and others are at a different level. This line of reasoning suggests that part of the γ_{jk} effect is due to α_k, the effect due to the level of A in which the group exists. Accordingly, we subtract the contribution of α_k:

$$\gamma_{jk} - \alpha_k = (\mu_{jk} - \mu) - (\mu_k - \mu)$$

(9.2) $$= \mu_{jk} - \mu_k$$

$$= \gamma_{j/k}$$

where $\gamma_{j/k}$ is the pure effect of the jkth group, uninflated by any contribution due to α_k. We can now substitute in Equation (9.1) for γ_{jk}, obtaining

(9.3) $$Y_{ijk} = \mu + \alpha_k + \gamma_{j/k} + \varepsilon_{ijk}$$

Each score is contributed to by a treatment effect, a group effect, and a residual component reflecting error of measurement and individual differences.

A common error in analyzing group designs is the failure to consider group effects in the model. In this case, the analysis proceeds as though the design were a completely randomized one-factor design with gn subjects in each of a treatment groups. This failure to separate the γ component from ε may result in an inflated F ratio as will be shown in the next section.

To complete the presentation of the underlying theory, and to arrive at the *EMS*, one must consider the nature of the effects. It is assumed that the levels of

A have been arbitrarily chosen by the experimenter, and consequently, that α_k is a fixed variable. Then, $\sum_k \alpha_k = 0$ and the variance component is defined as $\theta_A^2 = \sum_k^a \alpha_k^2/(a-1)$. The group effect $\gamma_{j/k}$ is viewed as a random variable, since the groups are clearly a random sample from the population of all possible groups of size n that could be composed. As usual, ε_{ijk} is also a random variable. The $\gamma_{j/k}$ and ε_{ijk} are assumed to be sampled from normally distributed populations with mean zero, and respective variances $\sigma_{G/A}^2$ and σ_e^2.

9.2.2 THE ANALYSIS OF VARIANCE.

Table 9–2 contains the *SV*, *df*, *SS*, *EMS*, and *F* for the *groups-within-treatments* design. One source exists for each term on the right-hand side of Equation (9.1). To facilitate the analysis, we have first divided our variability between two sources; ignoring the treatment variable A, we have between-groups and within-group variability. This breakdown corresponds to a model of the form of Equation (9.1). Subsequently, the between-groups source is further divided into the G/A and A components, as in the analysis of variance model just developed.

The *df* are easily obtainable once we have noted the sources. For the nested term G/A we can use the fact that the between-groups term is a composite of the A and G/A terms. Therefore, $df_{B.G} - df_A = df_{G/A}$, or

$$(ag-1)-(a-1)=a(g-1)$$

Alternatively, we note that G/A literally represents the summing, over all treatment levels, of the variability of group means about the mean of their treatment level. At each treatment level, we have $g-1$ *df* to represent the variability of g means about the treatment mean. Pooling over a levels, we again arrive at $a(g-1)$ *df*. Formulas for *SS* quantities are established as we have done previously.

The *EMS* follow the rules of thumb developed in Chapter 8. Considering the A line first, we immediately set down σ_e^2 and the null hypothesis term θ_A^2. Noting that the subscript G/A includes the letter A and that the essential letter G represents a random-effect variable, we include $\sigma_{G/A}^2$ in the expectation. The

Table 9–2 Analysis of variance for the group-within-treatment design

SV	df	SS	EMS	F
Total	$agn-1$	$\sum_i^n \sum_j^g \sum_k^a Y_{ijk}^2 - C$		
Between G	$ag-1$	$\dfrac{\sum_j^g \sum_k^a T_{\cdot jk}^2}{n} - C$		
A	$a-1$	$\dfrac{\sum_k^a T_{\cdot\cdot k}^2}{gn} - C$	$\sigma_e^2 + n\sigma_{G/A}^2 + ng\theta_A^2$	$\dfrac{MS_A}{MS_{G/A}}$
G/A	$a(g-1)$	$SS_{BG} - SS_A$	$\sigma_e^2 + n\sigma_{G/A}^2$	$\dfrac{MS_{G/A}}{MS_{S/G/A}}$
$S/G/A$	$ag(n-1)$	$SS_{tot} - SS_{BG}$	σ_e^2	

remaining two lines should pose no problems. Note that in this design, however, σ_e^2 is the sum of variances due to measurement errors and individual differences; that is, it includes $\sigma_{S/G/A}^2$. Only when there are between-subjects and within-subject terms in the analysis, in which case individual differences do not contribute to all sources, do we list separately the variance component due to individual differences.

9.2.3 POOLING.

On the basis of the *EMS*, it is clear that the treatment mean square should be tested against the mean square for groups within *A*. Consideration of the error *df* will indicate that this *F* test may often be lacking in power. Suppose that the experiment involves three four-man conference groups at each of three levels of *A*. The total of 36 subjects would generally be considered reasonable for a study involving three treatment levels, since 33 error *df* would be available if the design were a completely randomized one-factor design. The error *df* for the hierarchical design actually used are only $3(3-1)$, or six, however. This number of error *df* is not likely to make the experimenter feel secure if he fails to reject the null hypothesis. One possible solution is to assume a different model; specifically, the experimenter could assume that the group structures do not contribute to error variability, and consequently, that the $\sigma_{G/A}^2$ component can be deleted wherever it appears in the *EMS* column of Table 9–2. In this case, both $MS_{G/A}$ and $MS_{S/G/A}$ are estimates of σ_e^2 and can therefore be combined to provide a new error term. The combining, or *pooling* as it is generally called, takes the form $(SS_{G/A} + SS_{S/G/A})/(df_{G/A} + df_{S/G/A})$, an error mean square distributed on 33 *df*. For reasons that will be considered next, pooling should not be carried out unless the data strongly indicate that $\sigma_{G/A}^2$ is a negligible quantity.

Since the decision on pooling must await the collection of data, in designing the experiment one must proceed on the assumption that $MS_{G/A}$ will be the error term. Consequently, the investigator planning an experiment that involves a group of subjects as a unit should carefully evaluate $df_{G/A}$, and if it is too small to provide an adequate test of treatment effects, modify the design accordingly.

What are the consequences if the G/A and $S/G/A$ terms are pooled when $\sigma_{G/A}^2$ contributes to the data variability? To answer this, we first derive the expectation of our new error term, which will be labeled $MS_{S/A}$, and which is computed as

$$MS_{S/A} = \frac{SS_{G/A} + SS_{S/G/A}}{df_{G/A} + df_{S/G/A}}$$

$$= \frac{(df_{G/A})(MS_{G/A}) + (df_{S/G/A})(MS_{S/G/A})}{df_{G/A} + df_{S/G/A}}$$

Taking the expectations of the two sides yields

$$E(MS_{S/A}) = \left(\frac{df_{G/A}}{df_{G/A} + df_{S/G/A}}\right) E(MS_{G/A}) + \left(\frac{df_{S/G/A}}{df_{G/A} + df_{S/G/A}}\right) E(MS_{S/G/A})$$

Thus, $E(MS_{S/A})$ is a weighted average of $E(MS_{G/A})$ and $E(MS_{S/G/A})$, where the weights are proportions of *df* associated with the two terms. If we assume that the two expectations to the right of the equals sign are unequal, their average must lie between them. Thus, $E(MS_{S/A})$ will be less than $E(MS_{G/A})$, the larger of the two terms being averaged. But this means that the pooled error term has a smaller expectation than the error term that is generally appropriate for testing the *A* main effect (see Table 9–2). If the two terms being pooled do not have identical expectations—that is, if $\sigma^2_{G/A} \neq 0$—the *F* test of *A* against the pooled error term will be positively biased; under $H_0(\theta^2_A = 0)$, the *EMS* ratio will be greater than 1. Furthermore, the ratio of mean squares will not have an *F* distribution over replications of the experiment, since under H_0, we get $E(MS_A) \neq E(MS_{S/A})$.

Statisticians do not agree on the conditions under which pooling should be carried out. In the most extensive works available, Bozivich, Bancroft, and Hartley (1956) and Srivastava and Bozivich (1961) have investigated power and α under various violations of the model assumed in pooling, using computer populations in an approach like those of Section 4.3. Their work has been limited to hierarchical designs in which the components assumed to be zero are variances of random effects. The results are complicated; however, a rough rule of thumb would be to pool whenever there are a priori grounds for pooling and also a preliminary *F* test fails to be significant at the .25 level. The prior belief in the pooling model could be based on knowing the experimental situation or on an analysis of the results of preliminary tests in related experiments. In a particular social situation, for example, it may be unlikely that groups really do contribute to the variability in the data beyond the contribution due to individual differences. The preliminary test is a ratio of terms to be pooled; in the design under consideration we should test $MS_{G/A}$ against $MS_{S/G/A}$. If this *F* is nonsignificant at the .25 level, its numerator and denominator can be pooled.

The consequences of pooling inappropriately when fixed-effect components are to be neglected are less clear. I suspect that the same .25 level for the preliminary test will prove sufficiently conservative in most instances. Again, there should also be prior evidence. Pooling, when it is not warranted by the true state of affairs in the population, can generally result in false inferences, and when there is doubt about the appropriateness of pooling, it should not be used. Wherever possible, enough subjects should be run to assure reasonable power without pooling.

9.2.4 A NUMERICAL EXAMPLE.   Table 9–3 presents data for a group-within-treatment design. We first obtain the total sum of squares:

$$SS_{\text{tot}} = \sum_{i=1}^{4} \sum_{j=1}^{3} \sum_{k=1}^{2} Y_{ijk}^2 - C$$

$$= 5^2 + 6^2 + \cdots + 24^2 + 19^2 - \frac{(363)^2}{24}$$

$$= 6{,}671.00 - 5{,}490.38$$

$$= 1{,}180.62$$

Table 9–3   Data for a group-within-treatment design

	A_1				A_2	
G_{11}	G_{21}	G_{31}		G_{12}	G_{22}	G_{32}
5	7	16		24	9	17
6	18	5		21	23	26
18	4	9		12	28	24
12	11	14		16	19	19
$T_{.j1} =$ 41	40	44		$T_{.j2} =$ 73	79	86
$T_{..1} = 125$				$T_{..2} = 238$		

$$T_{...} = 363$$

The measure of between-groups variability is

$$SS_{\text{B.G}} = \frac{\sum_{j=1}^{3} \sum_{k=1}^{2} T_{.jk}^2}{4} - C$$

$$= \frac{(41)^2 + \cdots + (86)^2}{4} - \frac{(363)^2}{24}$$

$$= 6{,}045.75 - 5{,}490.38$$

$$= 555.37$$

Turning to the A source, we have

$$SS_A = \frac{\sum_{k=1}^{2} T_{..k}^2}{12} - C$$

$$= \frac{(125)^2 + (238)^2}{12} - \frac{(363)^2}{24}$$

$$= 6{,}022.42 - 5{,}490.38$$

$$= 532.04$$

Subtracting SS_A from SS_{BG}, we obtain the sum of squares for the nested group effect:

$$SS_{G/A} = SS_{\text{B.G}} - SS_A$$

$$= 555.37 - 532.04$$

$$= 23.33$$

The total variability can be partitioned into a between-groups and a within-group component. Therefore,

$$SS_{S/G/A} = SS_{\text{tot}} - SS_{\text{B.G}}$$

$$= 1{,}180.62 - 555.37$$

$$= 625.25$$

Table 9–4   Analysis of variance for the data of Table 9–3

SV	df	SS	MS	F
Total	23	1,180.62		
Between G	5	555.37		
A	1	532.04	532.04	92.98[a]
G/A	4	23.33	5.83	
$S/G/A$	18	625.25	34.74	

[a] $p < .01$

The results of the analysis are summarized in Table 9–4. The difference in performance under treatments A_1 and A_2 is clearly significant. Since the ratio of G/A to $S/G/A$ mean squares is less than 1, these two sources might be pooled to provide a new error term to test the A effect; the F test of Table 9–4 is so clearly significant, however, that this appears to be an unnecessary step.

9.3   A WITHIN-GROUP VARIABLE

We next consider the situation in which a variable is introduced into each group. There are again g groups at each of the a levels of the variable A. In addition, each group is divided into b levels of the variable B, with n subjects at each level. There may be several conference groups working under high stress, for example, and several others working under low stress. Within each conference group, half the subjects may be high-anxious (as measured by the Taylor Manifest Anxiety Scale), while the other half are low-anxious. Stress is the A variable and anxiety is the B variable.

This hierarchical design is represented in Table 9–5. The indices of notation are

$$i = 1, 2, \ldots, n$$
$$j = 1, 2, \ldots, g$$
$$k = 1, 2, \ldots, a$$
$$m = 1, 2, \ldots, b$$

The design involves bn subjects within each group, bgn subjects at each level of A, and $abgn$ subjects for the entire experiment.

9.3.1   THE ANALYSIS OF VARIANCE MODEL.   We first require some understanding of the structure of the design; in particular, we need some reasonable breakdown of the total variability. It helps to relate the present design to one we have previously considered. In this case, we have something analogous to the first design considered in Chapter 8, in which there was one between-subjects variable

Table 9-5 Data matrix for a hierarchical design containing a within-group variable

	A_1				A_k				A_a		
	G_{11} $\cdots$	G_{j1} $\cdots$	G_{g1}	$\cdots$	G_{1k} $\cdots$	G_{jk} $\cdots$	G_{gk}	$\cdots$	G_{1a} $\cdots$	G_{ja} $\cdots$	G_{ga}
B_1	Y_{1111}	Y_{1j11}	Y_{1g11}		Y_{11k1}	Y_{1jk1}	Y_{1gk1}		Y_{11a1}	Y_{1ja1}	Y_{1ga1}
	Y_{i111}	Y_{ij11}	Y_{ig11}		Y_{i1k1}	Y_{ijk1}	Y_{igk1}		Y_{i1a1}	Y_{ija1}	Y_{iga1}
	Y_{n111}	Y_{nj11}	Y_{ng11}		Y_{n1k1}	Y_{njk1}	Y_{ngk1}		Y_{n1a1}	Y_{nja1}	Y_{nga1}
B_m	Y_{111m}	Y_{1j1m}	Y_{1g1m}		Y_{11km}	Y_{1jkm}	Y_{1gkm}		Y_{11am}	Y_{1jam}	Y_{1gam}
	Y_{i11m}	Y_{ij1m}	Y_{ig1m}		Y_{i1km}	Y_{ijkm}	Y_{igkm}		Y_{i1am}	Y_{ijam}	Y_{igam}
	Y_{n11m}	Y_{nj1m}	Y_{ng1m}		Y_{n1km}	Y_{njkm}	Y_{ngkm}		Y_{n1am}	Y_{njam}	Y_{ngam}
B_b	Y_{111b}	Y_{1j1b}	Y_{1g1b}		Y_{11kb}	Y_{1jkb}	Y_{1gkb}		Y_{11ab}	Y_{1jab}	Y_{1gab}
	Y_{i11b}	Y_{ij1b}	Y_{ig1b}		Y_{i1kb}	Y_{ijkb}	Y_{igkb}		Y_{i1ab}	Y_{ijab}	Y_{igab}
	Y_{n11b}	Y_{nj1b}	Y_{ng1b}		Y_{n1kb}	Y_{njkb}	Y_{ngkb}		Y_{n1ab}	Y_{njab}	Y_{ngab}

and one within-subject variable. In the present instance, all variability is between subjects since there are no repeated measurements. There is one between-groups variable, however; there are different groups at each level of A. Furthermore, there is one within-group variable; all groups are represented at all levels of B. This suggests the SV of Table 9–6; it will be helpful to compare this with Table 8–2. Note how the Between G and Within G sources parallel the Between S and Within S sources of the earlier table. The sole discrepancy is the presence of an extra term $S/GB/A$, because we have an added level of nesting in the present design.

A model consistent with this view of the partitioning of the total variability is

(9.4) $$Y_{ijkm} = \mu + \alpha_k + \gamma_{j/k} + \beta_m + (\alpha\beta)_{km} + (\gamma\beta)_{jm/k} + \varepsilon_{ijkm}$$

Note that there are no interactions involving subjects, nor is there any AG or ABG effect. This is because subjects cross with none of the three other variables, and G does not cross with A. Note, as a further help in establishing models for designs involving nesting, that the interaction of a nested effect with another variable will also be nested. For example, the interaction of G/A with B is the nested interaction GB/A.

It is assumed that A and B are fixed-effect variables, but that the groups are a random sample from a large population of such groups. Consequently,

$$\sum_k \alpha_k = \sum_m \beta_m = \sum_k \sum_m (\alpha\beta)_{km} = 0$$

The variance components for the fixed effects are

$$\theta_A^2 = \frac{\sum_k \alpha_k^2}{a-1} \qquad \theta_B^2 = \frac{\sum_m \beta_m^2}{b-1}$$

$$\theta_{AB}^2 = \frac{\sum_k \sum_m (\alpha\beta)_{km}^2}{(a-1)(b-1)}$$

The terms $\gamma_{j/k}$, $(\gamma\beta)_{jm/k}$, and ε_{ijkm} comprise random samples from normally distributed populations with mean zero, and respective variances $\sigma_{G/A}^2$, $\sigma_{GB/A}^2$, and σ_e^2.

9.3.2   THE ANALYSIS OF VARIANCE.   Table 9–6 presents the pertinent aspects of the analysis of variance. The SV have already been discussed. The df column provides an excellent check on our breakdown of sources of variance. The initial division into between-groups and within-group sources can be checked by noting that $df_{\text{B.G}}$ reflects the variability of ag means about the grand mean; therefore, $ag-1$ is the required number. The $df_{\text{W.G}}$ reflects the variability of all bn scores within the group deviated about the group mean and then pooled over the ag groups; the result is $ag(bn-1)$. Adding the two df quantities, we obtain $abgn-1$, which is the correct total. The df for the A, B, and AB terms require no comment. The $df_{G/A}$ reflect that g group means have been subtracted from the

Table 9-6 Analysis of variance for a hierarchical design containing a within-group variable

SV	df	SS	EMS	F
Total	$abgn-1$	$\sum_i^n \sum_j^g \sum_k^a \sum_m^b Y_{ijkm}^2 - C$		
Between G	$ag-1$	$\dfrac{\sum_j^g \sum_k^a T_{.jk.}^2}{nb} - C$		
A	$a-1$	$\dfrac{\sum_k^a T_{..k.}^2}{ngb} - C$	$\sigma_e^2 + ng\sigma_{G/A}^2 + nbg\theta_A^2$	$\dfrac{MS_a}{MS_{G/A}}$
G/A	$a(g-1)$	$SS_{B.G} - SS_A$	$\sigma_e^2 + nb\sigma_{G/A}^2$	$\dfrac{MS_{G/A}}{MS_{S/GB/A}}$
Within G	$ag(bn-1)$	$SS_{tot} - SS_{B.G}$		
B	$b-1$	$\dfrac{\sum_m^b T_{...m}^2}{nga} - C$	$\sigma_e^2 + n\sigma_{GB/A}^2 + nga\theta_B^2$	$\dfrac{MS_B}{MS_{GB/A}}$
AB	$(a-1)(b-1)$	$\dfrac{\sum_k^a \sum_m^b T_{..km}^2}{ng} - C - SS_A - SS_B$	$\sigma_e^2 + n\sigma_{GB/A}^2 + ng\theta_{AB}^2$	$\dfrac{MS_{AB}}{MS_{GB/A}}$
GB/A	$a(b-1)(g-1)$	$\dfrac{\sum_j^g \sum_k^a \sum_m^b T_{.jkm}^2}{n} - \dfrac{\sum_k^a \sum_m^b T_{..km}^2}{ng} - \dfrac{\sum_j^g \sum_k^a T_{.jk.}^2}{nb} + \dfrac{\sum_k^a T_{..k.}^2}{ngb}$	$\sigma_e^2 + n\sigma_{GB/A}^2$	$\dfrac{MS_{GB/A}}{MS_{S/GB/A}}$
S/GB/A	$abg(n-1)$	$SS_{WG} - SS_B - SS_{AB} - SS_{GB/A}$	σ_e^2	

mean for the level of A in which the groups are nested (thus requiring $g-1$ df), that this process has been repeated at all a levels of A, and that the squared deviations have then been pooled. Similarly, GB/A represents the interaction of groups and B for each level of A $[(g-1)(b-1)$ $df]$ pooled over a levels of A. We note finally that there are $n-1$ df for the variability of scores within each level of B within each group; pooling over b levels and ag groups gives the $df_{S/GB/A}$. Adding the various df, we find that there are neither too few nor too many terms in the analysis; the individual terms sum to the appropriate total.

Once we have established the SV, and checked by partitioning the df to see whether this has been done correctly, SS and EMS follow. Remembering that A and B are fixed-effect variables and that groups are assumed to have random effects, we can readily verify that the entries in the EMS column follow the rules developed in Section 8.2.2. Once the EMS have been set down, appropriate F tests are immediately available.

9.3.3 A NUMERICAL EXAMPLE.

Table 9–7 presents data and cell totals for a hierarchical design with one within-group variable. The total variability is as usual,

$$SS_{tot} = \sum_{i}^{n} \sum_{j}^{g} \sum_{k}^{a} \sum_{m}^{b} Y_{ijkm}^2 - C$$

$$= (4)^2 + (5)^2 + \cdots + (22)^2 - \frac{(328)^2}{24}$$

$$= 5,502.00 - 4,482.67$$

$$= 1,019.33$$

Table 9–7   Data matrix for a hierarchical design including a within-group variable

	A_1		A_2		A_3	
	G_{11}	G_{21}	G_{12}	G_{22}	G_{13}	G_{23}
B_1	4	5	3	11	20	18
	6	9	10	6	23	17
	$T_{.j11}=10$	14	$T_{.j21}=13$	17	$T_{.j31}=43$	35
	$T_{..11}=24$		$T_{..21}=30$		$T_{..31}=78$	
B_2	8	10	12	14	19	24
	14	15	15	17	26	22
	$T_{.j12}=22$	25	$T_{.j22}=27$	31	$T_{.j32}=45$	46
	$T_{..12}=47$		$T_{..22}=58$		$T_{..32}=91$	
	$T_{..1.}=71$		$T_{..2.}=88$		$T_{..3.}=169$	
Group Totals						
	$T_{.jk.}=32$	39	40	48	88	81

This quantity is then partitioned into between-groups and within-group components:

$$SS_{\text{B.G}} = \frac{\sum_j^g \sum_k^a T_{.jk}^2}{nb} - C$$

$$= \frac{(32)^2 + \cdots + (81)^2}{4} - \frac{(328)^2}{24}$$

$$= 5,188.50 - 4,482.67$$

$$= 705.83$$

and

$$SS_{\text{W.G}} = SS_{\text{tot}} - SS_{\text{B.G}}$$

$$= 1,019.33 - 705.83$$

$$= 313.50$$

The A main effect contributes to the variability among groups. Therefore, we compute

$$SS_A = \frac{\sum_k^a T_{..k.}^2}{ngb} - C$$

$$= \frac{(71)^2 + (88)^2 + (169)^2}{8} - \frac{(328)^2}{24}$$

$$= 5,168.92 - 4,482.67$$

$$= 686.25$$

The residual variability due to differences among groups is

$$SS_{G/A} = SS_{\text{B.G}} - SS_A$$

$$= 705.83 - 686.25$$

$$= 19.53$$

We next analyze the within-group variability. The effect of B can be swiftly computed by using the single df formula:

$$SS_B = \frac{(T_{...1} - T_{...2})^2}{24}$$

$$= \frac{(132 - 196)^2}{24}$$

$$= 170.67$$

The AB interaction is investigated next:

$$SS_{AB} = \frac{\sum_k^a \sum_m^b (T_{..km}^2)}{ng} - C - SS_A - SS_B$$

$$= \frac{(24)^2 + \cdots + (91)^2}{4} - \frac{(328)^2}{24} - 686.25 - 170.67$$

$$= 5,353.50 - 4,482.67 - 686.25 - 170.67$$

$$= 13.91$$

Table 9–8   Analysis of variance for the data of Table 9–7

SV	df	SS	MS	F
Total	23	1,019.33		
Between G	5	705.83		
A	2	686.25	343.13	52.71[a]
G/A	3	19.53	6.51	
Within G	18	313.50		
B	1	170.67	170.67	46.89[b]
AB	2	13.91	6.96	1.91
GB/A	3	10.92	3.64	.37
$S/GB/A$	12	118.00	9.83	

[a] $p < .005$

[b] $p < .01$

Next, we have

$$SS_{GB/A} = \frac{\sum_j^g \sum_k^a \sum_m^b (T^2_{.jkm})^2}{n} - \frac{\sum_k^a \sum_m^b (T^2_{..km})^2}{ng}$$

$$- \frac{\sum_j^g \sum_k^a (T^2_{.jk.})^2}{nb} + \frac{\sum_k^a (T^2_{..k.})^2}{ngb}$$

$$= \frac{(10)^2 + (14)^2 + \cdots + (46)^2}{2} - 5,353.50 - 5,188.50 + 5,168.92$$

$$= 10.92$$

Finally, we have the residual variability among subjects:

$$SS_{S/GB/A} = SS_{W.G} - SS_B - SS_{AB} - SS_{GB/A}$$

$$= 313.50 - 170.67 - 13.91 - 10.92$$

$$= 118.00$$

Table 9–8 summarizes the analysis. Even on the small number of *df* provided in the example, the *A* and *B* main effects are highly significant. No other sources are significant.

9.4   REPEATED MEASUREMENTS IN GROUP-WITHIN-TREATMENT DESIGNS

The design of Section 9.2 and 9.3 can be extended by requiring several measures from each subject. For example, suppose we again have g groups under high stress A_1 and g groups under low stress A_2. Within each group there are n high-anxious subjects B_1 and n low-anxious subjects B_2. Each member of the group is tested on each of four trials; *trials* is the within-subject variable we shall label C in this section. We have g groups, generally, sampled randomly from a large population of such groups, at each of a levels of the independent variable A. Within each group are b arbitrarily chosen levels of the independent variable B; there are n different subjects at each of these levels. Thus, we have bn subjects in each of ag groups for a total of $abgn$ subjects, each providing one score at each of c levels of C. The analysis of variance is a direct extension of that presented in Table 9–6. The total variation of that table is now the between S variation; we add a within S source on $abgn(c-1)$ df. Note that $df_{\text{B.S}}$ and $df_{\text{w.s}}$ sum to $abcgn-1$, the appropriate df_{total} for the design under consideration. To partition the within S variability, first write C on $c-1$ df. The remaining SV are generated by crossing C with each of the between S sources of Table 9–6:

SV	df
C	$c-1$
AC	$(a-1)(c-1)$
GC/A	$a(g-1)(c-1)$
$\cdots$	
$SC/GB/A$	$abg(n-1)(c-1)$

Note the general form of interactions involving nested variables. The interaction of G/A and C, for example, is represented by GC/A, not G/AC; the df also suggest that within each of a levels of A, we have a nested interaction on $(g-1)(c-1)$ df. The SS and EMS should give no problems.

9.5 NESTING WITHIN-SUBJECT VARIABLES

To this point, we have considered designs in which subjects are nested within levels of an independent variable A, or within levels of some variable G, which in turn is nested within levels of A. We now turn to designs in which nesting is within, or in addition to, instead of between subjects. In such cases, the subject is faced with a randomly ordered series of stimuli or situations that are nested within levels of some variable. Examples are problems nested within difficulty levels, pictures nested within themes, and words nested within grammatical class. A measure such as response time, trials to criterion, or rating, furthermore, is obtained for each stimulus or situation. We shall now complicate matters by assuming that subjects are also nested within levels of some independent variable. We have, generally, n subjects at each of a levels of A for a total of an subjects. Each subject is tested with b different stimuli at each of c levels of C; that is, B is nested within levels of C (B/C). The design is very much like that of Table 8–10, except that here the levels of B are not the same at the various levels of C. We assume that the effects of *subjects* (S/A) and *stimuli* (B/C) are random and that the effects of A and B are fixed. In what follows, we shall try to develop an approach that is general enough to let us deal with variations of the designs in

Chapters 8 and 9 to any degree of complexity. Our first concern is determining SV and df for a general partitioning of the total variability.

STEP 1.   Partition the total variability into two main sources, between-subjects and within-subject variability:

SV	df
Total	$abcn - 1$
Between S	$an - 1$
Within S	$an(bc - 1)$

STEP 2.   Further partition the between-subjects variability:

SV	df
Between S	$an - 1$
A	$a - 1$
S/A	$a(n - 1)$

STEP 3.   We make a first try at partitioning the within-subject variability by viewing the design as involving bc levels of stimuli:

SV	df
Within S	$an(bc - 1)$
Stimuli	$bc - 1$
$A \times$ Stimuli	$(a - 1)(bc - 1)$
$S \times$ Stimuli$/A$	$a(n - 1)(bc - 1)$

Note that once we have written *stimuli*, we merely cross it with the sources generated in step 2.

STEP 4.   We now must partition the variability due to stimuli and its interactions:

SV	df
Stimuli	$bc - 1$
C	$c - 1$
B/C	$c(b - 1)$

Crossing each of the above with A yields:

SV	df
$A \times$ Stimuli	$(a - 1)(bc - 1)$
AC	$(a - 1)(c - 1)$
AB/C	$c(a - 1)(b - 1)$

Crossing C and B/C with S/A yields:

SV	df
$S \times \text{Stimuli}/A$	$a(n-1)(bc-1)$
SC/A	$a(n-1)(c-1)$
SB/AC	$ac(n-1)(b-1)$

Note that the interaction of S/A and B/C is SB/AC. This source corresponds to the *subject* $\times$ *stimuli* SS computed within each AC cell and then pooled over cells; note the correspondence between this verbal statement and the *df*.

In summary, we always first partition variability into between-subjects and within-subject terms. We then partition the between-subjects variability, following the lines developed in Chapter 5 and in Sections 9.2–9.4. In partitioning within-subject variability, we begin with our smallest experimental units, for example stimuli. The variability among those units is then further partitioned and interactions with between-subjects sources are then noted.

The end result for the design under consideration is presented in Table 9–9. We have omitted the cumbersome SS expressions; by this point, calculating any crossed or nested term should provide no difficulty. When in doubt, expanding *df* should lead to the proper expression. The rules of thumb (Chapter 8) have, as usual, provided the basis for generating the *EMS* of Table 9–9. Under the general model for getting the *EMS*, no single line in Table 9–9 provides an error term against which to test A, C, or AC, all terms of strong interest to the experimenter. If we were to test A against S/A, a significant result would be ambiguous; $\sigma^2_{AB/C}$ or θ^2_A, or both, might be contributing to the mean square. On the other hand, a significant result when A is tested against AB/C yields no greater assurance that θ^2_A is not zero; $\sigma^2_{S/A}$ may be making a sizable contribution to the mean square. We meet similar difficulty in trying to evaluate the contributions of C and AC. One possible solution to the problem is using preliminary tests of the sort described in our earlier discussion of pooling (Section 9.2.3). If we had prior grounds for believing that $\sigma^2_{AB/C} = 0$, for example, and a preliminary test of AB/C against SB/AC was not significant at the .25 level, it would be reasonable to delete $\sigma^2_{AB/C}$ wherever it appears in Table 9–9. In that case, S/A would provide an appropriate test of the hypothesis that $\theta^2_A = 0$. If this approach fails to provide an error term, quasi F ratios could be calculated (the nature of two such ratios, their *df*, and results of investigations of their properties have been presented in Section 7.6.4). Table 9–9 presents two alternative tests for each of the terms in question, as well as the general form of the *df* associated with the combination of mean squares used in such quasi F ratios.

That a subject is tested with several randomly sampled stimuli does not necessitate quasi F ratios, or a simplified model in which certain variance components are assumed to be zero. With many stimulus items, the experiment can be so designed that appropriate error terms are immediately available. Assume, for example, that we want to investigate how age of subjects and familiarity with words affect time of reading words aloud. We could use the

Table 9–9   Analysis of variance for a mixed design with subjects nested within levels of A and B nested within levels of C

SV	df	EMS
Total	$abcn - 1$	
Between S	$an - 1$	
(1) A	$a - 1$	$\sigma_e^2 + bc\sigma_{S/A}^2 + n\sigma_{AB/C}^2 + \sigma_{SB/AC}^2 + nbc\theta_A^2$
(2) S/A	$a(n - 1)$	$\sigma_e^2 + bc\sigma_{S/A}^2 \qquad\qquad + \sigma_{SB/AC}^2$
Within S	$an(bc - 1)$	
(3) C	$c - 1$	$\sigma_e^2 + b\sigma_{SC/A}^2 + na\sigma_{B/C}^2 \quad + \sigma_{SB/AC}^2 + nab\theta_c^2$
(4) AC	$(a - 1)(c - 1)$	$\sigma_e^2 + b\sigma_{SC/A}^2 + n\sigma_{AB/C}^2 + \sigma_{SB/AC}^2 + nb\theta_{AC}^2$
(5) SC/A	$a(n - 1)(c - 1)$	$\sigma_e^2 + b\sigma_{SC/A}^2 \qquad\qquad + \sigma_{SB/AC}^2$
(6) B/C	$c(b - 1)$	$\sigma_e^2 \qquad\qquad + na\sigma_{B/C}^2 + \sigma_{SB/AC}^2$
(7) AB/C	$c(a - 1)(b - 1)$	$\sigma_e^2 \qquad\qquad + n\sigma_{AB/C}^2 + \sigma_{SB/AC}^2$
(8) SB/AC	$ac(b - 1)(n - 1)$	$\sigma_e^2 \qquad\qquad\qquad\qquad + \sigma_{SB/AC}^2$

Quasi F (F') Ratios[a]

H_0 term	F_1'	F_2'
A	$\dfrac{(1)}{[(2) + (7) - (8)]}$	$\dfrac{[(1) + (8)]}{[(2) + (7)]}$
C	$\dfrac{(3)}{[(5) + (6) - (8)]}$	$\dfrac{[(3) + (8)]}{[(5) + (6)]}$
AC	$\dfrac{(4)}{[(5) + (6) - (8)]}$	$\dfrac{[(4) + (8)]}{[(5) + (6)]}$

$$df = \frac{(MS_1 \pm MS_2 \pm \cdots)^2}{MS_1/df_1 + MS_2/df_2 + \cdots}$$

[a] The digits in parentheses refer to the mean squares indexed by them. For example, (8) is shorthand for $MS_{SB/AC}$.

experimental design we have just finished analyzing; each subject would be tested on b words at each of c levels of familiarity. An alternative would be to create *acd* lists of words; that is, for each age by familiarity combination AC, we should have d different lists D/AC. Furthermore, we should have n subjects tested on each list of b words $S/D/AC$, with a resulting total of *acdn* subjects and *abcan* scores in all. Table 9–10 presents the appropriate analysis of variance, if A and C are assumed to have fixed effects, and B, D, and S to have random effects. The point of interest is that the A, C, and AC terms are tested against D/AC. By our example, if the null hypotheses for the first three lines are true (θ_A^2, θ_C^2, and θ_{AC}^2 all equal zero), the same components of variability contribute to them as to the variability among the different lists nested within AC cells. In using such a design, one must have a large enough number of lists to ensure adequate power against an error term distributed on $ac(d - 1)$ df.

The two designs whose analyses have been presented in Tables 9–9 and 9–10 are only two of several possible ways of dealing with variability due to

Table 9–10   Analysis of variance for a mixed design with subjects and B nested within stimulus sets D

SV	df	EMS
Total	$abcdn - 1$	
Between S	$acdn - 1$	
A	$a - 1$	$\sigma_e^2 + \sigma_{SB/D/AC}^2 + n\sigma_{B/D/AC}^2 + b\sigma_{S/D/AC}^2 + bn\sigma_{D/AC}^2 + bcdn\theta_A^2$
C	$c - 1$	$\sigma_e^2 + \sigma_{SB/D/AC}^2 + n\sigma_{B/D/AC}^2 + b\sigma_{S/D/AC}^2 + bn\sigma_{D/AC}^2 + abdn\theta_C^2$
AC	$(a-1)(c-1)$	$\sigma_e^2 + \sigma_{SB/D/AC}^2 + n\sigma_{B/D/AC}^2 + b\sigma_{S/D/AC}^2 + bn\sigma_{D/AC}^2 + bdn\theta_{AC}^2$
D/AC	$ac(d-1)$	$\sigma_e^2 + \sigma_{SB/D/AC}^2 + n\sigma_{B/D/AC}^2 + b\sigma_{S/D/AC}^2 + bn\sigma_{D/AC}^2$
$S/D/AC$	$acd(n-1)$	$\sigma_e^2 + \sigma_{SB/D/AC}^2 \qquad\qquad + b\sigma_{S/D/AC}^2$
Within S	$acdn(b-1)$	
$B/D/AC$	$acd(b-1)$	$\sigma_e^2 + \sigma_{SB/D/AC}^2 + n\sigma_{B/D/AC}^2$
$SB/D/AC$	$acd(b-1)(n-1)$	$\sigma_e^2 + \sigma_{SB/D/AC}^2$

randomly sampled items. Nevertheless, they provide an opportunity to note some of the factors involved in choosing among designs in this problematical area. On the one hand, we have an approximate F ratio whose properties are not entirely known. On the other hand, we require more stimulus material (*abcd* items against *bc* items) and F tests of C and AC, which will tend to be inefficient because variability among subjects contributes to the mean squares. As always, there is no simple rule for choosing among designs. Experimenters have to make the choice anew in each investigation, carefully weighing the relevant factors in light of their primary concerns and their knowledge of previous results in the area under study.

9.6   CONCLUDING REMARKS

Although there are countless variations of the hierarchical design, all yield to the same principles of analysis that have been earlier applied repeatedly (Chapters 8 and 9). The first step is to have a sound understanding of the layout of the design. Which variables are nested in which others? Which variables cross each other? Which are between-subjects and which are within-subject variables? The answers to these questions direct the partitioning of total variability and *df*. To generate *EMS*, we must determine which variables are to be viewed as random, which as fixed. This is not always an easy decision; the answer depends on how levels of the variable have been selected and on the range of generalization we intend. Once variables have been classified, applying the rules of thumb (Section 8.2.2) readily yields the *EMS*.

The nature of the data analysis is particularly sensitive to the choice of a structural model. If certain variables are assumed to have negligible effects, it will frequently be possible to have tests of greater power and to avoid quasi *F* ratios. Although such consequences of simplifying the model are desirable, we have espoused a more conservative approach to model construction, and therefore to data analysis. It is preferable to assume a general model, incorporating all the effects we can conceive. Sometimes prior information and preliminary tests of certain terms will let us delete certain parameters from the model and the corresponding variances from the *EMS* but assumptions alone are not sufficient grounds for such a procedure. Wishing some variance component to be zero will not make it so, and the price of wrongly assuming that the component is zero is ordinarily a Type I error in testing treatment effects of interest.

The presence of other random-effect variables besides subjects, a characteristic of the designs of this chapter, raises additional considerations in planning the experiment. One important point is that merely running many subjects will not ensure sufficient power to test null hypotheses of interest. In the designs of Sections 9.2–9.4, the value of *g*—the number of social groups, classes, litters, and so on—is the critical determinant of error *df* and thus of power. Unless there are grounds for pooling to obtain an error term on more *df*, there is little the experimenter can do after the data are collected. Thus it is important to work out the actual analysis of sources and *df* before collecting data and to modify the design in whatever ways seem necessary to obtain powerful tests of effects of interest. In the extreme case in which *g* = 1, there is not only a loss of power but a confounding of groups and levels of *A*. If one class is taught by one method and another by a second, is a difference in class means due to the different methods or to differences in the personal interactions within the two classes? To determine the effect of the treatment, we need some measure of variability among classes taught by the same method. Experimenters often do not realize that the failure to replicate groups within levels is not particularly different from running one subject at each level of *A* in a simple completely randomized one-factor design.

Similar comments hold for the designs of Section 9.5. When stimuli are a random sample from some population, considering test power requires that there be an adequate number of stimuli, or as in the case of the design of Table 9–10, an adequate number of sets of stimuli.

EXERCISES

9.1   Analyze the following data set.

	A_1		A_2		A_3	
G_{11}	G_{21}	G_{12}	G_{22}	G_{13}	G_{23}
5	8	11	32	12	18
21	23	15	18	22	31
14	10	16	26	36	38
12	17	23	25	18	37
8	15	27	26	34	28
16	20	31	17	41	32

9.2 There are three measures on each subject in the groups nested within levels of A. Analyze the data.

		B_1	B_2	B_3
	S_{111}	5	8	12
G_{11}	S_{211}	7	8	14
	S_{311}	8	10	17
	S_{411}	7	11	16
	S_{121}	10	12	19
G_{21}	S_{221}	8	10	20
	S_{321}	7	11	17
	S_{421}	8	2	18

A_1

		B_1	B_2	B_3
	S_{112}	18	21	28
G_{12}	S_{212}	17	19	23
	S_{312}	14	18	24
	S_{412}	16	21	27
	S_{122}	15	21	32
G_{22}	S_{222}	14	18	29
	S_{322}	15	23	28
	S_{422}	17	22	31

A_2

9.3 Five subjects are shown a series of 4 pleasant and 4 unpleasant pictures. Later, they are shown 16 pictures, including the old 8 pictures and 4 new pleasant and 4 new unpleasant pictures. At this time, they are asked to indicate their confidence that the picture is new or old on a scale of zero (very sure it is new) to nine (very sure it is old). Analyze the following data set. Note that A_1 = old, A_2 = new, B_1 = pleasant, and B_2 = unpleasant, and that the pictures are viewed as a random sample.

A_1

	P_{111}	B_1 P_{211}	P_{311}	P_{411}	P_{112}	B_2 P_{212}	P_{312}	P_{412}
S_1	5	6	4	8	5	8	8	9
S_2	5	7	6	5	7	8	9	7
S_3	6	7	5	8	8	5	9	6
S_4	4	7	6	6	9	7	8	7
S_5	6	7	5	6	6	9	6	8

A_2

	P_{121}	P_{221}	P_{321}	P_{421}	P_{122}	P_{222}	P_{322}	P_{422}
S_1	5	3	2	3	1	2	1	4
S_2	2	0	3	5	0	1	2	0
S_3	3	4	2	2	3	2	0	1
S_4	1	2	4	3	0	2	1	0
S_5	3	1	2	1	1	2	0	3

9.4 A study in economic bargaining involves thirty-six 3-man groups—a seller and two buyers. One of the buyers is a stooge who can side with the other buyer by keeping his own bids low, or with the seller by bidding the price of the commodity up. In one-third of the groups, the stooge aids the seller on 75 percent of the trials; in another 12 groups, 50 percent; in another 12 groups, 25 percent. An average concession score (distance from initial bid) was obtained from each S under two conditions: trials when stooge sided with S, trials when stooge was against S. These two measures were obtained from each of four blocks yielding eight measures per subject. Give SV, df, EMS.

9.5   An experiment was performed on the effects of socioeconomic status and intelligence on self-evaluation. Three school districts, each of a different social stratum, were chosen for participation in the experiment. Ten schools were selected at random from among the elementary schools in each district (that is, a total of 30 schools). Ten high and ten low IQ subjects were selected from each school. Each subject was given a self-evaluation scale, the dependent variable. The experimenter performed the following analysis.

SV	df
District D	2
Intelligence level I	1
$D \times I$	2
Within error	594
Total	599

Can you suggest an alternative analysis (including error terms)? What inference might be added? changed? Why?

9.6   Thirty-two subjects are randomly assigned to eight 4-man groups, four of which are task-oriented and four of which are ego-oriented. Each group is required to solve three problems under stress and three other problems under no stress. The score for each group is number of trials it takes the group to solve the problems. Give the *SV*, *df*, *EMS*, and error terms.

9.7   An analyst meets with twelve groups (*G*) for an hour each week. Each group consists of 3 males and 3 females (sex, *X*). Six of the groups are engaged in a directed therapy in which the analyst plays a visible and central role, and six of the groups are considerably more nondirective (therapy, *T*). Self-ratings are collected at the end of a year of therapy and analyzed.
(a)   What are the variables? Which are nested in which? Which are fixed? random?
(b)   What are *SV*, *df*, *EMS*, *F*?
(c)   Given that an evaluation of therapy and of the relative effectiveness of therapies for the two sexes are of primary interest, find the chief weakness of this design? If you had run this design as it stands, is there any possibility of remedying the weakness? Explain the possibility and its dangers.

9.8   A typical paradigm in cognitive psychology involves verifying visually presented statements. In an example, the *S* responds "Yes" to "A collie is a dog" and "No" to "A sparrow is a cat." Reaction time is recorded on each trial. In one such experiment, 60 true statements and 60 false statements were presented to all of 24 *S*s. In one-third of each kind of statement, subject (for example, collie) and predicate (for example, dog) are highly associated in the English language; associative strength (*A*) was moderate for another third of the sentences, and weak for the remaining third. For example, "A collie is a dog" is strongly associated and positive while "A bread is butter" is strongly associated and negative.
(a)   State *SV*, *df*, *EMS*, *F* ratios. That means quasi *F*s if needed.
(b)   An alternative design involving the same number of observations would have 16 *S*s who received only high *A* statements—30 true and 30 false, 16 *S*s with only moderate *A* statements, and 16 *S*s with only weak *A* statements. Now give the analysis of variance. Briefly discuss the possible pros and cons of (a) and (b).
(c)   Consider still another alternative. The design is exactly as in part (a) except that every *S* gets a different set of sentences. Now give the analysis of variance. Briefly discuss the merits of this design.

REFERENCES

Bozivich, H., Bancroft, T. A., and Hartley, H. O. Power of analysis of variance test procedures for certain incompletely specified models. *Annals of Mathematical Statistics* 27:1017–43 (1956).

Srivastava, S. R., and Bozivich, H. Power of certain analysis of variance test procedures involving preliminary tests. *Bulletin de l'Institut international statistique,* 33rd Session (1961).

10 LATIN SQUARE DESIGNS

10.1 INTRODUCTION

The basic Latin square design can be represented as

$$
\begin{array}{c}
\quad\quad C_1 \quad C_2 \quad C_3 \\
\begin{array}{c} B_1 \\ B_2 \\ B_3 \end{array}
\begin{bmatrix}
A_1 & A_2 & A_3 \\
A_2 & A_3 & A_1 \\
A_3 & A_1 & A_2
\end{bmatrix}
\end{array}
$$

where A, B, and C are variables in an experiment. Two aspects of this layout are worth noting. First, this is a Latin square because each level of A appears exactly once in each row (level of B) and in each column (level of C); the above layout is one of several that meet this definition. Secondly, the Latin square is an incomplete design in that it includes only 9 of the possible 27 combinations of levels of A, B, and C that would appear in a complete factorial design; more generally, assuming a rows and columns in the square, we have a^2 of the possible a^3 treatment combinations. We shall show later that this "incompleteness" has direct and important consequences for the applicability of the design.

The Latin square has several potential advantages over other designs. We have already touched on one, its ability to investigate several variables with less expenditure of time and subjects than complete factorial designs would involve. This is important in designing pilot studies or in case of a limited supply of available subjects.

Even more important is the efficiency of the Latin square relative to other designs. Consider, for example, a treatment × blocks design with a blocks and a treatment levels. The blocks might correspond to levels of performance on some measure of ability related to the dependent variable. We could introduce a second concomitant variable, perhaps a measure of motivation, with no additional experimental effort; the rows of our Latin square layout would be blocks based on level of intelligence and the columns would be blocks based on level of motivation. Since two sources of individual differences have been removed, we should have a very small error term and consequently a very precise test of the treatment variable A.

Perhaps the most prevalent use of the Latin square design is in situations in which it is desirable to test each subject under all levels of the treatment variable A. Then, the levels of B correspond to individual subjects or groups of subjects and the levels of C correspond to positions in time, for example, successive days of the experiment. The repeated-measurement design, in which the order of the A_j was independently randomized for each subject, permitted removing error variance due to individual differences. In the Latin square design, in which the orders of the A_j are chosen to meet the Latin square requirement, we are able to remove still another source of error variance, the variability due to temporal effects. Thus, if subjects were tested under each level of A on different days, and if the order was Latin-squared, variability due to days could be removed from the error variance. We should then have a more precise test of the A effects and also an independent evaluation of the variance due to days.

By this time, dazzled by the promised savings in effort and error variance, the reader may wonder why any investigator would contemplate any other design except the Latin square or some extension of it. There is a catch. The potential advantages, as usual, are accompanied by potential disadvantages. Information about interactions among the row, column, and treatment variables will frequently be impossible to extract. Furthermore, if such interactions exist, they may obscure inferences about the main effects of the treatments of interest. Whether the Latin square can reasonably be applied in a particular situation depends on whether interactions exist in the population, which variables interact, and which variables have random or fixed effects. Let us first describe how the actual design layout is selected.

10.2 SELECTING A LATIN SQUARE

Consider the arrangement of treatments

$$\begin{bmatrix} A_1 & A_2 & A_3 \\ A_2 & A_3 & A_1 \\ A_3 & A_1 & A_2 \end{bmatrix}$$

The characteristic that defines this arrangement as a Latin square is the occurrence of each treatment exactly once in each row and in each column. This square is only one of several possible. It is generally referred to as a *standard square*, since the first row and first column are in standard order (A_1, A_2, A_3). Interchanging rows and columns, we could get

$$\begin{bmatrix} A_1 & A_3 & A_2 \\ A_3 & A_2 & A_1 \\ A_2 & A_1 & A_3 \end{bmatrix}$$

This is one of $3!\,2! - 1$ nonstandard squares obtainable from the standard square. If there are four treatment levels, there are four possible standard squares. Since each of these can, by permutation of rows and columns, give rise to $4! \, 3! - 1$ nonstandard squares, the total number of 4×4 Latin squares is $4(4! \, 3! - 1) + 4$, or 576. As we add levels of A, even more marked increases occur in the possible number of squares.

Sets of standard squares are available in several sources. A very extensive presentation is the set of tables in Fisher and Yates (1955).

A standard square should be chosen at random from the complete set. Then randomly permute all the rows, all the columns, and for squares larger than 4×4, the letters. Suppose we choose at random the following 6×6 standard square:

$$\begin{bmatrix} A_1 & A_2 & A_3 & A_4 & A_5 & A_6 \\ A_2 & A_4 & A_5 & A_3 & A_6 & A_1 \\ A_3 & A_6 & A_1 & A_5 & A_4 & A_2 \\ A_4 & A_1 & A_2 & A_6 & A_3 & A_5 \\ A_5 & A_3 & A_6 & A_1 & A_2 & A_4 \\ A_6 & A_5 & A_4 & A_2 & A_1 & A_3 \end{bmatrix}$$

We note the order of appearance of the numbers 1 through 6 in a table of random numbers; we might obtain $\langle 1, 3, 2, 4, 6, 5 \rangle$. Then the rows are permuted so that the first row is still first, the third row is now second, the second row is now third, and so on. The result of the permutation of rows is

$$\begin{bmatrix} A_1 & A_2 & A_3 & A_4 & A_5 & A_6 \\ A_3 & A_6 & A_1 & A_5 & A_4 & A_2 \\ A_2 & A_4 & A_5 & A_3 & A_6 & A_1 \\ A_4 & A_1 & A_2 & A_6 & A_3 & A_5 \\ A_6 & A_5 & A_4 & A_2 & A_1 & A_3 \\ A_5 & A_3 & A_6 & A_1 & A_2 & A_4 \end{bmatrix}$$

We next permute all columns; we might have the random sequence $\langle 4, 1, 6, 2, 5, 3 \rangle$. Then the square becomes

$$\begin{bmatrix} A_4 & A_1 & A_6 & A_2 & A_5 & A_3 \\ A_5 & A_3 & A_2 & A_6 & A_4 & A_1 \\ A_3 & A_2 & A_1 & A_4 & A_6 & A_5 \\ A_6 & A_4 & A_5 & A_1 & A_3 & A_2 \\ A_2 & A_6 & A_3 & A_5 & A_1 & A_4 \\ A_1 & A_5 & A_4 & A_3 & A_2 & A_6 \end{bmatrix}$$

The random sequence dictating the permuting of letters might be $\langle 2, 4, 5, 1, 6, 3 \rangle$. The A_1s in the above square are replaced by A_2s, the A_2s by A_4s, the A_3s by

A_5s, and so on. Then we have

$$\begin{bmatrix} A_1 & A_2 & A_3 & A_4 & A_6 & A_5 \\ A_6 & A_5 & A_4 & A_3 & A_1 & A_2 \\ A_5 & A_4 & A_2 & A_1 & A_3 & A_6 \\ A_3 & A_1 & A_6 & A_2 & A_5 & A_4 \\ A_4 & A_3 & A_5 & A_6 & A_2 & A_1 \\ A_2 & A_6 & A_1 & A_5 & A_4 & A_3 \end{bmatrix}$$

This square can be considered a random selection from the population of 6×6 squares.

10.3 THE NONADDITIVITY PROBLEM

Suppose that an additive model provides a valid representation of the population of possible observations. Then population interaction effects are negligible and each score is viewed as the sum of μ, an error component, and main effects associated with the row, column, and treatment A variables. Throughout this chapter, we shall note situations in which such a model might be appropriate. We need only remark now that in such circumstances, efficient tests of main effects, readily interpretable, will be available.

Unfortunately, the best of all possible worlds does not always prevail. When treatment effects are not additive—that is, when interaction effects are present—it may be difficult to interpret F tests. Even under nonadditivity, this is not always so, but it is best to be aware of the possible problems. We begin by considering the simplest Latin square, a 2×2 design.

Suppose we had the set of cell totals

$$\begin{array}{cc} & \begin{array}{cc} C_1 & C_2 \end{array} \\ \begin{array}{c} B_1 \\ \\ B_2 \\ \end{array} & \begin{bmatrix} (A_1) & (A_2) \\ 5 & 8 \\ (A_2) & (A_1) \\ 6 & 11 \end{bmatrix} \end{array}$$

By the single df approach of Section 5.11, $SS_A = [(5 + 11) - (6 + 8)]^2/4 = 1$. Now consider SS_{BC}. The calculations are identical to those for SS_A. The finding that the SS_A equals the SS_{BC} is typical of the 2×2 Latin square. In fact, it can be shown that the sum of squares for the main effect of any one variable will always equal the sum of squares for the interaction effect of the other two variables. Thus, the B main effect is confounded with the AC interaction ($SS_B = SS_{AC}$) and the C effect cannot be distinguished from the AB interaction ($SS_C = SS_{AB}$). As a consequence of this confounding of main and interaction effects in the 2×2 design, the F test of a main effect can be interpreted only if it is assumed that the population variance due to the interaction of the other two variables is zero. The

Table 10–1   Data and subtotals from a 3^3 factorial design

	C_1			C_2			C_3		
	A_1	A_2	A_3	A_1	A_2	A_3	A_1	A_2	A_3
B_1	4	1	6	2	4	1	1	4	2
B_2	5	2	3	1	7	5	3	8	6
B_3	6	1	4	3	3	4	1	5	4
	15	4	13	6	14	10	5	17	12

C_1 Total = 32 C_2 Total = 30 C_3 Total = 34   Grand Total = 96

Subtotals for AB Cells

	A_1	A_2	A_3	B Totals
B_1	7	9	9	25
B_2	9	17	14	40
B_3	10	9	12	31
A Totals	26	35	35	96

SS_A can be interpreted to reflect variability due to differences in the effects of A_1 and A_2, for example, only if we assume that σ^2_{BC} is negligible.

Confounding exists in larger squares, but it is not complete. To examine the nature of the confounding, we shall compare the relation of main and interaction effects in a 3^3 factorial design with those in a 3×3 Latin square. We first consider the factorial arrangement, noting the example presented in Table 10–1. If $\frac{2}{9}$ is

Table 10–2   Data of Table 10–1 after adjustment for C effects

	C_1			C_2			C_3		
	A_1	A_2	A_3	A_1	A_2	A_3	A_1	A_2	A_3
B_1	4	1	6	$2\frac{2}{9}$	$4\frac{2}{9}$	$1\frac{2}{9}$	$\frac{7}{9}$	$3\frac{7}{9}$	$1\frac{7}{9}$
B_2	5	2	3	$1\frac{2}{9}$	$7\frac{2}{9}$	$5\frac{2}{9}$	$2\frac{7}{9}$	$7\frac{7}{9}$	$5\frac{7}{9}$
B_3	6	1	4	$3\frac{2}{3}$	$3\frac{2}{9}$	$4\frac{2}{9}$	$\frac{7}{9}$	$4\frac{7}{9}$	$3\frac{7}{9}$
	15	4	13	$6\frac{2}{3}$	$14\frac{2}{3}$	$10\frac{2}{3}$	$4\frac{1}{3}$	$16\frac{1}{3}$	$11\frac{1}{3}$

C_1 Total = 32 C_2 Total = 32 C_3 Total = 32   Grand Total = 96

Subtotals for AB Cells

	A_1	A_2	A_3	B Totals
B_1	7	9	9	25
B_2	9	17	14	40
B_3	10	9	12	31
A Totals	26	35	35	96

Table 10–3 Data from a Latin square design

	C_1	C_2	C_3	B Totals
B_1	(A_1) 4	(A_2) 4	(A_3) 2	10
B_2	(A_3) 3	(A_1) 1	(A_2) 8	12
B_3	(A_2) 1	(A_3) 4	(A_1) 1	6
A Totals	8	9	11	28

added to every score at C_2 and subtracted from every score at C_3, the C main effect has been removed (that is, SS_C is now zero); this can be seen in Table 10–2. Note, however, that the AB interaction effects are clearly unchanged. This is because C and AB effects are independent quantities in the complete factorial design.

We next turn to the Latin square approach. Table 10–3 presents a Latin square that involves three levels of each of three variables. If $\frac{4}{9}$ is added to all the C_1 scores, $\frac{1}{9}$ is added to all the C_2 scores, and $\frac{5}{9}$ is subtracted from all the C_3 scores, the C main effects are removed from the data matrix. The BC interaction effects have not been changed; the distance between any two rows in any column is exactly what it was before the adjustment. This is apparent in Table 10–4. The reader should verify that the adjustment has also not affected the variability due to the AC interaction effects. The AB interaction variability is changed, however. Retabling the data, before adjusting for C effects, we had

$$
\begin{array}{c}
 \begin{array}{ccc} B_1 & B_2 & B_3 \end{array} \\
\begin{array}{c} A_1 \\ \\ A_2 \\ \\ A_3 \end{array}
\left[
\begin{array}{ccc}
(C_1)\,4 & (C_2)\,1 & (C_3)\,1 \\
(C_2)\,4 & (C_3)\,8 & (C_1)\,1 \\
(C_3)\,2 & (C_1)\,3 & (C_2)\,4
\end{array}
\right]
\end{array}
$$

Table 10–4 Data of Table 10–3 after adjustment for C effects

	C_1	C_2	C	B Totals
B_1	$4\frac{4}{9}$	$4\frac{1}{9}$	$1\frac{4}{9}$	10
B_2	$3\frac{4}{9}$	$1\frac{1}{9}$	$7\frac{4}{9}$	12
B_3	$1\frac{4}{9}$	$4\frac{1}{9}$	$\frac{4}{9}$	6
A Totals	$9\frac{1}{3}$	$9\frac{1}{3}$	$9\frac{1}{3}$	28

We now have

$$
\begin{array}{c}
\begin{array}{ccc} B_1 & \qquad B_2 & \qquad B_3 \end{array} \\
\begin{array}{c} A_1 \\[18pt] A_2 \\[18pt] A_3 \end{array}
\begin{bmatrix}
(C_1)_{\;4\frac{4}{9}} & (C_2)_{\;1\frac{1}{9}} & (C_3)_{\;\frac{4}{9}} \\[14pt]
(C_2)_{\;4\frac{1}{9}} & (C_3)_{\;7\frac{4}{9}} & (C_1)_{\;1\frac{4}{9}} \\[14pt]
(C_3)_{\;1\frac{4}{9}} & (C_1)_{\;3\frac{4}{9}} & (C_2)_{\;4\frac{1}{9}}
\end{bmatrix}
\end{array}
$$

The sums of squares for the AB interaction are not the same for the two data sets. The situation is not quite the same as in the 2×2 Latin square. In that case, the SS_{AB} would be reduced to zero if variability due to C were removed from the matrix. In the larger squares, confounding is not complete. Nor are main and interaction effects completely independent, as they were in the factorial design considered earlier. In the factorial example, removing variability due to C left the variability due to the AB interaction completely unaffected.

The implication of the preceding discussion is that the expected mean square for a main effect in a Latin square design will have variance of interaction effects involving the other two variables as one of its components, unless such interaction effects are absent in the population. To understand why this happens, again consider the layout of Table 10–3. Assuming one score in each of the nine cells, and assuming a very general nonadditive model, we have

$$(10.1) \qquad Y_{jkm} = \mu + \alpha_j + \beta_k + \gamma_m + (\alpha\beta)_{jk} + (\alpha\gamma)_{jm} + (\beta\gamma)_{km} + (\alpha\beta\gamma)_{jkm} + \varepsilon_{jkm}$$

where α corresponds to A, β to B, and γ to C. The average score at B_1 is given by $\bar{Y}_{.1.} = (Y_{111} + Y_{212} + Y_{313})/3$. By Equation (10.1), we have

$$
\begin{aligned}
\bar{Y}_{.1.} = \tfrac{1}{3}[&(\mu + \alpha_1 + \beta_1 + \gamma_1 + (\alpha\beta)_{11} + (\alpha\gamma)_{11} + (\beta\gamma)_{11} + (\alpha\beta\gamma)_{111} + \varepsilon_{111}) \\
(10.2) \qquad &+ (\mu + \alpha_2 + \beta_1 + \gamma_2 + (\alpha\beta)_{21} + (\alpha\gamma)_{22} + (\beta\gamma)_{12} + (\alpha\beta\gamma)_{212} + \varepsilon_{212}) \\
&+ (\mu + \alpha_3 + \beta_1 + \gamma_3 + (\alpha\beta)_{31} + (\alpha\gamma)_{33} + (\beta\gamma)_{13} + (\alpha\beta\gamma)_{313} + \varepsilon_{313})]
\end{aligned}
$$

Assuming that α, β, and γ are fixed effects, we get

$$\alpha_1 + \alpha_2 + \alpha_3 = 0$$
$$\gamma_1 + \gamma_2 + \gamma_3 = 0$$
$$(\alpha\beta)_{11} + (\alpha\beta)_{21} + (\alpha\beta)_{31} = 0$$
$$(\beta\gamma)_{11} + (\beta\gamma)_{12} + (\beta\gamma)_{13} = 0$$

These four results can be proved algebraically once we assume fixed effects. Generally, however,

$$(\alpha\gamma)_{11} + (\alpha\gamma)_{22} + (\alpha\gamma)_{33} \neq 0$$
$$(\alpha\beta\gamma)_{111} + (\alpha\beta\gamma)_{212} + (\alpha\beta\gamma)_{313} \neq 0$$

The sum of interaction effects can be shown to be zero when we sum over all levels at one index *and hold all other indices constant*. The $(\alpha\beta)_{jk}$ and $(\alpha\beta\gamma)_{jkm}$ effects are varying relative to two indices as we sum them. The result of these developments is

(10.3)

$$\bar{Y}_{.1.} = \mu + \beta_1 + \tfrac{1}{3}[(\alpha\gamma)_{11} + (\alpha\gamma)_{22} + (\alpha\gamma)_{33}]$$
$$+ \tfrac{1}{3}[(\alpha\beta\gamma)_{111} + (\alpha\beta\gamma)_{212} + (\alpha\beta\gamma)_{313}]$$
$$+ \tfrac{1}{3}[\varepsilon_{111} + \varepsilon_{212} + \varepsilon_{313}]$$

The means at B_2 and B_3 are also contributed to by an average of three $\alpha\gamma$ components and an average of three $\alpha\beta\gamma$ components. Thus, the SS_B involves not only the variability of the β_k but also the variability among

$$(\tfrac{1}{3})[(\alpha\gamma_{11}) + (\alpha\gamma)_{22} + (\alpha\gamma)_{33}]$$
$$(\tfrac{1}{3})[(\alpha\gamma)_{31} + (\alpha\gamma)_{12} + (\alpha\gamma)_{23}], \text{ and}$$
$$(\tfrac{1}{3})[(\alpha\gamma)_{21} + (\alpha\gamma)_{32} + (\alpha\gamma)_{13}]$$

and the variability among

$$(\tfrac{1}{3})[(\alpha\beta\gamma)_{111} + (\alpha\beta\gamma)_{212} + (\alpha\beta\gamma)_{313}]$$
$$(\tfrac{1}{3})[(\alpha\beta\gamma)_{321} + (\alpha\beta\gamma)_{122} + (\alpha\beta\gamma)_{223}], \text{ and}$$
$$(\tfrac{1}{3})[(\alpha\beta\gamma)_{231} + (\alpha\beta\gamma)_{332} + (\alpha\beta\gamma)_{133}]$$

Related conclusions hold for SS_A and SS_C.

Confounding main and interaction effects is a potential source of trouble in using Latin square designs. But this problem does not preclude using the design. Transformations can be found that eliminate interaction effects, permitting a clear interpretation of tests of main effects. Even with interaction effects, reasonable interpretations of the F ratio will often be available.

10.4 BETWEEN-SUBJECTS DESIGNS

10.4.1 BASIC CALCULATIONS. Table 10–5 presents *SV*, *df*, and *SS* for an $a \times a$ Latin square with n subjects in each cell. Note that despite the abundance of terms in Equation (10.1), we distinguish only five sources of variance. In particular, we cannot separately extract independent terms representing each of the possible interactions. This is so because there are only $a^2 - 1$ *df* to account for the between-cells variability. We lose $3(a - 1)$ of these when we compute the sums of squares for the three main effects. The residual, $(a - 1)(a - 2)$, is less than $(a - 1)(a - 1)$, the *df* required to account for even a single interaction term.

Note that the $SS_{\text{B cells res}}$ can be viewed as the sum of squares for the interaction of any two Latin square variables further adjusted for the third

Table 10–5   Analysis of variance for a single Latin square, n subjects in a cell

SV	df	SS
Total	a^2n-1	$\sum_i \sum_j \sum_k \sum_m Y_{ijkm}^2 - C$
A	$a-1$	$\dfrac{\sum_j T_{.j..}^2}{an} - C$
B	$a-1$	$\dfrac{\sum_k T_{..k.}^2}{an} - C$
C	$a-1$	$\dfrac{\sum_m T_{...m}^2}{an} - C$
B cells res	$(a-1)(a-2)$	$SS_{cells} - C - SS_A - SS_B - SS_C$
S/cells	$a^2(n-1)$	$SS_{tot} - SS_A - SS_B - SS_C - SS_{B\ cells\ res}$

variable. That is,

$$SS_{B\ cells\ res} = SS_{AB} - SS_C$$
$$= SS_{AC} - SS_B$$
$$= SS_{BC} - SS_A$$

where the component sums of squares are computed in the usual manner.

Table 10–6 presents a data matrix for a 4×4 square; the entries are cell totals based on nine subjects each. We first calculate

$$C = \frac{(516)^2}{144} = 1,849$$

Table 10–6   Data for a single Latin square design

	C_1	C_2	C_3	C_4	$T_{.k.}$
B_1	(A_3) 12	(A_1) 24	(A_2) 9	(A_4) 48	93
B_2	(A_1) 36	(A_3) 18	(A_4) 57	(A_2) 21	132
B_3	(A_2) 27	(A_4) 69	(A_1) 33	(A_3) 15	144
B_4	(A_4) 51	(A_2) 24	(A_3) 27	(A_1) 45	147
$T_{..m} = $	126	135	126	129	$T_{...} = 516$

A Subtotals

A_1	A_2	A_3	A_4
138	81	72	225

The A, B, and C terms are obtained as in factorial designs:

$$SS_A = \frac{(138)^2 + (81)^2 + (72)^2 + (225)^2}{36} - C$$
$$= 412.5$$
$$SS_B = \frac{(93)^2 + (132)^2 + (144)^2 + (147)^2}{36} - C$$
$$= 51.5$$
$$SS_C = \frac{(126)^2 + (135)^2 + (126)^2 + (129)^2}{36} - C$$
$$= 1.5$$

The residual variability among the cell means is obtained as

$$SS_{B\text{ cells res}} = \frac{(12)^2 + (24)^2 + \cdots + (27)^2 + (45)^2}{9} - C - SS_A - SS_B - SS_C$$
$$= 35.5$$

and the within-cell variability is the difference between the SS_{tot} and the four terms above.

10.4.2   EXPECTED MEAN SQUARES AND F RATIOS.   Wilk and Kempthorne (1957) have considered the expected mean squares under the general nonadditive model represented by Equation (10.1). Their results are not simple and cannot be completely generated by the rules of thumb (Section 8.2.2). Nevertheless, they merit consideration because they provide the basis for decisions about using the Latin square design and interpretating results obtained from applying the design.

Fixed effects. Suppose that all three variables—A, B, and C—were fixed-effect variables. The expected mean squares would, as usual, contain an error-variance component σ_e^2, and a null hypothesis term θ_A^2 or θ_B^2 or θ_C^2. In addition, if all interactions were nonzero in the population sampled, they would also contribute to the expected mean squares. We should have an arrangement as shown.

SV	EMS
A	$\sigma_e^2 + na\theta_A^2 + n\theta_{BC}^2 + \dfrac{n(a-2)}{a}\,\theta_{ABC}^2$
B	$\sigma_e^2 + na\theta_B^2 + n\theta_{AC}^2 + \dfrac{n(a-2)}{a}\,\theta_{ABC}^2$
C	$\sigma_e^2 + na\theta_C^2 + n\theta_{AB}^2 + \dfrac{n(a-2)}{a}\,\theta_{ABC}^2$
B cells res	$\sigma_e^2 + n\theta_{AB}^2 + n\theta_{AC}^2 + n\theta_{BC}^2 + \dfrac{n(a-3)}{a}\,\theta_{ABC}^2$
S/ABC	σ_e^2

The interaction component in each main effect source is explained in Section 10.3, especially by the discussion following Equation (10.1). As we showed there, part of the sum of squares for any main effect is attributable to a portion of the variance due to the first-order interaction of the remaining two variables and part to a portion of the second-order variance attributable to interaction among all three variables. The remaining portion of all the first-order interaction variances, and also some portion of the second-order interaction variance, are all lumped together in the residual term. The strange-looking coefficients in the second-order interaction component is a result of the incompleteness of the design and falls out of the Wilk-Kempthorne derivations.

Consider the dilemma these expectations pose for the investigator. If the within-cell (S/ABC) term is used as a measure of error variance, a positive bias results; under the null hypothesis, the ratio of expected mean squares is greater than 1. The result will generally be too many Type I errors. Using the between-cell residual as an error term poses the opposite problem—negative bias. The ratio of expected mean squares will be less than 1 when H_0 is true. Treatment effects will have to be very large if they are to be detected as significant.

Although a transformation may be found that will remove nonadditivity, it is best to use the design in cases in which intuition and pilot data have already shown that unbiased F tests of the treatments of interest can be computed. This does not mean that complete nonadditivity is required. Let us consider a relevant example.

Suppose that we decide to increase the efficiency of a test of some treatment A by arbitrarily sorting subjects into 4 blocks by level of ability. We also use four experimenters who can run subjects concurrently, thereby reducing the number of weeks needed to complete the experiment. If we have four levels of A, a complete factorial design would involve 64 cells and take an unreasonable amount of time to complete. If we designate blocks as B and experimenters as C and use a 4×4 Latin square, however, we have a design that is potentially very efficient with respect to both error variance and running time. The F test of A is unbiased and efficient if the BC and ABC interactions are negligible, which is not an unreasonable assumption. Then $E(MS_A) = \sigma_e^2 + na\theta_A^2$ and the within-cell error term is completely adequate. If we can assume also that there is little variability due to experimenters, a partial test of the treatments $\times$ blocks AB interaction is available, for then we have $E(MS_C) = \sigma_e^2 + n\theta_{AB}^2$. A more powerful test is available if we also can assume that experimenters and treatments do not interact, for then $E(MS_{\text{B cells res}}) = \sigma_e^2 + n\theta_{AB}^2$ and $df_{\text{B cells res}}$ is greater than df_C.

Suppose that we want to investigate two treatment variables A and C. We might want to introduce experimenters as a third variable to reduce the time for each experimenter, or we could introduce blocks as a third variable to increase efficiency (Chapter 6). Or it may be that the experiment involves some third variable that while not of intrinsic interest is a necessary part of a well-planned experiment, perhaps position of the correct alternative in a multiple-choice discrimination task. In any event, if this variable B does not interact with either A or C, then S/ABC yields an unbiased error term for testing A and C effects.

Furthermore, a test of *B cells res* against the within-cell term provides a test of some portion of the *AC* interaction variance; again it is assumed that all other interactions are negligible.

In the preceding examples, we assumed strong a priori evidence that certain interaction terms are negligible. Occasionally, practical considerations will call for using the Latin square design when there is doubt about the status of critical interaction components. If the *F* test of *B cells res* is nonsignificant, significant tests of treatments can reasonably be assumed to reflect main and not interaction effects. If there is evidence of nonadditivity in the preliminary test of the residual term, and if it is feared that the interaction present is confounded with the treatment of interest, transformation to additivity should be sought. The criterion of the adequacy of the transformation will be the magnitude of the *F* test of *B cells res*; the smaller the *F*, the better the transformation.

Random effects. Frequently, the levels of one or more of our three variables can be assumed to have been randomly sampled from some population of levels. We want to generalize to the population of levels, which necessitates treating the variable, or variables, in question as random-effect variables. For example, first-grade schoolchildren are divided into classes on the basis of a readiness-to-read test. Designate this variable, level of readiness, as *C*. The symbol *A* might be methods of instruction in reading. Several different schools are involved in the study and these make up the levels of the random-effect variable *B*. Under a general nonadditive model that assumes the possibility of all interactions contributing variance, the expected mean squares (presented in the preceding section) must be extended. For each treatment source of variance, we again have an error component, a null hypothesis component, and first-order and second-order interaction components due to confounding. In addition, the rules of thumb (Chapter 8) are applied. Thus, in our example (*B* random), an *AB* component must be added to the expected mean square for *A* and a *BC* component to the expected mean square for *C*; both also contribute to *B* cells res. We should have the arrangement shown in the accompanying chart.

SV	EMS
A	$\sigma_e^2 + na\theta_A^2 + n\sigma_{BC}^2 + n\sigma_{AB}^2 + \dfrac{n(a-2)}{a}\sigma_{ABC}^2$
B (random)	$\sigma_e^2 + na\sigma_B^2 + n\theta_{AC}^2 + \dfrac{n(a-1)}{a}\sigma_{ABC}^2$
C	$\sigma_e^2 + na\theta_e^2 + n\sigma_{AB}^2 + n\sigma_{BC}^2 + \dfrac{n(a-2)}{a}\sigma_{ABC}^2$
B cells res	$\sigma_e^2 + n\sigma_{AB}^2 + n\theta_{AC}^2 + n\sigma_{BC}^2 + \dfrac{n(a-2)}{a}\sigma_{ABC}^2$
S/ABC	σ_e^2

If we can reasonably assume that θ^2_{AC} is the only interaction term that will contribute variance, then σ^2_{AB}, σ^2_{AC}, and σ^2_{ABC} drop out and the F tests of A and C against S/ABC are unbiased. Furthermore, under this assumption, B *cells res* can be viewed as a measure of the AC interaction and tested against S/ABC. Finally, B against B *cells res* provides an unbiased F test of the random-effect variable, though one considerably less powerful than the others since the error term is on fewer *df* and its variance will be larger than the variance of the S/ABC term if $\theta^2_{AC} > 0$.

Unbiased F tests of main effects can also be obtained if $\theta^2_{AC} = 0$ and any of the other interaction components are sizable. Under this assumption, the ratio of expected mean squares for A or C against B *cells res* will be 1 when H_0 is true. For reasons given in the preceding paragraph, however, this test will tend to be inefficient and lacking in power. A transformation to additivity would be preferred under these conditions.

Despite the possible problems, this is a very nice example of the usefulness of the Latin square. If there were a methods of instruction and a levels of reading readiness, we should require a^2 classes for each school to have a complete factorial design; this may be very impractical if not impossible. The Latin square gives us a way out with much less expenditure of effort. Even nonadditivity does not rule out the approach; as we indicated above, the bias of F tests depends on the particular combinations of interactions that are present. In the worst case, with all interactions suspect, transformations of the data can be sought. This loses information about the AC interaction, which may well be of interest, but provides a test of the important instructional variable A. Even if nonadditivity cannot be obtained, if instructions have a large enough effect to be of practical interest, they should prove significant when tested against B *cells res* despite the likely negative bias of the test.

10.4.3   A MODIFIED LATIN SQUARE DESIGN.   Suppose that we want to compare two *methods of psychotherapy P;* furthermore, we want to know whether the number of contact *hours per week H* is important, particularly whether differences in the effects of the therapies depend on hours of contact ($P \times H$ interaction). Subjects are chosen from the populations at four outpatient *clinics C* and are placed in one of four categories depending on the severity of *symptoms S*. Assessment of improvement by a panel of clinicians provides the dependent variable. We might lay out the design as

$$
\begin{array}{c}
\quad\quad C_1 \quad\quad C_2 \quad\quad C_3 \quad\quad C_4 \\
\begin{array}{c} S_1 \\ S_2 \\ S_3 \\ S_4 \end{array}
\left[
\begin{array}{cccc}
P_1 H_2 & P_1 H_1 & P_2 H_1 & P_2 H_2 \\
P_1 H_1 & P_1 H_2 & P_2 H_2 & P_2 H_1 \\
P_2 H_2 & P_2 H_1 & P_1 H_2 & P_1 H_1 \\
P_2 H_1 & P_2 H_2 & P_1 H_1 & P_1 H_2
\end{array}
\right]
\end{array}
$$

We have a 4×4 Latin square in which the subject populations at each of four clinics have been grouped according to level of S and then assigned to one of four *PH* combinations. It is a Latin square, because each of the four treatment combinations appears exactly once in each clinic and at each level of S. As usual with such designs, it is incomplete; the complete design would require 64 cells and each clinic would require enough patients to provide 16 combinations of S, P, and H.

This design permits us to study the main and interaction effects of two treatment variables, P and H in our examples. The design could be extended to more variables. There could be two levels of each of three treatment variables; then each row or column would contain eight treatment combinations. The seven *df* for treatment combinations would be partitioned into seven components, representing three main effects, three first-order interactions and one second-order interaction. Nor is the approach limited to two levels of the treatment variables. We could have a 12×12 square in which each row (or column) contained a complete 4×3 factorial design, for example, three levels of P and four levels of H. Then the eleven *df* for treatment combinations would be partitioned into two for P, three for H, and six for the *PH* interaction. The analysis of variance usually follows closely the analysis of variance of Table 10–5. In the general case, we label the row variable B, the column variable C, and the variables that make up our treatment combinations A and D. Table 10–7 presents an example with the variables completely labeled.

Numerical example. We use the data of Table 10–7 for the calculations. The entries are cell totals based on four subjects each. As usual, we first calculate the correction term:

$$C = \frac{(350)^2}{64} = 1{,}914.063$$

Table 10–7   Data for a Latin square of treatment combinations

	C_1	C_2	C_3	C_4	$T_{..k..}$
B_1	(A_1D_1)	(A_2D_1)	(A_2D_2)	(A_1D_2)	
	10	34	28	16	88
B_2	(A_2D_1)	(A_1D_2)	(A_1D_1)	(A_2D_2)	
	32	8	12	40	92
B_3	(A_2D_2)	(A_1D_1)	(A_1D_2)	(A_2D_1)	
	28	16	6	38	88
B_4	(A_1D_2)	(A_2D_2)	(A_2D_1)	(A_1D_1)	
	12	26	30	14	82
$T_{...m.} = 82$		84	76	108	$T_{.....} = 350$

AD Subtotals

A_1D_1	A_2D_1	A_1D_2	A_2D_2
52	134	42	122

Then

$$SS_B = \frac{(88)^2 + \cdots + (82)^2}{16} - C = 3.187$$

$$SS_C = \frac{(82)^2 + \cdots + (108)^2}{16} - C = 37.187$$

We next turn to the *ad* treatment combinations:

$$SS_{TC} = \frac{(52)^2 + \cdots + (42)^2}{16} - C = 417.687$$

This variability can be divided into three components: *A*, *D*, and *AD*. Taking advantage of the fact that each of these component terms is distributed on 1 *df*, we can use the short-cut formula of Section 5.11:

$$SS_A = \frac{(134 + 122 - 52 - 42)^2}{64} = 410.063$$

$$SS_D = \frac{(52 + 134 - 122 - 42)^2}{64} = 7.563$$

and

$$SS_{AD} = \frac{(52 + 122 - 134 - 42)^2}{64}$$

or

$$SS_{AD} = SS_{TC} - SS_A - SS_B$$
$$= .063$$

Next, we obtain

$$SS_{B \text{ cells res}} = \frac{10^2 + 34^2 + \cdots + 30^2 + 14^2}{4} - C - SS_B - SS_C - SS_{TC}$$

$$= 18.876$$

We have not bothered to provide the individual scores but the SS_{tot} is calculated in the usual way and

$$SS_{S/\text{cells}} = SS_{\text{tot}} - SS_B - SS_C - SS_{TC} - SS_{B \text{ cells res}}$$

10.4.4 USING SEVERAL LATIN SQUARES.   Suppose that in our example of the study of methods of psychotherapy *P* and hours of therapy *H* each week we wanted to investigate the interaction of *P* with severity of symptoms *S*. In the more general notation of the preceding section, we desire information on the interaction of *D* with *B* or *C* or both as well as with *A*; the previous design provided a direct test only of *AD*. One way of approaching the problem is to use the design of Section 10.4.2 as a building block. The variables *B* and *C* are again rows and columns and again only A is Latin-squared. There are a levels of A, B, and C. We randomly sample d squares from the population of $a \times a$ squares, where d may take on any value. Then we have one Latin square under treatment D_1, another at D_2, and so on. The layout might be

$$
D_1 \quad
\begin{array}{c}
B_1 \\ B_2 \\ B_3 \\ B_4
\end{array}
\begin{bmatrix}
A_4 & A_2 & A_1 & A_3 \\
A_1 & A_3 & A_2 & A_4 \\
A_3 & A_1 & A_4 & A_2 \\
A_2 & A_4 & A_3 & A_1
\end{bmatrix}
\quad
\begin{array}{cccc}
C_1 & C_2 & C_3 & C_4
\end{array}
$$

$$
D_2 \quad
\begin{array}{c}
B_1 \\ B_2 \\ B_3 \\ B_4
\end{array}
\begin{bmatrix}
A_2 & A_3 & A_1 & A_4 \\
A_3 & A_4 & A_2 & A_1 \\
A_1 & A_2 & A_4 & A_3 \\
A_4 & A_1 & A_3 & A_2
\end{bmatrix}
\quad
\begin{array}{cccc}
C_1 & C_2 & C_3 & C_4
\end{array}
$$

There are n subjects in each of the a^2d cells.

Assuming that the effects of the dimensions of the square (A, B, and C) are additive, the appropriate model would be

(10.4) $\qquad Y_{ijkmp} = \mu + \alpha_j + \beta_k + \gamma_m + \delta_p + (\alpha\delta)_{jp} + (\beta\delta)_{kp} + (\gamma\delta)_{mp} + \varepsilon_{ijkmp}$

Table 10–8 summarizes the analysis of variance. The pooled residual term represents summing over the levels of D of the between-cell residual for each square. In practice, we find the sum of squares among the a^2d cell means and then adjust for the main and interaction effects previously computed.

As in the past, we can test the additivity assumptions; in this case, we compute $MS_{\text{pooled res}}/MS_{S/\text{cells}}$. Nonsignificance would tend to support the model represented by Equation (10.4) and the expected mean squares of Table 10–8.

Table 10–8   Analysis of variance for several squares

SV	df	EMS[a]
Total	$a^2dn - 1$	
B	$a - 1$	$\sigma_e^2 + adn\theta_B^2$
C	$a - 1$	$\sigma_e^2 + adn\theta_C^2$
A	$a - 1$	$\sigma_e^2 + adn\theta_A^2$
D	$d - 1$	$\sigma_e^2 + a^2n\theta_D^2$
BD	$(a-1)(d-1)$	$\sigma_e^2 + an\theta_{BC}^2$
CD	$(a-1)(d-1)$	$\sigma_e^2 + an\theta_{CD}^2$
AD	$(a-1)(d-1)$	$\sigma_e^2 + an\theta_{AD}^2$
Pooled residual	$d(a-1)(a-2)$	σ_ξ^2
S/cells	$a^2d(n-1)$	σ_e^2

[a] Equation (10.4) is assumed.

Significance raises the usual problems. There is little to be added to our previous comments (in particular, see Section 10.4.2) except to note that not only will the interaction of two Latin square variables contribute to the main effect of the third, but also the interaction of two Latin square variables with D will be confounded with the interaction of the third variable with D. If BCD effects are not negligible in the population, for example, their variance will contribute to the AD mean square.

One advantage of the present design is that using many squares tends to minimize the danger due to interaction among Latin square variables. To see why this is so, again consider Section 10.3, particularly Equation (10.3) and the discussion following it. With a single $a \times a$ Latin square, the means at the different levels of B involve a distinct, nonoverlapping, sets of values of $(\alpha\gamma)_{jm}$. When we add more squares, the sets of $(\alpha\gamma)_{jm}$ values contributing to each $\bar{Y}_{.k.}$ tend to overlap; for example $(\alpha\gamma)_{23}$ might contribute to both $\bar{Y}_{.1.}$ and $\bar{Y}_{.2.}$. This reduces the relative contribution of σ^2_{AC} to the MS_B.

Despite the advantage just cited, it is occasionally profitable to replicate the *same* square at each level of D. Returning to our example of the investigation of several types of therapy, suppose we had strong prior grounds for believing that clinics would not interact with the other variables but that severity of symptoms might interact with P, the Latin square variable, and H, the between-squares variable; furthermore, the interaction is of some interest. In general terms, we assume that interactions with C are negligible but we are interested in possible interactions of A and D with B. If we use the same square at all levels of D, there is one change in the analysis of Table 10–8; the pooled residual can be partitioned into two terms, which under the current model, we designate AB' and ABD':

$$SS_{\text{pooled res}} = SS_{AB'} + SS_{ABD'}$$
$$d(a-1)(a-2) = (a-1)(a-2) + (a-1)(a-2)(d-1)$$

The prime sign indicates that only part of the variance is contained in the term in question. The remainder of the AB variability contributes to SS_C and the ABD variability is partly confounded with CD. The value $SS_{AB'}$ is computed as $SS_{B \text{ cells res}}$ was previously, where the cell in question contains dn scores.

10.4.5   THE BALANCED LATIN SQUARE DESIGN.   Suppose that we are interested in the main and interaction effects of three treatment variables A, B, and C. We are willing and able to run a complete factorial design involving a levels of all variables; that is, we propose a^3 cells with n subjects in each. There is a fourth variable D, however, which it is advisable or necessary to include in the study although its effects are not of interest. We might like to control for intelligence level by introducing a block variable, or we might wish to use several experimenters to shorten the duration of the experiment. Still another possibility is that we must use several schools or clinics to obtain sufficient subjects. Suppose that A and B are instructional variables in some classroom learning experiment, C is grade level, and D represents schools. To carry out the experiment as a complete four-factor design, we require a^3 cells for each school, presumably a^2 classes at each grade within a school. Generally, this will be impossible. We can compromise by having all combinations of A, B, and C but only $1/a$ of the total number of $ABCD$ cells. Assuming that $a = 3$, one possible layout is

$$
D_1 \begin{array}{c} \\ B_1 \\ B_2 \\ B_3 \end{array} \begin{array}{ccc} C_1 & C_2 & C_3 \\ \left[\begin{array}{ccc} A_2 & A_1 & A_3 \\ A_1 & A_3 & A_2 \\ A_3 & A_2 & A_1 \end{array}\right] \end{array} \quad D_2 \begin{array}{c} \\ B_1 \\ B_2 \\ B_3 \end{array} \begin{array}{ccc} C_1 & C_2 & C_3 \\ \left[\begin{array}{ccc} A_1 & A_3 & A_2 \\ A_3 & A_2 & A_1 \\ A_2 & A_1 & A_3 \end{array}\right] \end{array} \quad D_3 \begin{array}{c} \\ B_1 \\ B_2 \\ B_3 \end{array} \begin{array}{ccc} C_1 & C_2 & C_3 \\ \left[\begin{array}{ccc} A_3 & A_2 & A_1 \\ A_2 & A_1 & A_3 \\ A_1 & A_3 & A_2 \end{array}\right] \end{array}
$$

We achieve the balanced layout by first selecting an $a \times a$ square at random (see Section 10.2). Next we rotate the columns so that the mth column of the original square becomes the $m - 1$st column of the new square, the original first column becoming the ath column of the new square. We repeat this rotation once to generate still another square, and continue the process until we have a squares. The result is that at any level of one Latin square variable, all combinations of the other two Latin square variables are present. Then, when we compute the $\bar{Y}_{.k.}$, for example, each one is an average over all $(\alpha\gamma)_{jm}$ effects, and B is not confounded with AC.

There are still some potential sources of confounding that we should know. The means at the different levels of D will be based on different $(\alpha\beta\gamma)_{jkm}$ sets; thus D and ABC will be partially confounded. Consequently, the design should not be used if the main effect of D is of interest. Interactions of D with any of the Latin square variables could create problems in interpreting some of the F tests. For example, the B main effect could be viewed as a contrast among a sets of $(\alpha\gamma\delta)_{jmp}$ components; thus, if $\sigma^2_{ACD} > 0$, then ACD will be partially confounded with B. Other interactions with D, if sizable, will contribute to other main and interaction terms involving Latin square variables.

If interactions involving D are assumed negligible, the appropriate model is

(10.5)
$$
Y_{ijkmp} = \mu + \alpha_j + \beta_k + \gamma_m + \delta_p + (\alpha\beta)_{jk} + (\alpha\gamma)_{jm}
$$
$$
+ (\beta\gamma)_{km} + (\alpha\beta\gamma)_{jkm} + \varepsilon_{ijkmp}
$$

An analysis consistent with the model is presented in Table 10–9. Note that the

Table 10–9   Analysis of balanced Latin square design

SV	df	EMS[a]
Total	$a^3 n - 1$	
A	$a - 1$	$\sigma^2_e + a^2 n \theta^2_A$
B	$a - 1$	$\sigma^2_e + a^2 n \theta^2_B$
C	$a - 1$	$\sigma^2_e + a^2 n \theta^2_C$
AB	$(a - 1)^2$	$\sigma^2_e + a n \theta^2_{AB}$
AC	$(a - 1)^2$	$\sigma^2_e + a n \theta^2_{AC}$
BC	$(a - 1)^2$	$\sigma^2_e + a n \theta^2_{BC}$
D	$a - 1$	$\sigma^2_e + a^2 n \theta^2_D$
B cell res	$(a - 1)^3 - (a - 1)$	σ^2_e
S/cells	$a^3 (n - 1)$	σ^2_e

[a] Equation (10.5) is assumed.

between-cells residual can be taken as a measure of the *ABC* interaction, and that it is that portion that remains after we have adjusted for the component of *ABC* confounded with *D*.

10.5   REPEATED-MEASUREMENT DESIGNS

10.5.1   PRELIMINARY REMARKS.   Psychologists use the Latin square primarily as a repeated-measurement design. In many instances, subjects are few and it is essential that all subjects undergo all levels of the independent variable. In such a situation, the Latin square is potentially more efficient than the subjects × treatments design. They both permit us to extract variance due to individual differences as a separate source. The Latin square permits still greater efficiency, however, by also permitting us to account separately for variance due to temporal effects—for example, fatigue, practice, or boredom.

In the uses of this section each row, previously B_k, now represents one or more subjects that undergo the treatment levels, the A_j, in a specified sequence. Each column C_m represents a position within the sequence, a stage of practice. We view the rows as a random-effect variable and *C* and *A* as fixed-effect variables. It is important to distinguish between sequence and ordinal position effects. Consider five rats tested in a runway on five successive days, each day under a different drug. Each rat undergoes a different sequence of drugs, the set of five sequences forming a Latin square. Now, regardless of the sequence, we expect reduced running times over days due to the rats' increasing familiarity with the experiment. If additivity holds, the *C* source of variance will reflect these practice effects. To the extent that they are present, this design is more efficient than the subjects × treatments design, for such temporal effects are included in the error term in the latter design.

Presumably, the row means will also differ since they represent different subjects. There are other reasons why they might differ. Suppose that one rat receives a severe depressant on the first day and never wholly recovers; his average running time over the five days will be considerably longer than the time for a rat who receives the severe depressant on the last day of the experiment. Such effects are frequently referred to as carryover effects; when present, they distort our estimates of the treatment effects because the treatment effect will be either inflated or deflated by the carryover from the preceding treatment or treatments. In many experiments, such effects will be minimal. Consider a typical signal detection experiment, in which the subject is required to report the presence or absence of a tone against a noise background; I suspect that the signal-to-noise intensity ratio could be varied over trial blocks with little fear of carryover effects. In many experiments, a recovery period between presentations of the treatments will suffice; if no drug is lethal or does severe physical damage, our experiment with the rats could be carried out, free of carryover effects, with sufficiently long periods between tests under different drugs. There will be situations in which carryover effects persist for longer periods than can be practically incorporated into the experiment. Studies of learning, in which the treatment variable is type of material to be learned, may well fall into this category. There may even be situations for which the carryover effects are themselves of interest. Procedures have been developed for estimating carryover effects and adjusting treatment effects for their presence (Cochran and Cox, 1957, pp. 135–40).

We are assuming that designs are applied when carryover effects are negligible. Even so, there are still other sources of difficulty in interpreting the results of the analysis. We still have bias problems (Section 10.4.2). Because the Latin square is an incomplete design, interactions and main effects are partially confounded. Furthermore, interactions involving the random-effect variable, rows, will contribute to main effects—for example, $\sigma^2_{rows \times A}$, if nonzero, is a component of $E(MS_A)$. There are several possibilities, depending on the constellation of nonzero interactions. We shall refer to the F test as *unbiased* if with H_0 assumed true, the numerator and denominator expected mean squares are equal; *negatively biased* (α is deflated) if the denominator expectation is larger; and *positively biased* (α is inflated) if the numerator expectation is larger. Remember also that even in cases we designate unbiased, there may be positive bias due to heterogeneity of covariance (see Section 7.4.1). In these instances, the conservative adjustment for *df* could be used.

The designs we shall consider are mostly the designs of Section 10.4, with the difference that subjects are now nested in rows instead of cells. This will result in only slight changes in the breakdown of the sources of variance; most of the computations will proceed as before. In some ways, the analyses will be like those of Chapter 8; when several subjects undergo a particular sequence of treatments, we can view rows as a between-subjects variable and A and C as within-subject variables. In presenting the models we shall not bother to reiterate the definitions and assumptions used hitherto. Notation like η_i, α_j, and $(\eta\alpha)_{ij}$ are interpreted as usual, and the usual distributional assumptions still hold for random-effect variables.

10.5.2 A SINGLE LATIN SQUARE.   There will be situations in which so few subjects are available, or in which running time per subject is so great, that we must content ourselves with a subjects, one for each sequence of a treatments. We first assume an additive model in which all interactions are assumed negligible:

(10.6) $$Y_{ijk} = \mu + \eta_i + \alpha_j + \gamma_k + \varepsilon_{ijk}$$

The analysis of variance is summarized in Table 10–10. Under the restrictive assumptions represented by Equation (10.6), unbiased F tests are available.

Suppose that interactions involving subjects are present but treatments are not differentially affected by practice, that is, we still assume $\theta^2_{AC} = 0$. Because of the structural confounding in the Latin square, σ^2_{SC} contributes to $E(MS_A)$; so also does σ^2_{SA}, under the rules developed in Chapters 7 and 8. The F test of

Table 10–10   Analysis of variance for a single Latin square

SV	df	EMS	F
Total	a^2-1		
S	$a-1$	$\sigma_e^2 + n\sigma_S^2$	$\dfrac{MS_S}{MS_{res}}$
Columns	$a-1$	$\sigma_e^2 + n\theta_C^2$	$\dfrac{MS_C}{MS_{res}}$
A	$a-1$	$\sigma_e^2 + n\theta_A^2$	$\dfrac{MS_A}{MS_{res}}$
Residual	$(a-1)(a-2)$	σ_e^2	

treatments will nevertheless be unbiased in the sense that numerator and denominator have the same expectations *under* H_0. That is,

$$E(MS_A) = E(MS_{\text{B cells res}}) = \sigma_e^2 + \left(\frac{a-2}{a}\right)\sigma_{SAC}^2 + \sigma_{SA}^2 + \sigma_{SB}^2$$

Of course, the presence of interaction components in the error term will reduce efficiency, and nonadditivity introduces the possibility of positive bias from heterogeneity of covariance.

The situation is much worse if $\theta_{AC}^2 > 0$, regardless of the status of the row interaction components. The *AC* interaction variance contributes to the error variability but not to the *A* mean square. The result is a negatively biased *F* ratio and a possible consequent loss of power.

Tukey's single df test. If the single Latin square must be used, a test for nonadditivity should be used. Treatments and position in time are only too likely to interact, negatively biasing the *F* test. Tukey (1955) has provided a modification of the single *df* test described in Section 7.5.1. The following steps are involved:

1. Compute

$$\bar{Y}_{...} = \text{mean of all } a^2 \text{ scores}$$
$$\bar{Y}_{i..} = \text{mean of row } i$$
$$\bar{Y}_{.j.} = \text{mean for } A_j$$
$$\bar{Y}_{..k} = \text{mean for column } k$$
$$d_{ijk} = (\bar{Y}_{i..} - \bar{Y}_{...}) + (\bar{Y}_{.j.} - \bar{Y}_{...}) + (\bar{Y}_{..k} - \bar{Y}_{...})$$

2.

$$SS_{\text{nonadd}} = \left\{ \sum_{i,k} [(Y_{ijk} - \bar{Y}_{...}) - d_{ijk}] \, d_{ijk}^2 \right\}^2 / SS_{\text{B cell res}}$$

3.
$$F = \frac{SS_{\text{nonadd}}}{(SS_{\text{B cell res}} - SS_{\text{nonadd}})/[(a-1)(a-2)-1]}$$
$$df = 1, (a-1)(a-2)-1$$

The investigator would use the transformation that most reduces the magnitude of the F ratio computed in step 3.

Relative efficiency. If the additive model of Equation (10.7) is valid for the data, the Latin square will generally be more efficient than the subjects × treatments design (Chapter 7). Even under nonadditivity, the Latin square may prove more efficient, since some of the same interaction variability will also contribute to the subjects × treatments error term. We previously provided an intuitive argument for the Latin square's greater efficiency; we now derive an expression relating the two error terms, assuming nonadditivity in both designs.

Assume that there are a subjects and a treatment levels in both designs. Presumably, the variability within subjects is in the long run the same for both designs. That is,

$$E_{ST}(SS_{SA}) + E_{ST}(SS_A) = E_{LS}(SS_{\text{B cells res}}) + E_{LS}(SS_A) + E_{LS}(SS_C)$$

The subscripts ST and LS designate the two designs. Substituting components of variance, we have

$$(a-1)^2\sigma_{e_{ST}}^2 + (a-1)(\sigma_{e_{ST}}^2 + a\theta_A^2) = (a-1)(a-2)\sigma_{e_{LS}}^2$$
$$+ (a-1)(\sigma_{e_{LS}}^2 + a\theta_A^2)$$
$$+ (a-1)(\sigma_{e_{LS}}^2 + a\theta_C^2)$$

Then canceling the θ_A^2 terms and rearranging the remaining terms gives

$$\sigma_{e_{ST}}^2 = \sigma_{e_{LS}}^2 + \frac{1}{a}(\sigma_{e_{LS}}^2 + a\theta_C^2 - \sigma_{e_{LS}}^2)$$

or

(10.7)
$$MS_{ST} = MS_{\text{B cells res}} + \frac{1}{a}(MS_C - MS_{\text{B cells res}})$$

We can expect a smaller error term in the Latin square design if the variability due to C (position in time) exceeds its error term. The Latin square error term has fewer df than the subjects × treatments error term, however. Recall that Fisher (Section 6.2) has suggested the following measure of relative efficiency of design 1 to design 2:

$$RE = \left[\frac{MS_{\text{error 2}}}{MS_{\text{error 1}}}\right]\left[\frac{df_1+1}{df_1+3}\right]\left[\frac{df_2+3}{df_2+1}\right]$$

Note the adjustment for error df. In the present case, assuming that we have used the Latin square design and want to decide whether it will be profitable for future studies, we estimate the subjects × treatments error magnitude by Equation (10.7) and compute:

$$\textbf{(10.8)}\quad RE = \left[\frac{MS_{\text{B cells res}} + (1/a)(MS_C - MS_{\text{B cells res}})}{MS_{\text{B cells res}}}\right]$$
$$\times \left[\frac{(a-1)(a-2)+1}{(a-1)(a-2)+3}\right]\left[\frac{(a-1)^2+3}{(a-1)^2+1}\right]$$

Missing scores. Occasionally, a subject may miss a session and one cell in the square will be empty. We can estimate this missing value X_{ijk} by the method developed in Chapter 7. Following the additive model, we get

$$E(X_{ijk}) = \mu + \eta_i + \alpha_j + \gamma_k$$

and therefore

$$X = \left(\frac{T_{...} + X}{a^2}\right) + \left(\frac{T_{i..} + X}{a} - \frac{T_{...} + X}{a^2}\right) + \left(\frac{T_{.j.} + X}{a} - \frac{T_{...} + X}{a^2}\right)$$
$$+ \left(\frac{T_{..k} + X}{a} - \frac{T_{...} + X}{a^2}\right)$$

Simplifying and solving give

$$X = \frac{a(T_i + T_j + T_k) - 2T_{...}}{(a-1)(a-2)}$$

where

$$T_{...} = \text{obtained grand total}$$

$$T_{i..} = \text{obtained total for subject } i$$

$$T_{.j.} = \text{obtained total for } A_j$$

$$T_{..k} = \text{obtained total for } C_k$$

If several scores are missing, the iterative procedure of Chapter 7 can be used.

10.5.3   REPLICATING A LATIN SQUARE.   *The model.* It is assumed that a single $a \times a$ Latin square has been randomly sampled from the population of squares. A total of an subjects are randomly distributed among the a rows (sequences) of the square, with n subjects in each row; there are a scores for each subject. A representative score would be Y_{ijkm}, where

i indexes the subject within the row $(i = 1, 2, \ldots, n)$

j indexes the level of the treatment variable A $(j = 1, 2, \ldots a)$

k indexes the column in the square $(k = 1, 2, \ldots, a)$

m indexes the row within the square $(m = 1, 2, \ldots, a)$

As an example of the design, a might equal 4 and n might equal 2. Then, we could represent the design by

$$
\begin{array}{cc}
 & \begin{array}{cccc} C_1 & C_2 & C_3 & C_4 \end{array} \\
\begin{array}{cc} S_{11}, & S_{21} \\ S_{12}, & S_{22} \\ S_{13}, & S_{23} \\ S_{14}, & S_{24} \end{array} &
\begin{bmatrix} A_2 & A_4 & A_3 & A_1 \\ A_1 & A_3 & A_2 & A_4 \\ A_4 & A_2 & A_1 & A_3 \\ A_3 & A_1 & A_4 & A_2 \end{bmatrix}
\end{array}
$$

There are many models that could relate Y_{ijkm} and the population parameters. If carryover effects are negligible, we can ignore sequence and sequence interaction effects. If in addition, subjects do not interact with A or C, the appropriate model is

(10.9)
$$ Y_{ijkm} = \mu + \eta_{i/m} + \alpha_j + \gamma_k + (\alpha\gamma)_{jk} + \varepsilon_{ijkm} $$

As usual, $\eta_{i/m}$ and ε_{ijkm} are assumed to be normally distributed random variables; the other effects are fixed.

The analysis of variance. Table 10–11 summarizes the analysis of variance for the replicated square design; the model of Equation (10.9) is assumed.

The total variability is analyzed in a manner like that of Chapters 8 and 9. Since there are a scores for each of an subjects, there are a total of $a^2n - 1$ df. Part of this variability is due to variability among subjects; this accounts for $an - 1$ df. The remaining $an(a - 1)$ df represent the within-subject variability. Subjects may differ because they are in different rows in the square. There will be variability among subjects within rows also, due to individual differences. Thus the between-subjects variability is partitioned into two components, R and S/R.

Table 10–11   Analysis of variance for the replicated square design

SV	df	EMS	F
Total	$a^2n - 1$		
Between S	$an - 1$		
R	$a - 1$	$\sigma_e^2 + a\sigma_{S/R}^2 + n\theta_{AC}^2$	$\dfrac{MS_R}{MS_{S/R}}$
S/R	$a(n - 1)$	$\sigma_e^2 + a\sigma_{S/R}^2$	
Within S	$an(a - 1)$		
C	$a - 1$	$\sigma_e^2 + na\theta_e^2$	$\dfrac{MS_C}{MS_{res}}$
A	$a - 1$	$\sigma_e^2 + na\theta_A^2$	$\dfrac{MS_A}{MS_{res}}$
B cells res	$(a - 1)(a - 2)$	$\sigma_e^2 + n\theta_{AC}^2$	$\dfrac{MS_{\text{B cells res}}}{MS_{res}}$
Residual	$a(n - 1)(a - 1)$	σ_e^2	

(The interpretation of the test of R against S/R will be considered shortly when we deal with *EMS*.)

Partitioning the within-subject variability follows partitioning the single Latin square (Section 10.5.2), except that because of the replication within rows, there is an additional source, the (within-cell) *residual, $a(n-1)(a-1)$ df.*

The *EMS* follow directly from Equation (10.9). The source labeled R can be viewed as the *between-subjects component of the AC interaction*, since the test of R against S/R implies

$$\frac{E(MS_R)}{E(MS_{S/R})} = \frac{\sigma_e^2 + a\sigma_{S/R}^2 + n\theta_{AC}^2}{\sigma_e^2 + a\sigma_{S/R}^2}$$

and tests

$$H_0 : \theta_{AC}^2 = 0$$

Still assuming the validity of Equation (10.9), and therefore the *EMS* of Table 10–11, we have also unbiased and precise tests of A, C, and B *cells res.* The last term is intepreted as a test of the within-subject component of AC, since

$$\frac{E(MS_{\text{B cells res}})}{E(MS_{\text{W cells res}})} = \frac{\sigma_e^2 + n\theta_{AC}^2}{\sigma_e^2}$$

If strong prior evidence against the presence of other interactions is lacking, it is somewhat more difficult to draw inferences. If the two preliminary tests just described yield nonsignificant results, we shall assume that interaction variance is negligible, and that tests of A and C against the W *cells res* will be unbiased. In fact, if the preliminary test yields a small-enough result, the two residual terms might be pooled to provide more *df* for a more powerful test (the criterion for pooling will be discussed in the next chapter). If the preliminary test of the B *cells res* is significant, we can interpret it as reflecting AC variability alone (the model of Equation (10.9)) and tests of A and C will still be unbiased. This assumption is dangerous unless we have strong prior grounds for making it; if we proceed to test A and C against W *cells res* when R interactions are present, the F tests will be positively biased, since such sequence interactions will contribute to the numerator mean square but not to the error term. If sequence interactions are suspected, attempts should be made to find a transformation of the data that results in additivity; the criteria would be the preliminary F tests described above. Failing this, it is preferable to test A and C against B *cells res*; the tests may be inefficient (few *df*, large error variance) but unbiased if $\theta_{AC}^2 = 0$ or negatively biased if $\theta_{AC}^2 > 0$.

Numerical example. Table 10–12 shows the replicated square design. The analysis of variance follows that of the simple Latin square design with only a few exceptions. We calculate

$$C = \frac{(373)^2}{18} = 7,729.389$$

Table 10–12   Data for a replicated square design

		A_2	A_3	A_1	$T_{i\ldots m}$
R_1	S_{11}	9	23	27	59
	S_{21}	12	25	22	59
	$T_{.jk1}=$ 21	48	49	$T_{\ldots1}=118$
		A_1	A_2	A_3	
R_2	S_{12}	15	20	22	57
	S_{22}	8	26	29	63
	$T_{.jk2}=$ 23	46	51	$T_{\ldots2}=120$
		A_3	A_1	A_2	
R_3	S_{13}	16	27	33	76
	S_{23}	12	21	26	59
	$T_{.jk3}=$ 28	48	59	$T_{\ldots3}=135$
	$T_{.jk.}=$ 72	142	159	$T_{\ldots\ldots}=373$

A Subtotals

A_1	A_2	A_3
120	126	127

and

$$SS_{tot} = (9)^2 + (12)^2 + \cdots + (33)^2 + (26)^2 - C$$
$$= 887.611$$

The variability between subjects is given by

$$SS_{BS} = \frac{(59)^2 + \cdots + (76)^2 + (59)^2}{3} - C$$
$$= 82.944$$

This is partitioned into

$$SS_R = \frac{(118)^2 + (120)^2 + (135)^2}{6} - C = 28.778$$

and

$$SS_{S/R} = SS_{BS} - SS_R = 54.166$$

We next obtain a measure of the within-subject variability:

$$SS_{WS} = SS_{tot} - SS_{BS} = 804.667$$

Table 10–13　Analysis of variance for the data of Table 10–12

SV	df	SS	MS	F
Total	17	887.611		
Between S	5	82.944		
R	2	28.778	14.389	.797
S/R	3	54.166	18.055	
Within S	12	804.667		
A	2	4.778	2.389	.210
C	2	708.778	354.389	31.117[a]
Error	8	91.112	11.389	

[a] $p < .001$

This can be further partitioned into several components:

$$SS_A = \frac{(120)^2 + (126)^2 + (127)^2}{6} - C = 4.778$$

$$SS_C = \frac{(72)^2 + (142)^2 + (159)^2}{6} - C = 708.778$$

$$SS_{B\ \text{cells res}} = \frac{(21)^2 + (48)^2 + \cdots + (48)^2 + (59)^2}{2} - C - SS_R - SS_A - SS_C = 8.777$$

and

$$SS_{\text{res}} = SS_{\text{WS}} - SS_A - SS_C - SS_{B\ \text{cells res}} = 82.335$$

The *df* for B cells res are $(a-1)(a-2)$, or 2, and those for the residual are $a(a-1)(n-1)$, or 6. Therefore,

$$\frac{MS_{B\ \text{cells res}}}{MS_{\text{res}}} = \frac{4.389}{13.723}$$

Since the *F* is clearly not significant, it seems reasonable to pool the *B cells res* and residual terms to obtain increased power. The result is

$$MS_{\text{error}} = \frac{82.335 + 8.777}{6 + 2} = 11.389$$

The analysis is summarized in Table 10–13.

10.5.4　USING SEVERAL LATIN SQUARES.　Using several Latin squares reduces the possibility of confounding main with interaction effects. For this reason, it is frequently preferable to randomly sample *n* squares and assign one subject to each of the *an* sequences rather than to replicate one square *n* times. The latter approach (preceding section) would be most useful when the *AC* interaction is of

Table 10–14 Analysis of several squares with repeated measurements

SV	df	EMS
Total	$a^2n - 1$	
S	$an - 1$	$\sigma_e^2 + a\sigma_S^2$
A	$a - 1$	$\sigma_e^2 + na\theta_A^2$
C	$a - 1$	$\sigma_e^2 + na\theta_C^2$
Residual	$(an - 2)(a - 1)$	σ_e^2

especial interest and evidence available before the experiment supports the validity of Equation (10.9).

Assuming an additive model, we have the SV, df, and EMS of Table 10–14. Since the calculations are simple, we shall omit a numerical example. The one point to note is that if AC interaction effects are not negligible, they will contribute to the pooled residual mean square but not to MS_A. The result will be negative bias and a consequent loss of power. Subject interactions contribute to both the treatment and error mean squares and will cause no problem.

10.5.5 INVESTIGATING ADDITIONAL INDEPENDENT VARIABLES. *Latin squaring treatment combinations.* Suppose that we are interested in the main and interaction effects of two treatment variables A and D. Several variations of the Latin square design merit consideration. One approach is to test each subject under all *ad* combinations; the layout is similar to that of Section 10.4.3. This approach is particularly useful when we have few subjects although they are available long enough to be tested under all conditions. We could use a single $ad \times ad$ Latin square, or we could replicate the square n times or select n such squares each replicated only once. The analyses follow those of the preceding three sections; the only difference is that where we had an A source of variance before, we now have a *treatment combination* source of variance on $ad - 1$ df. This variability is then further partitioned in an A, a D, and an AD source.

Replicating one square. The replicated square design (Section 10.5.3) can be extended to permit investigating additional treatment variables. As before, there could be exactly a sequences. However, n of the subjects in a particular sequence would be at D_1, n more at D_2, and so on, giving a total of adn subjects, dn in each of a sequences. Consider as an example of the application of this design the measurement of reaction time under each of four levels of stress A. Three experimental groups of eight subjects each are tested under all stress levels; the groups differ in level of anxiety D. In this case, $a = 4$, $d = 3$, and $n = 2$. In general,

    i indexes the subjects within a row at a level of D
    j indexes the level of A
  m indexes the level of D
    k indexes the level of C (position in time)
    p indexes the row in a level of D

The layout of the design in our example might be

$$
\begin{array}{cccc}
 & C_1 & C_2 & C_3 & C_4
\end{array}
$$

$$
D_1 \quad
\begin{array}{l}
S_{111}, S_{211} \\
S_{112}, S_{212} \\
S_{113}, S_{213} \\
S_{114}, S_{214}
\end{array}
\begin{bmatrix}
A_4 & A_2 & A_1 & A_3 \\
A_2 & A_3 & A_4 & A_1 \\
A_1 & A_4 & A_3 & A_2 \\
A_3 & A_1 & A_2 & A_4
\end{bmatrix}
$$

$$
D_2 \quad
\begin{array}{l}
S_{121}, S_{221} \\
S_{122}, S_{222} \\
S_{123}, S_{223} \\
S_{124}, S_{224}
\end{array}
\begin{bmatrix}
A_4 & A_2 & A_1 & A_3 \\
A_2 & A_3 & A_4 & A_1 \\
A_1 & A_4 & A_3 & A_2 \\
A_3 & A_1 & A_2 & A_4
\end{bmatrix}
$$

$$
D_3 \quad
\begin{array}{l}
S_{131}, S_{231} \\
S_{132}, S_{232} \\
S_{133}, S_{233} \\
S_{134}, S_{234}
\end{array}
\begin{bmatrix}
A_4 & A_2 & A_1 & A_3 \\
A_2 & A_3 & A_4 & A_1 \\
A_1 & A_4 & A_3 & A_2 \\
A_3 & A_1 & A_2 & A_4
\end{bmatrix}
$$

The design under discussion allows evaluating main effects and interactions of A and D with fewer subjects than a completely randomized design, fewer measurements per subject than a simple repeated-measurement design, and a potentially more efficient test of A and AD than the mixed design (Chapter 8).

The analysis of variance. Table 10–15 presents the *SV, df, EMS,* and *F* ratios. The between-subjects variability is partitioned in the same way as in Section 8.3, where we dealt with two between-subjects variables. The within-subject effects are generated if one remembers that there is confounding between any main effect associated with the square (that is, R, C, A) and the interaction of the other two variables. Note, however, that D can interact with each of the main effects associated with the square and with B cells res as well.

As in previous tables in this chapter, we have omitted the clumsy *SS* formulas. The only terms that might cause any problem are *B cells res* and *B cells res* $\times D$. To calculate the former, sum over the *dn* scores in each of the a^2 combinations of rows and columns in the square. The *SS* based on these a^2 units will be referred to as $SS_{\overline{RC}}$. Then,

$$
SS_{\text{B cells res}} = SS_{\overline{RC}} - SS_R - SS_C - SS_A
$$

Note that this corresponds to

$$
df_{\text{B cells res}} = (a^2 - 1) - 3(a - 1)
$$
$$
= (a - 1)(a - 2)
$$

For $SS_{\text{B cells res}} \times D$, the basic unit is each combination of row, column, and level of D. The *SS* based on these $a^2 d$ means, which is a term distributed on $a^2 d - 1$ *df*, is

Table 10–15 Analysis of variance for the replicated square design with a between-squares factor

SV	df	EMS	F
Total	$a^2dn - 1$		
Between S	$adn - 1$		
R	$a - 1$	$\sigma_e^2 + a\sigma_{S/RD}^2 + adn\theta_{AC}^2$	$\dfrac{MS_R}{MS_{S/RD}}$
D	$d - 1$	$\sigma_e^2 + a\sigma_{S/RD}^2 + a^2n\theta_D^2$	$\dfrac{MS_D}{MS_{S/RD}}$
RD	$(a-1)(d-1)$	$\sigma_e^2 + a\sigma_{S/RD}^2 + an\theta_{ADC}^2$	$\dfrac{MS_{RD}}{MS_{S/RD}}$
S/RD	$ad(n-1)$	$\sigma_e^2 + a\sigma_{S/RD}^2$	
Within S	$adn(a-1)$		
A	$a - 1$	$\sigma_e^2 + adn\theta_A^2$	$\dfrac{MS_A}{MS_{res}}$
C	$a - 1$	$\sigma_e^2 + adn\theta_C^2$	$\dfrac{MS_C}{MS_{res}}$
AD	$(a-1)(d-1)$	$\sigma_e^2 + an\theta_{AD}^2$	$\dfrac{MS_{AD}}{MS_{res}}$
DC	$(d-1)(a-1)$	$\sigma_e^2 + an\theta_{DC}^2$	$\dfrac{MS_{DC}}{MS_{res}}$
B cells res	$(a-1)(a-2)$	$\sigma_e^2 + dn\theta_{AC}^2$	$\dfrac{MS_{\text{B cells res}}}{MS_{res}}$
B cells res $\times D$	$(a-1)(a-2)(d-1)$	$\sigma_e^2 + n\theta_{ADC}^2$	$\dfrac{MS_{\text{B cells res} \times D}}{MS_{res}}$
Residual	$ad(a-1)(n-1)$	σ_e^2	

designated $SS_{\overline{RCD}}$ and

$$SS_{\text{B cells res}\times D} = SS_{\overline{RCD}} - SS_R - SS_C - SS_D - SS_{RD}$$
$$- SS_{CD} - SS_{AD} - SS_{\text{B cells res}}$$

The model that has been chosen is an extension of Equation (10.9). Carryover effects are assumed negligible; therefore we have the equation

(10.10) $Y_{ijkmp} = \mu + \eta_{i/kp} + \alpha_j + \delta_k + \gamma_m + (\alpha\delta)_{jk} + (\alpha\gamma)_{jm}$

$$+ (\delta\gamma)_{km} + (\alpha\delta\gamma)_{jkm} + \varepsilon_{ijkmp}$$

This equation generates the *EMS* for Table 10–15, which in turn indicate the appropriate error terms. Furthermore, it should be apparent that if Equation

(10.10) is valid, R can be interpreted as a between-subjects measure of AC, RD as a between-subjects measure of ADC, B *cells res* as a within-subject measure of AC, and B *cells res* $\times D$ as a within-subject measure of ADC. As in Section 10.5, the F tests of A and C effects against the residual will be positively biased if the sequence of treatment presentations interacts with A or C. If there is reason to suspect such carryover effects, the following preliminary tests can be carried out:

$$F = \frac{MS_R}{MS_{S/RD}} \qquad\qquad F = \frac{MS_{B\ \text{cells res}}}{MS_{\text{res}}}$$

$$F = \frac{MS_{RD}}{MS_{S/RD}} \qquad\qquad F = \frac{MS_{B\ \text{cells res} \times D}}{MS_{\text{res}}}$$

If any of these are significant, a transformation to additivity should be sought.

Note that if sequence interactions are suspect, our recommended procedure provides no way of testing for AC or ADC effects. There is no test that permits determining which interaction component is present when, for example, B *cells res* is significant. Such a result only reveals that *some* interaction component is present. A significant B *cells res* can be interpreted to be AC (or B *cells res* $\times D$ to be ADC) if there is a strong a priori reason for assuming Equation (10.10) to be valid. Such an assumption might be founded on previous experimentation and on knowing the independent and dependent variables, that is, knowing the experimental situation.

A numerical example. Table 10–16 presents data from a 3×3 Latin square replicated four times, twice at each of two levels of B. Subtotals that are required in calculating sums of squares have also been computed. The correction term is

$$C = \frac{(416)^2}{36} = 4,807.111$$

The total variability is

$$SS_{\text{tot}} = (12)^2 + (4)^2 + \cdots + (18)^2 + (16)^2 - C$$

$$= 5,814.000 - 4,807.111$$

$$= 1,006.889$$

We next obtain the between-subjects variability:

$$SS_{\text{B.S}} = \frac{(26)^2 + (28)^2 + \cdots + (37)^2 + (43)^2}{3} - C$$

$$= 5,218.000 - 4,807.111$$

$$= 410.889$$

Table 10–16 Data for a replicated Latin square design with a between-squares factor

D_1	A_1	A_3	A_2	$\sum Y_{ij1mp}$
S_{111}	12	4	10	26
S_{211}	14	6	8	28
$T_{.jk11}=26$	10	18		$T_{..k11}=54$
S_{112}	6 (A_2)	9 (A_1)	3 (A_3)	18
S_{212}	11	17	5	33
$T_{.jk12}=17$	26	8		$T_{..k12}=51$
S_{113}	5 (A_3)	12 (A_2)	18 (A_1)	35
S_{213}	7	6	10	23
$T_{.jk13}=12$	18	28		$T_{..k13}=58$
$T_{.jk1.}=55$	54	54		$T_{...1.}=163$

D_2	A_1	A_3	A_2	$T_{i.2mp}$
S_{121}	19	11	20	50
S_{221}	21	11	20	52
$T_{.jk21}=40$	22	40		$T_{..k21}=102$
S_{122}	16 (A_2)	18 (A_1)	8 (A_3)	42
S_{222}	14	12	3	29
$T_{.jk22}=30$	30	11		$T_{..k22}=71$
S_{123}	6 (A_3)	17 (A_2)	14 (A_1)	37
S_{223}	9	18	16	43
$T_{.jk23}=15$	35	30		$T_{..k23}=80$
$T_{.jk2.}=85$	87	81		$T_{...2.}=253$

A Subtotals

	A_1	A_2	A_3	
D_1	80	53	30	163
D_2	100	105	48	253
	180	158	78	416

C Subtotals

	C_1	C_2	C_3	
D_1	55	54	54	163
D_2	85	87	81	253
	140	141	135	416

This is now further partitioned:

$$SS_R = \frac{(54+102)^2 + (51+71)^2 + (58+80)^2}{12} - C$$

$$= 4{,}855.333 - 4{,}807.111$$

$$= 48.222$$

$$SS_D = \frac{(163-253)^2}{36} = 225.000$$

$$SS_{RD} = \frac{(54)^2 + \cdots + (80)^2}{6} - C - SS_R - SS_D$$

$$= 5{,}121.000 - 4{,}807.111 - 48.222 - 225.000$$

$$= 40.667$$

and

$$SS_{S/RD} = SS_{BS} - SS_R - SS_D - SS_{RD} = 370.222$$

The within-subject variability is computed next:

$$SS_{W.S} = SS_{tot} - SS_{BS} = 596.000$$

Partitioning this, we obtain

$$SS_C = \frac{(140)^2 + (141)^2 + (135)^2}{12} - C$$

$$= 4{,}808.833 - 4{,}807.111$$

$$= 1.722$$

$$SS_A = \frac{(180)^2 + (158)^2 + (78)^2}{12} - C$$

$$= 5{,}287.333 - 4{,}807.111$$

$$= 480.222$$

$$SS_{DC} = \frac{(55)^2 + (54)^2 + \cdots + (81)^2}{6} - C - SS_D - SS_C$$

$$= 5{,}035.333 - 4{,}807.111 - 225.000 - 1.722$$

$$= 2.500$$

$$SS_{AD} = \frac{(80)^2 + (53)^2 + \cdots + (48)^2}{6} - C - SS_A - SS_D$$

$$= 5{,}573.000 - 4{,}807.111 - 480.222 - 225.000$$

$$= 60.667$$

$$SS_{\text{B cells res}} = \frac{(26+40)^2 + (10+22)^2 + \cdots + (28+30)^2}{4}$$
$$- C - SS_R - SS_C - SS_A$$
$$= 5{,}338.000 - 4{,}807.111 - 48.222 - 1.722 - 480.222$$
$$= .723$$

$$SS_{\text{B cells res} \times D} = \frac{(26)^2 + (10)^2 + \cdots + (35)^2 + (30)^2}{2} - C - SS_D$$
$$- SS_{\text{B cells res}} - SS_{RD} - SS_R - SS_{DC} - SS_C - SS_{AD} - SS_A$$
$$= 5{,}668.000 - 4{,}807.111 - 225.000 - .723 - 40.667$$
$$- 48.222 - 2.500 - 1.722 - 60.667 - 480.222$$
$$= 1.166$$

and

$$SS_{\text{res}} = SS_{\text{w.s}} - SS_C - SS_A - SS_{DC} - SS_{AD} - SS_{\text{B cells res}} - SS_{\text{B cells res} \times D}$$
$$= 48.000$$

Table 10–17 summarizes the results of the analysis.

Using several squares. The design using several Latin squares can also be extended to handle an additional independent variable. Suppose that we randomly sample dn squares. Squares are randomly assigned to levels of D so that there are n squares at each D_k. A total of adn subjects are then randomly distributed among the adn sequences of the selected squares, one subject in each

Table 10–17 Analysis of variance for the data of Table 10–16

SV	df	SS	MS	F
Total	35	1,006.889		
Between S	11	410.889		
R	2	48.222	24.111	1.490
D	1	225.000	225.000	13.906[a]
RD	2	40.667	20.334	1.257
S/RD	6	97.080	16.180	
Within S	24	596.000		
C	2	1.722	.861	.215
A	2	480.222	240.111	60.028[b]
DC	2	2.500	1.250	.313
AB	2	60.667	30.334	7.584[a]
B cells res	2	.723	.362	.091
B cells res $\times D$	2	1.166	.583	.146
Residual	12	48.000	4.000	

[a] $p < .01$
[b] $p < .001$

Table 10–18　Analysis with several squares and a between-squares factor

SV	df	EMS
Total	a^2dn-1	
Between S	$adn-1$	
D	$d-1$	$\sigma_e^2 + a\sigma_{S/D}^2 + a^2n\theta_D^2$
S/D	$d(an-1)$	$\sigma_e^2 + a\sigma_{S/D}^2$
Within S	$adn(a-1)$	
A	$a-1$	$\sigma_e^2 + adn\theta_A^2$
C	$a-1$	$\sigma_e^2 + adn\theta_C^2$
AD	$(a-1)(d-1)$	$\sigma_e^2 + an\theta_{AD}^2$
BC	$(a-1)(d-1)$	$\sigma_e^2 + an\theta_{CD}^2$
Residual	$b(an-2)(a-1)$	σ_e^2

sequence. If we assume that the dimensions of the square—subjects, positions in time, A—do not interact, the following model is appropriate:

$$(10.11) \qquad Y_{ijkm} = \mu + \eta_{i/k} + \alpha_j + \gamma_m + \delta_k + (\alpha\delta)_{jk} + (\delta\gamma)_{km} + \varepsilon_{ijkm}$$

Table 10–18 presents the analysis of variance. The interaction effects AC and ACD, if present, will contribute to the residual error term and not to A or C, and will thus cause negative bias in those tests and also in tests of AD and ACD. The advantage of the design lies in the likelihood that confounding will be minimized since several squares are used.

10.6   CONCLUDING REMARKS

The assets of the Latin square design are clear—fewer subjects are needed; systematic treatment biases are reduced through counterbalancing; and error variance is reduced since variability due to two factors, the rows and columns of the design, is removed. The dangers are less immediately obvious but we should all find them familiar by now. Fixed-effect interactions involving the treatment variable contribute to the between-cells residual term but not to the treatment source; an F ratio using these two sources under such nonadditivity is negatively biased. When random interaction effects are present, they contribute to the treatment source but not to the within-cell error term; an F ratio of these two terms under these conditions is positively biased. When repeated measurements are involved, there is the danger of carryover effects; the effect of a treatment may depend on which of several treatments preceded it. Also, heterogeneity of covariance may cause positive bias.

　　Despite the problems that we have noted, the advantage of the Latin square design, or some variation of it, will frequently outweigh its disadvantages. Carryover effects will be minimal in many experiments; in others they can be reduced by introducing sufficiently long intervals between treatment levels; and in still others we can statistically account for them (Cochran and Cox, 1957, pp.

135–40). If we have prior knowledge of the types of interactions we may encounter in our data, we can still achieve unbiased F tests. If an AC interaction is likely, and no other, we should be wise to replicate a single square and to use the within-cell term as our measure of error. In the contrary case—if for example, random-effect interactions are likely—the between-cells residual makes most sense. Where we suspect nonadditivity but do not know the specific sources, transformations that reduce nonadditivity can be sought. Also, confounding interaction with main effects becomes less serious when squares are larger and more are used.

Even the possible presence of negative bias in the test of treatments does not rule out using the Latin square design. Such bias means that whenever the null hypothesis is true, the ratio of *EMS* will be less than 1; presumably, the probability of a Type I error will be lower than the nominal α level. When the null hypothesis is false, however, the test of treatments may be more powerful than what other designs provide. If variability due to columns is large relative to variability due to interactions, its removal from numerator and denominator of the F test of treatments will generally increase the expected F ratio. The situation is analogous to that depicted in Figure 3–1, in which we similarly distinguished between bias and efficiency in discussing properties of estimators.

Finally, it is important to realize that there are situations in which the supply of subjects is severely restricted; the experiment may be unpleasant or dangerous and volunteers consequently few, or the experiment may require a particular type of subject, perhaps a particular clinical case that it is difficult to obtain. Then, the choice of design is quickly reduced to a choice among repeated-measurement designs. The subjects × treatments design, with an independent randomization for each subject, has many of the same problems that plague the Latin square—heterogeneity of covariance or carryover effects. Thus the Latin square, with its potentially greater efficiency, may be preferred.

EXERCISES

10.1 A study is designed to evaluate three methods A of teaching first-year French to high school students. Three tracks B (ability levels) are included from each of three schools C. The design, with cell totals based on 10 subjects each, is given here. Present a reasonable model, and the corresponding analysis of variance. Assume $MS_{S/cells}$

$$
\begin{array}{c}
\quad\quad C_1 \quad\quad\quad C_2 \quad\quad\quad C_3 \\
\begin{array}{c} B_1 \\ \\ B_2 \\ \\ B_3 \end{array}
\left[
\begin{array}{ccc}
(A_1) & (A_2) & (A_3) \\
85 & 77 & 75 \\
(A_3) & (A_1) & (A_2) \\
64 & 75 & 76 \\
(A_2) & (A_3) & (A_1) \\
60 & 59 & 69
\end{array}
\right]
\end{array}
$$

10.2 Each of 36 subjects was tested on four signal detection problems, which varied in location of the target. Subjects differed in sequence of presentations of the problems (four sequences were used) and in the instructions read to them (three sets of instructions). Present the appropriate analysis of variance table.

10.3 A clinical investigator is concerned with how alcohol affects reaction time *rt* to perceptual inputs. He runs three experimental groups of 20 *S*s each on four tasks, each task on a different day. The groups differ in alcohol intake. The tasks involve determining (a) simple *rt* to an auditory stimulus, (b) simple *rt* to a visual stimulus, (c) choice *rt* to auditory stimuli, and (d) choice *rt* to visual stimuli. The tasks are presented in 60 sequences composing 15 Latin squares; median *rt*s are obtained for each *S* on each task. Present the analysis of variance.

10.4 Five highly practiced *S*s were each required to learn different lists on five different days; the lists varied in type of item—for example, words, digits, nonsense syllables. Since prior practice in the task should minimize learning-to-learn and proactive inhibition should be slight with the different materials used, the sequence of lists was Latin-squared. Each list had seven items; the dependent measure was errors per position in the list. Present the *SV* and *df*.

10.5 Assume that three-man groups are required to solve four problems that vary in difficulty. The purpose is to investigate the number of communications from each individual as a function of problem difficulty. The presentation is Latin-squared and four groups are run through each sequence. Present *SV* and *df*.

10.6 We are interested in gambling behavior under variations in initial stake (three levels), payoffs (three levels), and probability of winning (three levels). Suggest several alternative designs and discuss their relative merits. Assume that 81 subjects are available.

10.7 Consider a $2 \times 2 \times 2$ experiment. Suppose we have four experimenters, each of whom runs two of the eight cells. In the notation of Chapter 5,

E_1	E_2	E_3	E_4
(1), *bc*	*a, abc*	*b, c*	*ab, ac*

Several effects are confounded among experimenters. Which effects are confounded, and in what way?

REFERENCES

Cochran, W. C., and Cox, G. M. *Experimental Designs*, 2nd ed. New York: John Wiley & Sons, 1957.

Fisher, R. A., and Yates, F. *Statistical Tables for Biological, Agricultural and Medical Research.* Edinburgh: Oliver & Boyd, 1955, pp. 80–82.

Tukey, J. W. Test for nonadditivity in the Latin square. *Biometrics* 11:111–13 (1965).

Wilk, M. B., and Kempthorne, O. Nonadditivities in a Latin square. *Journal of the American Statistical Association* 52:218–36 (1957).

SUPPLEMENTARY READING

Readers who have trouble with the notation and derivations of the Wilk–Kempthorne (1957) article might profit from the following treatment based on their paper:

Gaito, J. The single Latin square design in psychological research. *Psychometrika* 23:369–78 (1958).

A different model for the replicated square design from the one presented in this chapter, but one that leads to similar conclusions about the effects of nonadditivity, is discussed in

Gourlay, N. *F* test bias for experimental designs of the Latin square type. *Psychometrika* 20:237–87 (1955).

Gaito has also discussed the relation between the designs of Chapters 7 and 10.

Gaito, J. Repeated measurement designs and counterbalancing. *Psychological Bulletin* 58:46–54 (1961).

Several papers have introduced ways of balancing the Latin square for carryover effects—among these are (see also the pages cited in Cochran and Cox, 1957):

Alimena, B. A. A method of determining unbiased distribution in Latin square. *Psychometrika* 27:315–18 (1962).

Atkinson, G. F. Designs for sequences of treatments with carryover effects. *Biometrics* 22:292–309 (1966).

Benjamin, L. S. A special Latin square for the use of each subject as his own control. *Psychometrika* 30:499–513 (1966).

Bradley, J. V. Complete counterbalancing of immediate sequential effects in a Latin square design. *Journal of the American Statistical Association* 53:525–28 (1958).

Williams, R. N. Experimental designs for serially correlated observations. *Biometrika* 39:151–67 (1952).

Numerous additional incomplete designs that involve confounding but hold the potential for efficient analyses can be found in Cochran and Cox (1957).

11 FURTHER DATA ANALYSES: QUALITATIVE INDEPENDENT VARIABLES

11.1 INTRODUCTION

We have so far been concerned with examining the total variability among treatment population means. The null hypothesis that the variance of the entire set of means is zero has been tested, and estimates of the variance of the entire set of means have been established. Such procedures are often only the beginning of a complete analysis of the data. Left unanswered are questions about the slope and shape of performance curves; and differences among means within subsets smaller than the entire treatment set are still unexplored.

If the independent variable is quantitative (amount of reward, for example, or length of time in therapy), the overall F test is often only preliminary, useful in determining whether or not to proceed further with the analysis. Knowing that there is significant variability among the means, we may ask whether performance shows a general improvement over the levels of the independent variable, and whether several apparent changes in the direction of the function are significant. In short, the overall F indicates that the means do not fall on a straight line with slope of zero. We then wish to know more about the slope and shape of the best-fitting function.

In the case of a qualitative variable (say type of reward or type of therapy), the F test is again only a first step. Consider, for example, an experiment in which motivation is manipulated. There is one group in which correct responses are rewarded and errors are not punished (group R), a second group that is punished for errors and also not rewarded for correct responses (group P), a third group that is both rewarded for correct responses and punished for incorrect responses (group RP), and a fourth group that receives neither reward for correct responses nor punishment for incorrect responses (group NRP). A significant overall F shows that differences exist among the treatment population means. Why? Are reward and punishment different in effect? Is the combination RP more effective than either alone? Does the RP effect differ from the average effect for the R and P groups? Is the effect of the NRP treatment significantly different from the effect of the average of the three incentive groups? These are all reasonable questions, and others could be asked.

Analyses relevant to quantitative independent variables will be considered in Chapter 17. For the present, the discussion is restricted to the sort of comparisons among means, as just indicated. These comparisons can also be made with quantitative variables (Do the effects of one and two food pellets of reward differ?) but for these variables the analyses of Chapter 17 will generally be more fruitful.

To clarify the analyses that follow, we return to the example of the experiment on reward and punishment. Consider each of the questions raised. Are reward and punishment different in effect? implies the null hypothesis

$$\mu_R - \mu_P = 0$$

Is the combination *RP* more effective than either alone? implies the null hypotheses

$$\mu_{RP} - \mu_R = 0$$

and

$$\mu_{RP} - \mu_P = 0$$

Does the *RP* effect differ from the average effect for the *R* and *P* groups? implies the null hypothesis

$$\mu_{RP} - \tfrac{1}{2}(\mu_R + \mu_P) = 0$$

Is the effect of the *NRP* treatment significantly different from that of the average of the three incentive groups? implies the null hypothesis

$$\mu_{NRP} - \tfrac{1}{3}(\mu_R + \mu_P + \mu_{RP}) = 0$$

We can rewrite the five null hypotheses:

$$(1)\mu_R + (-1)\mu_P + (0)\mu_{RP} + (0)\mu_{NRP} = 0$$
$$(-1)\mu_R + (0)\mu_P + (1)\mu_{RP} + (0)\mu_{NRP} = 0$$
$$(0)\mu_R + (-1)\mu_P + (1)\mu_{RP} + (0)\mu_{NRP} = 0$$
$$(-\tfrac{1}{2})\mu_R + (-\tfrac{1}{2})\mu_P + (1)\mu_{RP} + (0)\mu_{NRP} = 0$$
$$(-\tfrac{1}{3})\mu_R + (-\tfrac{1}{3})\mu_P + (-\tfrac{1}{3})\mu_{RP} + (1)\mu_{NRP} = 0$$

A close examination of these equations indicates the general form of the null hypotheses to be considered in this chapter (and in Chapter 17 also; the difference is in the rationale for selecting the multipliers of the μs). The general null hypothesis is

$$\psi = 0$$

where $\psi = \sum_j w_j \mu_j$, and the w_j are the coefficients that multiply the μ_j; they vary as a function of the specific hypotheses being tested. When the sum of the coefficients is zero (this will be the case through this chapter), ψ is referred to as a linear contrast. Estimation is generally the problem of estimating ψ or of obtaining a confidence interval for it.

Given a set of a means, we find many possible contrasts that could be tested or estimated. If we assume the truth of the overall null hypothesis ($H_0: \mu_1 = \cdots = \mu_j = \cdots = \mu_a$), the probability of at least one Type I error within such a set of tests may be considerably greater than α. In the next section, we will elaborate on this observation. In Section 11.3, we will consider several approaches to dealing with this inflation of α in the special case in which null hypotheses of the form $H_0: \mu_j - \mu_{j'} = 0$ are tested. In Section 11.4, we will turn to more complex contrasts: those in which more than two means have nonzero weights. In Section 11.5, we will extend the discussion to contrasts based on interaction. In Section 11.6, we will summarize our recommendations for testing sets of contrasts.

11.2   MULTIPLE CONTRASTS AND ERROR RATES

Contrasts are tested under one or both of two sets of circumstances: (1) before collecting data, the experimenter raises certain questions that dictate specific sets of contrasts; (2) after the data have been collected, the usual F test has proved significant and the investigator wants to examine the results further to detect those contrasts that cause the overall null hypothesis to be rejected. In either case, we have a situation that can be likened to a series of tosses of a coin. Although the probability of a head on any one toss is .5, the probability of *at least one head* in a series of tosses is greater than .5, and the magnitude of this probability is a direct function of the total number of tosses. Similarly, any single test of a comparison has probability α of a Type I error. As the number of comparisons increases, however, the probability of *at least one Type I error* increases. If we did 10 independent significance tests, for example, each at the .05 level, the probability of at least one Type I error would be $1 - (.95)^{10} \approx .40$. The same problem exists if the largest observed contrast is selected for testing. This is equivalent to testing all contrasts, since the probability that the largest observed contrast is significant is the probability that at least one contrast is significant.

It will now be helpful to distinguish between two types of error rates. The usual α level, the probability that a single comparison results in a Type I error, will be referred to as the *error rate per comparison* (*EC*). The probability that an entire set of comparisons contains at least one Type I error will be referred to as the *error rate per family* (*EF*). In the coin toss example, $EC = .05$ but $EF = .40$. The distinction may be clarified by picturing 2,000 replications of an experiment. Five contrasts, all of which are true for the population, are tested in each experiment. Thus, we have 2,000 sets of five significance tests. A single experiment can result in 0 to 5 erroneous rejections of H_0. The actual results might resemble the accompanying chart (page 293).

The total number of Type I errors is $402 + (2)(43) + (3)(5) + (4)(2) = 511$. Thus, the *EC* equals $511/10,000 \approx .05$. However, 452 experiments (with each experiment viewed as a family of five contrasts) contain one or more Type I errors. Thus, the *EF* equals $452/2,000 = .226$.

Number of experiments with X errors	X
1,548	0
402	1
43	2
5	3
2	4
0	5
2,000	

There are two extreme views on the discrepancy between *EF* and *EC*. On the one hand, we could ignore the *EF* and test every contrast at a fixed α level chosen independent of the total number of contrasts to be tested. The objection to this approach is that the likelihood of a Type I error increases with the number of contrasts tested. Significant findings that are not readily replicable, because they are erroneous, are more likely to be obtained by investigators who test relatively many contrasts. If the experiment is a pilot study, not to be published but used as a source of leads for subsequent studies, the damage may be slight. But if the results are to be published or put to practical application, the use of the *EC* can be dangerous.

One could move to another extreme. Energetic researchers test thousands of null hypotheses in their lifetimes. Should they guard against error rate based on those many hypothesis tests, carrying out each at some infinitesimally low α level? If they did, Type II error rates would soar to undesirable heights.

The approach ordinarily taken falls somewhere between these two extremes. Our proposed compromise is to use some set of contrasts involved in a single experiment as the unit for controlling error rate. We shall choose a criterion for significance that will give a constant *EF* regardless of the number of treatment groups. In effect, a rule is required for adjusting the *EC* downward as the total number of contrasts increases, and adjusting so that the change in the number of contrasts does not alter the *EF*.

One issue that arises is the proper choice of the family of contrasts, the unit for controlling error rate. This choice usually is between the total set of possible contrasts in an experiment and the set of contrasts associated with a given source of variance. In a one-factor design, these are equivalent. Consider, however, a two-factor design. For simplicity, assume that we are interested in testing all null hypotheses of the form $\mu_{j.} - \mu_{j'.} = 0$ and all those of the form $\mu_{.k} - \mu_{.k'} = 0$. Assume that *EF* is .05. If the experiment is the basis for the family of contrasts, the criterion for significance will be such that the probability of at least one Type I error is less than or equal to .05 for the set of $\frac{1}{2}a(a-1) + \frac{1}{2}b(b-1)$ contrasts. If, however, the family is that set of contrasts associated with a source of variance, there are two families of contrasts in our example, one consisting of $\frac{1}{2}a(a-1)$ contrasts and the other of $\frac{1}{2}b(b-1)$ contrasts. Now the criterion for the significance of any *A* contrast would be set so that the probability of at least one Type I error in the set of $\frac{1}{2}a(a-1)$ contrasts is no more than .05; similarly, the criterion for the significance of any B contrasts would be set so that the probability of at least one Type I error in the set of $\frac{1}{2}b(b-1)$ contrasts is no more than .05.

We recommend defining the family as the set of contrasts associated with a source of variance rather than with the experiment as a whole. Since the recommended approach involves smaller families, we shall not need as large a critical value for significance of any contrast at some designated EF, and the test will therefore be more powerful. Furthermore, investigators often precede tests of contrasts by some overall F test. It will be easier to relate the results of a family of contrasts with the results of the preceding F tests if the overall test and the family are based on the same unit, a single source of variance.[1]

11.3   COMPARISONS BY PAIRS

Rejecting the overall null hypothesis $\mu_1 = \cdots = \mu_j = \cdots = \mu_a$ implies that there exists at least one pair of treatment population means μ_j and $\mu_{j'}$ such that $\mu_j \neq \mu_{j'}$. Ordinarily, the investigator will want to determine which pairs are responsible for the variability among the a treatment population means. Several procedures exist for testing these pairwise differences while controlling the Type I error rate for the family of $\frac{1}{2}a(a-1)$ possible tests of this form. We consider three approaches (alternatives that for various reasons I find less satisfactory are included in reviews by Miller, 1966, and Games, 1971). These approaches are Tukey's (1953) *Wholly Significant Difference* (*WSD*) test, the Newman-Keuls (1939, 1952) test, and the Bonferroni t test (Dunn, 1961; Perlmutter and Myers, 1972). We shall also give Dunnett's (1955) test for controlling EF for the set of contrasts of a experimental group means with a control group mean.

11.3.1   TUKEY'S *WSD* PROCEDURE.   The significance tests and confidence intervals that are part of Tukey's method are based on the sampling distribution of q, the Studentized range statistic. This statistic is defined as the range of a set of observations (the range is the largest minus the smallest value) divided by an estimate of the standard deviation of the population from which the sample of observations has been drawn. It is assumed that the observations are drawn independently from a normally distributed population. In applying the Studentized range statistic to the analysis of data from a one-factor completely randomized design, there are a observations, the treatment group means. The assumption that they are independently sampled from the same normal population implies the overall null hypothesis and homogeneity of variance. Our best estimate of the sampling error of the observations would be

(11.1)
$$\hat{\sigma}_{\bar{Y}} = \sqrt{\frac{MS_{S/A}}{n}}$$

[1] In previous editions, I used Ryan's (1959) terminology, referring to *error rate experimentwise* (*EW*). Unfortunately, this implies that the unit for controlling error rate will be the entire set of contrasts performed in connection with an experiment. Because this is exactly counter to the preceding recommendation, I have now adopted Tukey's (1953) more neutral term, *family*, to refer to the set of contrasts to which the error rate is referred. Thus, EF replaces the EW of the earlier edition.

More generally, $\hat{\sigma}_{\bar{Y}}^2$ will be the MS_{error} calculated for the overall F test, divided by the number of observations on which each mean is based. (The situation is somewhat different when contrasts are performed on a within-subject variable (Section 11.3.6).) Contrasts among the $\bar{Y}_{..k}$ in a two-factor design, for example, would require $\hat{\sigma}_{\bar{Y}} = \sqrt{MS_{S/AB}/an}$.

The sampling distribution of q depends on a (the number of means ranged over) and on the df associated with $\hat{\sigma}_{\bar{Y}}$. Critical values of q are presented in the Appendix, Table A–9, for the .01 and .05 (two-tailed) significance levels for various values of a and errors df. We shall refer to the critical value as q_α, the value of q needed for significance at the α level. To test the significance of any contrast of two means $(\bar{Y}_{.j} - \bar{Y}_{.j'})$, we could calculate $(\bar{Y}_{.j} - \bar{Y}_{.j'})/\hat{\sigma}_{\bar{Y}}$ (where $\bar{Y}_{.j} > \bar{Y}_{.j'}$) and evaluate it against q_α to determine whether the result is significant. An equivalent alternative is to calculate $q_\alpha \hat{\sigma}_{\bar{Y}}$. Any difference between two means greater than this product is significant at the α level.

For reasons explained earlier (Chapter 3), we feel that estimation, particularly interval estimation, is an important and too often neglected inferential tool. The Tukey method readily establishes confidence intervals on contrasts between two means. Such confidence intervals, based on the EF concept, are somewhat different in interpretation from the intervals previously encountered in the text, which are based on the EC concept. When EC is our error rate, we are concerned with whether a single interval does or does not contain the parameter; confidence is the proportion of intervals that in the long run would contain the parameter. In the present context, in which we wish to control EF, the experiment falls into one of two categories: (1) either the intervals computed for all k contrasts contain the parameters being estimated, or (2) at least one interval does not contain its parameter; confidence is the proportion of experiments that in the long run fall in the first category.

By Tukey's WSD approach, the confidence interval for any contrast of two treatment population means $\mu_j - \mu_{j'}$ is $\hat{\psi} \pm q_\alpha \hat{\sigma}_{\bar{Y}}$, where $\hat{\psi} = \bar{Y}_{.j} - \bar{Y}_{.j'}$. In other words, the probability is at least $1 - \alpha$ that

(11.2)
$$\hat{\psi} - q_\alpha \hat{\sigma}_{\bar{Y}} \leq \psi \leq \hat{\psi} + q_\alpha \hat{\sigma}_{\bar{Y}}$$

for all ψ of the form $\mu_j - \mu_{j'}$. Values of q_α depend on the total number of means in the family and on error df, and $\hat{\sigma}_{\bar{Y}}$ depends on the design, as noted earlier.

A numerical example. To see how both significance tests and confidence interval are constructed, consider the following set of ordered group means:

$\bar{Y}_1$	$\bar{Y}_2$	$\bar{Y}_3$	$\bar{Y}_4$	$\bar{Y}_5$	$\bar{Y}_6$
4	7	8	10	12	16

Assume that each mean is based on six subjects and that $MS_{S/A} = 13.5$. Then, $\hat{\sigma}_{\bar{Y}} = \sqrt{13.5/6} = 1.5$. The critical q value for a set of six means and 30 error df is 4.30. Therefore, we regard as significant all differences by pairs among means that exceed (1.5) (4.30), or 6.45. The result can be displayed graphically:

$$\bar{Y}_1 \quad \bar{Y}_2 \quad \bar{Y}_3 \quad \bar{Y}_4 \quad \bar{Y}_5 \quad \bar{Y}_6$$

Any means that lie above the same line do not differ significantly from each other; all other contrasts are significant. Thus, $\bar{Y}_1$ differs significantly from $\bar{Y}_5$ and $\bar{Y}_6$; and $\bar{Y}_2$ and $\bar{Y}_6$, and $\bar{Y}_3$ and $\bar{Y}_6$, also form significant contrasts. In confidence intervals, the .95 interval for any $\mu_j - \mu_{j'}$ is $\bar{Y}_{.j} - \bar{Y}_{.j'} \pm 6.45$.

11.3.2 THE NEWMAN-KEULS TEST.

The procedure of Newman and Keuls also uses the Studentized range statistic (Table A–9). The difference between it and the Tukey procedure is that the Newman-Keuls test is sequential; we test the largest range first and test successively smaller ranges only if the previous test has a significant outcome. As we range over successively smaller subsets of means, moreover, the criterion for the significance shifts. More specifically, we label the group means $\bar{Y}_1, \bar{Y}_2, \ldots, \bar{Y}_a$, from smallest to largest. Then we first compare $\bar{Y}_a - \bar{Y}_1$ with $\hat{\sigma}_{\bar{Y}} q_{\alpha;a,df_2}$ and reject $H_a(\mu_a = \mu_1)$ if the difference between means exceeds the critical value. If the result is not significant, we stop testing; if the result is significant, we evaluate $\bar{Y}_{a-1} - \bar{Y}_1$ and $\bar{Y}_a - \bar{Y}_{2'}$ each against $\hat{\sigma}_{\bar{Y}} q_{\alpha;a-1,df_2}$. Note that the q value is now obtained from the $a - 1$ column of Table A–9. Testing proceeds as in the first stage. If we reject $H_{a-1}: \mu_{a-1} = \mu_1$, we now evaluate $\bar{Y}_{a-2} - \bar{Y}_1$ and $\bar{Y}_{a-1} - \bar{Y}_2$, this time using $q_{\alpha;a-2,df_2}$.

A numerical example. The Newman-Keuls test will be illustrated with the data of the preceding section. First, calculate the criterion statistic for each possible subset of means. When we are ranging over the full ordered set of four means, we need $\hat{\sigma}_{\bar{Y}} q_{\alpha,a-1,df_2} = (1.5)(4.30) = 6.45$, as in the Tukey test. The critical values for tests over smaller subsets of means are:

Set size =	2	3	4	5	6
Criteria =	4.34	5.24	5.76	6.15	6.45

Note that each value is a product of $\hat{\sigma}_{\bar{Y}}$ (1.5 in our example) and a q value appropriate for the error df (30 in our example) and the number of means ranged over.

The test proceeds as follows. First, it is easier to keep track of the testing sequence if we establish a 6×6 table, as in Table 11–1. We then compare the largest range against the critical value, 6.45. Because $\bar{Y}_6 - \bar{Y}_1 = 12$, we place an asterisk in the upper right-hand corner of that table. The circle indicates that the WSD test also resulted in significance for this contrast. We next consider the two contrasts on the diagonal immediately to the left of the (1, 6) cell. $\bar{Y}_6 - \bar{Y}_2$ and $\bar{Y}_5 - \bar{Y}_1$ are both larger than 6.15, the criterion difference for sets of five means. The next set of tests are performed on the diagonal set: (1, 4), (2, 5), and (3, 6). These three differences are evaluated against the criterion value 5.76. Only $\bar{Y}_5 - \bar{Y}_2$ is not significant.

Table 11–1 A summary of tests of pairwise contrasts of an illustrative set of six means ($EF = .05$, $n = 6$, $MS_{error} = 13.5$)

		SET SIZE				
	2	3	4	5	6	
Newman-Keuls criterion	4.34	5.24	5.76	6.15	6.45	
WSD criterion	6.45	6.45	6.45	6.45	6.45	
Bonferroni criterion	6.76	6.76	6.76	6.76	6.76	
$\bar{Y}_j$	4	7	8	10	12	16

	1	2	3	4	5	6
1	...	...	...	*	⊛	⊛
2		...	...	...	...	⊛
3			...	...	...	⊛
4				...	...	*
5					...	...
6						...

* $\bar{Y}_j - \bar{Y}_{j'}$ exceeds $N - K$ criterion.
⊛ $\bar{Y}_j - \bar{Y}_{j'}$ exceeds all three criteria.

There may now be some question about which additional tests are permissible. The failure to obtain a significant $(2, 5)$ contrast implies that $\mu_2 = \mu_3 = \mu_4 = \mu_5$ and argues against further testing within that set of means bounded by $\bar{Y}_2$ and $\bar{Y}_5$. On the other hand, if we were to consider only that the $(1, 4)$ contrast was significant, we should proceed to test the $(2, 4)$ contrast. It is the nonsignificant $(2, 5)$ contrast that takes precedence. We immediately place dashes (to indicate nonrejection) in all cells within the triangle that has the $(2, 5)$ cell in its upper right-hand corner. Thus, the only cells on the next diagonal on which tests will be permitted are the $(1, 3)$ and $(3, 6)$. Of these, only $\bar{Y}_6 - \bar{Y}_3$ exceeds the current criterion, the value 5.24. Therefore, tests on the $(1, 2)$ and $(2, 3)$ cells are ruled out; $\bar{Y}_6 - \bar{Y}_5$ is tested, however, and proves significant against the criterion value, 4.34. Using the same underscoring system as in the *WSD* test, we can display our final results as

$$\bar{Y}_1 \quad \bar{Y}_2 \quad \bar{Y}_3 \quad \bar{Y}_4 \quad \bar{Y}_5 \quad \bar{Y}_6$$

It should be evident that the Newman-Keuls test is more powerful than the *WSD*. Although the two procedures provide identical power against the overall null hypothesis—that is, they are identical tests for the widest range—the Newman-Keuls redefines the family of contrasts as a smaller set when preceding tests have yielded significant results. The Newman-Keuls appears as a result to be a more reasonable compromise between control of the *EF* and the desire for power against false null hypotheses. A confidence interval for the family of contrasts is not available with the Newman-Keuls approach, however, because the family does undergo redefinition depending on the outcome of the sequence of tests.

11.3.3   THE BONFERRONI t STATISTIC.   If one has a set of k independent contrasts, controlling the EF is fairly simple. Under these conditions,

(11.3)
$$EF = 1 - (1 - EC)^k$$

where EC is the significance level for each contrast. We choose EC equal to EF/k. Then,

$$1 - \left(1 - \frac{EF}{k}\right)^k = 1 - \left[1 - k\left(\frac{EF}{k}\right) + \binom{k}{2}\left(\frac{EF}{k}\right)^2 - \cdots - \left(\frac{EF}{k}\right)^k\right]$$

If EF is small (say .10 or less), $(EF/k)^j$ for $j > 1$ will be close to zero, and therefore,

$$1 - \left(1 - \frac{EF}{k}\right)^k \approx 1 - \left[1 - k\frac{EF}{K}\right] = EF$$

Thus, for orthogonal (independent) sets of contrasts, we can get approximately any desired EF by performing a t test at the EF/k level. The contrasts under consideration in the present section, the $\frac{1}{2}a(a-1)$ comparisons by pairs, are not orthogonal (criteria for orthogonality will be defined in Section 11.4). The Bonferroni procedure can still be applied, though with the understanding that it is conservative. It can be used with nonindependent tests since under those conditions the true EF will never be more than $k(EC)$; that is,

(11.4)
$$EF \leq 1 - (1 - EC)^k$$

The equality holds (approximately) only for k independent tests.
    The test statistic will be

(11.5)
$$t = \frac{\bar{Y}_j - \bar{Y}_{j'}}{\sqrt{\dfrac{2MS_{\text{error}}}{n}}}$$

As usual, MS_{error} is the error term for the F test of the overall null hypothesis (see Sections 11.3.5 and 11.3.6 for possible exceptions).
    If we wished an EF of .05 for a family of five contrasts, and had $df_{\text{error}} = 20$, the criterion for significance would be that t must exceed 2.845, the critical value for 20 df and $\alpha = .01$ (two-tailed); see Table A-3, the t distribution table. (The reader should be aware that t^2 is distributed like F with numerator df equal to 1; thus, an equivalent test would be to require the calculated F for the contrast to exceed 8.10, the critical value for $\alpha = .01$ and 20 df in Table A-5; 8.10 is the square of 2.845.) An equivalent test procedure would be to regard as significant

all pairwise contrasts having

$$\bar{Y}_j - \bar{Y}_{j'} > t_{EF/k}\sqrt{\frac{2MS_{error}}{n}}$$

Frequently, the desired significance level will not be available in Tables A–3 or A–5; for example, if $EF = .10$ and $k = 8$, each contrast must be tested at the $.0125\alpha$ level. Table A–12 provides values of t ($= \sqrt{F}$ when $df_1 = 1$) for various combinations of k and error df at $EF = .01$, $.05$, and $.10$. If other combinations are involved in an experiment, find z_α, the normal deviate significant at the EF/k level, using a two-tailed test. The quantity

(11.6)
$$\left[z_\alpha + \frac{z_\alpha^3 + z_\alpha}{4(df_{error} - 2)} \right]$$

is distributed approximately like t; if its value is exceeded, reject H_0. Be careful to note that the required z score is the one exceeded with probability equal to $(\frac{1}{2})(EF/k)$.

The use of $\frac{1}{2}$ in the preceding calculation is related to the general point that all tests of contrasts described in this section are two-tailed: differences in either direction that exceed some critical magnitude will be rejected.

A further point in considering significance tests of orthogonal contrasts is that they need not be equally weighted. Let us assume that we desire an EF of $.10$ and that k equals 3. One of the contrasts is of greater interest than the other two and we are willing to have a higher Type I error rate in testing it in order to achieve greater power. We might carry out this test at the $.05$ level and the other two at the $.025$ level. So long as the sum of the ECs is EF, the actual error rate per family will approximately equal EF.

We can readily obtain confidence intervals on the magnitudes of the contrast. On the assumption that k tests are equally weighted, the probability is at least $1 - \alpha$ that

(11.7)
$$\hat{\psi} - t_{EF/k}\sqrt{\frac{2MS_{error}}{n}} \le \psi \le \hat{\psi} + t_{EF/k}\sqrt{\frac{2MS_{error}}{n}}$$

for the k contrasts of interest.

A numerical example. Again consider the data of Table 10–1, which were previously analyzed by the Tukey, *WSD*, and Newman-Keuls approaches. Once again we want to evaluate all contrasts by pairs; accordingly, k is 15. Turning to Table A–12, for EF of $.05$, k of 15, and 30 error df, we get the critical value of t as 3.1869.[2] Therefore, $\bar{Y}_j - \bar{Y}_{j'}$ must exceed $(3.1869)\sqrt{(\frac{2}{6})(13.5)} = 6.76$ for $H_0: \mu_j = \mu_{j'}$ to be rejected. With this particular data set, the Bonferroni approach and the *WSD* find the same contrasts to be significant. Nevertheless, the Tukey procedure has the greater power since for this situation, it requires a smaller critical value of

---

[2] Roughly the same critical value of t would result from applying Formula (11.6). We need the z score that is exceeded with probability $.05/30$, or $.0017$. Precise tables of the normal curve give us 2.93 as the required value. If we substitute this value for z in Formula (11.6), our approximate t is 3.1807, which is reasonably close to the true value 3.1869.

$\bar{Y}_j - \bar{Y}_{j'}$ for significance than the Bonferroni t does. Another way of stating this is to say that the confidence interval for the family of 15 contrasts is narrower with the Tukey procedure. This observation suggests a general method for selecting one of two methods for simultaneously testing a set of contrasts (the approach is not applicable with successive testing procedures like the Newman-Keuls): before analyzing (or even collecting) the data, calculate the ratio of the widths of the confidence intervals. The ratio of interval widths for the Tukey to the Bonferroni test, with EF equal to α, is $q_\alpha/(\sqrt{2})t_{\alpha/k}$. For the data set of our example, this ratio is $4.30/(3.19\sqrt{2})$, or .95.

Dunn (1961) has considered this ratio for various combinations of k (number of contrasts), a (number of means in the total set), df_{error}, and α. When all possible contrasts by pairs are performed, the interval width will always be narrower for the Tukey procedure. For fixed a, the advantage of the Tukey procedure declines as α decreases, df increases, or k decreases. As k decreases from the total number of comparisons by pairs, a point will be reached at which the Bonferroni t will be the preferred test.

It is important to emphasize that our discussion of the Bonferroni t is predicated on the assumption that the set of k tests has been planned before the experiment. To understand why this must be so, consider the following situation. Assume that after viewing the data in our example, we choose to test only $\bar{Y}_6 - \bar{Y}_1$ and $\bar{Y}_6 - \bar{Y}_2$, the two largest differences; each of the two tests is carried out at the .025 level. The value EF is considerably greater than .05 in this case. We have essentially tested all fifteen differences because if any two of them are significant, the two largest will be. If our decision to test these two contrasts had been made before looking at the data, there would be only two contrasts in the family and we should be justified in claiming that EF was less than or equal to .05.

Although our concern here is with comparisons of means, we note the general applicability of the Bonferroni approach to a wide variety of situations involving multiple significance tests. For example, to test the significance of each of k correlation coefficients, or to evaluate k chi square tests, we could set the EC for each test at EF/k.

11.3.4   DUNNETT's (1955) TEST: COMPARING A CONTROL GROUP WITH EXPERIMENTAL GROUPS.

If one of the a groups is a control group, there are $a - 1$ nonindependent comparisons that may be of interest. So the error rate can be kept at α for the entire set of comparisons, the statistic

$$\frac{\bar{Y}_{.j} - \bar{Y}_c}{\sqrt{MS_{S/A}[(1/n_j) + (1/n_c)]}}$$

is evaluated against the statistic d, whose distribution depends on the value of a and the error df. Significant values of d are presented in Table A–10. The null hypothesis test can also be carried out by determining whether

(11.8) $$\bar{Y}_{.j} - \bar{Y}_c > \left[\sqrt{MS_{S/A}\left(\frac{1}{n_j} + \frac{1}{n_c}\right)} \right] d_{\alpha; a, a(n-1)}$$

where $\bar{Y}_c$ is the mean of the control group, n_j and n_c are the numbers of observations in the jth experimental group and the control group, and $d_{\alpha;a,a(n-1)}$ is the value of d required for significance at the α level when there are a groups (including group C), with n measurements in a group. For other designs besides the completely randomized one-factor, $MS_{S/A}$ and $a(n-1)$ are replaced by the appropriate error term and error df, and the ns are always the numbers of observations on which the two means are based.

Note that Table A–10 presents one-tailed probabilities. Therefore, if we are concerned with detecting differences in either direction at the .05 level, we require the value of d significant at the .025 level. Similarly, if we require a 95 percent confidence interval, the d to use would be the value appropriate to the .025 level of significance.

The probability is $1 - \alpha$ that all $a - 1$ confidence intervals include the true difference $(\mu_{.j} - \mu_c)$ if the confidence interval is computed as

$$(\bar{Y}_{.j} - \bar{Y}_c) - d\hat{\sigma}\sqrt{\frac{1}{n_j} + \frac{1}{n_c}} \le \mu_j - \mu_c \le (\bar{Y}_{.j} - \bar{Y}_c) + d\hat{\sigma}\sqrt{\frac{1}{n_j} + \frac{1}{n_c}}$$

where $\hat{\sigma}^2 = MS_{\text{error}}$ and d is the d required for significance at the α level.

11.3.5 UNEQUAL GROUP SIZES. The *WSD* and Newman-Keuls procedures, but not the Bonferroni, assume that all ns are equal. Two proposals for modifying these procedures have been considered for the unequal n case. It has been suggested that n be replaced by n', the harmonic mean of the a values of n, or by n'', the harmonic mean of the two values of n on which the contrasted means are based, that is,

(11.9)
$$n' = a\left(\frac{1}{n_1} + \cdots + \frac{1}{n_j} + \cdots + \frac{1}{n_a}\right)^{-1}$$

and

(11.10)
$$n'' = 2\left(\frac{1}{n_j} + \frac{1}{n_{j'}}\right)^{-1}$$

Then, assuming a one-factor completely randomized design gives

(11.11)
$$\hat{\sigma}_{\bar{Y}} = \sqrt{\frac{MS_{S/A}}{n'}}$$

or

(11.12)
$$\hat{\sigma}_{\bar{Y}} = \sqrt{\frac{MS_{S/A}}{n''}}$$

When $a = 2$, the last two expressions are equivalent. The denominator of the usual two-group t test is just $(\sqrt{2})\hat{\sigma}_{\bar{Y}}$; the value $\hat{\sigma}_{\bar{Y}}$ has been defined above.

There have been few investigations of how adequate these adjustments are for unequal ns. Fortunately, there is some available evidence on their relative merits. Keselman, Murray, and Rogan (1975) established computer populations with equal means and variances. They drew 1000 samples for each of several combinations of group sizes. By the *WSD* procedure, the proportion of experiments that resulted in one or more Type I errors was generally close to the designated *EF*. The largest discrepancy between the obtained result and the theoretical *EF* occurred when the ratio of n_is was $1:1:21:21$ and n' was used in place of n. In this case, when the theoretical *EF* was .05, the proportion of experiments having at least one Type I error was .077. In general, the n' adjustment resulted in a proportion of experiments with Type I error greater than the nominal *EF*. With the n'' adjustment, the observed probability usually was lower than the theoretical value and the discrepancies were smaller than when n' was used. The amount of data is limited but it appears that the harmonic mean of n_j and $n_{j'}$ provides an adequate replacement for n in testing the difference between $\bar{Y}_j$ and $\bar{Y}_{j'}$.

In the investigation described above, population variances were homogeneous. The complementary case, in which ns are equal but variances are not, does not appear to be a problem; the *WSD* (and presumably the Newman-Keuls) test has been shown to be robust to violations of the homogeneity of variance assumption (Keselman and Toothaker, 1974; Keselman, 1975). Problems arise, however, when both ns and population variances vary—not a surprising result after our discussion of the assumptions underlying the *F* test (Section 4.3). Keselman and Toothaker used variances .4, .8, 1.2, and 1.6 with ns of 8, 9, 11, and 16. If we consider a family consisting of the six possible contrasts by pairs, with *EF* nominally at .01, the observed *EF*s were .003 for positive correlations of n and σ, and .036 for negative correlation; for *EF* theoretically at .05, the observed results were .030 and .110. Although the range of conditions investigated is limited, it is reasonable to expect these results to hold up for other combinations of n and σ. After all, the calculated q equals $(\sqrt{2})t$ and its sampling distribution therefore should be sensitive to violations of assumptions in much the way that of t is. You may recall that the t test shows the same pattern of α inflation with negatively correlated ns and variances, and deflation with positively correlated ns and variances.

In view of the preceding discussion, we recommend a modified Bonferroni t procedure when ns and variances are unequal across conditions. Specifically, we suggest that Welch's t (Equation (4.19)) with its corrected df (Equation (4.20)) be used in place of the usual t test. The sampling distribution of Welch's statistic is well known, and with the appropriate adjustment for df, it closely approximates the distribution of the t. As usual in the Bonferroni procedure, the operative significance level for each test will be EF/k. A possible loss of power, relative to statistics based on the Studentized range q, seems a small price to pay to ensure that the sampling distribution of the test statistic conforms closely to the tabled values.

11.3.6 REPEATED-MEASUREMENT DESIGNS.   In using the Studentized range statistic (the *WSD* and Newman-Keuls tests), it is assumed that the treatment means are independently distributed with homogeneous variances. If all subjects are tested under all a treatment levels, it is doubtful that these assumptions hold. Scheffé (1959) has pointed out that the test statistic $(\bar{Y}_{.j} - \bar{Y}_{.j'})/\sqrt{MS_{AS}/n}$ will be distributed like q if variances and covariances are homogeneous. In the more general nonadditive case in which this condition is not met, the above statistic only approximates the Studentized range statistic in distribution, and I know of no evidence about the quality of that approximation. Unless there is clear evidence to support the assumption of homogeneous variances and covariances, therefore, the safest procedure would be to use the Bonferroni t test with denominator based solely on the pair of conditions being compared. For each contrast of the form $\bar{Y}_{.j} - \bar{Y}_{.j'}$, calculate the n difference scores of the form $Y_{ij} - Y_{ij'}$. If we label these the $d_{i(j,j')}$, the denominator of the t test is

$$(11.13) \qquad \hat{\sigma}_{\bar{d}(j,j')} = \sqrt{\frac{\sum_i d_{i(j,j')}^2 - n\bar{d}_{.(j,j')}^2}{n(n-1)}}$$

This is equivalent to using $\sqrt{MS_{SA}/n}$ as an error term but with the levels of A defined as only those two involved in the contrast being tested. As always, each test is evaluated at the EF/k level. The Bonferroni approach might also be used instead of the Dunnett when each subject is tested under several experimental conditions and one control condition.

11.3.7 *F* TESTS AND CONTRASTS BY PAIRS. Experimenters who are primarily interested in testing differences by pairs are often concerned about the role of the F test of the overall null hypothesis. Among the questions that arise are: Must the overall F test be significant before I can do the Newman-Keuls (or the *WSD*, the Bonferroni, or the Dunnett)? Is there any need to do an F test at all? Should I do both and reject the overall null hypothesis if either is significant? These questions will be somewhat easier to deal with in a specific example in which the relation between the F and the q statistic is considered.

Assume three groups of 11 subjects each, and further assume that $MS_{S/A}$ equals 11. This assumption simplifies things by causing $\hat{\sigma}_{\bar{Y}}$ to be equal to 1, but it is not necessary. Let us further place the restriction that $\bar{Y}_1$ is the smallest of the three means and is zero, and $\bar{Y}_3$ is the largest mean. Then, for the Newman-Keuls or the *WSD* test to reject the largest pairwise difference at the .05 level, $\bar{Y}_3 - \bar{Y}_1$, or $\bar{Y}_3$, must be greater than 3.49, the value of q required for significance with $a = 3$ and 30 error df. This statement is represented by the horizontal line at $\bar{Y}_3 = 3.49$ in Figure 11-1. Any point representing a pair of values of $\bar{Y}_2$ and $\bar{Y}_3$, where $\bar{Y}_3$ is greater than 3.49, will lie above the line and will cause to be rejected the hypothesis that the largest pairwise difference is zero.

The value of F required for significance at the .05 level with $df_1 = 2$ and $df_2 = 30$ is 3.32. The set of points $(\bar{Y}_2, \bar{Y}_3)$ that yield values of F at least this large are those lying above the curved line in Figure 11-1. (We terminate the function at $\bar{Y}_2 = 3.16$ because of the stipulation that $\bar{Y}_3$ will be the larger of the two values.)

Figure 11–1   Rejection regions for F and q with $MS_{error}/n = 1$, $df_{error} = 30$.

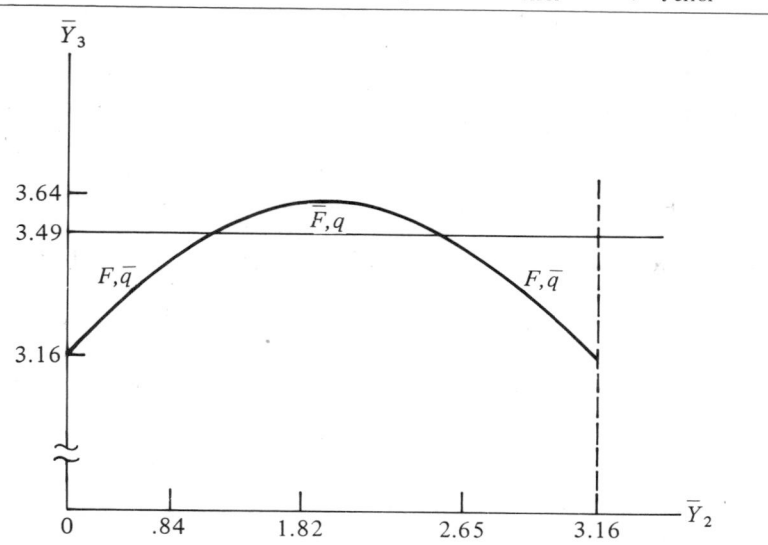

There are two fundamental characteristics of Figure 11–1 that should be appreciated. First, if the overall null hypothesis ($H_0: \mu_1 = \mu_2 = \mu_3$) is assumed true, the probability of sampling a point within its rejection region is .05 for either test. Second, the rejection regions overlap largely, but not completely. There are certain patterns of means that will yield a significant F but not a significant q value (the areas labeled $F, \bar{q}$) and others that will result in significant values of q, but not of F (the area labeled $\bar{F}, q$). In general, F will be the more sensitive test when most of the means fall close together.

These two observations shed some light on the questions we raised. Since the two tests provide equivalent control of Type I error rates, there is no logical necessity that the Studentized range test be preceded by an F test, and certainly not by a significant F. To demand a significant F before the WSD or the Newman-Keuls test is to adjust the actual Type I error rate to some level below the nominal value, and consequently to lose power. To see why this is so, consider Figure 11–1. The significant-F-first strategy removes $\bar{F}, q$ from the rejection region for the Studentized range test, thus reducing the Type I error rate below .05. The loss in power by such a strategy may not be great, but there is no apparent gain to warrant this approach.

Some investigators reject the overall null hypothesis if either F or q is significant. By Figure 11–1, the rejection region for this strategy includes all points above whichever is lower, the curve or the line. Increasing the size of the rejection region in this way will increase power to detect some departures from the overall null hypothesis, but at the cost of an increased Type I error rate. The increase in error rate will probably not be more than 1 or 2 percent. Whether the potential gain in power warrants this approach is uncertain until we have more information about its power against various alternatives to the null hypothesis. It currently seems preferable to perform only the Newman-Keuls test if interest is solely in contrasts by pairs; if the test of the largest difference between two means is not significant, the overall null hypothesis would not be rejected.

The situation is different when we consider some subset of contrasts planned before the experiment. Again, if the *EF* is controlled by the Bonferroni procedure, or by the Dunnett test, there is no reason to require a significant *F* before the test of contrasts. On the other hand, we strongly recommend, although it is not necessary, that the *F* test accompany the Bonferroni or Dunnett tests. It is quite possible that the overall null hypothesis is false but that the particular contrasts planned do not represent those means that differ. With this possibility, a small inflation in Type I error rate seems a small price to pay for the potential added information from the *F* test. Note the distinction between this situation and the case in which *all* comparisons by pairs are tested. In the case under discussion, in which a subset of contrasts is selected a priori, rejecting a contrast implies rejection of the overall null hypothesis. Failure to reject a contrast, however, does not imply overall equality of the treatment population means as it does when the largest difference is selected for testing by the Studentized range tests.

11.4 CONTRASTING MEANS: THE GENERAL CASE

11.4.1 CALCULATIONS.

In the preceding section, we considered tests of the null hypothesis $H_0 : \mu_j - \mu_{j'} = 0$. This is a special although very common contrast. Now we consider the more general null hypothesis $H_0 : \psi = 0$, in which $\psi = \sum_j w_j \mu_j$; the only restriction is that $\sum_j w_j = 0$. Interest in such contrasts arises naturally in many experiments. We noted (Section 11.1), for example, that the question whether the combined reward-only R and punishment-only P groups differed in average performance from a reward and punishment RP group implies $H_0 : \frac{1}{2}(\mu_R + \mu_P) - \mu_{RP} = 0$. In what follows, we shall develop computational formulas for testing linear contrasts of the general form. After those mechanics are behind us, we shall consider controlling *EF*.

The *t* statistic provides a test of the general null hypothesis that $\psi = 0$. The numerator of the *t* is always a random variable minus its expectation under H_0, in this case, $\hat{\psi}$. The denominator is our best estimate of the standard deviation of the sampling distribution of our random variable, in this case, $\hat{\sigma}_{\hat{\psi}}$. The formula for this measure of sampling variation is a generalization of the denominator of the more common *t* test for the difference between two means. We have

$$\hat{\sigma}_{\hat{\psi}}^2 = \hat{\sigma}_{\sum_j w_j \bar{Y}_j}^2$$

Assuming that scores are independently distributed, the variance of the sum is the sum of the variances. That is,

$$\hat{\sigma}_{\hat{\psi}}^2 = \sum_j \hat{\sigma}_{w_j \bar{Y}_j}^2$$

But the sampling variance of $w_j \bar{Y}_j$ is w_j^2 times the sampling variance of Y_j. We therefore have

$$\hat{\sigma}_{\hat{\psi}}^2 = \sum_j w_j^2 \hat{\sigma}_{\bar{Y}_j}^2 \quad \text{or} \quad \hat{\sigma}_{\hat{\psi}}^2 = \sum_j w_j^2 \left(\frac{\hat{\sigma}_j^2}{n_j} \right)$$

Under the assumption of homogeneity of variance, we may let $\hat{\sigma}_1^2 = \cdots = \hat{\sigma}_j^2 = \cdots = \hat{\sigma}_a^2 = MS_{S/A}$, yielding

(11.14)
$$\hat{\sigma}_{\hat{\psi}} = \sqrt{(MS_{S/A}) \left(\sum_j \frac{w_j^2}{n_j} \right)}$$

If the treatment populations are normally distributed, then under H_0,

(11.15)
$$\frac{\hat{\psi}}{\sqrt{(MS_{S/A})[\sum_j (w_j^2/n_j)]}}$$

is distributed as t on $(\sum_j n_j - a)$ df. In multifactor designs, n_j is the number of observations on which $\bar{Y}_j$ is based and $MS_{S/A}$ is replaced by the error term for the overall test of A effects (see Section 11.4.4 for a possible exception in repeated-measurement designs).

Formula (11.15) is convenient for establishing confidence intervals on ψ; the appropriate interval is $\hat{\psi} \pm (t_{\alpha, df_{\text{error}}}) \hat{\sigma}_{\hat{\psi}}$, where $\hat{\sigma}_{\hat{\psi}}$ is defined by Equation (11.14). If the investigator wants only a significance test, there is no need for calculating square roots. Remembering that t^2 is an F ratio with one numerator df, we have

(11.16)
$$F = \frac{\hat{\psi}^2 / (\sum_j w_j^2 / n_j)}{MS_{S/A}} = \frac{SS_{\hat{\psi}}}{MS_{S/A}}$$

With equal ns, an equivalent formula for the numerator of F is

(11.17)
$$SS_{\hat{\psi}} = \frac{(\sum_j w_j T_j)^2}{n \sum w_j^2}$$

Note that multiplying all weights by a constant leaves $SS_{\hat{\psi}}$ unchanged. This observation enables us to avoid dealing with fractions in testing hypotheses about contrasts. If we want to get the confidence interval, however, multiplying the weights by a constant has the effect of multiplying the interval width by that constant.

It may be helpful to note the relation between the calculations for $SS_{\hat{\psi}}$ and other calculations previously met in this test. The quantity defined in Equation (11.17) and its equivalent for unequal n [numerator of Equation (11.16)] are general formulas for any sum of squares distributed on 1 df. For example, there is no real difference between the above formula and the quantity on the right of Equation (5.39). In Chapter 5 we had a 2×2 design, and we provided a shortcut formula for SS_{AB}:

$$SS_{AB} = \frac{(T_{11} + T_{22} - T_{12} - T_{21})^2}{4n}$$

This can be rewritten

$$SS_{AB} = \frac{[(1)(T_{11}) + (1)(T_{22}) + (-1)(T_{12}) + (-1)(T_{21})]^2}{[(1)^2 + (1)^2 + (-1)^2 + (-1)^2]n}$$

which is clearly of the form of the right-hand quantity of Equation (11.17). In general, the formula for any sum of squares on 1 *df* can be represented by

(11.18) $$SS_{1df} = \frac{\text{(sum of the weighted cell totals)}^2}{\text{(sum of the squared weights)} \times \text{(cell frequencies)}}$$

and if the *n*s are unequal,

(11.19) $$SS_{1d_f} = \frac{\text{(sum of the weighted cell means)}^2}{\text{sum of the ratios of squared weights to cell frequencies}}$$

A numerical example. An example of the application of Equation (11.17) is the reward and punishment experiment. Assume 10 subjects in each of the 4 treatment groups. The total number of errors in 20 trials for each group is

R	P	RP	NRP
42	34	25	88

To determine whether the average effect of the *R* and *P* treatments differs from the average of the *RP* treatment ($H_0: \frac{1}{2}(\mu_R + \mu_P) - \mu_{RP} = 0$), we compute

$$SS_{\hat{\psi}} = \frac{[42 + 34 - 2(25)]^2}{(10)(6)} = \frac{676}{60} = 11.27$$

Using Equation (11.16), which would also apply if the cell frequencies were unequal, we have

$$SS_{\hat{\psi}} = \frac{[4.2 + 3.4 - 2(2.5)]^2}{.1 + .1 + .4} = \frac{6.76}{.6} = 11.27$$

Note that to simplify calculations, we have used the coefficients 1, 1, and -2 instead of $\frac{1}{2}$, $\frac{1}{2}$, and -1. This does not change the value of $SS_{\hat{\psi}}$.

Suppose that we wanted a .95 confidence interval on ψ. Assume that $MS_{S/A}$ is 15. Our best estimate of ψ is

$$\hat{\psi} = \frac{1}{2}(\bar{Y}_R + \bar{Y}_P) - \bar{Y}_{RP} = 1.3$$

The expression for $\hat{\sigma}_{\hat{\psi}}$ is provided by Equation (11.14):

$$\hat{\sigma}_{\hat{\psi}} = \sqrt{(15)\left[\frac{(\frac{1}{2})^2 + (\frac{1}{2})^2 + (-1)^2}{10}\right]} = 1.5$$

Because the error term is based on 4 groups of 10 subjects, $df_{\text{error}} = 36$ and the critical value of *t* is approximately 2.029. Then, the desired bounds on ψ are $1.3 \pm (2.029)(1.5)$, or -1.74 and 4.34. Since the interval contains zero, the null hypothesis cannot be rejected.

The basic calculations having been established, it is time to turn to the problem of controlling *EF* for those situations in which several contrasts of the general class are performed for one set of means.

11.4.2   PLANNED COMPARISONS.   First is a general case, in which the experimenter designates *k* contrasts of interest before the experiment. This involves two situations, orthogonal and nonorthogonal contrasts.

Orthogonal contrasts. There are many possible contrasts of the form

$$\psi = \sum w_j \mu_j$$

if the sole restriction is that $\sum w_j = 0$. Many of these contrasts will be correlated with each other; they will carry redundant information. To take an extreme example, suppose

$$\bar{Y}_3 - \bar{Y}_2 = 3 \qquad \text{and} \qquad \bar{Y}_2 - \bar{Y}_1 = 7$$

Then, the value of $\bar{Y}_3 - \bar{Y}_1$ must be 10, and if the first two contrasts are significant, then the third must be. Sets of independent contrasts can be obtained, however; each is called an *orthogonal set*. If *a* treatment means are assumed contrasted, each set of orthogonal contrasts will have $a - 1$ members. The sum of squares for each term will be distributed on 1 *df*, and the total of the $a - 1$ sums of squares will equal the SS_A distributed on $(a - 1)$ *df*.

Two contrasts ψ_p and $\psi_{p'}$ are orthogonal if (1) the sum of the coefficients for each contrast is zero, that is, $\sum_j w_{jp} = 0$ and $\sum_j w_{jp'} = 0$; and (2) the sum of cross-products of coefficients is zero, that is, $\sum_j w_{jp} w_{jp'} = 0$. To illustrate orthogonality, we again return to the example of the experiment on reward and punishment. One possible set of three orthogonal contrasts is

	R	P	RP	NRP
$w_{j1} - 1$	+1	0	0	
$w_{j2} - 1$	-1	+2	0	
$w_{j3} - 1$	-1	-1	+3	

To verify that the three contrasts are independent, we note that

$$\sum_j w_{j1} w_{j2} = (-1)(-1) + (+1)(-1) + (0)(+2) + (0)(0) = 0$$

$$\sum_j w_{j1} w_{j3} = (-1)(-1) + (+1)(-1) + (0)(-1) + (0)(+3) = 0$$

$$\sum_j w_{j2} w_{j3} = (-1)(-1) + (-1)(-1) + (+2)(-1) + (0)(+3) = 0$$

By the data of Section 11.4.1

$$SS_A = \frac{(42)^2 + (34)^2 + (25)^2 + (88)^2}{10} - \frac{(189)^2}{40}$$

$$= 1,128.9 - 893.025$$

$$= 235.875$$

For the three contrasts, we have

$$SS_{\psi_1} = \frac{(42-34)^2}{(2)(10)} = 3.2$$

$$SS_{\psi_2} = \frac{[42+34-(2)(25)]^2}{(6)(10)} = 11.267$$

$$SS_{\psi_3} = \frac{[42+34+25-(3)(88)]^2}{(12)(10)} = \frac{26,569}{120} = 221.408$$

Adding yields

$$\sum_p SS_{\psi_p} = 3.2 + 11.267 + 221.408$$

$$= 235.875$$

$$= SS_A$$

It is not necessary to test all members of an orthogonal set. Suppose that there were six levels of A and only two contrasts of interest, orthogonal to each other. We might divide the SS_A into three quantities corresponding to the two contrasts and a residual $(SS_A - SS_{\psi_1} - SS_{\psi_2})$.

If the contrasts are orthogonal, control of the EF is fairly simple. If each test is carried out with α equal to the desired EF divided by k (the number of tests), the probability of one or more Type I errors is approximately the nominal EF (Section 11.3.3).

Nonorthogonal contrasts. Orthogonal contrasts are easily interpreted because they are not redundant, and a close approximation to the desired EF is readily obtained. In many or most cases, however, the contrasts of interest will not be orthogonal. After all, contrasts are tested because they are of psychological import, not because they are independent of each other.

It can be proved from elementary probability theory that the probability of at least one Type I error for a set of k nonindependent tests, each carried out at the EF/k level, is less than EF. Therefore we recommend the procedures described above for the orthogonal case, even for contrasts that are not orthogonal. Note that with correlated contrasts the approach is conservative for Type I error rate; the probability of one or more Type I errors over the set of k contrasts is at most EF. Furthermore, the test tends to be more conservative as k increases. One should realize, therefore, that testing nonorthogonal contrasts and adding more contrasts involves compromise. We gain more information in the sense of investigating more questions, or questions that are not part of a restricted orthogonal set. The quality of the information is somewhat degraded, however, in the sense that our test becomes conservative, more so with more comparisons; we lose power and have wider confidence intervals. In any specific instance, it may be worth trying to divide the questions of interest into two sets for level of importance. Frequently, certain contrasts are basic to the study; others would be nice to know about but are not fundamental. Then the total error rate could be divided so that the *EC* is higher for the primary contrasts. These will then be tested with somewhat more power and the confidence intervals will be narrower.

11.4.3 POST HOC ANALYSIS: SCHEFFÉ'S MULTIPLE COMPARISON METHOD.   Despite careful planning, certain differences among means that have not been considered will attract attention when the data have been collected. It would be foolish to ignore these merely because they are unforeseen. The contrasts in question might well provide insight into the processes under investigation. On the other hand, picking out the largest contrasts after the experiment has been completed is tantamount to investigating the universe of possible contrasts. We are obligated to pay a price in power and confidence interval width for this additional information, for we are controlling error rate over a larger set of contrasts than what would be the case if the contrasts had been planned. We shall enlarge on one approach to controlling *EF* for post hoc comparisons.

Assume a levels of the treatment variable A in a completely randomized one-factor design. To test the null hypothesis that the pth contrast is zero, that is that $\sum_j w_{jp}\mu_j = 0$, according to Scheffé (1959), the obtained F statistic should be evaluated against $(a-1)F_{\alpha, a-1, a(n-1)}$, where the criterion F is the one required for significance at the α level on $a-1$ and $a(n-1)$ df. (With unequal n, the denominator df are $\sum n_j - a$.) The basis for this null hypothesis test is the following theorem (proved by Scheffé in Section 3.5 of his text): the probability is $1-\alpha$ that the values of all contrasts simultaneously lie within the confidence intervals of the form

$$(\textbf{11.20}) \quad \hat\psi_P - \hat\sigma_{\hat\psi_P}\sqrt{(a-1)F_{\alpha, a-1, a(n-1)}} < \psi_P < \hat\psi_P + \hat\sigma_{\hat\psi_P}\sqrt{(a-1)F_{\alpha, a-1, a(n-1)}}$$

where $\hat\psi_P = \sum_j w_{jp}\bar Y_{.j}$ and $\hat\sigma_{\hat\psi_P}$ is defined in Equation (11.14). We are again controlling the error rate per family, this time for the set of all possible contrasts.

Experimenters who have used the Scheffé procedure have sometimes been perplexed to find that a significant test of the overall main effect has not been followed by at least one significant contrast. If there is some difference within the set of μ_j, why is this difference not reflected within one of the subsets of μ_j subsequently considered? The answer has as a basis that if the overall test is significant at the α level, at least the maximum possible contrast will also be significant at the α level. Unfortunately, the maximum possible contrast may have been of little interest and therefore may not have been computed. There is no guarantee that the obvious contrasts (say differences within pairs of means) or the contrasts most interesting for the psychologist will be significant when the overall F test is.

Related to the point just discussed is that the Scheffé test equals the overall F test in power only when the detection of the maximum possible contrast is at issue. The power to detect the significance of other contrasts is lower than with the main effect test because the EF is held constant over the entire set of possible contrasts. Scheffé, recognizing the power problem, suggests that the EF be set at 10 percent. This may strike some as heresy; but, it should be noted, EF and EC are different concepts, and there is no reason to demand that traditional EC levels of significance be applied to the *EF*. Even with the *EF* at 10 percent, the *EC* will generally be quite low.

As with other multiple contrast procedures we have considered, there is no logical necessity that the Scheffé test be preceded by a significant *F* test of the overall null hypothesis. On the other hand, that the power to test the maximum contrast equals the power of the overall *F* test means that no contrast will be significant unless the overall *F* test is. There seems little point in spending energy on a series of post hoc tests by the Scheffé method without first determining, by the overall *F* test, whether there is any likelihood of a significant contrast.

A numerical example. In Section 11.4.1, we presented a sample data set and calculated the confidence interval for $\psi = \frac{1}{2}(\mu_R + \mu_P) - \mu_{RP}$; the interval was based on the ordinary *t* test with $\alpha = .05$. The necessary quantities of our calculations were $a = 4$, $n = 10$, $MS_{S/A} = 15$, and $\hat{\psi} = \frac{1}{2}(4.2 + 3.4) - 2.5 = 1.3$. If we want to find the .95 confidence interval by the Scheffé procedure, we require the value $(3)F_{.05,3,36}$, or approximately 8.61. From Equation (11.14), we calculate $\hat{\sigma}^2_{\hat{\psi}} = 1.5$ (the computational details for this are in Section 11.4.1), and by Equation (11.20), we have $1.3 \pm (1.5)\sqrt{8.61}$, or -3.10 and 5.70.

Dunn (1961) and Perlmutter and Myers (1973) have compared the Scheffé and Bonferroni *t* intervals and have noted that in many cases, it is preferable to designate every contrast of any possible interest before the experiment. This permits using the Bonferroni method, which even for large *k*, will often give smaller confidence intervals than the Scheffé. If the investigator has some set of *k* contrasts in mind before viewing the data, the choice between the Scheffé and Bonferroni procedures can be made by calculating the ratio of the width of the Bonferroni confidence interval to the width of the Scheffé; the ratio reduces to a ratio of critical values of the test statistics: $(t_{EF/k,df_{error}})/\sqrt{(a-1)F_{EF,a-1,df_{error}}}$. If this ratio is less than one, the Bonferroni approach will provide more precise estimation of ψ and more powerful tests of null hypotheses. (Keselman, 1974, has performed computer sampling studies that support the conclusion about power.)

11.4.4 REPEATED-MEASUREMENT DESIGNS.

Frequently *n* subjects are tested under all *a* levels of some treatment variable and several contrasts among means are then assessed. If σ^2_j, the variance for the population of subjects under treatment A_j, equals σ^2 for all *j*, and if population variances of difference scores for all A_j and $A_{j'}$ are all equal (for conditions on the variance-covariance matrix, see Chapter 7), we can prove that

$$E(MS_{\hat{\psi}}) = \sigma^2_e + (1-\rho)\sigma^2 + n\theta^2_\psi \qquad \theta^2_\psi = \frac{(\sum_j w_j \alpha_j)^2}{\sum w^2_j}$$

Since we have already proved in Chapter 7 that

$$E(MS_{SA}) = \sigma^2_e + (1-\rho)\sigma^2$$

it is clear that the error term against which *A* is tested is also the error term against which various contrasts should be tested.

If the variances and covariances are not homogeneous, each contrast requires a different error term (which we could call $p(A) \times S$, the variability of the pth contrast over subjects) if the ratio of *EMS* is to be one under the null hypothesis. Using the usual Scheffé procedure with MS_{SA} as the error term is an approximation under these conditions. The quality of the approximation is not clear although it obviously will depend on just how much the assumption of homogeneous variances of differences is violated.

An appropriate error term $MS_{p(A) \times S}$ can be calculated for each contrast ψ_p (see next section). Accordingly, we prefer to designate in advance of the experiment those contrasts that could conceivably be of any psychological interest and then to compute Bonferroni t statistics, each with its own denominator depending on the contrast being tested. There is a loss in error df (we now have $n - 1$ instead of $(a - 1)(n - 1)$ df), but this must be considered in light of the knowledge that the appropriate error term is used in the Bonferroni procedure. Furthermore, the Bonferroni often provides more power than the Scheffé procedure even for fairly large numbers of contrasts.

11.5   ANALYSIS OF INTERACTION

Suppose that we have sampled 30 subjects from each of 4 clinical populations P. Furthermore, each set of 30 subjects is further divided so that 10 subjects are randomly assigned to each of 2 experimental conditions and the remaining 10 to a control condition C. Thus, we have 120 subjects in a 4×3 design. We might want to determine whether the difference between the average of the combined experimental groups and the average of the control group is a function of the clinical population sampled. That is, we are interested in the variability of $\psi = \frac{1}{2}(\mu_{E_1} + \mu_{E_2}) - \mu_C$ as a function of clinical population. This suggests that the interaction of P and C is of interest, which is true but does not completely describe our goal. A significant interaction merely suggests that the spread among the three conditions varies with different clinical populations; we want to know whether the particular contrast cited—that between the experimental conditions and the control condition—varies over populations. Is it at least one source of the overall interaction variability? The question involves the interaction of the pth component of C with P, or $p(C) \times P$. In general, $p(A) \times B$ refers to the variation of ψ_p as a function of B.

In the preceding section, we noted that the appropriate error term against which to test the contrast ψ_p will often be $MS_{p(A) \times S}$. The calculations for this term are exactly the same as for $p(A) \times B$ except that we now have one score in each cell of the basic two-factor design.

Perhaps the simplest approach to calculating $SS_{p(A) \times B}$ is first to calculate the sum of squares for $p(A)$ at each level of B. Applying Equation (11.17) gives

$$SS_{p(A) \text{ at } B_k} = \frac{(\sum_j w_{jp} T_{.jk})^2}{n \sum_j w_{jp}^2}$$

Summing the b values of the above term and subtracting the "correction term" SS_{ψ_p} yields

(11.21)
$$SS_{p(A)\times B} = \frac{\sum_k (\sum_j w_{jp} T_{.jk})^2}{n \sum_j w_{jp}^2} - \frac{(\sum_j w_{jp} T_{.j.})^2}{bn \sum_j w_{jp}^2}$$

This sum of squares is distributed on $b-1$ df. We are essentially calculating the variability of b contrasts about their average. That the quantity of Equation (11.21) consists of $b-1$ squared terms also indicates the df.

Table 11–2 presents cell totals for a 3×3 design, 4 subjects in each cell. Assume that we are interested in testing $p(A)\times B$, where the contrast is between the average of A_2 and A_3 against A_1; the coefficients are therefore 1, $-\frac{1}{2}$, and $-\frac{1}{2}$. Since the SS are unchanged by multiplication by a constant, we shall use 2, -1, and -1. We have calculated the sums of squares for this contrast at each level of B. For example,

$$SS_{p(A) \text{ at } B_1} = \frac{[(2)(23)-31-55]^2}{(4)(6)} = 66.67$$

Adding the values for each level of B and subtracting SS_{ψ_p} gives

$$SS_{p(A)\times B} = 812.71 - 715.68 = 97.03$$

Suppose we had three subjects tested at all three levels of A. We could consider only the cell totals of Table 11–2. The first subject would have the scores 23, 31, and 55, and so on. We wish to test whether $SS_{\hat{\psi}_p}$ is significant; in other words, we are interested in the null hypothesis $H_0: \frac{1}{2}(\mu_2 + \mu_3) - \mu_1 = 0$. An appropriate error term is $SS_{p(A)\times S}$, which we should calculate exactly as we calculated $SS_{p(A)\times B}$ except that the totals are now based on one score instead of four.

Table 11–2   Data for a numerical example

	A_1	A_2	A_3	$SS_{p(A) \text{ at } B_k}$
B_1	4	6	12	
	5	9	14	
	6	8	13	
	8	8	16	
$T_{.j1}=23$		31	55	66.67
B_2	5	14	15	
	7	12	18	
	6	16	15	
	4	16	17	
$T_{.j2}=22$		58	65	260.04
B_3	6	14	24	
	4	14	22	
	5	17	20	
	8	15	28	
$T_{.j3}=23$		60	94	486.00
$w_{jp}=$	2	-1	-1	

It is possible to calculate sums of squares for other types of contrasts, for example, $p(A) \times B \times C$, or even $p(A) \times q(B)$. The first of these involves a component of the ABC interaction sum of squares. The question is much like that in a BC interaction. Instead of asking whether the variability in the means at the levels of B changes over levels of C, we are concerned with variability in the contrast $p(A)$. Do B and C interact relative to this parameter? The calculations are a straightforward extension of those just presented. We obtain a sum of squares for $p(A)$ at each combination of A and B, then subtract $SS_{\hat{\psi}_p}$, $SS_{p(A) \times B}$, and $SS_{p(A) \times C}$.

Contrasts like $p(A) \times q(B)$ are only slightly more complicated. Suppose we wished to test the null hypothesis that the $(2, -1, -1)$ contrast we previously considered was equal for B_1 and B_3. Our weights are:

$$w_{1p} = 2, \qquad w_{2p} = -1, \qquad w_{3p} = -1$$
$$w_{1q} = 0, \qquad w_{2q} = 1, \qquad w_{3q} = -1$$

In general, the null hypothesis is

$$H_0 : \sum_j^a \sum_k^b w_{jp} w_{kq} \mu_{jk} = 0$$

The appropriate sum of squares, distributed on one df, is

(11.22)
$$SS_{p(A) \times q(B)} = \frac{\left(\sum_j^a \sum_k^b w_{jp} w_{kq} T_{.jk} \right)^2}{n \sum_j \sum_k (w_{jp} w_{kq})^2}$$

To illustrate, we apply Equation (11.22) to the data of Table 11–2:

$$SS_{p(A) \times q(B)} = \frac{[(0)(2)(23) + (0)(-1)(31) + \cdots + (-1)(-1)(94)]^2}{(4)(12)}$$

$$= 17.52$$

11.6   CONCLUDING REMARKS

In this summary of our inferences about contrasts, what follows should be considered guidelines, not rules. There are no firm rules either for defining the family of contrasts or choosing a method for controlling error rate for the family. These decisions are very much like those we face in determining α. By defining the family more broadly, or by choosing procedures that call for larger contrasts for significance, we sacrifice power for tighter control of Type I error rates.

In multifactor designs, the family should consist of those contrasts of interest that are associated with a variance source. Second, to test all contrasts by pairs when means are independently distributed and are based on equal number of scores, the Newman-Keuls procedure provides the best balance between power and control of Type I errors. If interval estimates are desired, however (and they should be), the Tukey *WSD* procedure should be used. When *n*s are unequal, the harmonic mean of the two ns involved in the contrast should be used in place of n. If variances seem to be heterogeneous and the ns differ, Welch's t should be calculated for each contrast. Then, following the Bonferroni approach, each t should be evaluated at an α level equal to the desired EF divided by the total number of contrasts by pairs. The Bonferroni approach should also be applied to data from repeated-measurement designs when heterogeneity of covariance is suspected.

Third, when other contrasts than contrasts by pairs, or additional ones, are decided on after viewing the data, Scheffé's procedure is generally appropriate. If the set of contrasts can be planned before viewing the data, however, even with a large family the Bonferroni t test may be more powerful. The investigator should usually choose the procedure with the narrower confidence interval; the ratio of intervals does not depend on the values of the means and it is therefore valid to use this approach.

The Bonferroni approach is a useful procedure for controlling EF in a wide variety of situations. It is true that contrasts must be planned before viewing the data. Nevertheless, even when many contrasts are planned, Bonferroni intervals will often be narrower and tests more powerful than in alternative procedures. In any event, the Bonferroni approach often can be applied in situations in which assumptions underlying alternative procedures are violated, or in which alternative procedures are not available.

EXERCISES

11.1 Consider the following anova table and group means (each mean is the sum of scores for the group divided by 75–15 subjects, five measurements each).

SV	df	SS	MS	F
Total	374	32,086.80		
Between S	74	27,938.15		
Groups	4	7,189.87	1,797.47	6.04[a]
S/G	70	20,820.99	297.44	
Within S	300	4,148.65		
Time	4	1,846.20	461.55	60.49[a]
$T \times G$	16	166.24	10.39	1.36
$S \times T/G$	280	2,136.21	7.63	
A	B	C	D	E
20.546	14.560	13.332	11.160	7.200

[a] $p < .001$

Perform the Tukey, Newman-Keuls, and Bonferroni tests on the group means, making all comparisons of the form A versus B, A versus C, and so on.

11.2 (a) *GSR* measures are taken on parachutists. Measures are obtained on groups two weeks before the jump (*BJ*-2), one week before (*BJ*-1), on the day of the jump before jumping (*DJ*-*P*), and on the day of the jump after jumping (*DJ*-*A*). Also, a control group of normal (nonparachuting) cowards *C* is tested. The mean scores for the groups are:

BJ-2	BJ-1	DJ-A	DJ-P	C
5	5	7	9	2

One hypothesis is that parachutists two weeks before the jump behave like the controls. Determine whether these two groups differ significantly from the other three groups. There are six subjects in each group, and the MS_W was 4.0.

(b) Suppose the above contrast was one of many being tested. What would your criterion of significance then be? Find the Scheffé confidence interval.

11.3 Compute the *SS* for numerator and denominator of each of the following contrasts, assuming that all subjects go through all levels of *B*:

(a) the mean for treatments B_1 and B_2 versus the mean for treatment B_3,

(b) the variability in the above contrast over the levels of *A*.

		B_1	B_2	B_3
A_1	S_{11}	4	5	3
	S_{21}	1	6	2
A_2	S_{12}	5	4	2
	S_{22}	3	7	4

11.4 Each cell contains a group total based on five subjects. Compute the SS_{AB}, given that the *AB* interaction can be viewed as a pool of two orthogonal sums of squares.

	A_1	A_2	A_3
B_1	21	16	9
B_2	4	12	7

11.5 Five rats are each tested in a Skinner box under treatments: four drugs and a placebo. Tests are separated in time to minimize carryover effects and the orders of presentation are Latin-squared. The error *MS* is 11.0. The means are

D_1	D_2	D_3	D_4	C
1	6	8	13	7

Carry out both Dunnett's test for the D_j versus *C* and Tukey's test for all pairs at the 5 percent level. What conclusions do you reach on the relative power to test D_j versus *C*? Why do you think this happens? How does the Bonferroni *t* compare?

11.6 Post Ph.D. productivity measures are obtained on random samples of size 10 of clinicians, experimentalists, and social psychologists from the University of California (Berkeley), Stanford, Minnesota, Northwestern, Penn State, and Yale. Lay out a data matrix for the appropriate design. Then set up contrasts to test the following hypotheses:

H_1: Midwestern Ph.D.s are less likely to perish than non-Midwestern Ph.D.s are.

H_2: This regional difference is more marked for the public institutions than for the private.

H_3: This regional difference is more marked among clinicians than among nonclinicians.

H_4: Minnesota clinicians outpublish all other clinicians.

H_5: Minnesota's superiority is more marked in clinical areas than in other areas.

11.7 Consider the following data set:

	E						C			
A_1	10,	14,	17,	17,	12	10,	2,	5,	6,	7
A_2	19,	19,	23,	20,	18	10,	4,	1,	9,	10

where E and C refer to experimental and control conditions.

(a) We want to test $H_0: \psi = [(\mu_{1,E} - \mu_{1,C}) - (\mu_{2,E} - \mu_{2,C})] = 0$. Calculate $F = \hat{\psi}^2 / \hat{\sigma}_{\hat{\psi}}$. Is this significant at the .05 level?

(b) An experimenter who ran this experiment failed to obtain significance. He analyzed the data again as follows. He subtracted $\bar{Y}_{1,C}$ from each of the $Y_{i1,E}$ and $\bar{Y}_{2,C}$ from each of the $Y_{i2,E}$, effectively reducing the data to two groups of "corrected" scores. Carry out the adjustment and compute the F for the two groups of adjusted scores.

(c) The revised F test yielded a significant result. Does the experimenter really have a legitimate, more powerful test of interaction? If not, what has he done wrong? Why does this procedure yield a larger F? Under what conditions would the procedure be acceptable?

11.8 The following may help you to understand orthogonality of contrasts better.

(a) For the set of cell means presented below (assume $n = 10$), calculate SS_{AB}, using the contrast formula. Then estimate α_1 and α_2 and adjust the data for these values. Recalculate SS_{AB} for the adjusted data. Next state the weights for each of the four cells for the A and AB contrasts. Are the two sets of weights orthogonal? What is one implication of orthogonality, according to the preceding exercise?

$$
\begin{array}{c c c}
 & B_1 & B_2 \\
A_1 & \begin{bmatrix} 4 \\ 9 \end{bmatrix} & \begin{matrix} 8 \\ 8 \end{matrix} \\
A_2 & &
\end{array}
$$

(b) Suppose that we have the following three groups. Each of the cell means is based on 10 scores.

A_1	A_2	A_3
18	11	6

Calculate the sum of squares for each of the following three contrasts:

$$\psi_1 = \mu_1 - \mu_2$$

$$\psi_2 = \frac{\mu_1 + \mu_2}{2} - \mu_3$$

$$\psi_3 = \mu_1 - \mu_3$$

Now adjust the means so that there is no effect due to ψ_1. Recalculate the SS for the other two contrasts, using the adjusted means. Compare the results with the original values of the SS. How does the outcome of the comparison relate to whether or not the contrast is orthogonal to ψ_1 in each of the two cases?

11.9 We have five experimental groups in a problem-solving study. Two groups are given instructions to facilitate solving; call these groups F_1 and F_2. There are also two interference conditions; these groups are I_1 and I_2. There is also a control group C, given neutral instructions. As the investigator, I want to know whether the difference between the average facilitation score and the control average differs significantly from the difference between the control mean and the average inhibition score. There are 10 subjects in a group and the cell totals are:

F_1	F_2	C	I_1	I_2
46	49	38	18	17

To achieve my purpose, I could:

1. Subtract 3.8 from all scores in the four experimental groups and then compute a $(+1, +1, -1, -1)$ contrast on the adjusted group data.
2. Subtract 3.8 from the F_1 and F_2 scores and add 3.8 to the I_1 and I_2 scores and then do the $(+1, +1, -1, -1)$ contrast.
3. Do a $(+1, +1, -4, +1, +1)$ contrast.

One of these is correct. For each choice, indicate whether it is right or wrong and justify your response.

REFERENCES

Dunn, O. J. Multiple comparisons among means. *Journal of the American Statistical Association* 56: 52–64 (1961).

Dunnett, C. W. A multiple comparison procedure for comparing several treatments with a control. *Journal of the American Statistical Association* 50:1096–1121 (1955).

Games, P. A. Multiple comparisons of means. *American Educational Research Journal* 8: 531–65 (1971).

Keselman, H. J. The statistic with the smaller critical value. *Psychological Bulletin* 81: 130–31 (1974).

Keselman, H. J. A power investigation of the Tukey multiple comparison statistic. *Educational and Psychological Measurement* 36: 97–104 (1975).

Keselman, H. J., Murray, R., and Rogan, J. Effect of very unequal group sizes on Tukey's multiple comparison test. *Educational and Psychological Measurement*, in press.

Keselman, H. J., and Toothaker, L. E. Comparison of Tukey's *T*-method and Scheffé's *S*-method for various numbers of all possible differences of averages contrasts under violations of assumptions. *Educational and Psychological Measurement* 34: 511–20 (1974).

Miller, R. G., Jr. *Simultaneous Statistical Inference.* New York: McGraw-Hill, 1966.

Perlmutter, J., and Myers, J. L. A comparison of two procedures for testing multiple contrasts. *Psychological Bulletin* 79: 181–84 (1973).

Ryan, T. A. Multiple comparisons in psychological research. *Psychological Bulletin* 56: 26–47 (1959).

Scheffé, H. *The Analysis of Variance,* New York: John Wiley & Sons, 1959.

Tukey, J. W. The problem of multiple comparisons. Unpublished manuscript.

12 SIMPLE LINEAR REGRESSION

12.1 INTRODUCTION

The design models presented thus far are special cases of an equation usually referred to as the general linear model; and the analyses of variance considered to this point are special cases of a more general approach to data, linear regression analysis. A knowledge of the more general system of data analysis is important because with this concept you can better understand several additional topics. These include estimation and hypothesis testing when cell frequencies are disproportionate, the analysis of covariance, the analysis of functional relations when independent variables are quantitative ("trend" analysis), and analyses of data when several dependent variables are observed (multivariate analysis of variance). Viewing all these, as well as the usual analyses of variance, as special cases within the general domain of regression analysis not only permits us to see some interesting relations among them, but provides a general set of calculations for testing hypotheses and estimating within all these areas.

In examining linear regression, we first take up the simplest case, the linear relation between two variables X and Y. After having defined a simple linear function between the two variables, derived formulas for estimating the parameters of the function, and established tests of hypotheses about the parameters, we shall show that the ordinary F test of the equality of two treatment population means is equivalent to a test for a linear relation between two variables. This should support our assertion that the usual analysis of variance is a special case of regression analysis, clarifying what that assertion implies.

12.2 DESCRIBING SIMPLE LINEAR RELATIONS

12.2.1 THE EQUATION FOR A LINE.

Consider the following pairs of numbers:

$$X = 2 \quad 3 \quad 6 \quad 9$$
$$Y = 7 \quad 9 \quad 15 \quad 21$$

If we were to plot the four points these pairs represent, they would fall on a straight line. Furthermore, knowing any X value, we can immediately generate the Y value by multiplying by 2 and adding 3; that is, $Y = 3 + 2X$. As we probably all know from elementary school days, a straight line can generally be represented by the function

(**12.1**) $$Y = b_0 + b_1 X$$

In the specific example we have just looked at, b_0 is 3 and b_1 is 2. We refer to b_0 as the *intercept*; it is the value of Y when X equals zero. The constant b_1 is the *linear regression coefficient*, often referred to as the *slope* of the line. Its value is the amount of change in Y for each change of one unit in X.

It would be surprising to find a perfect linear relation between any two variables. Nevertheless, Equation (12.1) is useful because it often provides a reasonable approximation to data. Figure 12–1 should help to make this statement clearer. In panel (a), the data are perfectly described by a straight line; knowledge of X and the quantities b_0 and b_1 yield the individual's true Y value. Unfortunately, this is a situation that exists in mathematics textbooks—rarely if ever in statistics textbooks or in the real world. Panel (b) is considerably more

Figure 12–1 Y as a linear function of X for four data sets.

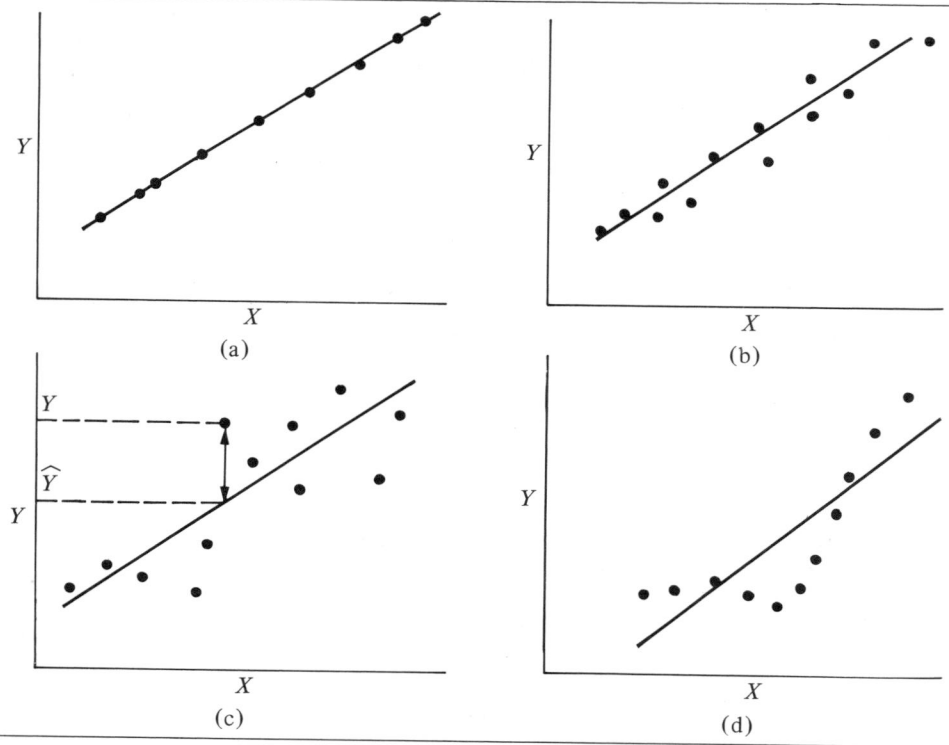

interesting. The points no longer fall on the straight line. They fall close to the line, however. In this case, the line captures the one clear trend in the data, the apparently systematic increase in Y with increases in X. Furthermore, because the points lie close to the line, the value of Y obtained by inserting an individual's X value into the equation for the line will be close to the individual's actual Y value; that is, X is a good predictor of Y in panel (b). The main difference between panel (b) and panel (c) is simply that the Y values are scattered more widely about the line. Still, while the straight line does not yield very good predictions of Y from X, it does a better job than many other functions of X would. It is much like using the mean of a set of scores as an index of central tendency. By any one of several criteria, it may be the most appropriate index, but if the variance in the data is large, it will not precisely represent any individual score. Panel (d) represents a somewhat different situation. Here, the data depart systematically from a straight line. As a consequence, there are functions of X (for example, $Y = b_0 + b_1 X + b_2 X^2$) that will describe the data better than the line will.

Reconsider panel (c) for a moment. For one point, we have indicated the vertical distance between the point and the line. That vertical distance corresponds to the difference between the individual's observed Y value and the value of Y that would be predicted if we inserted the individual's X value into Equation (12.1). The line that we have drawn in each panel has been drawn to minimize the sum of the squared distances of observed from predicted values of Y. In what follows, we shall present formulas for b_0 and b_1 that result in such *least-squares* lines. (A general discussion of least-squares estimation can be found in Section 3.6.)

12.2.2 THE LEAST-SQUARES LINEAR FUNCTION. Because we now recognize that the points will not fall exactly on the line, we replace Equation (12.1) by

(12.2)
$$\hat{Y} = b_0 + b_1 X$$

The diacritical mark is used to indicate that we have an estimate of an individual's Y value, which is based on knowledge of his value of X and the linear relation. To describe the observed score, we have

(12.3)
$$Y = \hat{Y} + d$$

where $d = Y - \hat{Y}$. Finding the best-fitting line according to a least-squares criterion is the problem of finding values of b_0 and b_1 that allow $\sum d^2$ to be a minimum.

Least-squares values of b_0 and b_1 are obtained by taking partial derivatives of $\sum d^2$ with respect to the two coefficients; substituting for $\hat{Y}$,

(12.4)
$$\sum d^2 = \sum (Y - b_0 - b_1 X)^2$$

Setting the derivatives equal to zero results in what are referred to as *normal equations:*

(12.5)
$$\sum Y - n b_0 - b_1 \sum X = 0$$
$$\sum XY - b_0 \sum X - b_1 \sum X^2 = 0$$

Solving simultaneously, we obtain

(**12.6**) $$b_0 = \bar{Y} - b_1 \bar{X}$$

and

(**12.7**) $$b_1 = \frac{\sum XY - n\bar{X}\bar{Y}}{\sum X^2 - n\bar{X}^2}$$

While Equation (12.7) is convenient for calculating, the following form is often used:

(**12.8**) $$b_1 = \frac{\sum xy}{\sum x^2}$$

where $x = X - \bar{X}$ and $y = Y - \bar{Y}$. If we divide numerator and denominator by n, we can see clearly that the least-squares formula for b_1 is the ratio of the sample covariance to the sample variance of the Xs.

Substituting Equation (12.6) into Equation (12.2) eliminates b_0, yielding an equivalent expression that is frequently seen:

(**12.9**) $$\hat{Y} = \bar{Y} + b_1 x$$

Equation (12.2), or (12.9), describes the best-fitting straight line, best-fitting in the sense of minimizing the sum of squared deviations of observed from predicted values of Y. "Best-fitting" does not necessarily mean a good fit; the points may be widely scattered about the line. Panel (c) of Figure 12.1 is one such case.[1] Clearly, we need a measure of error variance that can be an index of goodness of fit.

12.2.3   VARIABILITY ABOUT THE THE BEST-FITTING LINE.   Figure 12–2 is helpful in understanding the relation between the usual variability about the mean and the variability we have minimized in selecting the best-fitting line for a sample of data. Note that the deviation of a score from the grand mean, $Y - \bar{Y}$, can be viewed as consisting of two components: the deviation of the score from the value predicted given X, $Y - \hat{Y}$, and the deviation of the prediction from the mean, $\hat{Y} - \bar{Y}$. This partitioning of the deviation from the mean is simply expressed by the algebraic identity

(**12.10**) $$(Y - \hat{Y}) = (Y - \hat{Y}) + (\hat{Y} - \bar{Y})$$

Squaring both sides and summing over all n individuals in the sample, we obtain

$$\sum (Y - \bar{Y})^2 = \sum (Y - \hat{Y})^2 + \sum (\hat{Y} - \bar{Y})^2 + 2 \sum (Y - \hat{Y})(\hat{Y} - \bar{Y})$$

Substituting in terms of Equations (12.9) and (12.7), and simplifying, we get the

---

[1] The best-fitting straight line may not be a good fit in another sense; some other function may capture the data better, as in panel (d). In Section 12.3, we shall give an inferential model that makes explicit the assumption that a straight line describes the relation between X and Y in the population better than any other function does. In Chapter 17, we shall give tests of this assumption.

Figure 12–2 Partitioning $Y - \bar{Y}$ into component parts.

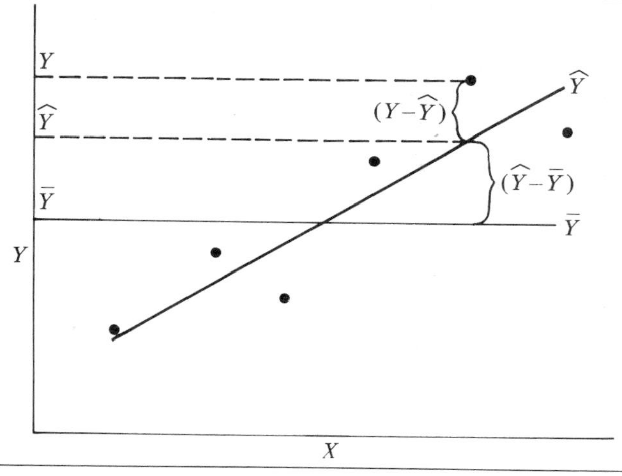

rightmost term equal to zero, and the result can be written

(12.11)
$$\sum y^2 = \left[\sum y^2 - \frac{(\sum xy)^2}{\sum x^2}\right] + \frac{(\sum xy)^2}{\sum x^2}$$

$$SS_{\text{total}} = SS_{\text{error}} + SS_{Y}$$

Note that what we have labeled SS_{error} is identical to $\sum d^2$; the expression is equivalent to the right-hand side of Equation (12.4). Equation (12.11) succinctly expresses that the total sum of squares can be partitioned into two components. Part of the total variability can be viewed as error variability. This part is the numerator of a variance about a line—not about a mean, the more usual case. The remainder of the total sum of squares $SS_{\hat{Y}}$ is the sum of squared deviations of predictions about their average (which incidentally is $\bar{Y}$, the average of the observed scores).[2] The partitioning is reminiscent of the partitioning of the total sum of squares into a within-group and a between-groups component in the usual analysis of variance for a one-factor completely randomized design. As in the analysis of variance, the partitioning of total variability is fundamental to calculations involved in tests of hypotheses.

One other analogy to material we have presented earlier may be worth considering. If we replace x by w in Equation (12.11), the expression for $SS_{\hat{Y}}$ is equivalent to the expression for $SS_{\hat{\psi}}$ (Chapter 11). Thus, the sum of squares for a single df contrast can be regarded as the sum of squares for the regression of individual scores on the weights used in a contrast for a set of means. The point

---

[2] The proof that the average predicted score equals the average score follows:

$$\bar{\hat{Y}} = \frac{\sum_{i=1}^{n}(a + bX)}{n} = a + b\bar{X} = (\bar{Y} - b\bar{X}) + b\bar{X} = \bar{Y}$$

that we shall continue to develop is that the tests presented earlier are special cases of linear regression analysis.

12.2.4   A NUMERICAL EXAMPLE.   Table 12–1 presents a set of values of X and Y, and also calculations based on these. Part (a) of the table corresponds to Equation (12.11). We have used deviations about $\bar{X}$ and $\bar{Y}$ to calculate the regression coefficients and the variance of the scores about the line that they determine. We could get the same results without finding deviations about the

Table 12–1   An Example of Linear Regression Analysis

(a) X	Y	x	y	x^2	y^2	xy
1	11	-2.5	1	6.25	1	-2.5
2	3	-1.5	-7	2.25	49	10.5
3	7	$-.5$	-3	.25	9	1.5
4	9	.5	-1	.25	1	$-.5$
5	9	1.5	-1	2.25	1	-1.5
6	21	2.5	11	6.25	121	27.5
21	60	0.0	0	17.50	182	35.0

$$b_1 = \frac{\sum xy}{\sum x^2} = \frac{35}{17.5} = 2$$

$$b_0 = \bar{Y} - b\bar{X} = 10 - (2)(3.5) = 3$$

$$\hat{Y} = 3 + 2X$$

$$SS_{error} = \sum y^2 - \frac{(\sum xy)^2}{\sum x^2} = \frac{182 - (35)^2}{17.5} = 112$$

$$SS_{\hat{Y}} = \frac{(\sum xy)^2}{\sum x^2} = b_1^2 \sum x^2 = \frac{(35)^2}{17.5} = 70$$

$$SS_{total} = \sum y^2 \qquad\qquad = 182$$

Table 12-1 (*continued*)

(b)	X	Y	$\hat{Y}$	$d = Y - \hat{Y}$	d^2	$\hat{Y} - \bar{Y}$	$(\hat{Y} - \bar{Y})^2$
	1	11	5	6	36	-5	25
	2	3	7	-4	16	-3	9
	3	7	9	-2	4	-1	1
	4	9	11	-2	4	1	1
	5	9	13	-4	16	3	9
	6	21	15	6	36	5	25
	21	60	60	0	112	0	70

$$SS_{\text{error}} = \sum (Y - \hat{Y})^2 = 112$$

$$SS_{\hat{Y}} = \sum (\hat{Y} - \bar{Y})^2 = 70$$

$$SS_{\text{total}} = \sum (Y - \bar{Y})^2 = 182$$

mean, using the raw score formulas for sums of squares and sums of products (Chapter 2). Part (b) of the table uses a different, but equivalent, approach to calculating sums of squares, one based on deviations of predicted scores. To obtain the error sum of squares $SS_{Y|X}$ the deviations of observed data from predicted are squared and summed. Note that Y and $\hat{Y}$ have the same mean; this will always be the case. The quantity $\sum (\hat{Y} - \bar{Y})^2$ is the variability of the predicted scores, which together with the error sum of squares accounts for the total sum of squares.

The proportion of total variability that is predictable by knowing X is one possible index of how well the straight line describes the data. After all, if all the points fall exactly on the line, observed values equal predicted values, and $SS_{\hat{Y}}$ equals SS_{total}. On the other hand, if b_1 equals zero, all values of $\hat{Y}$ equal $\bar{Y}$; knowing X provides no better prediction than knowing $\bar{Y}$ alone and $SS_{\hat{Y}}/SS_{\text{total}}$ equals zero. This ratio can be shown to equal r^2, the square of the correlation coefficient. It equals $\frac{70}{182}$, or .38, in the example.

12.3   THE LINEAR REGRESSION MODEL

Thus far, we have considered describing a set of n pairs of scores within a linear regression framework. Although description is important, we ordinarily want to raise questions about the population from which we have sampled. What are the regression coefficients for the function relating Y and X in the population? What is our best estimate of the variance about that function? These are only a few of the inferential questions we shall confront henceforth. To draw inferences about the population, we shall need a model of regression for the population. This means a more detailed set of assumptions than we needed for developing our descriptive statistics. In what follows, we present such a model. Later we shall develop formulas for estimation and tests of significance that will incorporate many of the statistics presented in the preceding section.

We begin by viewing a single individual's Y score as a sum of population parameters, much as we did in developing the analysis of variance model for the one-factor completely randomized design. We assume that the observed score equals some component that is constant for all individuals who have the same value of X, plus an error component that is due to individual differences and error of measurement. More precisely,

(12.12)
$$Y = \beta_0 + \beta_1 X + \varepsilon$$

where β_0 and β_1 are the intercept and slope for the least-squares linear function for the population, and ε is the error component. The regression coefficients are expressed like Equations (12.6) and (12.7) for the sample regression coefficients:

(12.13)
$$\beta_0 = \mu_Y - \beta_1 \mu_X$$

(12.14)
$$\beta_1 = \frac{E(XY) - \mu_X \mu_Y}{\sigma_X^2}$$

The symbol ε represents the vertical displacement of a point from the regression line; distributional assumptions about ε will be presented shortly. The parameters μ_Y and μ_X are expected values for the populations of Y and X scores.

Equation (12.12) can be rewritten in a form that is like Equation (12.9) for sample data:

(12.15)
$$Y = \mu_Y + \beta_1 (X - \mu_X) + \varepsilon$$

In addition to Equation (12.12) (or equivalently (12.15)), we require the following assumptions:

1.   *X is selected.* In the language developed earlier, we view X as a fixed-effect variable. We assume that the same set of values of X would be present in any replication of the study. This implies that we can replace μ_X by $\bar{X}$ in Equation (12.15). This assumption is often not correct. The investigator will frequently draw a sample of individuals and measure both Y and X; thus, successive samples will give rise to somewhat different distributions of X. Fortunately, formulas to be presented for estimation and hypothesis testing will still be

appropriate so long as distributional assumptions about ε (assumptions 3 to 5) are met.

2. *X is measured without error.* Even if we select values of X in advance—say we choose subjects having certain scores on a personality scale—there is always the possibility that the scores are in error due to the unreliability of the measuring instrument. A possible result of such errors of measurement is that the usual formulas provide biased results in estimation and tests of significance. The situation is not too bad if X, its errors of measurement, and the ε are independently and normally distributed and if X and Y have a joint distribution known as the bivariate normal; in large samples, normality of X is not required. In this case, the degree of bias in estimating β_1 depends on the magnitude of the error in measuring X relative to the variance of the true values of X; even with unreliable data this ratio will often be quite small. Snedecor and Cochran (1967, pp. 164–66) provide a more detailed discussion of this assumption.

3. *Linearity.* Corresponding to each value of X, there is a population of Y values having mean $\mu_{Y|X}$. The set of conditional means lies in a straight line described by

(12.16)
$$\mu_{Y|X} = \mu_Y + \beta_1(X - \mu_X)$$

Considering this equation together with Equation (12.15), it follows that the error component can be defined as

(12.17)
$$\varepsilon = Y - \mu_{Y|X}$$

This assumption implies that any deviation of scores from the best-fitting population line is chance variability. If the dispersion of points in the population resembled the dispersion of panel (d) of Figure 12–1, the inferential model now being developed would be inappropriate. To apply such a model would be to treat systematic departures from the straight line incorrectly, as though they happened by chance. One consequence would be that tests of null hypotheses about the value of β_1 would have spuriously inflated error terms and would be therefore negatively biased.

4. *Normality.* Each population of values of ε is independently and normally distributed with mean zero. The value of the mean follows from Equation (12.17) and the fact that the average of all deviations about a mean can be proved to be zero.

5. *Homogeneity of variance.* Each population of values of ε has the same variance σ_e^2. Textbooks on regression usually refer to this as homoscedasticity, but it seems preferable to emphasize the similarity to the analysis of variance model by using similar terms. One example of the problems that may stem from heterogeneity of variance is evident when we consider establishing confidence intervals on the conditional expectation $\mu_{Y|X}$. The usual formula involves pooling squared deviations from linearity for the entire set of X values. With heterogeneous variances, this might provide a poor estimate of the variance of the population of interest.

Before proceeding further, let us compare this model with the model developed earlier for a one-factor completely randomized design (Chapter 4). In the earlier case, we assumed a treatment populations, each having a normal distribution with mean μ_j and variance σ_e^2. Within the context set by the regression model, we again have a populations (letting a be the number of selected values of X), each having a normal distribution with mean $\mu_{Y|X}$ and variance σ_e^2. The chief distinction is that the model for the one-factor design is more general; no restriction is placed on the relation among the population means. The model currently under consideration requires that the means lie along a straight line (assumption 3).

12.4   INFERENCES ABOUT POPULATION PARAMETERS

12.4.1   ESTIMATING THE POPULATION ERROR VARIANCE σ_e^2.

To develop confidence intervals and tests of significance for population parameters, we need an estimate of the population error variance. The parameter to be estimated is defined as

$$\sigma_e^2 = E(Y - \mu_{Y|X})^2$$

Intuitively, the appropriate estimate might be the average squared deviation of observed from predicted values of Y for the sample of n pairs of scores. This is almost correct, but not quite. We shall show that an unbiased estimate of σ_e^2 requires that we divide SS_{error} by $n-2$ rather than by n. That is,

(12.18)
$$\hat{\sigma}_e^2 = \frac{\sum (Y - \hat{Y})^2}{n-2} = \frac{\sum d^2}{n-2}$$

Estimating the regression line involves estimating two parameters, β_0 and β_1, and thus the loss of two *df*. It is not surprising, therefore, that the denominator of Equation (12.18) is $n-2$. It has always been the case in previous chapters that the denominators of unbiased estimates of population variances have been the *df* associated with the numerator. A more formal proof follows.

PROOF. We begin with an identity, in which $Y - \mu_{Y|X}$, or ε, is viewed as a sum of three components:

$$Y - (\mu_Y + \beta_1 x) = [Y - (\bar{Y} + b_1 x)] + (\bar{Y} - \mu_Y) + (b_1 - \beta_1)x$$
$$= (y - b_1 x) + (\bar{Y} - \mu_Y) + (b_1 - \beta_1)x$$

Squaring both sides and summing over the n values for the sample yields

$$\sum [Y - (\mu + \beta_1 x)]^2 = \sum (y - b_1 x)^2 + n(\bar{Y} - \mu_Y)^2 + (b_1 - \beta_1)^2 \sum x^2$$
$$+ 2(\bar{Y} - \mu_Y) \sum (y - b_1 x) + 2(b_1 - \beta_1) \sum x(y - b_1 x)$$
$$+ 2(\bar{Y} - \mu_Y)(b_1 - \beta_1) \sum x$$

It is a happy fact that each of the three cross-product terms (the last three terms on the right-hand side of the equation) equals zero and therefore can be ignored. Note that for the first such term,

$$\sum (y - b_1 x) = \sum y - b_1 \sum x = 0 - (b_1)(0)$$

and for the next,

$$\sum x(y - b_1 x) = \sum xy - b_1 \sum x^2 = \sum xy - \sum xy = 0$$

and for the last,

$$\sum x = 0$$

We therefore have

$$\sum \varepsilon^2 = \sum d^2 + n(\bar{Y} - \mu_Y)^2 + (b_1 - \beta_1)^2 \sum x^2$$

Taking the expectations of both sides, and rearranging terms, yields

$$E\left(\sum d^2\right) = E\left(\sum \varepsilon^2\right) - E[n(\bar{Y} - \mu)^2] - E\left[(b_1 - \beta_1)^2 \sum x^2\right]$$

Since the expected squared error is the error variance,

$$E\left(\sum \varepsilon^2\right) = \sum [E(\varepsilon^2)] = n\sigma_e^2$$

Also, since the variance of a mean is the population variance divided by the sample size,

$$E[n(\bar{Y} - \mu)^2] = n\left(\frac{\sigma_e^2}{n}\right) = \sigma_e^2$$

Finally, $\sigma_{b_1}^2$, or $E(b_1 - \beta_1)^2$, equals $\sigma_e^2/\sum x^2$ (as we shall demonstrate in the next section). Also, because x is assumed fixed over repeated samples, we have

$$E\left[(b_1 - \beta_1)^2 \sum x^2\right] = \sum x^2 E(b_1 - \beta_1)^2$$

$$= \sum x^2 \left(\frac{\sigma_e^2}{\sum x^2}\right)$$

$$= \sigma_e^2$$

Putting these results together gives

$$E\left(\sum d^2\right) = (n - 2)\sigma_e^2$$

and

(12.19)
$$E\left(\frac{\sum d^2}{n - 2}\right) = \sigma_e^2$$

This concludes the proof that the statistic defined in Equation (12.18) is an unbiased estimate of the population error variance.

12.4.2   INFERENCES ABOUT REGRESSION COEFFICIENTS.   In many situations, although there are no prior grounds for adopting the linear model, the data will be well fitted by a linear function. In other situations, theory developed before the data collection will not only predict a linear function but provide a psychological interpretation of the regression coefficients. Suppose we hypothesized that trying to decide whether a particular target letter was in a word involved a letter-by-letter examination of the word. Then reaction time would be represented as a linear function of word length; b_1 represents the time spent scanning each letter of the word and b_0 represents the remainder of the total reaction time, those components of time (for example, time to push the response button) that are independent of word length. Whether our motivation for assuming a linear model lies solely in the data or in a combination of theory and data, our ultimate goal is a statement about the function relating X and Y in the population. We want to go from b_0 and b_1 to β_0 and β_1. This suggests such questions as: Are the sample regression coefficients unbiased estimates of the population coefficients? How variable are the sample regression coefficients?

We first consider the slope of the line. We have

$$b_1 = \frac{\sum xy}{\sum x^2} = \frac{\sum xY}{\sum x^2}$$

because $\sum xy = \sum xY - \sum x\bar{Y}$; but $\sum x\bar{Y} = \bar{Y}\sum x = 0$. Substituting for Y in terms of Equation (12.15) gives

$$b_1 = \frac{\sum x(\mu_Y + \beta_1 x + \varepsilon)}{\sum x^2}$$

Because $\sum x\mu_Y = 0$,

(**12.20**)
$$b_1 = \beta_1 + \frac{\sum x\varepsilon}{\sum x^2}$$

Because x is assumed to be fixed over samples,

$$E\left(\frac{\sum x\varepsilon}{\sum x^2}\right) = \frac{\sum [xE(\varepsilon)]}{\sum x^2}$$

But $E(\varepsilon) = 0$. Therefore, $E(b_1) = \beta_1$; the value b_1 is an unbiased estimator of β_1. We can also show that b_0 is an unbiased estimator of β_0:

(**12.21**)
$$\begin{aligned} b_0 &= \bar{Y} - b_1\bar{X} \\ &= (\beta_0 + \beta_1\bar{X} + \bar{\varepsilon}) - b_1\bar{X} \\ &= \beta_0 - (b_1 - \beta_1)\bar{X} + \bar{\varepsilon} \end{aligned}$$

Since $E(\bar{\varepsilon}) = 0$ and we have just shown that $E(b_1) = \beta_1$, we find that $E(b_0) = \beta_0$.

To calculate confidence intervals and tests of significance for hypothesized values of the regression coefficients, we require estimates of the variances of their sampling distributions. A formula for $\hat{\sigma}^2_{b_1}$ is readily derived from Equation (12.20):

$$\sigma^2_{b_1} = E(b_1 - \beta_1)^2 \qquad \text{by definition of a variance}$$

$$= E\left(\frac{\sum x\varepsilon}{\sum x^2}\right)^2 \qquad \text{from Equation (12.20)}$$

$$= \left(\frac{1}{\sum x^2}\right)^2 \left[\sum x^2 E(\varepsilon^2)\right] \qquad \text{because the } xs \text{ are fixed}$$

$$= \frac{E(\varepsilon^2)}{\sum x^2}$$

$$= \frac{\sigma^2_e}{\sum x^2}$$

Therefore,

(12.22)
$$\hat{\sigma}^2_{b_1} = \frac{\hat{\sigma}^2_e}{\sum x^2} = \frac{\sum d^2}{(n-2)\sum x^2}$$

where, as previously, $d = Y - \hat{Y}$.

It should be apparent from this development that the sampling variability of the slope depends on two things. First, as we should expect, the smaller the variability of Y about the population line, the more reliable the estimation of β_1. Second, and perhaps less expected, estimates of β_1 are more reliable when the selected values of X are more variable; $\sigma^2_{b_1}$ decreases as $\sum x^2$ increases. It may be easier to understand this if we consider some possible data sets. Suppose $X_1 = 1$, $X_2 = 3$, $Y_1 = 2$, and $Y_2 = 4$. For these data, $b_1 = 1$. We can create another data set for which b_1 is 1, but one that has larger value of $\sum x^2$. Let X_1 again equal 1 and Y_1 again equal 3, but set X_2 and Y_2 at 5 and 7. For the first data set, a decrease of one unit in Y_2 reduces b_1 from 1 to .5, but for the second data set (with the more variable X), a reduction in Y_2 of one unit reduces the slope from 1 to only .75. The more widely spread the values of X, the smaller the effect on the slope of an error in any Y. Another way to increase $\sum x^2$ is to add a point to an existing set of data points (unless the added value of X is placed at the exact mean of the other values). Intuitively, the more numerous the points on which the line is based, the smaller the effect on the slope of a change in any one value of Y.

Given Equation (12.22), we are now able to test hypotheses about the value of β_1. If our regression model is correct, it can be shown that $(b_1 - \beta_1)/\hat{\sigma}_{b_1}$ is distributed as t on $n-2$ df; alternatively, we recognize that the square of this quantity is distributed as F on 1 and $n-2$ df. To understand the test, consider the data of Table 12–1. A frequent question about such data is whether Y and X are related in the population from which the data have been sampled. This translates into $H_0: \beta_1 = 0$. Recalling that $\hat{\sigma}^2_e = \sum d^2/(n-2)$, we have

$$t = \frac{2-0}{\sqrt{(\frac{1}{4})(112)/(17.5)}} = \frac{2}{1.265} = 1.58$$

which, on 4 *df*, is not significant at the .05 level. Of course, we are really not enthusiastic about significance tests with so few *df*; they do tend to lack power.

Computing a confidence interval on the slope will often be of interest. The $1 - \alpha$ interval on β_1 is $b_1 \pm t_{n-2,\alpha} \hat{\sigma}_{b_1}$, where $t_{n-2,\alpha}$ is the value of t required for significance at the α level against a two-tailed alternative. Suppose that we want a .95 confidence interval for β_1. The critical value of t is 2.776. Therefore, the desired limits are $2 \pm (2.776)(1.265)$, or -1.51 and 5.51.

Starting with Equation (12.21), we can derive an estimate of the variance of the sampling distribution of the intercept b_0:

$$\sigma_{b_0}^2 = E(b_0 - \beta_0)^2$$

$$= \bar{X}^2 E(b_1 - \beta_1)^2 + E(\bar{\varepsilon})^2 - 2\bar{X}E[\bar{\varepsilon}(b_1 - \beta_1)]$$

$$= \bar{X}^2 \sigma_{b_1}^2 + \frac{\sigma_e^2}{n} - 0$$

Therefore,

(12.23)
$$\hat{\sigma}_{b_0}^2 = \hat{\sigma}_e^2 \left(\frac{\bar{X}^2}{\sum x^2} + \frac{1}{n} \right)$$

The quantity $(b_0 - \beta_0)/\hat{\sigma}_{b_0}^2$ is distributed as t on $n - 2$ *df*. Accordingly, significance tests and confidence intervals can be established just as in the case of the slope. Again using the data of Table 12–1, we test whether the best-fitting line goes through the origin; $H_0 : \beta_0 = 0$. For these data,

$$t = \frac{3 - 0}{\sqrt{(\frac{6}{4})(18.67)[(3.5)^2/17.5] + \frac{1}{6}}} = \frac{3}{4.93} = .61$$

which is clearly not significant on four *df*. The .95 confidence interval is $3 \pm (2.776)(4.93)$, or -10.69 and 16.69, a wide interval.

12.4.3 INFERENCES ABOUT THE POPULATION REGRESSION LINE.

Having calculated estimates of the regression coefficients, we shall probably want to use them to estimate the conditional population means $\mu_{Y|X}$. It seems as though the value of Y predicted for some X from the sample, that is, $\hat{Y}$, would be a reasonable estimate of the mean of the population of values of Y at that value of X. Let us see whether this is so. First, we note that $\hat{Y}$ is an unbiased estimator of $\mu_{Y|X}$:

(12.24)
$$\hat{Y} - \mu_{Y|X} = (\bar{Y} + b_1 x) - (\mu_Y + \beta_1 x)$$

$$= (\bar{Y} - \mu_Y) + (b_1 - \beta_1)x$$

Since we know that $E(\bar{Y}) = \mu_Y$ and $E(b_1) = \beta_1$, the expectation for the right-hand side is zero. Therefore, $E(\hat{Y} - \mu_{Y|X}) = 0$ and $\hat{Y}$ is an unbiased estimate of $\mu_{Y|X}$.

To establish confidence intervals for some $\mu_{Y|X}$, we need an estimate of the variance of the sampling distributions of $\hat{Y}$. To derive this, we first express $\bar{Y}$ by population parameters. From Equation (12.15), it follows that $\bar{Y} = \mu_Y + \bar{\varepsilon}$, and Equation (12.24) is now

$$\hat{Y} - \mu_{Y|X} = \bar{\varepsilon} + (b_1 - \beta_1)x$$

The sampling variance of $\hat{Y}$, $\sigma_{\hat{Y}}^2$, is therefore

(12.25)
$$E(\hat{Y} - \mu_{Y|X})^2 = E[\bar{\varepsilon} + (b_1 - \beta_1)x]^2$$

Because the ε are assumed to be independently distributed, $E[\varepsilon(b_1 - \beta_1)x] = 0$. Furthermore, the developments leading to Equation (12.22) allow us to express $E(b_1 - \beta_1)^2$ as $\sigma_e^2 / \sum x^2$. Also, $E(\bar{\varepsilon})^2 = \sigma_e^2 / n$. Integrating these facts, we get the variance estimate:

(12.26)
$$\hat{\sigma}_{\hat{Y}}^2 = \hat{\sigma}_e^2 \left[\frac{1}{n} + \frac{x^2}{\sum x^2} \right]$$

Before using this result, we might note that $\hat{Y}$ is not only an unbiased estimator of $\mu_{Y|X}$ but also a least-squares one. So you can see how this is, consider any other estimator, which we shall call $\hat{Y} + C$. Replacing $\hat{Y}$ by $\hat{Y} + C$ in Equation (12.25) causes the right-hand side to be increased by the amount C^2. (The expectations of the products of C and the other terms can easily be shown to equal zero.) Thus the sampling variance is minimal when $C = 0$, that is, when we use $\hat{Y}$ as our estimator.

Having arrived at Equation (12.26), we can readily obtain confidence intervals. The $1 - \alpha$ interval for $\mu_{Y|X}$ is

(12.27)
$$\hat{Y} \pm t_{n-2, \, \alpha} \hat{\sigma}_{\hat{Y}}$$

We should note that several factors influence the precision of our estimation. As always, with intervals of this sort, the width of the interval increases as confidence increases (α decreases), error variance increases, or n decreases. Also, the variance of the slope is a factor, and this varies inversely with $\sum x^2$ (for reasons noted in Section 12.4.2). One other factor is of interest; the greater the distance of X from $\bar{X}$, the wider the interval about $\mu_{Y|X}$. To understand why this is so, consider the situation in which $X = \bar{X}$. Then, $\hat{Y} = \bar{Y}$ regardless of the value of b_1. For this value of X, therefore, the variance of the slope should not contribute to the variance of $\hat{Y}$, and it does not; since x is zero, Equation (12.26) reduces to an estimate of the variance of $\bar{Y}$, namely $\sigma_{\bar{Y}}^2 = \sigma_e^2 / n$. It should be clear from Equation (12.9) that the contribution of b_1 to $\hat{Y}$ increases as the absolute value of x does; thus, the variance of a prediction obtained at some value of X will depend on how far X is from its mean.

To illustrate the calculations, we again use the data of Table 12–1. When $X = 2$, $\hat{Y} = 7$. This is our best estimate of $\mu_{Y|X=2}$. The .95 confidence interval about this parameter is

$$7 \pm (2.776) \sqrt{(28) \left[\frac{1}{6} + \frac{(1.5)^2}{17.5} \right]}$$

which yields bounds of $-.98$ and 14.98.

In Section 12.4.2, we calculated a confidence interval for β_0 using these same data. Such an interval is merely a special case (in which $X = 0$) of the current developments. In view of our discussion of the relation of the interval width to the value of x^2, it is interesting to note that the interval just calculated is much narrower than the one obtained at $X = 0$.

12.4.4 PREDICTING AN INDIVIDUAL'S SCORE.

A common application of the developments for simple linear regression is the following. A regression line, together with an estimated error variance, is calculated for some sample of n individuals. We now wish to predict the Y value for some individual who was not in the original sample. Note the difference between this and the prediction generated in Section 12.4.3. There, we predicted the mean for a population having a certain value of X; here, we want to predict Y, which is the mean plus a component of error. Because the expectation of that error is zero, $\hat{Y}$ is again the unbiased estimator, just as in estimating $\mu_{Y|X}$. Because of the variance of this error component, however, the variance of $\hat{Y}$ about Y will equal the variance of $\hat{Y}$ about $\mu_{Y|X'}$ plus σ_e^2. Our estimator error variance is now

(12.28)
$$\hat{\sigma}_{\hat{Y}}^2 = \hat{\sigma}_e^2 \left(1 + \frac{1}{n} + \frac{x}{\sum x^2} \right)$$

The formula for the confidence interval for an individual Y value is still (12.27), but with Equation (12.28) instead of (12.26) substituted for $\hat{\sigma}_{\hat{Y}}^2$. If $X = 2$, the .95 confidence interval for the score of an individual not in the sample is

$$7 \pm (2.776) \sqrt{(28) 1 + \frac{1}{6} + \frac{(1.5)^2}{17.5}}$$

or -9.72 and 23.72. There is much less precision in estimating an individual score than there was in estimating a population mean.

Confidence intervals for Y are sometimes misinterpreted. We do not mean to imply that .95 of the population of scores at $X = 2$ lie between the computed boundaries. The correct interpretation is the following. Assume that we compute a new regression line for each of many samples of size n; as usual X is fixed. Further assume that for each sample, we estimate a score and its confidence interval for some additional individual at a value of X that will be the same for all samples. Following each prediction, we observe the score for the additional individual. In .95 of the cases, the observed value of Y for the individual who was not in the sample will lie within the computed confidence interval.

12.5   THE ONE-FACTOR DESIGN WITH TWO GROUPS: A SPECIAL CASE OF REGRESSION ANALYSIS

Although the material in this chapter is of interest in its own right, the primary reason for its presentation has been to introduce a regression framework within which we might relate the usual analyses of variance to other analyses. We are now ready for the first step in that integration. We will take the simplest F test, that for a two-group design, and show that it is equivalent to the squared t test of the hypothesis that a linear regression coefficient is zero.

We first consider the relation between the usual experimental design model and the regression model. The experimental design model asserts:

$$Y_{ij} = \mu + \alpha_j + \varepsilon_{ij}$$

with the usual assumptions that the ε_{ij} are normally and independently distributed with mean zero and variance σ_e^2 within each population. To consider the two-group experiment as a problem in regression analysis, we assign the value X_1 to the n_1 subjects in one group and the value X_2 to the n_2 subjects in the second group. The only restriction on the values of X is that $X_1 \neq X_2$. The regression model (Equation 12.12) is

$$Y_{ij} = \beta_0 + \beta_1 X_{ij} + \varepsilon_{ij}$$

where $X_{i1} = X_1$ and $X_{i2} = X_2$ for all i. Distributional assumptions about the ε_{ij} are the same as for the analysis of variance model. Because there are only two populations of Y values, the population means must lie on a straight line. Consequently, $\mu_j = \beta_0 + \beta_1 X_j$, and

(12.29) $$\mu_1 - \mu_2 = \beta_1(X_1 - X_2)$$

Since X_1 cannot equal X_2, the null hypothesis that $\mu_1 = \mu_2$ is equivalent, within the regression framework, to the null hypothesis that $\beta_1 = 0$. We know from Section 12.4.2 that this hypothesis can be tested by

(12.30) $$F = \frac{b_1^2}{\hat{\sigma}_{b_1}^2} = \frac{b_1^2}{\hat{\sigma}_e^2 / \sum x^2} = \frac{b_1^2 \sum x^2}{\hat{\sigma}_e^2}$$

where F is distributed on 1 and $n - 2$ df; in the present case, $n = n_1 + n_2$. With only two values of X, $b_1 = (\bar{Y}_1 - \bar{Y}_2)/(X_1 - X_2)$, and algebraic manipulation demonstrates that

$$\sum x^2 = \frac{(X_1 - X_2)^2}{(1/n_1) + (1/n_2)}$$

From these developments, the numerator of Equation (12.30) can be rewritten as

(12.31) $$b_1^2 \sum x^2 = \frac{(\bar{Y}_{.1} - \bar{Y}_{.2})^2}{(1/n_1) + (1/n_2)}$$

which is exactly the expression for $SS_{\hat{\psi}}$ in Equation (11.16) and equals SS_A when $a = 2$. Turning now to the denominator of Equation (12.30), recall that

$$\hat{\sigma}_e^2 = \frac{\sum (Y - \hat{Y})^2}{n - 2}$$

From Equation (12.11),

$$\sum (Y - \hat{Y})^2 = SS_{error} = \sum y^2 - \frac{(\sum xy)^2}{\sum x^2}$$

By definition, $\sum y^2 = SS_{total}$; also, if b_1 is replaced by its least-squares expression,

$b_1^2 \sum x^2 = (\sum xy)^2 / \sum x^2$. Therefore,

$$\sum (Y - \hat{Y})^2 = SS_{total} - b_1^2 \sum x^2 = SS_{total} - SS_A$$

for the special case of two values of x. It follows that

$$\hat{\sigma}_{error}^2 = \frac{SS_{S/A}}{n-2} = MS_{S/A}$$

and Equation (12.30) is algebraically identical to the usual F test of $H_0: \mu_1 = \mu_2$. Note that the result is completely independent of the values of X_1 and X_2. On the other hand, this is not true in establishing a confidence interval for $\mu_1 - \mu_2$. Since $\beta_1 = (\mu_1 - \mu_2)/(X_1 - X_2)$, the interval for β_1 would have to be multiplied by $X_1 - X_2$ to get the interval for $\mu_1 - \mu_2$. The situation is the same as the one in which the F test of a contrast is unaffected by multiplication of the weights by a constant (Chapter 11); however, the confidence interval is affected.

The preceding material was not given to provide an alternative to the usual significance test for comparing two treatment population means. We do hope, however, that it provides an alternative view of the usual significance test. In any design, we can establish a relation between the experimental design model and some linear regression model and therefore between the parameters of the two models. Tests of hypotheses about regression parameters will often be clumsier than equivalent tests about population means, but there will be situations in which the regression approach has the advantage. It will be useful, therefore, to understand the more general approach, regression analysis, so that our choice between procedures will be based on the dictates of the situation rather than on limitations in our knowledge. In chapters which follow, we shall extend the development of the regression approach, tying it to design models more complex than the one considered here. Before doing so, we shall digress briefly (Chapter 13) to present some mathematics that will help the development of regression analysis.

EXERCISES

12.1 (a) Find the regression equation for the following data set:

$$X = 2, 3, 3, 5, 8, 6, 4, 4, 9, 8$$
$$Y = 10, 19, 16, 26, 41, 33, 24, 21, 48, 40$$

(b) Estimate σ_e^2.
(c) Test $H_0: \beta_1 = 0$.
(d) Calculate the .95 confidence interval for $\mu_{Y|X=6}$.

12.2 We frequently want to predict Y for some X that was not in the original sample. Discuss what the precision of estimation is (a) when X is interpolated between two values included in the original sample (for example, $X = 7$ in Exercise 12.1); (b) when we extrapolate beyond the original range of values of X (for example, $X = 10$ in Exercise 12.1).

12.3 Subjects are required to check for the presence of some target letter in a set of letters. There are 4 groups of 10 subjects each. Each group is assigned to a different list length; $L = 2, 4, 6,$ or 8 letters. The value RT is measured and the results are:

$$L_j = \quad 2 \quad\quad 4 \quad\quad 6 \quad\quad 8$$
$$\bar{Y}_{.j} = 480 \quad 520 \quad 540 \quad 540$$
$$\hat{\sigma}_j^2 = 360 \quad 315 \quad 324 \quad 333$$

Joe Anova considers the usual analysis of variance design model: $Y_{ij} = \mu + \alpha_j + \varepsilon_{ij}$.

(a) Calculate all appropriate terms and test $H_0 : \mu_1 = \mu_2 = \mu_3 = \mu_4$.

Jim Regress sees the study very differently. He says that this is a regression problem. We have 40 subjects, each with an L score and a Y score. He assumes the regression (linear) model: $Y_{ij} = \mu + \beta(L_{.j} - \bar{L}_{..}) + \varepsilon_{ij}$. He tests $H_0 : \beta = 0$ (that is, there is no linear relation between Y and L).

(b) Are Joe and Jim testing equivalent hypotheses? Briefly, justify your answer. If your answer is No, how are their two null hypotheses related? If Joe's is false, should Jim's also be? Or if Jim's is false, should Joe's be?

(c) Calculate $\sum\sum x^2$, $\sum\sum y^2$, and $\sum\sum xy$, given $x = x_{ij} - \bar{X}_{..}$ and $y = Y_{ij} - \bar{Y}_{..}$. (Note that $\sum\sum y^2$ is SS_{total} in the analysis of variance.)

(d) Using Equation (12.11), and the results from (c), partition SS_{total} into SS_{error} and $SS_{\hat{Y}}$.

(e) Calculate the F test of $H_0 : \beta_1 = 0$.

12.4 Joe Anova and Jim Regress (Exercise 12.3) calculate the same SS_{total}. The partitioning of this variability is different. If your calculations are correct in Exercise 12.3, SS_A is greater than $SS_{\hat{Y}}$ and $SS_{S/A}$ is smaller than SS_{error}.

(a) Will these inequalities hold for any data set from a one-factor design? Explain your answer.

(b) Can you think of an alternative to the two models presented in the previous problem?

REFERENCE

Snedecor, G. W., and Cochran, W. G. *Statistical Methods*, 6th ed. Ames: Iowa University Press, 1967.

13 MATRIX ALGEBRA

13.1 INTRODUCTION

A *matrix* is a rectangular array of numbers, or symbols representing numbers. Certainly we are not unfamiliar with matrices, for we have encountered many data matrices in this work and we have discussed a special matrix (Chapter 7) that contained the variances and covariances for conditions in a repeated-measurement design. In short, the raw elements of statistical analyses are entries in data matrices and in matrices containing such summary statistics as means, variances, and covariances. Therefore, it is not surprising that statisticians have found useful the notation and operations of a branch of mathematics known as matrix algebra. While matrix algebra has not been absolutely necessary in our work so far, it will become increasingly useful as we further consider regression analysis and its applications. This chapter is only an introduction to matrix algebra and therefore is limited. Despite this, it should provide some useful computations and a sense of some of the advantages of using matrix algebra. It should prove useful even to those who are already familiar with matrix algebra. Our examples are from regression analysis and we have derived several general results that will be important subsequently. (Some additional aspects of matrix algebra will be presented in Chapter 18, where they will be needed for understanding the material on multivariate analysis of variance.)

13.2 DEFINITIONS

A *matrix* is a rectangular array of numbers, or symbols that represent numbers. Thus a specific data set obtained by testing each of four subjects under these conditions can be represented by the 4×3 matrix:

$$\mathbf{Y} = \begin{bmatrix} 5 & 2 & 7 \\ 6 & 1 & 3 \\ 2 & 1 & 4 \\ 8 & 9 & 6 \end{bmatrix}$$

In general, data from such a study would be represented by the $n \times a$ matrix:

$$\mathbf{Y} = \begin{bmatrix} Y_{11} & \cdots & Y_{1j} & \cdots & Y_{1a} \\ \cdot & & \cdot & & \cdot \\ \cdot & & \cdot & & \cdot \\ \cdot & & \cdot & & \cdot \\ Y_{n1} & \cdots & Y_{nj} & \cdots & Y_{na} \end{bmatrix}$$

The brackets signify that the collection of numbers, or symbols, is to be viewed as a matrix and is subject to the operations of matrix algebra. The symbol that represents the matrix, $\mathbf{Y}$, is printed in boldface type. We shall follow both of these conventions throughout. We are sometimes interested in both the matrix $\mathbf{Y}$ and its *transpose*, symbolized $\mathbf{Y}'$. The transpose $\mathbf{Y}'$ is derived from $\mathbf{Y}$ by interchanging the rows and columns of $\mathbf{Y}$. Thus, the 4×3 matrix presented above has the 3×4 transpose

$$\mathbf{Y}' \begin{bmatrix} 5 & 6 & 2 & 8 \\ 2 & 1 & 1 & 9 \\ 7 & 3 & 4 & 6 \end{bmatrix}$$

Any matrix with equal numbers of rows and columns is called a *square matrix*. Among these, *symmetric matrices* are of frequent interest in statistics. A matrix $\mathbf{Y}$ is symmetric if $\mathbf{Y} = \mathbf{Y}'$. *Equality* holds only if two matrices of the same dimensionality (numbers of rows and columns) are identical for every entry. The variance-covariance matrix (defined in Chapter 7) must be symmetric because the covariance of A_j and $A_{j'}$ is identical to the covariance of $A_{j'}$ and A_j; thus the cell in the jth row and the j'th column contains the same entry as the cell in the j'th row and the jth column. An example of a symmetric matrix is

$$\begin{bmatrix} 7 & 2 & 8 \\ 2 & 9 & 3 \\ 8 & 3 & 4 \end{bmatrix}$$

A matrix of special interest is the *identity matrix*, which is a square matrix with ones in all diagonal cells and zeros elsewhere; for example,

$$\mathbf{I} = \begin{bmatrix} 1 & 0 & 0 \\ 0 & 1 & 0 \\ 0 & 0 & 1 \end{bmatrix}$$

We shall always use the symbol $\mathbf{I}$ to denote matrices of this kind. The identity matrix is important because multiplication by $\mathbf{I}$ leaves a matrix unchanged. Formally, $\mathbf{I}$ is said to be an identity operator for matrix multiplication and $\mathbf{IY} = \mathbf{Y}$.

We frequently deal with matrices that consist of only a single row or column. Such matrices are *row* or *column vectors*. The set of treatment population means, for example, can be represented by the column vector

$$\boldsymbol{\mu} = \begin{bmatrix} \mu_i \\ \cdot \\ \cdot \\ \cdot \\ \mu_j \\ \cdot \\ \cdot \\ \cdot \\ \mu_a \end{bmatrix}$$

The transpose $\boldsymbol{\mu}'$ is a row vector.

These terms just presented will suffice for us to consider some simple operations on matrices.

13.3   BASIC OPERATIONS

Just as ordinary algebra involves addition, subtraction, multiplication, and division of individual numbers, matrix algebra involves analogous manipulations of matrices. Consider addition first. For the matrices $\mathbf{X}$ and $\mathbf{Y}$, both of the same dimensionality, $r \times c$, addition is defined as

$$(\textbf{13.1}) \qquad \mathbf{X} + \mathbf{Y} = \begin{bmatrix} X_{11} + Y_{11} & \cdots & X_{1c} + Y_{1c} \\ & \cdot & \\ & \cdot & \\ & \cdot & \\ & X_{ij} + Y_{ij} & \\ & \cdot & \\ & \cdot & \\ & \cdot & \\ X_{r1} + Y_{r1} & \cdots & X_{rc} + Y_{rc} \end{bmatrix}$$

Addition of matrices is commutative: $\mathbf{X} + \mathbf{Y} = \mathbf{Y} + \mathbf{X}$. Furthermore, it is associative: $(\mathbf{X} + \mathbf{Y}) + \mathbf{Z} = \mathbf{X} + (\mathbf{Y} + \mathbf{Z})$.

As we might expect, subtraction of matrices is much like addition. The matrices $\mathbf{X}$ and $\mathbf{Y}$ must have the same dimensions. Then,

$$(\textbf{13.2}) \qquad \mathbf{X} - \mathbf{Y} = \begin{bmatrix} X_{11} - Y_{11} & \cdots & X_{1c} - Y_{1c} \\ & \cdot & \\ & \cdot & \\ & \cdot & \\ & X_{ij} - Y_{ij} & \\ & \cdot & \\ & \cdot & \\ & \cdot & \\ X_{r1} - Y_{r1} & \cdots & X_{rc} - Y_{rc} \end{bmatrix}$$

Matrix multiplication is different from what the preceding developments may have led us to expect. First, for a single number k (often called a *scalar*), and a matrix $\mathbf{Y}$, the operation multiplying k times $\mathbf{Y}$ has meaning. In contrast, $k+\mathbf{Y}$ or $k-\mathbf{Y}$ is not defined. The product $k\mathbf{Y}$ is defined as a new matrix, in which all entries are multiplied by k. That is,

(13.3)

$$k\mathbf{Y} = \begin{bmatrix} kY_{11} \cdots kY_{1c} \\ \cdot \\ \cdot \\ \cdot \\ kY_{ij} \\ \cdot \\ \cdot \\ \cdot \\ kY_{r1} \cdots kY_{rc} \end{bmatrix}$$

A second distinction between addition and subtraction on the one hand, and multiplication on the other, is that $\mathbf{X}$ and $\mathbf{Y}$ need not be of equal dimensionality for the operation $\mathbf{X} \times \mathbf{Y}$ to be carried out. As we shall see, however, there is a restriction on the dimensions of the two matrices. We shall also find that unlike multiplication of numbers, matrix multiplication is not ordinarily commutative (usually, $\mathbf{XY} \neq \mathbf{YX}$).

To understand these points, we must first understand how we multiply matrices. Consider an example in which $\mathbf{X}$ is a 2×3 matrix and $\mathbf{Y}$ is 3×2. Then,

$$\mathbf{XY} = \begin{bmatrix} X_{11} & X_{12} & X_{13} \\ X_{21} & X_{22} & X_{23} \end{bmatrix} \times \begin{bmatrix} Y_{11} & Y_{12} \\ Y_{21} & Y_{22} \\ Y_{31} & Y_{32} \end{bmatrix}$$

and multiplying each row by each column, and adding the three products, we get

$$\mathbf{XY} = \begin{bmatrix} X_{11}Y_{11} + X_{12}Y_{21} + X_{13}Y_{31} & X_{11}Y_{12} + X_{12}Y_{22} + X_{13}Y_{32} \\ X_{21}Y_{11} + X_{22}Y_{21} + X_{23}Y_{31} & X_{21}Y_{12} + X_{22}Y_{22} + X_{23}Y_{32} \end{bmatrix}$$

Inserting numbers may make multiplication easier to understand. Let

$$\mathbf{X} = \begin{bmatrix} 2 & 5 & 6 \\ 3 & 1 & 4 \end{bmatrix} \quad \text{and} \quad \mathbf{Y} = \begin{bmatrix} 8 & 1 \\ 4 & 6 \\ 2 & 7 \end{bmatrix}$$

Then,

$$\mathbf{XY} = \begin{bmatrix} (2)(8) + (5)(4) + (6)(2) & (2)(1) + (5)(6) + (6)(7) \\ (3)(8) + (1)(4) + (4)(2) & (3)(1) + (1)(6) + (4)(7) \end{bmatrix}$$

$$= \begin{bmatrix} 48 & 74 \\ 36 & 37 \end{bmatrix}$$

It is important to recognize that the entry in the ith row and the jth column of the product matrix is the result of summing the products by pairs of the ith row of **X** and the jth column of **Y**. As a consequence, row i of **X** must have the same number of entries as column j of **Y**. That is, **X** must have as many columns as **Y** has rows. This is why we require that for **X** of dimensionality $a \times b$, the matrix **Y** must have dimensionality $b \times c$. Note that another consequence of the way in which we define matrix multiplication is that the product matrix has dimensionality $a \times c$. Thus, when **X** is a 2×3 matrix and **Y** is a 3×2 matrix, **XY** is 2×2. If **X** were a 2×3 matrix and **Y** were a 3×6 matrix, the product matrix **XY** would have two rows and six columns.

Still another consequence of this type of multiplication is that it is not generally commutative. There will be some cases in which we cannot even form the product **YX**. For example, if **X** is 2×3 and **Y** is 3×6, we cannot define **YX**, even though **XY** is defined. This is because the number of columns in **Y** does not equal the number of rows in **X**. Even when we can define **YX**, it will not usually equal **XY**. For example, if **X** is 2×3 and **Y** is 3×2, the product **XY** is 2×2. In contrast, the product **YX** will have three rows and three columns. Since matrices are equal only if they are of identical dimensionality and have identical entries in all cells, the two products cannot be equal. Finally, consider the product of two square matrices. Let

$$\mathbf{X} = \begin{bmatrix} 2 & 3 \\ 4 & 5 \end{bmatrix} \quad \text{and} \quad \mathbf{Y} = \begin{bmatrix} 6 & 7 \\ 8 & 9 \end{bmatrix}$$

The product matrix **XY** is

$$\mathbf{XY} = \begin{bmatrix} (2)(6)+(3)(8) & (2)(7)+(3)(9) \\ (4)(6)+(5)(8) & (4)(7)+(5)(9) \end{bmatrix} = \begin{bmatrix} 36 & 41 \\ 64 & 73 \end{bmatrix}$$

and **YX** is

$$\mathbf{YX} = \begin{bmatrix} (6)(2)+(7)(4) & (6)(3)+(7)(5) \\ (8)(2)+(9)(4) & (8)(3)+(9)(5) \end{bmatrix} = \begin{bmatrix} 40 & 53 \\ 52 & 69 \end{bmatrix}$$

Once again, commutativity fails to hold. Because multiplication is commutative only in special cases, reference to the product of **X** and **Y** can be confusing. Often, to ensure clarity, the product **XY** will be described as premultiplication of **Y** by **X** or as postmultiplication of **X** by **Y**.

The associative principle does apply to multiplication. The product of three matrices, for example, can be obtained as **W(XY)** or **(WX)Y**. Furthermore, matrix multiplication is distributive. Thus, **W(X+Y) = WX + WY**.

The use of matrix notation, together with the operations defined in this section, permits us to express sets of equations in very simple ways. Consider the development of the linear regression model for the bivariate case in Section 12.3.

We define the following matrices:

$$\textbf{(13.4)} \qquad \mathbf{Y} = \begin{bmatrix} Y_1 \\ \cdot \\ \cdot \\ \cdot \\ Y_i \\ \cdot \\ \cdot \\ \cdot \\ Y_n \end{bmatrix} \quad \mathbf{X} = \begin{bmatrix} 1 & X_1 \\ \cdot & \cdot \\ \cdot & \cdot \\ \cdot & \cdot \\ 1 & X_i \\ \cdot & \cdot \\ \cdot & \cdot \\ \cdot & \cdot \\ 1 & X_n \end{bmatrix} \quad \boldsymbol{\beta} = \begin{bmatrix} \beta_0 \\ \beta_1 \end{bmatrix} \quad \boldsymbol{\varepsilon} = \begin{bmatrix} \varepsilon_1 \\ \cdot \\ \cdot \\ \cdot \\ \varepsilon_i \\ \cdot \\ \cdot \\ \cdot \\ \varepsilon_n \end{bmatrix}$$

By the rules of matrix multiplication,

$$\mathbf{X}\boldsymbol{\beta} = \begin{bmatrix} (1)\beta_0 + \beta_1 X_1 \\ \cdot \\ \cdot \\ \cdot \\ (1)\beta_0 + \beta_1 X_i \\ \cdot \\ \cdot \\ \cdot \\ (1)\beta_0 + \beta_1 X_n \end{bmatrix}$$

Adding $\mathbf{X}\boldsymbol{\beta} + \boldsymbol{\varepsilon}$, we have

$$\begin{bmatrix} \beta_0 + \beta_1 X_1 + \varepsilon_1 \\ \cdot \\ \cdot \\ \cdot \\ \beta_0 + \beta_1 X_i + \varepsilon_i \\ \cdot \\ \cdot \\ \cdot \\ \beta_0 + \beta_1 X_n + \varepsilon_n \end{bmatrix}$$

Referring to Equation (12.12), we find that the ith row of the preceding matrix provides the linear regression equation for Y_i. Then, in matrix notation, the population regression model, Equation (12.12), is

$$\textbf{(13.5)} \qquad \mathbf{Y} = \mathbf{X}\boldsymbol{\beta} + \boldsymbol{\varepsilon}$$

In the more general, multivariate case we have several independent variables. We might be interested in some performance measure Y as a function of intelligence, age, educational level, and sex; note that a variable such as sex can be included merely by coding it as one or zero, depending on the sex of the subject. Column vectors are added to $\mathbf{X}$ so that if there are a independent variables, $\mathbf{X}$ is an $n \times (a+1)$ matrix; also $\boldsymbol{\beta}$ is an $(a+1) \times 1$ column vector whose entries are β_0 and β_j, the regression coefficients on the $X^{(j)}$. With these quantities defined, Equation (13.5) applies unchanged.

Matrix algebra also provides a useful approach to estimating parameters in regression analysis. Recall the normal equations, (12.5); these can be rewritten

(13.6)
$$n\hat{\beta}_0 + \left(\sum X\right)\hat{\beta}_1 = \sum Y$$

$$\left(\sum X\right)\hat{\beta}_0 + \left(\sum X^2\right)\hat{\beta}_1 = \sum XY$$

If we had more independent variables, we should have to add a normal equation for each in order to estimate each added regression coefficient. Therefore, it will prove useful to have a single equation in matrix notation that represents any number of normal equations. Using the bivariate case (Equations 13.6) as an example, we can develop such a matrix equation. First, from the definitions of Equations (13.4) and the rules of matrix multiplication,

$$\mathbf{X'X} = \begin{bmatrix} 1 & \cdots & 1 & \cdots & 1 \\ X_1 & \cdots & X_i & \cdots & X_n \end{bmatrix} \begin{bmatrix} 1 & X_1 \\ \cdot & \cdot \\ \cdot & \cdot \\ \cdot & \cdot \\ 1 & X_i \\ \cdot & \cdot \\ \cdot & \cdot \\ \cdot & \cdot \\ 1 & X_n \end{bmatrix}$$

$$= \begin{bmatrix} (1)(1) + \cdots + (1)(1) & (1)(X_1) + \cdots + (1)(X_n) \\ (X_1)(1) + \cdots + (X_n)(1) & X_1^2 + \cdots + X_n^2 \end{bmatrix}$$

$$= \begin{bmatrix} n & \sum X_i \\ \sum X_i & \sum X_i^2 \end{bmatrix}$$

Furthermore,

$$(\mathbf{X'X})\hat{\beta} = \begin{bmatrix} n & \sum X_i \\ \sum X_i & \sum X_i \end{bmatrix} \begin{bmatrix} \hat{\beta}_0 \\ \hat{\beta}_1 \end{bmatrix}$$

$$= \begin{bmatrix} n\hat{\beta}_0 + \left(\sum X\right)\hat{\beta}_1 \\ \left(\sum X\right)\hat{\beta}_0 & \left(\sum X^2\right)\hat{\beta}_1 \end{bmatrix}$$

Thus, $(\mathbf{X'X})\hat{\beta}$ represents the left-hand terms of Equations (13.6). Also,

$$\mathbf{X'Y} = \begin{bmatrix} 1 & \cdots & 1 & \cdots & 1 \\ X_1 & \cdots & X_i & \cdots & X_n \end{bmatrix} \begin{bmatrix} Y_1 \\ \cdot \\ \cdot \\ \cdot \\ Y_i \\ \cdot \\ \cdot \\ \cdot \\ Y_n \end{bmatrix}$$

$$= \begin{bmatrix} (1)(Y_1) + \cdots + (1)(Y_n) \\ (X_1)(Y_1) + \cdots + (X_n)(Y_n) \end{bmatrix} = \begin{bmatrix} \sum Y \\ \sum XY \end{bmatrix}$$

Thus, $\mathbf{X'Y}$ represents the right-hand terms of Equations (13.6). In general, any set of normal equations can be represented by

(13.7) $$(\mathbf{X'X})\hat{\beta} = \mathbf{X'Y}$$

If $\mathbf{X}$ is an $n \times (a+1)$ matrix with ones in the first column and $\hat{\beta}$ is a column vector of length $a+1$, Equation (13.7) provides a simple and general system for representing any number of normal equations. The introduction of one more operation, matrix inversion, lets us solve Equation (13.7) for $\hat{\beta}$. We then have least-squares numerical estimates of the population regression coefficients. Furthermore, we can obtain estimates of the variances and covariances of the regression coefficients as part of solving Equation (13.7). In Section 13.5, we shall develop these points. First, we need to have some sense of the meaning of matrix inversion.

13.4 INVERSE OF A MATRIX

Consider some square matrix $\mathbf{A}$. This matrix is said to have an *inverse* $\mathbf{A}^{-1}$ if

(13.8) $$\mathbf{A}\mathbf{A}^{-1} = \mathbf{A}^{-1}\mathbf{A} = \mathbf{I}$$

where $\mathbf{I}$ is the identity matrix (defined in Section 13.2). Only square matrices have inverses, and not all of these do. Matrices that can be inverted are often referred to as *nonsingular*, in contrast to *singular* matrices, which do not have an inverse. (We shall pursue this distinction in Section 13.7.) Note now that multiplication by $\mathbf{A}^{-1}$ is the matrix algebra analogue to division by some number a in ordinary arithmetic, since division by a is equivalent to multiplication by its inverse (reciprocal). Furthermore, $a \times (1/a) = 1$ is analogous to Equation (13.8), $\mathbf{I}$ being the matrix algebra analogue to the ordinary number 1.

One of the most important applications of the concept of the inverse is in solving sets of simultaneous equations. If a set of equations can be expressed as

$\mathbf{AX} = \mathbf{Y}$, then $\mathbf{A}^{-1}\mathbf{AX} = \mathbf{A}^{-1}\mathbf{Y}$. Since $\mathbf{A}^{-1}\mathbf{A} = \mathbf{I}$, this can be rewritten

(13.9) $\mathbf{X} = \mathbf{A}^{-1}\mathbf{Y}$

and we have a solution to the original set of equations. For a numerical example, consider

$$3X_1 + 4X_2 = 14$$
$$-X_1 + 5X_2 = 27$$

In the way usually taught in high school, we find the solution set to be $X_1 = -2$ and $X_2 = 5$. In matrix notation, we have $\mathbf{AX} = \mathbf{Y}$, where

$$\mathbf{A} = \begin{bmatrix} 3 & 4 \\ -1 & 5 \end{bmatrix}, \quad \mathbf{X} = \begin{bmatrix} X_1 \\ X_2 \end{bmatrix}, \quad \text{and} \quad \mathbf{Y} = \begin{bmatrix} 14 \\ 27 \end{bmatrix}$$

By methods to be developed shortly, we find that

$$\mathbf{A}^{-1} = \left(\frac{1}{19}\right)\begin{bmatrix} 5 & -4 \\ 1 & 3 \end{bmatrix}$$

To check this result, we write

$$\mathbf{A}^{-1}\mathbf{A} = \left(\frac{1}{19}\right)\begin{bmatrix} (5)(3)+(-4)(-1) & (5)(4)+(-4)(5) \\ (1)(3)+(3)(-1) & (1)(4)+(3)(5) \end{bmatrix} = \begin{bmatrix} 1 & 0 \\ 0 & 1 \end{bmatrix}$$

Premultiplying $\mathbf{Y}$ by $\mathbf{A}^{-1}$, as required by Equation (13.9), we have

$$\mathbf{X} = \left(\frac{1}{19}\right)\begin{bmatrix} 5 & -4 \\ 1 & 3 \end{bmatrix}\begin{bmatrix} 14 \\ 27 \end{bmatrix}$$

$$= \left(\frac{1}{19}\right)\begin{bmatrix} (5)(14)+(-4)(27) \\ (1)(14)+(3)(27) \end{bmatrix}$$

$$= \begin{bmatrix} -2 \\ 5 \end{bmatrix}$$

13.5   SOME STATISTICAL APPLICATIONS

13.5.1   SOLVING FOR $\hat{\boldsymbol{\beta}}$.   We now have the tool we need to solve Equation (13.7) for estimates of regression coefficients. The process is exactly like that just shown. We first note that $\mathbf{X'X}$ is a square matrix, $(a+1) \times (a+1)$; furthermore, we limit discussion to cases in which $\mathbf{X'X}$ is nonsingular. Then, premultiplying both sides of Equation (13.7) by $(\mathbf{X'X})^{-1}$, we get

$$(\mathbf{X'X})^{-1}(\mathbf{X'X})\hat{\boldsymbol{\beta}} = (\mathbf{X'X})^{-1}(\mathbf{X'Y})$$

(13.10) $\mathbf{I}\hat{\boldsymbol{\beta}} = (\mathbf{X'X})^{-1}(\mathbf{X'Y})$

$$\hat{\boldsymbol{\beta}} = (\mathbf{X'X})^{-1}(\mathbf{X'Y})$$

As we have indicated, Equation (13.10) holds for any number of independent variables; a big advantage of matrix notation is the generality and simplicity of the representation and consequently of most derivations and proofs. Nevertheless, it may be helpful to see the application of Equation (13.10) to a special case with which we have some familiarity, the case in which we have only one independent variable. Then, the definitions of Equation (13.4) apply. In Section 13.3, we showed that

(13.11)
$$\mathbf{X'X} = \begin{bmatrix} n & \sum X \\ \sum X & \sum X^2 \end{bmatrix}$$

and the inverse can be shown to be

(13.12)
$$(\mathbf{X'X})^{-1} = \frac{1}{n\sum X^2 - (\sum X)^2} \begin{bmatrix} \sum X^2 & -\sum X \\ -\sum X & n \end{bmatrix}$$

You should verify this last result by multiplying the above two matrices to obtain the identity matrix.

In Section 13.3, we also showed that

(13.13)
$$\mathbf{X'Y} = \begin{bmatrix} \sum Y \\ \sum XY \end{bmatrix}$$

Applying Equation (13.10) (and using the notation of Chapter 12, b_j instead of $\hat{\beta}_j$) gives

(13.14)
$$\hat{\boldsymbol{\beta}} = (\mathbf{X'X})^{-1}(\mathbf{X'Y}) = \begin{bmatrix} b_0 \\ b_1 \end{bmatrix} = \begin{bmatrix} \dfrac{(\sum X^2)(\sum Y) - (\sum X)(\sum XY)}{n\sum X^2 - (\sum X)^2} \\ \dfrac{n\sum XY - (\sum X)(\sum Y)}{n\sum X^2 - (\sum X)^2} \end{bmatrix}$$

The two entries in the right-hand vector are algebraically equivalent to the formulas for b_0 and b_1 given by Equations (12.6) and (12.7).

13.5.2 UNBIASED ESTIMATES.   We have earlier showed that b_0 and b_1 were unbiased estimates of the population parameters (Chapter 12), carrying out two separate proofs, one for each coefficient. The whole process of establishing lack of bias in an estimator is considerably simplified by using matrix notation. We can show that for any number of independent variables, Equation (13.10) provides a set of unbiased estimates of the $a+1$ population regression coefficients: β_0, $\beta_1, \ldots, \beta_a$. From Equation (13.10),

$$E(\hat{\boldsymbol{\beta}}) = E[(\mathbf{X'X})^{-1}(\mathbf{X'Y})] = [(\mathbf{X'X})^{-1}\mathbf{X'}]E(\mathbf{Y})$$

because $\mathbf{X}$ is fixed over replications of the study. From Equation (13.5), we can replace $\mathbf{Y}$ by $\mathbf{X\beta} + \boldsymbol{\varepsilon}$; then,

$$E(\hat{\boldsymbol{\beta}}) = [(\mathbf{X'X})^{-1}\mathbf{X'}]E(\mathbf{X\beta} + \boldsymbol{\varepsilon}) = (\mathbf{X'X})^{-1}(\mathbf{X'X})\boldsymbol{\beta}$$

because $E(\mathbf{X\beta}) = \mathbf{X\beta}$ and $E(\mathbf{\varepsilon}) = \mathbf{0}$. Since a matrix times its inverse equals $\mathbf{I}$, then $(\mathbf{X'X})^{-1}(\mathbf{X'X}) = \mathbf{I}$, and therefore,

$$E(\hat{\mathbf{\beta}}) = \mathbf{I\beta} = \mathbf{\beta}$$

and the proof is complete.

13.5.3　VARIANCE OF PARAMETER ESTIMATES.　　We have also earlier separately derived estimates of the variances of the sampling distributions of b_0 and b_1 (Chapter 12). In what follows, we shall show that the variances and covariances of any set of estimators of $a+1$ population regression coefficients is a simple function of the inverse of $\mathbf{X'X}$. First, consider

$$(\mathbf{13.15}) \quad E(\hat{\mathbf{\beta}} - \mathbf{\beta})(\hat{\mathbf{\beta}} - \mathbf{\beta})' = E \begin{bmatrix} \hat{\beta}_0 - \beta_0 \\ \hat{\beta}_1 - \beta_1 \\ \vdots \\ \hat{\beta}_a - \beta_a \end{bmatrix} [\hat{\beta}_0 - \beta_0 \quad \hat{\beta}_1 - \beta_1 \cdots \hat{\beta}_a - \beta_a]$$

$$= \begin{bmatrix} E(\hat{\beta}_0 - \beta_0)^2 & E(\hat{\beta}_0 - \beta_0)(\hat{\beta}_1 - \beta_1) \cdots \\ E(\hat{\beta}_1 - \beta_1)(\hat{\beta}_0 - \beta_0) & E(\hat{\beta}_1 - \beta_1)^2 \qquad \cdots \\ E(\hat{\beta}_a - \beta_a)(\hat{\beta}_0 - \beta_0) & E(\hat{\beta}_a - \beta_a)(\hat{\beta}_1 - \beta_1) \cdots \end{bmatrix}$$

The entries on the diagonal are variances of the $\hat{\beta}_j$, the estimates of the population regression coefficients. The off-diagonal entries are covariances of the estimates. We have already seen how the variances of the sampling distributions of $\hat{\beta}_0$ and $\hat{\beta}_1$ (b_0 and b_1) are used in formulas for significance tests and confidence intervals; in certain problems, the covariances are also of interest because their magnitudes tell something about the dependencies among tests of hypotheses for different parameters.

　　We shall refer to the above variance-covariance matrix as cov $(\hat{\mathbf{\beta}})$. Then

$$(\mathbf{13.16}) \qquad \text{cov}(\hat{\mathbf{\beta}}) = E(\hat{\mathbf{\beta}} - \mathbf{\beta})(\hat{\mathbf{\beta}} - \mathbf{\beta})'$$

$$= E[(\mathbf{X'X})^{-1}(\mathbf{X'Y}) - \mathbf{\beta}][(\mathbf{X'X})^{-1}(\mathbf{X'Y}) - \mathbf{\beta}]'$$

with Equation (13.10) used for substituting for $\hat{\mathbf{\beta}}$. Equation (13.5) permits substitution for $\mathbf{Y}$:

$$(\mathbf{X'X})^{-1}(\mathbf{X'Y}) = (\mathbf{X'X})^{-1}\mathbf{X'}(\mathbf{X\beta} + \mathbf{\varepsilon})$$

$$(\mathbf{13.17}) \qquad\qquad = (\mathbf{X'X})^{-1}(\mathbf{X'X})\mathbf{\beta} + (\mathbf{X'X})^{-1}\mathbf{X'\varepsilon}$$

$$= \mathbf{\beta} + (\mathbf{X'X})^{-1}\mathbf{X'\varepsilon}$$

Substituting Equation (13.17) into (13.16) gives

$$(\mathbf{13.18}) \qquad\qquad \text{cov}(\hat{\mathbf{\beta}}) = E[(\mathbf{X'X})^{-1}\mathbf{X'\varepsilon}][(\mathbf{X'X})^{-1}\mathbf{X'\varepsilon}]'$$

At this point it is helpful to recognize that $(\mathbf{AB})' = \mathbf{B'A'}$ for any $\mathbf{A}$ and $\mathbf{B}$ (see

Exercise 13.2). Therefore, we can rewrite:

$$[(\mathbf{X'X})^{-1}\mathbf{X'}\boldsymbol{\varepsilon}]' = \boldsymbol{\varepsilon}'\mathbf{X}[(\mathbf{X'X})^{-1}]'$$

However, $\mathbf{X'X}$ is symmetric and therefore its inverse must be also. This means that $[(\mathbf{X'X})^{-1}]' = (\mathbf{X'X})^{-1}$. Equation (13.18) can now be rewritten as

(13.19) $$\text{cov}(\hat{\boldsymbol{\beta}}) = E[(\mathbf{X'X})^{-1}\mathbf{X'}(\boldsymbol{\varepsilon}\boldsymbol{\varepsilon}')\mathbf{X}(\mathbf{X'X})^{-1}]$$

Rearranging terms, we have

(13.20) $$\text{cov}(\hat{\boldsymbol{\beta}}) = (\mathbf{X'X})^{-1}(\mathbf{X'X})(\mathbf{X'X})^{-1}E(\boldsymbol{\varepsilon}\boldsymbol{\varepsilon}')$$
$$= (\mathbf{X'X})^{-1}E(\boldsymbol{\varepsilon}\boldsymbol{\varepsilon}')$$

Assuming independence of errors, and noting that $E(\varepsilon_1^2) = E(\varepsilon_2^2 = \cdots = E(\varepsilon_n^2) = \sigma_e^2$, yields

$$E(\boldsymbol{\varepsilon}\boldsymbol{\varepsilon}') = \sigma_e^2\mathbf{I}$$

Thus, the final expression for the matrix of Equation (13.15) is

(13.21) $$\text{cov}(\hat{\boldsymbol{\beta}}) = \sigma_e^2(\mathbf{X'X})^{-1}$$

Let us again consider the simple case with which we are most familiar, assuming only one independent variable and matrices defined by Equation (13.4). Then, Equation (13.12) defines $(\mathbf{X'X})^{-1}$. Substituting into Equation (13.21), we get

$$\text{cov}(\hat{\boldsymbol{\beta}}) = \left[\frac{\sigma_e^2}{n\sum X^2 - (\sum X)^2}\right]\begin{bmatrix} \sum X^2 & -\sum X \\ -\sum X & n \end{bmatrix}$$

We have stated that the diagonal elements are variances of sample regression coefficients. Therefore, we claim that

(13.22) $$\hat{\sigma}_{b_0}^2 = \hat{\sigma}_e^2\left(\frac{\sum X^2}{n\sum X^2 - (\sum X)^2}\right)$$

It is of some interest to compare this with Equation (12.23), the expression derived previously for $\hat{\sigma}_{b_0}^2$. There we had

$$\hat{\sigma}_{b_0}^2 = \hat{\sigma}_e^2\left(\frac{\bar{X}^2}{\sum x^2} + \frac{1}{n}\right) = \hat{\sigma}_e^2\left(\frac{n\bar{X}^2 + \sum x^2}{n\sum x^2}\right)$$

Note that $n\sum x^2 = n\sum X^2 - (\sum X)^2$. Then the expression immediately above is identical to Equation (13.22).

What of $\sigma_{b_1}^2$, the second diagonal entry in the variance-covariance matrix? We have

$$\hat{\sigma}_{b_1}^2 = \hat{\sigma}_e^2\left(\frac{n}{n\sum X^2 - (\sum X)^2}\right)$$

Since the denominator equals $n \sum x^2$, this expression is algebraically equivalent to the result derived earlier as Equation (12.22).

The point is not that the same result can be derived in several different ways. The point is that the matrix approach provides a general derivation for the entire set of variances and covariances. It accomplishes this with no more effort or space than what is required for deriving the variance of any one regression coefficient by the methods used in Chapter 12. We have a simple and elegant, but powerful tool for representing our model and the data, and for deriving the consequences of the model. We shall better realize the full power of the approach in considering multiple regression. There, we shall face larger systems of normal equations and larger variance-covariance matrices with the same matrix equations developed in this chapter.

13.6   A COMPUTATIONAL METHOD

Let us solve the equation $\mathbf{X'X\beta} = \mathbf{X'Y}$ for numerical estimates of β. Although we can bypass the computation of $(\mathbf{X'X})^{-1}$ in this process, we shall not. As we showed in the preceding section, the elements of this matrix are components of variance and covariance estimates. If values of only $\hat{\beta}$ are required, the additional calculations that yield $(\mathbf{X'X})^{-1}$ can be omitted from the general computation.

In working with real data, we shall more often than not have the analyses carried out at the nearest computing center. Despite this, it seems useful to provide a way of calculating results by hand (or at least by a hand-held calculator). This permits us to understand some of the points that will arise by working through some sample data. Although the method we shall give is not the most efficient possible, it is relatively easy to learn and to relate to the computing algorithms often taught in elementary algebra courses. It takes advantages of the fact that $\mathbf{X'X}$ is a symmetric matrix.

The *Doolittle method* is a technique for solving the set of normal equations for estimates of the population regression coefficients. We illustrate it by analyzing a simple numerical example. Suppose that Y is performance on a memory task and we want to analyze it as a function of age $X^{(1)}$ and intelligence test score $X^{(2)}$. (We shall develop the multiple linear regression model in the next chapter.) We define these quantities:

$$\mathbf{Y} = \begin{bmatrix} Y_1 \\ \cdot \\ \cdot \\ \cdot \\ Y_n \end{bmatrix} \qquad \mathbf{X} = \begin{bmatrix} 1 & X_1^{(1)} & X_1^{(2)} \\ 1 & X_2^{(1)} & X_2^{(2)} \\ \cdot & \cdot & \cdot \\ \cdot & \cdot & \cdot \\ \cdot & \cdot & \cdot \\ 1 & X_n^{(1)} & X_n^{(2)} \end{bmatrix} \qquad \mathbf{\beta} = \begin{bmatrix} \beta_0 \\ \beta_1 \\ \beta_2 \end{bmatrix}$$

We assume that these quantities are related in the manner described by Equation (13.6). Then, numerical entries for $\hat{\beta}$ are obtained by solving Equation (13.7).

This requires numerical values for $\mathbf{X'X}$ and $\mathbf{X'Y}$. Suppose

$$\mathbf{X'X} = \begin{bmatrix} 2 & 2 & 3 \\ 2 & 1 & 4 \\ 3 & 4 & 2 \end{bmatrix} \qquad \mathbf{X'Y} = \begin{bmatrix} 10 \\ 13 \\ 9 \end{bmatrix}$$

Then we should write Equation (13.7) as

(13.23)
$$\begin{aligned} &(1) \quad 2\hat{\beta}_0 + 2\hat{\beta}_1 + 3\hat{\beta}_2 = 10 \\ &(2) \quad 2\hat{\beta}_0 + \ \hat{\beta}_1 + 4\hat{\beta}_2 = 13 \\ &(3) \quad 3\hat{\beta}_0 + 4\hat{\beta}_1 + 2\hat{\beta}_2 = \ 9 \end{aligned}$$

Alternatively,

(13.23′)
$$\begin{bmatrix} 2 & 2 & 3 \\ 2 & 1 & 4 \\ 3 & 4 & 2 \end{bmatrix} \begin{bmatrix} \beta_0 \\ \beta_1 \\ \beta_2 \end{bmatrix} = \begin{bmatrix} 10 \\ 13 \\ 9 \end{bmatrix}$$

There are many possible sequences of operations on these equations, all of which lead to the same solution. We shall consider one such sequence, which generalizes to problems involving more independent variables.

STAGE 1.   We first operate on the equations in such a way that the coefficients of $\hat{\beta}_0$ will be 1, 0, and 0 as we go from the first to the third line of Equation (13.23). Therefore,
  a. Multiply line (1) by one-half.
  b. Replace line (2) by line (2) minus line (1).
  c. Replace line (3) by line (3) minus $\frac{3}{2}$ times line (1).
The result of these steps is

(13.24)
$$\begin{aligned} &(1) \quad \hat{\beta}_0 + \hat{\beta}_1 + \tfrac{3}{2}\hat{\beta}_2 = 5 \\ &(2) \qquad\quad -\hat{\beta}_1 + \hat{\beta}_2 = 3 \\ &(3) \qquad\qquad \hat{\beta}_1 - \tfrac{5}{2}\hat{\beta}_2 = -6 \end{aligned}$$

In matrix notation, we have

(13.24′)
$$\begin{bmatrix} 1 & 1 & \tfrac{3}{2} \\ 0 & -1 & 1 \\ 0 & 1 & -\tfrac{5}{2} \end{bmatrix} \begin{bmatrix} \hat{\beta}_0 \\ \hat{\beta}_1 \\ \hat{\beta}_2 \end{bmatrix} = \begin{bmatrix} 5 \\ 3 \\ -6 \end{bmatrix}$$

Steps a through c are equivalent to premultiplying $\mathbf{X'X}$ and $\mathbf{X'Y}$ by a matrix that we shall label $\mathbf{O}_1$ ($\mathbf{O}$ for "operator"), where

$$\mathbf{O}_1 = \begin{bmatrix} \tfrac{1}{2} & 0 & 0 \\ -1 & 1 & 0 \\ -\tfrac{3}{2} & 0 & 1 \end{bmatrix}$$

You should verify that $\mathbf{O}_1(\mathbf{X'X})$ equals

$$\begin{bmatrix} 1 & 1 & \frac{3}{2} \\ 0 & -1 & 1 \\ 0 & 1 & -\frac{5}{2} \end{bmatrix}$$

and $\mathbf{O}_1(\mathbf{X'Y})$ equals

$$\begin{bmatrix} 5 \\ 3 \\ -6 \end{bmatrix}$$

STAGE 2.   The next set of operations will convert the coefficients of $\hat{\beta}_1$ to one and zero in lines (2) and (3) of Equation (13.24).

    a. Multiply line (2) by -1.

    b. Replace line (3) by the sum of lines (2) and (3). This yields

(13.25)
$$\begin{aligned} (1)\quad & \hat{\beta}_0 + \hat{\beta}_1 + \tfrac{3}{2}\hat{\beta}_2 = 5 \\ (2)\quad & \hat{\beta}_1 - \hat{\beta}_2 = -3 \\ (3)\quad & -\tfrac{3}{2}\hat{\beta}_2 = -3 \end{aligned}$$

STAGE 3.   We convert the coefficient of $\hat{\beta}_2$ to one in line (3) of Equation (13.25) by multiplying the line by $-\frac{2}{3}$. The result is

(13.26)
$$\begin{aligned} (1)\quad & \hat{\beta}_0 + \hat{\beta}_1 + \tfrac{3}{2}\hat{\beta}_2 = 5 \\ (2)\quad & \hat{\beta}_1 - \hat{\beta}_2 = -3 \\ (3)\quad & \hat{\beta}_2 = 2 \end{aligned}$$

At this point, we have concluded what is ordinarily called the "forward solution." It is now simple to solve for $\hat{\beta}_1$ in line (2) of Equation (13.26) and then for $\hat{\beta}_0$ in line (1). The result is

$$\hat{\boldsymbol{\beta}} = \begin{bmatrix} 3 \\ -1 \\ 2 \end{bmatrix}$$

    Another way of obtaining $\hat{\boldsymbol{\beta}}$ yields $(\mathbf{X'X})^{-1}$ as part of the process. We continue through the "backward solution" to accomplish this.

STAGE 4.   We move back through the lines in Equation (13.26). Our purpose is to achieve a zero multiplier of $\hat{\beta}_2$ in lines (2) and (1) of the equation:

    a. Replace line (2) of Equation (13.26) by the sum of lines (2) and (3).

    b. Replace line (1) of Equation (13.26) by line (1) minus $\frac{3}{2}$ times line (3). We now have

(13.27)
$$\begin{aligned} (1)\quad & \hat{\beta}_0 + \hat{\beta}_1 = 2 \\ (2)\quad & \hat{\beta}_1 = -1 \\ (3)\quad & \hat{\beta}_2 = 2 \end{aligned}$$

STAGE 5.   We wish to eliminate $\hat{\beta}_1$ in line (1) of Equation (13.27). Therefore, replace line (1) by line (1) minus line (2). We have the solution that was arrived at following Stage 3.

Recall that when we presented Stage 1 we noted that the transformations in Equation (13.24) were equivalent to premultiplying both sides of the matrix representation by a matrix designated $\mathbf{O}_1$. We could demonstrate that each stage corresponds to premultiplication by some operator; in essence, the stages are equivalent to premultiplying $(\mathbf{X'X})\hat{\boldsymbol{\beta}}$ by the product of a series of "operator" matrices, $\mathbf{O}_5\,\mathbf{O}_4\,\mathbf{O}_3\,\mathbf{O}_2\,\mathbf{O}_1$. Call this product $\mathbf{\Pi}$. We know that

$$\mathbf{\Pi}(\mathbf{X'X})\hat{\boldsymbol{\beta}} = \hat{\boldsymbol{\beta}}$$

because our stages have isolated $\hat{\boldsymbol{\beta}}$. Then, $\mathbf{\Pi}(\mathbf{X'X})$ must equal the identity matrix $\mathbf{I}$. Therefore, $\mathbf{\Pi}$ must equal $(\mathbf{X'X})^{-1}$.

This line of reasoning implies that we should find the values of each of the $\mathbf{O}_k$, as we did for $\mathbf{O}_1$. Then, the inverse we seek would be their product. Table 13–1 presents a simple approach and summarizes our procedure. Columns C_1 through C_3 represent the coefficients of the $\hat{\boldsymbol{\beta}}$s at each stage. In Stage 5, we have the identity matrix. This is because the series of operations finally corresponds to multiplication of $\mathbf{X'X}$ (the matrix formed by the first three rows and first three

Table 13–1   An example of the application of the Doolittle method

OPERATION		CODE	C_1	C_2	C_3	C_0	E_1	E_2	E_3	SUM
		R_1	2	2	3	10	1	0	0	18
		R_2	2	1	4	13	0	1	0	21
		R_3	3	4	2	9	0	0	1	19
	$\frac{1}{2}R_1$	R_4	1	1	$\frac{3}{2}$	5	$\frac{1}{2}$	0	0	9
Stage 1	$R_2 - R_1$	R_5	0	-1	1	3	-1	1	0	3
	$R_3 - \frac{3}{2}R_1$	R_6	0	1	$-\frac{5}{2}$	-6	$-\frac{3}{2}$	0	1	-8
	R_4	R_7	1	1	$\frac{3}{2}$	5	$\frac{1}{2}$	0	0	9
Stage 2	$(-1)R_5$	R_8	0	1	-1	-3	1	-1	0	-3
	$R_6 + R_5$	R_9	0	0	$-\frac{3}{2}$	-3	$-\frac{5}{2}$	1	1	-5
	R_7	R_{10}	1	1	$\frac{3}{2}$	5	$\frac{1}{2}$	0	0	9
Stage 3	R_8	R_{11}	0	1	-1	-3	1	-1	0	-3
	$-\frac{2}{3}R_9$	R_{12}	0	0	1	2	$\frac{5}{3}$	$-\frac{2}{3}$	$-\frac{2}{3}$	$\frac{10}{3}$
	$R_{10} - \frac{3}{2}R_{12}$	R_{13}	1	1	0	2	-2	1	1	4
Stage 4	$R_{10} + R_{11}$	R_{14}	0	1	0	-1	$\frac{8}{3}$	$-\frac{5}{3}$	$-\frac{2}{3}$	$\frac{1}{3}$
	R_{12}	R_{15}	0	0	1	2	$\frac{5}{3}$	$-\frac{2}{3}$	$-\frac{2}{3}$	$\frac{10}{3}$
	$R_{13} - R_{14}$	R_{16}	1	0	0	3	$-\frac{14}{3}$	$\frac{8}{3}$	$\frac{5}{3}$	$\frac{11}{3}$
Stage 5	R_{14}	R_{17}	0	1	0	-1	$\frac{8}{3}$	$-\frac{5}{3}$	$-\frac{2}{3}$	$\frac{1}{3}$
	R_{15}	R_{18}	0	0	1	2	$\frac{5}{3}$	$-\frac{2}{3}$	$-\frac{2}{3}$	$\frac{10}{3}$

columns) by its inverse $(\mathbf{X'X})^{-1}$. The quantity C_0 initially contains the vector $\mathbf{X'Y}$. By Stage 5, we have $(\mathbf{X'X})^{-1}(\mathbf{X'Y})$, the solution set.

Columns E_1 through E_3 initially make up an identity matrix. Because the operations correspond to premultiplication by $(\mathbf{X'X})^{-1}$, that matrix is to be found in R_{16} through R_{18} of these columns. This can be verified by

$$(\mathbf{13.28}) \qquad \underbrace{\begin{bmatrix} -\frac{14}{3} & \frac{8}{3} & \frac{5}{3} \\ \frac{8}{3} & -\frac{5}{3} & -\frac{2}{3} \\ \frac{5}{3} & -\frac{2}{3} & -\frac{2}{3} \end{bmatrix}}_{(\mathbf{X'X})^{-1}} \underbrace{\begin{bmatrix} 2 & 2 & 3 \\ 2 & 1 & 4 \\ 3 & 4 & 2 \end{bmatrix}}_{(\mathbf{X'X})} = \underbrace{\begin{bmatrix} 1 & 0 & 0 \\ 0 & 1 & 0 \\ 0 & 0 & 1 \end{bmatrix}}_{\mathbf{I}}$$

If we had defined $\mathbf{O}_k$ at each stage and found the product $\mathbf{\Pi}$, it would have equalled $(\mathbf{X'X})^{-1}$.

The last column provides a check on our work. The "sum" entry, for example, in R_9 is $(\frac{1}{2})(18)$, as demanded by the operation for that row. It is also the sum of all the entries to its left. If these computations disagree, an error has been made. Furthermore, with "real" data, the last column checks rounding-off errors. It is usually wise to carry out accuracy to six decimal places since errors in rounding off often accumulate over the stages.

The approach of Table 13–1 generalizes to large sets of equations. In the general case, there will be k equations with k parameters to be estimated. The goal of the forward solution (Stages 1 through 3 in our example) is to convert the diagonal entries of $\mathbf{X'X}$ (C_1 through C_k) into ones and the entries below the diagonal into zeros. The desired conversion is accomplished in k successive stages, in a direction left to right through the columns. The goal of the backward solution is to move from the upper triangular matrix to an identity matrix, that is, to convert the entries above the diagonal in the first k columns to zeros. This is accomplished in $k-1$ additional stages, as one goes from right to left. When the identity matrix has been formed in C_1 through C_k, the last k rows of C_0 (originally $\mathbf{X'Y}$) will contain $\hat{\beta}$ and the last k rows of the E columns will contain the inverse $(\mathbf{X'X})^{-1}$.

We have chosen to present the Doolittle method because both the mechanics and the underlying rationale are relatively easy to understand. Abbreviated versions that provide the same information with less labor are described in many sources, some of which are listed at the end of this chapter.

13.7 RANK OF A MATRIX

Consider the following set of equations:

$$\begin{array}{ll} (1) & 4\hat{\beta}_0 + 4\hat{\beta}_1 + 6\hat{\beta}_2 = 20 \\ (2) & 4\hat{\beta}_0 + 6\hat{\beta}_1 + 8\hat{\beta}_2 = 24 \\ (3) & 6\hat{\beta}_0 + 8\hat{\beta}_1 + 11\hat{\beta}_2 = 34 \end{array}$$

Let us carry out the first few stages of the forward solution of the Doolittle method. Since the matrix of coefficients is symmetric, this is possible. In the first stage, (a) divide line (1) by four; (b) replace line (2) by line (2) minus line (1); and (c) replace line (3) by line (3) minus $\frac{3}{2}$ times line (1). The result is

$$\hat{\beta}_0 + \hat{\beta}_1 + \tfrac{3}{2}\hat{\beta}_2 = 5$$
$$2\hat{\beta}_1 + 2\hat{\beta}_2 = 4$$
$$2\hat{\beta}_1 + 2\hat{\beta}_2 = 4$$

In the second stage, we (a) multiply line (2) by $\frac{1}{2}$, and (b) replace line (3) by line (3) minus (2). The result is somewhat strange:

(13.29)

$$\begin{aligned}
&(1) \quad \hat{\beta}_0 + \hat{\beta}_1 + \tfrac{3}{2}\hat{\beta}_2 = 5 \\
&(2) \quad \hat{\beta}_1 + \hat{\beta}_2 = 2 \\
&(3) \quad 0 + 0 + 0 = 0
\end{aligned}$$

The third line has been eliminated. The reason for this is that line (3) of Equation (13.28) was not linearly independent of the other two lines. In fact,

$$\text{line } (3) = [\tfrac{1}{2} \times \text{line } (1)] + \text{line } (2)$$

When an equation in the original set can be expressed as a linear function of one or more of the remaining equations, the set can be reduced to a set of linearly independent equations, as in the preceding example.[1] The number of linearly independent rows or columns in a matrix is called the *rank* of the matrix. The matrix $\mathbf{X'X}$, which corresponds to Equation (13.28), has rank 2 because only the first two rows are independent. (It should be noted that if a matrix has r linearly independent rows, there will be exactly r linearly independent columns.) In the example developed in Section 13.6, the matrix had a rank of three.

The concept of rank is important because if a matrix has a rank less than the number of rows, there is no unique solution for the set of parameter estimates. Basically, we have more unknowns than we have independent equations. In such cases, the matrix is often said to be of *deficient rank*. Deficiency in rank results in the inability to calculate an inverse; the matrix is singular. The implications may be clearer if we consider Equation (13.29) more closely.

Let $\hat{\beta}_2 = 1$; then $\hat{\beta}_1$ must also equal 1. From line (1), $\hat{\beta}_0 = \frac{5}{2}$. Now let $\hat{\beta}_2 = 0$; in this case $\hat{\beta}_1 = 2$ and $\hat{\beta}_0 = 3$. In fact, the same two equations will yield solutions no matter what our initial choice of $\hat{\beta}_2$ is; there are an infinite number of possible solution sets. You should convince yourself of this by choosing any initial value

---

[1] A more precise statement of linear independence is the following. In a set of vectors

$$\mathbf{V}_1, \mathbf{V}_2, \ldots, \mathbf{V}_i, \ldots, \mathbf{V}_k$$

the members are linearly independent if and only if there exists no set of weights c_i (at least one must be nonzero) such that $\sum c_i \mathbf{V}_i = 0$. In Equation (13.28),

$$\mathbf{V}_1 = [4\,4\,6\,20], \qquad \mathbf{V}_2 = [4\,6\,8\,24], \qquad \mathbf{V}_3 = [6\,8\,11\,34]$$

Since $(\frac{1}{2})\mathbf{V}_1 + (1)\mathbf{V}_2 + (-1)\mathbf{V}_3 = 0$, there is a dependency.

for one of the parameter estimates and then using Equation (13.29) to solve for the other estimates.

Matrices of deficient rank are of interest because they do occur in regression approaches to analysis of variance. Consider as an example two treatment groups with n_1 and n_2 subjects; the total number of subjects is n. Define a matrix

$$\mathbf{X} = \begin{bmatrix} 1 & 1 & 0 \\ 1 & 1 & 0 \\ \cdot & \cdot & \cdot \\ \cdot & \cdot & \cdot \\ \cdot & \cdot & \cdot \\ 1 & 0 & 1 \\ 1 & 0 & 1 \end{bmatrix}$$

This matrix has $n.$ ones in the first column, n_1 ones and n_2 zeros in the second column, and n_1 zeros and n_2 ones in the third column. We also let

$$\boldsymbol{\beta} = \begin{bmatrix} \mu \\ \alpha_1 \\ \alpha_2 \end{bmatrix}$$

Then, $\mathbf{Y} = \mathbf{X}\boldsymbol{\beta} + \boldsymbol{\varepsilon}$ corresponds to the structural equation of Chapter 4:

$$Y_{ij} = \mu + \alpha_j + \varepsilon_{ij}$$

If we want to estimate the analysis of variance parameters, the normal equations are $\mathbf{X'X}\hat{\boldsymbol{\beta}} = \mathbf{X'Y}$, or

$$n.\hat{\mu} + n_1\hat{\alpha}_1 + n_2\hat{\alpha}_2 = T_{..}$$
$$n_1\hat{\mu} + n_1\hat{\alpha}_1 + 0 = T_{1.}$$
$$n_2\hat{\mu} + 0 + n_2\hat{\alpha}_2 = T_{2.}$$

Since $n_1 + n_2 = n.$ and $T_{.1} + T_{.2} = T_{..}$, the first equation is the sum of the second two; $\mathbf{X'X}$ is of deficient rank. This does not mean that we shall not be able to estimate parameters when we view analysis of variance within a regression framework. (We shall provide solutions to this particular problem in Chapter 15.) It does mean, however, that matrices of deficient rank can crop up in statistical work; and since we should like to be able to estimate our population parameters, we should be aware of this possibility.

EXERCISES

13.1 Let

$$\mathbf{X} = \begin{bmatrix} 4 & 3 & 6 \\ 2 & 1 & 8 \end{bmatrix} \qquad \mathbf{Y} = \begin{bmatrix} 5 & 1 \\ 7 & 9 \end{bmatrix}$$

State whether each of the following operations can be executed; if so, calculate the result. (a) $\mathbf{X}+\mathbf{Y}$; (b) $\mathbf{XY}$; (c) $\mathbf{X'Y}$; (d) $\mathbf{YX}$; (e) $\mathbf{Y'X}$.

13.2 Let

$$\mathbf{X} = \begin{bmatrix} X_{11} & X_{12} \\ X_{21} & X_{22} \\ X_{31} & X_{32} \end{bmatrix} \qquad \mathbf{Y} = \begin{bmatrix} Y_{11} & Y_{12} & Y_{13} & Y_{14} \\ Y_{21} & Y_{22} & Y_{23} & Y_{24} \end{bmatrix}$$

Verify that $(\mathbf{XY})' = \mathbf{Y}'\mathbf{X}'$.

13.3 Let $\mathbf{x}$ and $\mathbf{y}$ be $n \times 1$ column vectors of deviations about the means. Express the variances and covariance in terms of $\mathbf{x}$ and $\mathbf{y}$.

13.4 Use the Doolittle method to solve the following system of equations:

$$\begin{bmatrix} 3 & 1 & 6 & 1 \\ 1 & 4 & 2 & 2 \\ 6 & 2 & 5 & 4 \\ 1 & 2 & 4 & 3 \end{bmatrix} \begin{bmatrix} X_1 \\ X_2 \\ X_3 \\ X_4 \end{bmatrix} = \begin{bmatrix} 30 \\ 3 \\ 28 \\ 11 \end{bmatrix}$$

13.5 Verify that $\mathbf{X}$ has a rank of two where

$$\mathbf{X} = \begin{bmatrix} 3 & 1 & 5 & 7 \\ 2 & 4 & 2 & 6 \\ 7 & 4 & 11 & 17 \end{bmatrix}$$

The approach involves operating on the rows (as in the Doolittle method) and demonstrating that one can be eliminated.

SUPPLEMENTARY READING

Several recent textbooks on multivariate statistics have included chapters, or appendices, that give fuller treatments than the current chapter. Among these are:

Bock, R. D. *Multivariate Statistical Methods in Behavioral* Research. New York: McGraw-Hill, 1975.

Harris, R. J. *A Primer of Multivariate Statistics.* New York: Academic Press, 1975.

Morrison, D. F. *Multivariate Statistical Methods*, 2nd ed. New York: McGraw-Hill, 1976.

Lengthier treatments are in:

Horst, P. *Matrix Algebra for Social Scientists.* New York: Holt, Rinehart, and Winston, 1963.

Searle, S. R. *Matrix Algebra for the Biological Sciences.* New York: John Wiley & Sons, 1966.

A clear treatment of the abbreviated Doolittle procedure can be found in

Graybill, F. A. *An Introduction to Linear Statistical Models*, vol. 1. New York: McGraw-Hill, 1961.

14 MULTIPLE REGRESSION

14.1 INTRODUCTION

The simple linear model just investigated can be extended to multiple regression, the functional relation between the dependent variable Y and a set of independent variables (sometimes called "predictors") $X_1, X_2, \ldots, X_p$. There are several reasons for our interest in multiple regression. First, in such applications as clinical diagnosis and personnel selection, it is desirable to select some subset of independent variables that provide the most information about Y and weight them to minimize errors of prediction. A second reason for interest in multiple regression is that the approach not only includes the calculations of the first eleven chapters as special cases, but also has the flexibility for being directly applicable to studies in which cell frequencies are disproportionate. As we have noted (Chapter 5), data from such studies are troublesome within the usual analysis of variance framework. The multiple regression approach to disproportionate n will be presented in Chapter 15; however, the basic computations for estimation and significance testing are developed here.

A third reason for our interest in regression analysis is that it allows us to treat continuous variables as continuous. Often, investigators arbitrarily divide subjects into groups on the basis of variables such as income or intelligence. Similarly, stimuli are sometimes categorized; for example, words can be classified into high and low frequency groups. Such categorization loses information about the functional relation between the independent and the dependent variables. In some cases, the categorization approach may have greater efficiency, power, robustness, or ease of analysis (for example, see the discussions of the extreme groups versus the correlational approaches in Chapter 6, and of blocking versus analysis of covariance in Chapter 16), but the regression approach will be better in others. Knowing about regression analysis permits an intelligent choice.

A fourth advantage of knowing multiple regression is that certain important and common forms of data analyses are regression analyses. If we recognize this about analysis of covariance (Chapter 16) and trend analysis (Chapter 17), the concepts and computations should cause little trouble; they will have already been developed in Chapters 12 through 14. We can then concentrate on interpreting the results of such analyses.

A final reason for our concern with multiple regression is that it provides a more complete framework for dealing with problems that arise with the repeated-measurement design. Repeated measurements on subjects are almost never independent of each other. We have discussed this fact, its implications, and some modifications of the usual analysis of variance in response to it. Chapter 18 provides an alternative treatment within the regression framework.

14.2 A MODEL FOR MULTIPLE REGRESSION

We already have been introduced to a detailed model for the regression of Y on a single independent variable X (Section 12.3). The multiple regression model is a straightforward extension to the case in which there are several predictors.[1] We begin by assuming that for each combination of values of the p independent variables, there is a population of Y scores whose mean is denoted by $\mu_{Y|X_1X_2\cdots X_p}$ (the mean of Y scores given a particular set of values of the p measures). Then, an individual's Y score can be viewed as equal to this conditional mean plus some component of error ε. That is,

(14.1)
$$Y = \mu_{Y|X_1\cdots X_p} + \varepsilon$$
$$= \beta_0 + \beta_1 X_1 + \cdots + \beta_p X_p + \varepsilon$$

If we take the expectation for both sides of Equation (14.1),

$$E(Y) = \mu = \beta_0 + \beta_1 \mu_{X_1} + \cdots + \beta_p \mu_{X_p}$$

(as usual, $E(\varepsilon) = 0$), and solving for β_0 yields

(14.2)
$$\beta_0 = \mu - \sum_{j=1}^{p} \beta_j \mu_{X_j}$$

To understand the remaining regression coefficients, assume that we have two independent variables and that the function relating $\mu_{Y|X_1X_2}$ to them is $\mu_{Y|X_1X_2} = 5 + 2X_1 + 3X_2$. Then, holding X_2 constant, we have

X_1	$\mu_{Y|X_1X_2}$
0	$5 + 0 + 3X_2$
1	$5 + 2 + 3X_2$
2	$5 + 4 + 3X_2$
3	$5 + 6 + 3X_2$

The value β_1 is the slope of a line, the change in $\mu_{Y|X_1X_2}$ for each unit change in

[1] Although the discussion in this chapter is in terms of the linear equation, (14.1), interactive (for example, $\beta_j X_k X_{k'}$) or polynomial ($\beta_j X_k^2$) terms can be included. Some of the X_j in Equation (14.1) could be viewed as representing cross-products or powers of independent variables.

X_1 with X_2 held constant. Some writers use the notation $\beta_{Y_{1\cdot2}}$ to make explicit the idea that regression is on X_1 with X_2 held constant. In general, β_j is the regression of the population mean on X_j with all other Xs held constant.

It is convenient to rewrite Equation (14.1) in matrix notation:

(14.3) $$\mathbf{Y} = \mathbf{X}\boldsymbol{\beta} + \boldsymbol{\varepsilon}$$

As in Chapter 13, $\mathbf{Y}$ is an $n \times 1$ vector of observations, $\boldsymbol{\beta}$ is also a column vector consisting of the $p + 1$ values of β_j. Also, $\boldsymbol{\varepsilon}$ is an $n \times 1$ vector of error components, and $\mathbf{X}$ is an $n \times (p + 1)$ matrix. The first column of $\mathbf{X}$ consists of ones and the jth column consists of the values of X_{ij}. We assume that:

1. The matrix $\mathbf{X}$ has full rank. That is, no row (column) is a linear combination of any of the other rows (columns). Therefore, the rank of $\mathbf{X}$ is $p + 1$.[2] (Rank was discussed in Section 13.7.)
2. The value $\mathbf{X}$ is fixed over replications of the experiment.[3]
3. The value $\mathbf{X}$ is measured without error.
4. The value $\boldsymbol{\varepsilon}$ is normally distributed with $E(\boldsymbol{\varepsilon}) = 0$ and $E(\boldsymbol{\varepsilon}\boldsymbol{\varepsilon}') = \sigma_e^2 I$. Thus, for each combination of X_js there is a population of values of ε that are normally and independently distributed with mean zero and variance σ_e^2.

14.3   BASIC STATISTICS

14.3.1   REGRESSION COEFFICIENTS.   Whether our purpose is only to obtain the best-fitting function for a sample of data, or to compute confidence intervals and test statistics for population parameters, we have to calculate certain quantities based on $\mathbf{X}$ and $\mathbf{Y}$. To illustrate these calculations, as well as certain inferential procedures, we use an example to which we shall refer throughout this chapter. Suppose that we are interested in the effects of previous performance and self-esteem on expectations of performance on some new task. In studies on this kind of problem, subjects are typically divided into high and low self-esteem groups on the basis of scores on some test. Then, the first of two experimental sessions is rigged so that a random half of each self-esteem group experiences success and a random half experiences failure; this is accomplished by manipulating either the task or the evaluation given to subjects. Finally, measures such as expectancy about performance on the next task are subjected to a two-factor analysis of variance.

The experimental procedure just described has some drawbacks. Some investigators object to misleading subjects about their performances in the first session. Others may be concerned that the deception will be unsuccessful. Furthermore, the requirement that the outcome in the first session be manipulated may narrow the range of tasks that can be used. Finally, the procedure

[2] When a matrix is not square but of full rank, that rank is the smaller of the two dimensions. Since $n \geq p + 1$ (we must have at least one point for each parameter to be estimated), the rank is $p + 1$.

[3] As noted in Chapter 12, formulas for significance tests and interval estimates are not affected if the Xs are random instead of fixed variables. Results for the random case have been derived by Graybill (1961, Chapter 10).

Table 14–1 Data for a numerical example of regression analysis

SUBJECT	PROBLEMS COMPLETED, X_1	SELF-ESTEEM SCORE, X_2	EXPECTANCY FOR NEW TASK, Y
1	14	6	14
2	8	15	5
3	9	19	8
4	13	33	11
5	10	39	15
6	11	38	14
7	14	74	18
8	15	74	19
9	11	11	7
10	16	78	17

ignores considerable available information because it fails to use the actual self-esteem and performance scores that the subjects obtain.

An alternative approach to the research avoids these problems. Subjects could be told their true performance scores; indeed, they could calculate them themselves to minimize mistrust of the investigator; and they could be told group averages to give them some reference for evaluating their own performances. Then, a regression analysis could be carried out with first-task performance as X_1 and self-esteem score as X_2. Moreover, any number of other measures— additional indices of first-task performance or other personality correlates—could also be included in the regression analysis.

Table 14–1 presents data that have been made up for our hypothetical study. Assume that X_1 is the number of problems correctly completed in some fixed period, X_2 is some measure of self-esteem, and Y is the number of problems the subject expects to correctly complete on some second task. If Equation (14.3) is assumed, the least-squares estimates of the vector of regression coefficients β is calculated by

(**14.4**) $$\mathbf{b} = (\mathbf{X'X})^{-1}(\mathbf{X'Y})$$

These estimates are unbiased, as we proved in Section 13.5.2. To obtain numerical values, we must calculate

(**14.5**) $$\mathbf{X'X} = \begin{bmatrix} n & \sum X_1 & \sum X_2 \\ \sum X_1 & \sum X_1^2 & \sum X_1 X_2 \\ \sum X_2 & \sum X_1 X_2 & \sum X_2^2 \end{bmatrix} = \begin{bmatrix} 10 & 121 & 387 \\ 121 & 1,529 & 5,127 \\ 387 & 5,127 & 21,833 \end{bmatrix}$$

The inverse is

(**14.6**) $$(\mathbf{X'X})^{-1} = \begin{bmatrix} 2.866608 & -.265652 & .011571 \\ -.265652 & .027695 & -.001795 \\ .011571 & -.001795 & .000262 \end{bmatrix}$$

We also need

(14.7)
$$(\mathbf{X'Y}) = \begin{bmatrix} \sum Y \\ \sum X_1 Y \\ \sum X_2 Y \end{bmatrix} = \begin{bmatrix} 128 \\ 1,641 \\ 5,932 \end{bmatrix}$$

Then, from Equation (14.4),

(14.8)
$$\mathbf{b} = \begin{bmatrix} b_0 \\ b_1 \\ b_2 \end{bmatrix} = \begin{bmatrix} -.372526 \\ .797523 \\ .091025 \end{bmatrix}$$

Therefore, the least-squares linear regression equation for the data of Table 14–1 is

(14.9)
$$\hat{Y} = -.372526 + (.797523)X_1 + (.091025)X_2$$

The coefficient of X_2 is so small that it is tempting to disregard it. Its contribution may not be trivial, however. We shall later consider a significance test of whether β_2 is zero; we now merely caution the reader not to rely on visual judgments and to carry sufficient decimal places to ensure that seemingly small coefficients are not dropped due to rounding-off errors.

We have noted that the values of the b_j are unbiased estimates of the β_j. This assumes that all influences on Y have been included in the regression equation. Suppose that contrary to this assumption, Y depends on some variable X_3, which has not been included in the regression equation. Furthermore, X_3 is correlated with X_1 and X_2. Then, estimates of β are positively or negatively biased to an unknowable degree. Assume, for example, that β_2 is zero; X_2 has no influence on Y. Furthermore, assume that β_3 is greater than zero but X_3 is not included in the regression equation that we calculate. Finally, assume that X_2 and X_3 are correlated. In our numerical example, X_3 might be intelligence. If the purpose of the research is to predict Y accurately, the correlation of self-esteem and intelligence is of little concern. If interest is centered on the theoretical proposition that self-esteem influences expectancy, however, there is a real issue. If b_2 differs significantly from zero, we may infer a relation between self-esteem and expectancy when intelligence is actually what is influencing expectancy. Note that the usual "experimental" procedure of dividing subjects into high and low self-esteem groups does not solve the problem. A significant main effect (in the analysis of variance) may imply an influence of self-esteem, or intelligence, or both.

Snedecor and Cochran (1967, pp. 394–97) have discussed in some detail this problem of bias in estimating the regression coefficients. One strategy they suggest is studying a population in which there is a limited range of values of those X variables that are not of direct interest. This can be shown to reduce the correlations between the predictors included in the analysis and those that have not been included; in this way the bias in estimation is reduced. Snedecor and Cochran also suggest another alternative—including in the regression analysis any X variable that might have a marked effect on Y, even if the variable is not of interest in itself. There is some tradeoff involved in adding variables in this way. If the additional variables correlate highly with the original set of X_j, the sampling variability of the b_j will increase. (We shall return to this point in Section 14.5.2.)

14.3.2 DO THE X VARIABLES HAVE AN EFFECT?   Continuing to use the data of Table 14–1, we shall develop a test of $H_0: \beta_1 = \beta_2 = 0$. This hypothesis implies that $Y = \beta_0 + \varepsilon$, and from Equation (14.2), this is equivalent to $Y = \mu + \varepsilon$. In other words, the null hypothesis is an assertion that there is no systematic variation in Y as a function of X_1 and X_2; these variables have no explanatory power (or predictive, if that is the goal).

The approach to testing this hypothesis follows the one first met in analyzing variance for a simple one-factor design. In that case, we partitioned the total sum of squares into a component due to the treatment variable and a component due to error. Under the null hypothesis, the group means differed only because of error, and therefore MS_A and $MS_{S/A}$ were independent estimates of σ_e^2. An F test enabled us to evaluate the null hypothesis. In the current situation, we partition the total sum of squares into a component due to the variability of the $\hat{Y}$ and a component due to error. Under the null hypothesis that $\beta_1 = \beta_2 = 0$, the variability of predicted scores is due solely to error. Accordingly, $MS_{\hat{Y}}/MS_{\text{error}}$ tests the null hypothesis.

Starting with the identity $(Y - \bar{Y}) = (\hat{Y} - \bar{Y}) + (Y - \hat{Y})$, we can readily demonstrate that

(14.10)
$$\sum (Y - \bar{Y})^2 = \sum (\hat{Y} - \bar{Y})^2 + \sum (Y - \hat{Y})^2$$

or

$$SS_{\text{total}} = SS_{\hat{Y}} + SS_{\text{error}}$$

Equation (14.10) can be written in a more convenient form for calculating sums of squares:

(14.11)
$$\sum Y^2 - C = (\sum \hat{Y}^2 - C) + (\sum Y^2 - \sum \hat{Y}^2)$$

where C is, as in earlier chapters, T^2/n. In matrix notation,

(14.12)
$$\mathbf{Y'Y} - \mathbf{C} = (\hat{\mathbf{Y}}'\hat{\mathbf{Y}} - \mathbf{C}) + (\mathbf{Y'Y} - \hat{\mathbf{Y}}'\hat{\mathbf{Y}})$$

There are at least two ways of calculating $\sum \hat{Y}^2$. We have

$$\sum \hat{Y}^2 = \hat{\mathbf{Y}}'\hat{\mathbf{Y}} = (\mathbf{Xb})'(\mathbf{Xb}) = \mathbf{b'X'Xb}$$

Alternatively, recall that

(14.13)
$$\mathbf{b} = (\mathbf{X'X})^{-1}(\mathbf{X'Y})$$

Therefore,

(14.14)
$$\sum \hat{Y}^2 = \mathbf{b'X'Xb} = (\mathbf{b'X'})\mathbf{X}[(\mathbf{X'X})^{-1}(\mathbf{X'Y})] = \mathbf{b'(X'Y)}$$

Table 14–2   Analysis of variance to test $H_0: \beta_1 = \beta_2 = 0$

SV	df	SS	MS	F
$\hat{Y}(12)$	$p = 2$	$\mathbf{b}'(\mathbf{X}'\mathbf{Y}) - \mathbf{C} = 1801.01 - 1638.4 = 162.61$	81.31	11.62[a]
Error	$n - 1 - p = 7$	$\mathbf{Y}'\mathbf{Y} - \mathbf{b}'(\mathbf{X}'\mathbf{Y}) = 211.60 - 162.21 = 48.99$	7.00	
Total	$n - 1 = 9$	$\mathbf{Y}'\mathbf{Y} - \mathbf{C} = 1850 - 1638.4 = 211.60$		

[a] $p < .01$

The preceding formulas have been applied to the data of Table 14–1 and the numerical results are presented in Table 14–2. The degrees of freedom associated with each term merit comment. As usual, SS_{total} has $n - 1$ associated *df*. The $\hat{Y}$ are determined by $p + 1$ parameter estimates $(\hat{\beta}_0, \hat{\beta}_1, \ldots, \hat{\beta}_j, \ldots, \hat{\beta}_p)$; one *df* is lost from this total because $SS_{\hat{Y}(12)}$ represents variation of the predicted values about the grand mean.[4] The remaining $n - 1 - p$ *df* are associated with SS_{error}. Note the parallel to earlier developments in this book. The error *df* have generally been $N - k$, where N is the total number of scores and k is the number of parameters estimated. We have $n - 1$ *df*, for example, in estimating the variability about a single mean; $an - a$ *df* when the error variance is variability about a parameter estimates (the $\hat{\mu}_j$); and $n - 2$ *df* for variability of scores about a line determined by estimates of β_0 and β_1. Also, as in previous developments, division of SS_{error} by df_{error} yields an unbiased estimate of σ_e^2, under the assumptions of Section 14.2. (The normality assumption is not needed to prove that the estimate is unbiased; it is needed, however, to prove that the ratio of mean squares is distributed as F.)

The F test of the data of Table 14–1 proves to be significant at the .01 level. We reject the null hypothesis that both regression coefficients are zero. Of course, this leaves open the question whether both X_1 and X_2 are required to account for the variability in Y. In the next section, we shall consider separate tests of β_1 and β_2. Before doing so, we shall try to draw some additional information from the calculations carried out to this point.

One index of the importance of X_1 and X_2 in accounting for the variability in Y is $SS_{\hat{Y}(12)}/SS_{\text{total}}$, which for our example is .77. This can be shown to equal the squared multiple correlation coefficient $R^2_{Y \cdot 12}$. Another way of thinking about this measure is that it is also identical to the squared correlation between Y and $\hat{Y}$. The statistic R^2 generally overestimates the proportion of population variance for which the X variables account. To understand why, consider the case where ρ, the population correlation, is zero. Because of sampling error, it is very unlikely that R would equal zero; consequently, R^2 will be greater than zero. In view of this inflation, Wherry (1931) has suggested an alternative formulation for R. The

---

[4] We use the notation $\hat{Y}(12)$ to indicate that β_1 and β_2 have been estimated to obtain the predictions. This is useful because in subsequent sections we shall ignore one or more predictors; the b coefficient will be arbitrarily set to zero. Thus, $SS_{\hat{Y}(1)}$ would represent the variability of values predicted by the reduced equation $\hat{Y} = b_0 + b_1 X_1$.

development is very similar to that for ω^2 in Chapter 4. We want an estimate of

$$\rho^2 = \frac{\sigma_{\hat{Y}}^2}{\sigma_{\text{total}}^2} = 1 - \frac{\sigma_e^2}{\sigma_{\text{total}}^2}$$

Replacing the population variances by unbiased estimates gives

(14.15)

$$\tilde{R}^2 = 1 - \frac{SS_e/(n-1-p)}{SS_{\text{total}}/(n-1)}$$

$$= 1 - \left(\frac{SS_e}{SS_{\text{total}}}\right)\left(\frac{n-1}{n-1-p}\right)$$

$$= 1 - (1-R^2)\left(\frac{n-1}{n-1-p}\right)$$

Darlington (1968) has discussed some of the properties of this statistic, and also properties of other measures of the validity of the regression equation. For the data in our example, the "shrunken" $R^2(\tilde{R}^2)$ is .70.

The general form of the confidence interval for the β_j is

(14.16)

$$b_j \pm t_{1-\alpha}\sigma_{b_j}$$

The value of t is that required for significance at the α level (assuming a $1-\alpha$ confidence interval) with $n-1-p\ df$. If we desire .95 confidence, the required t is 2.365 for our example in which we have seven error df. The standard error is

(14.17)

$$\hat{\sigma}_{b_j} = \hat{\sigma}_e\sqrt{c_{jj}}$$

where $\hat{\sigma}_e$ is the error mean square of Table 14–2 and c_{jj} is the jth diagonal entry of the inverse matrix $(\mathbf{X'X})^{-1}$; note that $j = 0, 1,$ or 2. From Equations (14.16) and (14.17), we can calculate the three confidence intervals. For the intercept β_0, the boundaries are $-.372526 \pm (2.365)(\sqrt{7.00})(\sqrt{2.866608})$; the interval extends from -10.97 to 10.22. This is a wide interval but not surprising in view of the small number of observations and the variability of the data. The interval for the regression of Y on X_1 extends from $-.24$ to 1.84 and the interval for β_2 extends from $.05$ to $.13$.

14.4   TESTING A SUBSET OF THE β_j

In our example of the study of expectancies Y as a function of previous performance X_1 and self-esteem X_2, we assumed the regression equation

(14.18)

$$Y = \beta_0 + \beta_1 X_1 + \beta_2 X_2 + \varepsilon$$

Whether our interest is in predicting Y or in understanding factors that influence its variability, we may wish to determine whether both of the X variables are important. We may ask, for example, whether the equation

(14.19)

$$Y = \beta_0 + \beta_1 X_1 + \varepsilon$$

provides as good an account of the variability in Y as Equation (14.18). In other words, if X_1 is already included in the regression equation, does X_2 add significantly to our ability to predict Y? In general, we can ask whether adding any set of r independent variables significantly adds to the variability already accounted for by the other $p - r$ independent variables. Note that in the preceding section, the test of H_0:$\beta_1 = \beta_2 = 0$ represents the limiting case where $r = p$.

We shall again use the sample data of Table 14–1 to illustrate calculations, initially investigating whether X_2 is important, given that X_1 is included in the regression equation. Subsequently, we shall turn things around and examine whether X_1 makes a significant contribution beyond what X_2 makes. The first step is to assume the complete regression equation (Equation 14.18) and calculate the variability of the Y predicted on the basis of estimates of all three β coefficients. The quantity $SS_{\hat{Y}(12)}$ was calculated in the preceding section and is 162.61 (see Table 14–2). The second step is to compute the variability of the $\hat{Y}$ under the hypothesis that $\beta_2 = 0$; call this quantity $SS_{\hat{Y}(1)}$. The numerator of the F test of the null hypothesis that β_2 equals zero will be a mean square based on the difference between $SS_{\hat{Y}(12)}$ and $SS_{\hat{Y}(1)}$. Before carrying out the calculations, consider why this should be so.

The quantity $SS_{\hat{Y}(12)}$ represents the variability predictable from estimates of β_0, β_1, and β_2. The quantity $SS_{\hat{Y}(1)}$ represents variability predictable from estimates of a reduced set of parameters, in this case, β_0 and β_1. If the difference between the two sums of squares is small, adding X_2 to the regression equation has not notably increased our ability to account for the variability in Y. This would suggest that Equation (14.19) should be preferred to Equation (14.18) as a model of the data. On the other hand, if $SS_{\hat{Y}(12)}$ is considerably larger than $SS_{\hat{Y}(1)}$, knowing X_2 improves predictability of Y, and the preferred model would be Equation (14.18). Thus, $SS_{\hat{Y}(12)} - SS_{\hat{Y}(1)}$ reflects the contribution X_2 makes beyond that already due to X_1. We shall refer to this difference in sums of squares as the sum of squares due to X_2 after adjustment for X_1. This can be represented by

(14.20) $$SS_{\hat{Y}(2|1)} = SS_{\hat{Y}(12)} - SS_{\hat{Y}(1)}$$

Since $SS_{\hat{Y}(12)}$ is already available, we need only calculate $SS_{\hat{Y}(1)}$. To do this, we shall ignore the X_2 data (as Equation (14.19) implies) and get a new set of estimates of β_0 and β_1. Designate this 2×1 vector of coefficients by the symbol $\mathbf{b}^*$, to contrast it with $\mathbf{b}$ and to indicate that it is based on a reduced set of independent variables. Also, let $\mathbf{X}^*$ be the $n \times 2$ matrix consisting of $\mathbf{X}_0$ (all ones) and $\mathbf{X}_1$. Then,

(14.21) $$\mathbf{b}^* = (\mathbf{X}^{*\prime}\mathbf{X}^*)^{-1}\mathbf{X}^{*\prime}\mathbf{Y}$$

and

(14.22) $$SS_{\hat{Y}^*} = \mathbf{b}^{*\prime}\mathbf{X}^{*\prime}\mathbf{Y} - C$$

Note the resemblance of these equations to Equations (14.13) and (14.14); only the asterisk indicates that we are dealing with a reduced set of independent variables.

The actual calculations for $SS_{\hat{Y}(1)}$ are as follows. We have

$$\mathbf{X}^{*\prime}\mathbf{X}^{*} = \begin{bmatrix} 10 & 121 \\ 121 & 1{,}529 \end{bmatrix} \quad \text{and} \quad \mathbf{X}^{*\prime}\mathbf{Y} = \begin{bmatrix} 128 \\ 1{,}641 \end{bmatrix}$$

Inverting $\mathbf{X}^{*\prime}\mathbf{X}^{*}$ gives

$$(\mathbf{X}^{*\prime}\mathbf{X}^{*})^{-1} = \begin{bmatrix} 2.355932 & -.186441 \\ -.186441 & .015408 \end{bmatrix}$$

Applying Equations (14.21) and (14.22), we have

$$\mathbf{b}^{*} = \begin{bmatrix} -4.389834 \\ 1.420641 \end{bmatrix}$$

and

$$\begin{aligned} \mathbf{SS}_{\hat{Y}(1)} &= b^{*\prime}X^{*\prime}Y - C \\ &= 1{,}769.37 - 1{,}638.40 \\ &= 130.97 \end{aligned}$$

Applying Equation (14.20), we find that the adjusted contribution of X_2 is 31.64.

To compute the F ratio, we must divide $SS_{\hat{Y}(2|1)}$ by the proper df. In general, this is $p - (p - r)$, or r. In the specific example under consideration, there are two df associated with $SS_{\hat{Y}(12)}$ and one associated with $SS_{\hat{Y}(2)}$. Then, $SS_{\hat{Y}(2|1)}$ is distributed on one df.

The numerical results of two analyses of variance of the data are presented in Table 14–3. The upper portion of the table presents a test of the contribution of X_2 after the variability predictable from knowledge of X_1 has been removed. In essence, $SS_{\hat{Y}(2|1)}$ is the basis for a test of $H_0: \beta_2 = 0$. Although this test is now the focus of our interest, we should note that the computations required for it enable us to test two other null hypotheses. First, the $\hat{Y}(12)$ term provides a test of

Table 14–3 Testing the contributions of X_1 and X_2 to the data of Table 14–1

SV	df	SS	MS	F
Total	$n - 1 = 9$	211.60		
$\hat{Y}(12)$	$p = 2$	162.61	81.31	11.62[a]
$\hat{Y}(1)$	$p - r = 1$	130.97	130.97	18.71[a]
$\hat{Y}(2\|1)$	$r = 1$	31.64	31.64	4.52
Error	$n - 1 - p = 7$	48.99	7.00	
Total	9	211.60		
$\hat{Y}(12)$	2	162.61	81.31	11.62[a]
$\hat{Y}(2)$	1	139.78	139.78	19.97[a]
$\hat{Y}(1\|2)$	1	22.83	22.83	3.26
Error	7	48.99	7.00	

[a] $p < .01$

$H_0: \beta_1 = \beta_2 = 0$. This test was performed in the previous section and the results given in Table 14.2. The significance of the F ratio permits us to conclude that estimating β_1 and β_2 increases the variation in Y we can predict. The $\hat{Y}(1)$ term permits a test of $H_0: \beta_1 = 0$, assuming that X_2 is ignored (that is, β_2 is set to zero). This F test is also significant; when X_2 is omitted from the regression equation, X_1 accounts for a significant proportion of the observed variance in Y.

The next line of the table represents the contribution of X_2 after the contribution of X_1 is adjusted for. This term $\hat{Y}(2|1)$ is not significant at the .05 level. The results of the series of significance tests, lead to the inference that including X_1 in the regression equation accounts for a significant part of the variability in Y but subsequently adding X_2 does not significantly improve the account of the data.

In the lower part of Table 14–3, we consider whether adding X_1 to a regression equation that already includes X_2 significantly improves our ability to predict the data. Essentially, we try to choose between Equation (14.18) and

$$(\textbf{14.23}) \qquad\qquad Y = \beta_0 + \beta_2 X_2 + \varepsilon$$

Before considering this issue, note that $\hat{Y}(2)$ is a significant source of variance; this indicates that knowing X_2 alone significantly enhances our ability to account for the variability in Y. Turning next to the test of $\hat{Y}(1|2)$, we again find that there is no significant increase in the variability accounted for when one of the independent variables (X_1 this time) is added to the regression equation.

To summarize the results from the two parts of Table 14–3: Including either X_1 or X_2 in the regression equation (in other words, estimating either β_1 or β_2) significantly improves our ability to predict Y. Once either variable has been taken into account, however, the additional improvement due to the second independent variable is not large enough to be statistically significant. This is because X_1 and X_2 have a sizable positive correlation ($r_{12} = .66$) and therefore convey redundant information about Y. It is important to realize that other patterns of correlations among the X and the Y variables could exist with considerably different consequences. Consider, for example, two X variables, each of which has a low positive correlation with Y and a relatively large negative correlation with each other. With this pattern of correlations, adding either independent variable to a regression equation that already includes the other variable is likely to result in a larger contribution to the predicted variability than what either variable alone makes. Cohen and Cohen (1975, pp. 84–91) have a good discussion of this and other patterns of correlations by pairs and their likely consequences.

The computational approach underlying the results in Table 14–3 can be used to test any hypothesis of the form $H_0: \beta_1 = \beta_2 = \cdots = \beta_r = 0$, where $r \leq p$. The general approach consists of the following steps:

1. Compute $SS_{\hat{Y}(12\cdots p)}$ on p df. The formula is $\mathbf{b'X'Y} - \mathbf{C}$.
2. Compute $SS_{\hat{Y}(r+1\cdots p)}$ on $p - r$ df. The formula is $\mathbf{b^*X^{*'}Y} - \mathbf{C}$, where $\mathbf{b^{*'}} = [b_0^* \; b_{r+1}^* \cdots b_p^*]$ (the * indicates that these values will generally differ from the

b_j calculated under the full model) and $\mathbf{X}^{*\prime}$ is a $(p-r+1) \times n$ matrix whose rows are the vectors $\mathbf{X}_0'$ (n ones), $\mathbf{X}_{r+1}', \ldots, \mathbf{X}_p'$.

3.   Subtract $SS_{Y(r+1 \cdots p)}$ from $S_{Y(12) \cdots p)}$; the result is distributed on $r\, df$. Test this against

$$MS_{error} = \frac{\mathbf{Y'Y} - \mathbf{b'X'Y}}{n-1-p}$$

If the result is significant, one or more of the set of r parameters specified under H_0 is not zero. Stated somewhat differently, a significant F implies that the ability to predict Y is improved by adding one or more of the variables $X_1, X_2, \ldots, X_r$ to a regression equation that already includes the other $p - r$ variables.

One implication of the developments in this section is that the contribution of X depends on when it is entered into the regression equation. In our example, when X_1 was entered first, it accounted for a sum of squares of 130.97; when X_1 was entered after X_2, it accounted for a sum of squares of only 22.83. We have also noted that the patterns of relations among the Xs and Y may be such that a variable can contribute more when entered later. In any event, the conclusions we reach about the relative importance of different X variables will generally be a consequence of the order in which they are considered. An example is the hypothetical study that motivated our numerical example. Depending on whether we test β_1 or β_2 first, we conclude that either X_1 or X_2 (but not both) contributes significantly to the predictable variation in Y. Since there are such inferential difficulties, some discussion of strategies for determining the order of entering variables into the regression equation is warranted.

14.5   APPROACHES TO REGRESSION ANALYSIS

14.5.1   A PRIORI ORDERING OF TESTS.   The interpretative difficulties just encountered exist since self-esteem and performance, the two predictors in our example, were correlated. The result is a nonzero covariance, a contribution to the variability in Y that is not unique to either X_1 or X_2. The sequence of testing in the top part of Table 14–3 results in the assignment of this contribution to X_1, and the sequence in the bottom part of the table attributes the covariation to X_2. Sometimes, the choice between such possible sequences may be determined before viewing the data; logical relations among the independent variables, or the questions of interest, may dictate the order in which variables are considered and therefore the attribution of the covariation.

In the example of the preceding sections, there is no logical way in which task performance could influence self-esteem scores, provided that those scores were obtained before the performance tests. On the other hand, it is reasonable to hypothesize that people who view themselves able to perform better do perform better; self-esteem X_2 influences performance X_1. From this standpoint, the analysis in the bottom part of Table 14–3 is preferred. The variability accounted for by the presumed causal factor when the other factor is ignored $SS_{\hat{Y}(2)}$ is assessed first, then the unique contribution due to the other factor $SS_{\hat{Y}(1|2)}$. Of course, it is possible that both self-esteem and performance scores are largely determined by some X_3, perhaps intelligence, which we have omitted from the regression analysis. If this is the case, the argument for attributing the covariation to self-esteem is fallacious. We should have included a measure of intelligence in the study. Then, the appropriate order of testing would have been *intelligence* X_3, *self-esteem* X_2, and *performance* X_1. The sums of squares would have been computed in the order $SS_{\hat{Y}(3)}$, $SS_{\hat{Y}(2|3)}$, and $SS_{\hat{Y}(1|23)}$; note that these quantities add up to $SS_{\hat{Y}(123)}$, the variability of the points predicted by the complete equation in which all β coefficients are estimated.

The issue how to derive causal inferences from regression analyses is not easily resolved. Readers who are interested in the problem would do well to consider methodologies that have been developed to evaluate causal models. Several treatments of a technique called path analysis, and of related techniques, are available. Introductions are presented by Darlington (1968, pp. 166–68) and Kerlinger and Pedhazur (1973; pp. 306–31). Both contain numerous references to primary sources.

14.5.2 STEPWISE REGRESSION.   When no theory exists to guide the order in which independent variables are entered into the regression equation, the data must determine that order. The usual procedure is first to consider the proportion of variance due to each variable when all the other variables are ignored. That X_j for which $SS_{\hat{Y}(j)}$ is largest is entered first into the regression coefficient. The variable entered next would be the one that accounted for the largest proportion of the variability in Y after adjustment for X_j. In other words, the variable entered second would be that X_k $(k \neq j)$ that would give the largest $SS_{\hat{Y}(k|j)}$. The process continues until including additional terms does not produce enough of an increment in predicted variability to some criterion designated before the analysis. This is usually an F ratio, perhaps an F of 1 or perhaps the value required for significance. Kerlinger and Pedhazur (1973) present a numerical example and Harris (1975) examines one available computer program in detail.[5]

The chief problem with stepwise regression is that the results are often unreliable—for several reasons. First, when the number of independent variables is large, many significance tests will be performed. The *EF* can be huge (cf. Chapter 11). Second, trivial differences in sums of squares will dictate the weights given to variables. In our example, self-esteem X_2 would have been entered first because it accounted for 66 percent of the total sum of squares whereas performance X_1 accounted for only 62 percent. The difference might well have been chance. Nevertheless, if we needed to include a significant contribution at the .05 level in the final regression equation, the adjusted X_1 contribution would fall

---

[5] The approach described here is often referred to as the forward approach. An alternative, backward approach is sometimes used, in which the analysis starts with the full set of predictors. Those that make the smallest contribution to the predictable variability are dropped first from the regression equation. This approach will not necessarily give the same results as the forward approach.

short of the criterion. This particular problem is severe when the X variables have high correlations. A third problem is that the sampling variability of the b_i tends to be high. This variability of b_i depends on the proportion of variance in X_i that is predictable from the other X variables. Finally, it is quite possible that a variable entered early contributes nothing at all after other variables have been entered.

In view of these comments, sample sizes should be large. How large is not clear; recommendations vary from several hundred subjects (Kerlinger and Pedhazur, 1973) through 50 more than the value of p (Harris, 1975) to a $40:1$ ratio of n to p (Cohen and Cohen, 1975). The appropriate n will depend on the number of predictors and the degree to which they are correlated with each other. Pilot work will help provide estimates of those quantities that are components of the calculations for $\hat{\sigma}_{b_i}$. Cohen and Cohen (1975) have provided a useful form for expressing the standard error of the regression coefficient:

(14.24)
$$\hat{\sigma}_{b_i} = \frac{\sum_i (Y_i - \bar{Y})^2}{\sum_i (X_{ij} - \bar{X}_{.j})^2} \sqrt{\frac{1 - R_Y^2}{n - p - 1}} \sqrt{\frac{1}{1 - R_j}}$$

where R_j^2 is the proportion of $\sum (X_{ij} - \bar{X}_{.j})^2$ predictable by a regression equation in which the other $p - 1$ independent variables are entered.

Another precaution to take in doing stepwise regression is to use a program that reviews what a variable contributes after other variables have been entered. In this way, variables can be eliminated if they correlate highly with Y when other variables are ignored but contribute little unique variability when their contribution is adjusted for the contribution of other variables.

Finally, cross-validation is an excellent idea. This is a procedure in which the weights derived from one sample of subjects are used to predict scores in another sample; observed and predicted scores are then correlated. The same procedure can be applied to a single sample that has been split in half for this purpose. All too often, the correlations are low, attesting to the instability of estimates of β, about which we commented above. Incidentally, cross-validation coefficients almost as high as those obtained through least-squares estimation of the β coefficients have frequently been obtained by simpler methods of constructing regression equations. If we want only accuracy of prediction, one effective approach is to convert all measures, including the Y values, to standardized scores $((X_{ij} - \bar{X}_{.j})/\hat{\sigma}_j)$, and then merely weight the variables equally (Dawes and Corrigan, 1974).

Stepwise regression has been used primarily to predict behavior rather than to develop or test explanations of processes that underlie behavior. The type of causal analysis outlined in the preceding section better lends itself to theoretical endeavors. Nevertheless, theorists have turned to the stepwise approach in research areas in which there are many highly correlated, independent variables and in which there is no strong theory to guide the sequence of analysis. To note just two examples from the field of cognitive psychology, Loftus and Suppes (1972) and Anderson and Reder (1974) have considered such variables as word length and word familiarity (among many others; Loftus and Suppes had 12 predictors and Anderson and Reder used 22) in theorizing how information is stored and retrieved in memory. The same problems exist in doing this research as in trying to establish a regression equation for predictive purposes, and the precautions discussed earlier should be taken. Indeed, there are probably greater risks in using stepwise regression as an exploratory tool for suggesting variables for subsequent study and as a source of speculation about underlying processes. If two subsets of independent variables yield the same R^2, which subset is incorporated into the regression equation is of little consequence if prediction is the sole concern. In more theoretically motivated work, including a variable attaches explanatory significance to it and choosing between two such subsets of variables may mean choosing between theoretical positions. It is one thing to make such a choice knowingly, quite another to leave it to a program for stepwise regression.

14.5.3   ORTHOGONAL VARIABLES.   By this time, it may have occurred to the reader that independent variables in regression analysis are rarely independent. (Nor are they truly variables if we have selected the values. Unfortunately, we have not been able to find any term that suited us better.) Variables that are orthogonal (by the definition presented in Chapter 11) have at least two advantages. First, estimates of population parameters are much more stable when the X variables are orthogonal. In that case, R_j^2 goes to zero for all j in Equation (14.24); accordingly, the standard error decreases. The second, and perhaps more important, consequence of orthogonality is interpretative simplicity. Estimates of regression coefficients do not change as a function of whatever other coefficients are estimated. Furthermore, the sum of squares attributable to some variable is the same after adjustment for other variables. If X_1 and X_2 were othogonal in the example used in this chapter, $SS_{\hat{Y}(1)}$ would equal $SS_{\hat{Y}(1|2)}$. As a result, we should not need to worry about the order in which variables enter the regression equation, and the preceding material on approaches to regression analysis would be totally unnecessary.

There is one class of analyses in which orthogonality prevails. Tests of main and interaction effects are orthogonal in analyzing variance for complete factorial designs with equal or proportional cell frequencies. In such designs, removing effects of one variable has no influence on the estimation of the effects of the other variables.

With the evident advantages of having orthogonal X variables, it is tempting to select people, or stimuli, so that the resulting design is factorial with equal cell frequencies. We shall next consider the merits of such a factorial approach against those of a random sampling approach.

14.6   RESEARCH STRATEGIES

Much psychological research is not experimental. Variables are not manipulated; instead, individuals are selected for characteristics like age, intelligence, or score on some personality scale. Sometimes stimuli, like words, are selected for characteristics such as length, pleasantness, or familiarity. A common approach to such research has been to divide subjects or stimuli into groups, to categorize by what is ordinarily a continuous measure. Typically, the element is characterized as high or low for the attribute in question. Often when more than one such attribute is considered, categorization is carried out so as to produce equal cell frequencies, rather than by some criterion determined before collecting data. Consider, for example, a study of expectancy about performance in some laboratory task Y as a function of degree of chronic depression X_1 and score on a measure of externalization X_2. Assume that after a median split on the two variables, 100 subjects fall so as to produce the following cell frequencies:

$$
\begin{array}{cc}
 & \begin{array}{cc} X_2 \\ \text{High} & \text{Low} \end{array} \\
X_1 \begin{array}{c} \text{High} \\ \text{Low} \end{array} & \begin{bmatrix} 35 & 15 \\ 15 & 35 \end{bmatrix}
\end{array}
$$

The researcher might invite 15 subjects from each cell to the laboratory and obtain both expectancy and performance measures from this subset of subjects. The usual multifactor analysis of variance would then be carried out on the data from this two-factor design.

The pervasive use of such "designs" stands as testimony to the computational simplicity of analysis of variance for equal cell frequencies. It probably also reflects an ignorance of the alternatives by many researchers. This is not to say that categorization has no other apparent advantages besides computational simplicity. The procedure does result in orthogonal X variables and consequently, more reliable parameter estimates and relatively straightforward interpretation of the results of significant tests. For several reasons, however, these advantages are illusory. First, estimates of the marginal population parameters, the $\mu_{j.}$ and $\mu_{.k}$, are almost certain to be biased. In obtaining such estimates from designs on which equal cell frequencies have been forced, equal weight is given to both cell means in the row or column in question. In the example considered above, the relative weights should be 35 and 15.

Furthermore, the original pattern of cell frequencies indicates a positive correlation between X_1 and X_2. Thrusting orthogonality on the experimental design does not resolve the interpretative problems associated with nonorthogonal variables; it merely avoids them. Using the design procedure we have illustrated is equivalent to asking what the effects of variables would be if the variables were not related as they are related. We are asking questions about a population that does not exist.

A second possible approach to the research in question would also involve categorization, but the distribution of cell frequencies would more closely approximate the relative sizes of populations. If we wished, for example, to collect expectancy data from only 60 subjects (the number used when there were 15 in a cell), they could be sampled at random. Assuming the same positive correlation between the X variables that we assumed in discussing the equal-n procedure, this random sampling would yield disproportionate cell frequencies. The appropriate analysis will be discussed in the next chapter; in essence, it involves transforming the X scores to one of two values and then carrying out a least-squares regression analysis of the form presented in this chapter. There seems little point to this procedure. It has all the disadvantages of multiple regression with nonorthogonal variables, and at the same time, it loses power and information about functional relations among the X variables and Y.

It seems most sensible to retain the original set of X scores and to carry out the regression analysis. Such an approach more truly represents relations among variables in the population, and because all the data are used, it gives a clearer sense of those relations and more power for statistical tests. Admittedly, there are very real problems with regression analysis when variables are not orthogonal; and often the variables are not orthogonal. We must choose between stepwise analyses and several possible prior orderings, according to the state of our knowledge and theory. Interpretation will usually be difficult when X variables are correlated, unless we are willing to assume that one variable influences the second, or that one variable has no influence on Y. Nevertheless, if we understand the possible problems, we at least can be aware of the constraints on our inferences. Furthermore, if there is a well-developed theory that postulates causal relations among X variables, methods based on the regression approach, like path analysis, can further our understanding of behavior. In summary, random sampling coupled with regression analysis is an imperfect approach to inference but it uses all the data and meets our variables as they are in the universe from which we sample.

The preceding discussion should not be misconstrued as advocating that we never block on continuous X variables. We have already considered two sets of circumstances in which categorization of subjects is appropriate (Chapter 6). We do feel that investigators should be comfortable with traditional analysis of variance *and* multiple regression, and therefore be able to base their approach on the research problem. When the independent variables are attributes of people or stimuli, rather than manipulable treatment conditions, the regression approach often will make more sense.

14.7   CONCLUDING REMARKS

We have assumed a linear relation between the dependent variable Y and a set of p independent variables. In matrix notation, that relation takes the form

$$\mathbf{Y} = \mathbf{X}\boldsymbol{\beta} + \boldsymbol{\varepsilon}$$

The regression coefficients were estimated by solving a set of $p + 1$ simultaneous equations of the form

$$\mathbf{X}'\mathbf{X}\hat{\boldsymbol{\beta}} = \mathbf{X}'\mathbf{Y}$$

and having the least-squares solution

$$\hat{\boldsymbol{\beta}} = \mathbf{b} = (\mathbf{X}'\mathbf{X})^{-1}(\mathbf{X}'\mathbf{Y})$$

These estimates yield a set of predicted Y values:

$$\hat{\mathbf{Y}} = \mathbf{Xb}$$

or

$$\hat{Y} = b_0 + b_1 X_1 + b_2 X_2 + \cdots + b_p X_p$$

The sum of squared deviations of the $\hat{Y}$ about $\bar{Y}$ is

$$SS_{\hat{Y}(12\cdots p)} = \mathbf{b}'\mathbf{X}'\mathbf{Y} - C$$

where C is the usual correction term for the mean, T^2/n. The SS_{error} is readily obtained by subtracting the preceding term from SS_{total}. To obtain a confidence interval for any particular coefficient b_j, we calculate $\hat{\sigma}^2_{b_j} = c_{jj} MS_{\text{error}}$, where c_{jj} is a diagonal element in $(\mathbf{X}'\mathbf{X})^{-1}$.

The F ratio $MS_{\hat{Y}(12\cdots p)}/MS_{\text{error}}$ (on p and $n - p - 1$ df) tests whether *any* of the p independent variables contributes significantly to the predicted variability. Significance indicates that one or more of the X variables is important. To evaluate the contribution of any one variable X_j, we set β_j to zero and estimate the other p coefficients again. The difference between the resulting sum of squares and sums of squares for the complete model is the contribution of X_j adjusted for the remaining variables and is the basis for an F test of whether X_j makes a significant additional contribution to the variability of Y. The contribution of X_j after adjustment for some subset of the remaining predictors could also be calculated. Unless the independent variables are orthogonal, the nature of the adjustment influences the outcome of the data analysis and conclusions about the relative importance of the independent variables. The proper adjustment in assessing the contribution of a variable will be an issue of special interest in the discussion of nonorthogonal factorial designs (next chapter).

There are several other types of null hypotheses that we have not considered. The interested reader can consult Graybill (1961) or Searle (1971) for tests of the null hypothesis that r of the β_j have other specific values besides zero. For example, a theory that makes the statement

$$Y = \beta_0 + \beta_1 X_1 + X_2$$

is equivalent to $H_0: \beta_2 = 1$. Tests of linear contrasts of the β_j are also of interest. An investigator might want, for example, to test $H_0: \beta_1 - \beta_2 = 0$, or $H_0: \beta_1 + \beta_2 - 2\beta_3 = 0$. Procedures for doing so are presented in the Graybill and Searle books.

The primary purpose of this chapter has been to prepare you for reconsidering analysis of variance as a special case of regression analysis. As a consequence, we have either omitted, or considered only briefly, important and interesting issues. A good beginning for the reader who wants to know more about such issues is a paper by Darlington (1968), in which several fundamental questions are raised and careful consideration is given to possible solutions.

EXERCISES

14.1   Assuming the complete model

$$Y = \beta_0 + \beta_1 X_1 + \beta_2 X_2 + \beta_3 X_3 + \varepsilon$$

calculate least-squares estimates of the β_i for the following data set.

Subject	1	2	3	4	5	6	7	8	9	10
$Y =$	24	16	14	22	25	18	21	22	8	18
$X_1 =$	30	2	22	23	13	10	2	1	0	1
$X_2 =$	10	4	7	13	6	12	9	9	7	10
$X_3 =$	20	17	13	11	17	12	12	15	16	16

14.2   We want to determine whether X_1 contributes significantly to the variability in the data set of Exercise 14.1. (a) Carry out the analysis, ignoring X_2 and X_3; (b) carry out the analysis after adjusting for the other two X variables.

14.3   Let $\mathbf{X}$ be an $n \times 3$ matrix whose column vectors are $\mathbf{X_0}$ (all ones), and $\mathbf{X_1}$ and $\mathbf{X_2}$, two predictor variables. We solve for

$$\mathbf{b} = \begin{bmatrix} b_0 \\ b_1 \\ b_2 \end{bmatrix}$$

by $\mathbf{b} = (\mathbf{X'X})^{-1}(\mathbf{X'Y})$. Suppose that some other variable, X_3, also influences Y; that is,

$$\mathbf{Y} = \mathbf{X\beta} + \mathbf{X_3\beta_3} + \mathbf{\varepsilon}$$

Show that $E(\mathbf{b})$ is a biased estimator of $\mathbf{\beta}$, specifically, that

$$E(\mathbf{b}) = \mathbf{\beta} + (\mathbf{X'X})^{-1}\mathbf{X'X_3\beta_3}$$

14.4   We have the following data set:

Subject	1	2	3	4	5
$Y =$	4	1	5	7	12
$X_1 =$	2	−1	−2	−1	2
$X_2 =$	−2	−1	0	1	2

(a) Show that X_1 and X_2 are orthogonal variables. What are the implications for $(\mathbf{X'X})$?

(b) Calculate $(\mathbf{X'X})^{-1}$. What have you discovered about the inverse of any matrix with zeros in the off-diagonal cells?

(c) Calculate $\mathbf{b}$ under the model $Y = \beta_0 + \beta_1 X_1 + \beta_2 X_2 + \varepsilon$.

(d) Calculate $SS_{\hat{Y}(12)}$

(e) Calculate b_1 and $SS_{\hat{Y}(1)}$ under the reduced model $Y = \beta_0 + \beta_1 X_1 + \varepsilon$. How does the estimate of b_1 compare with the estimate of b_1 calculated in (c)?

(f) Calculate $SS_{\hat{Y}(1|2)}$. How does this compare with $SS_{\hat{Y}(1)}$?

REFERENCES

Anderson, J. R., and Reder, L. M. Negative judgments in and about semantic memory. *Journal of Verbal Learning and Verbal Behavior* 13:664–81 (1974).

Cohen, J., and Cohen, P. *Applied Multiple Regression/Correlation Analysis for the Behavioral Sciences.* Hillsdale, N.J.: Lawrence Erlbaum Associates, 1975.

Darlington, R. B. Multiple regression in psychological research and practice. *Psychological Bulletin* 69:161–82 (1968).

Dawes, R. M., and Corrigan, B. Linear models in decision making. *Psychological Bulletin* 81:95–106 (1974).

Graybill, F. A. *An Introduction to Linear Statistical Models, Vol. 1.* New York: McGraw-Hill, 1961.

Harris, R. J. *A Primer of Multivariate Statistics.* New York: Academic Press, 1975.

Kerlinger, F. N., and Pedhazur, E. J. *Multiple Regression in Behavioral Research.* New York: Holt, Rinehart, and Winston, 1973.

Loftus, E. F., and Suppes, P. Structural variables that determine the speed of retrieving words from long-term memory. *Journal of Verbal Learning and Verbal Behavior,* 11:770–77 (1972).

Searle, S. R. *Linear Models.* New York: John Wiley & Sons, 1971.

Snedecor, G. W., and Cochran, W. G. *Statistical Methods,* 6th ed. Ames: Iowa State University Press, 1967.

Wherry, R. J. A new formula for predicting the shrinkage of the coefficient of multiple correlation. *Annals of Mathematical Statistics* 2:440–57 (1931).

15 REGRESSION ANALYSIS AND FACTORIAL DESIGNS

15.1 INTRODUCTION

The analysis of variance is an elegant system for partitioning variability within factorial designs. The system breaks down when confronted by data from multifactor designs with disproportionate cell frequencies (Chapter 5). The usual sums of squares are no longer orthogonal, as we can see in a two-factor design, when SS_A, SS_B, and SS_{AB} no longer add to the between-cells sum of squares. Multiple regression analyses can be applied to such situations. Since multiple regression is a system for evaluating the contributions of variables that are not orthogonal to each other, it seems appropriate to bring the theory and computational machinery of regression analysis to bear on data obtained from factorial studies in which the factors are not orthogonal.

Some writers have implied that regression analysis is a panacea for the ills engendered by disproportionate n. We should be forewarned that this is not the case. The exact nature of the analysis, or of its interpretation, is rarely clear-cut when variables are correlated. This holds for the application to factorial designs. In considering the contribution of A to the variability in the data, do we ignore B and AB? adjust for the contribution of B? adjust for the contribution of AB? adjust for the contribution of B and AB? These are parallel to questions raised about the general multiple regression analysis of Chapter 14. Here, as there, the answers are not easy. (We shall reconsider the issue later in this chapter.)

There are other, new concerns when regression analysis is applied to data from factorial designs. If the variables are qualitative, coding of groups is an issue. It is not immediately obvious what values of X are to be assigned to each group. We shall consider several alternative coding systems, the similarities and differences in results obtained using them, and relations among them.

A more fundamental issue is that experimental design models are of deficient rank; the system of equations required for getting parameter estimates (the *normal equations*) are not linearly independent and therefore do not yield a unique set of parameter estimates. This situation can be remedied by placing constraints on the parameter estimates, reducing the system of equations to one of full rank (that is, one in which the equations are linearly independent of each other). We shall take up the relation of such constraints to coding the X variables, and to assumptions about population parameters. (Before you read the main body of this chapter, you may benefit by reviewing Section 13.7, in which rank is considered.)

Although the interesting problems in applying regression analysis to factorial designs occur with multifactor designs, we shall start with one-factor designs. This will enable us to consider such matters as rank, constraints, and coding systems, and also computational formulas, within a simpler context.

15.2   THE ONE-FACTOR DESIGN

15.2.1   THE DESIGN MATRIX.   Consider three treatment groups with n_j subjects in A_j and $n. = \sum n_j$. The usual experimental design model is

(15.1)
$$Y_{ij} = \mu + \alpha_i + \varepsilon_{ij}$$

In matrix notation, this is

(15.2)
$$\mathbf{Y} = \mathbf{X}\boldsymbol{\phi} + \boldsymbol{\varepsilon}$$

where

(15.3) $\mathbf{Y} = \begin{bmatrix} Y_{11} \\ \cdot \\ \cdot \\ \cdot \\ Y_{n_1 1} \\ Y_{12} \\ \cdot \\ \cdot \\ \cdot \\ Y_{n_2 2} \\ Y_{13} \\ \cdot \\ \cdot \\ \cdot \\ Y_{n_3 3} \end{bmatrix}$ $\mathbf{X} = \begin{bmatrix} 1 & 1 & 0 & 0 \\ \cdot & \cdot & \cdot & \cdot \\ \cdot & \cdot & \cdot & \cdot \\ 1 & 1 & 0 & 0 \\ 1 & 0 & 1 & 0 \\ \cdot & \cdot & \cdot & \cdot \\ \cdot & \cdot & \cdot & \cdot \\ 1 & 0 & 1 & 0 \\ 1 & 0 & 0 & 1 \\ \cdot & \cdot & \cdot & \cdot \\ \cdot & \cdot & \cdot & \cdot \\ 1 & 0 & 0 & 1 \end{bmatrix}$ $\boldsymbol{\phi} = \begin{bmatrix} \mu \\ \alpha_1 \\ \alpha_2 \\ \alpha_3 \end{bmatrix}$ $\boldsymbol{\varepsilon} = \begin{bmatrix} \varepsilon_{11} \\ \cdot \\ \cdot \\ \varepsilon_{n_1 1} \\ \varepsilon_{12} \\ \cdot \\ \cdot \\ \varepsilon_{n_2 2} \\ \varepsilon_{13} \\ \cdot \\ \cdot \\ \varepsilon_{n_3 3} \end{bmatrix}$

We use $\boldsymbol{\phi}$ rather than $\boldsymbol{\beta}$ to designate the parameter set to avoid confusion with two-factor designs, in which the β_j represent effects of the treatment variable B.

Since Equation (15.2) resembles the multiple regression equation of Chapter 14, it is logical to try to estimate the parameters, and to test the null hypothesis $H_0: \alpha_1 = \alpha_2 = \alpha_3 = 0$ by the methods of that chapter. Recall that parameters can be estimated by solving the set of simultaneous equations:

(15.4)
$$\mathbf{X}'\mathbf{X}\hat{\boldsymbol{\phi}} = \mathbf{X}'\mathbf{Y}$$

With three groups, Equation (15.4) corresponds to

$$n_.\hat{\phi}_0 + n_1\hat{\phi}_1 + n_2\hat{\phi}_2 + n_3\hat{\phi}_3 = T_{..}$$

(15.5)

$$n_1\hat{\phi}_0 + n_1\hat{\phi}_1 \qquad\qquad = T_{.1}$$

$$n_2\hat{\phi}_0 \qquad\quad + n_2\hat{\phi}_2 \qquad = T_{.2}$$

$$n_3\hat{\phi}_0 \qquad\qquad\quad n_3\hat{\phi}_3 = T_{.3}$$

We have noted that such systems of equations are of deficient rank (Section 13.7); the first equation can be obtained by summing the other three. Because of this lack of linear independence, an infinite number of possible ϕ vectors will satisfy Equations (15.5). Any particular vector is not really an estimate of the population parameters of Equation (15.3), but one of many possible solutions. To remind us of this, we shall designate the solution vector as ϕ^0, and not as $\hat{\phi}$; the vector ϕ^0 has elements μ_1^0, α_1^0, α_2^0, α_3^0. In general, the experimental design model involves p parameters (four in this example), and therefore p simultaneous equations. Only r of these are linearly independent; r is the number of groups and is less than p. To test the null hypothesis of equality of treatment population means, and to estimate and test various contrasts among the μ_j, the p normal equations must be transformed into an equivalent set of r linearly independent equations. One way to accomplish this is to place $p - r$ constraints on the elements of ϕ^0. There are several kinds of constraints for the one-factor design.

15.2.2   CONSTRAINING THE SOLUTION SET. $\alpha_3^0 = 0$. A very simple approach is to set one of the elements of ϕ^0 at zero; we arbitrarily have selected α_3^0. This is equivalent to redefining $\mathbf{X}$ and ϕ^0 so that

(15.6)
$$\mathbf{X} = \begin{bmatrix} 1 & 1 & 0 \\ \cdot & \cdot & \cdot \\ \cdot & \cdot & \cdot \\ \cdot & \cdot & \cdot \\ 1 & 0 & 1 \\ \cdot & \cdot & \cdot \\ \cdot & \cdot & \cdot \\ \cdot & \cdot & \cdot \\ 1 & 0 & 0 \end{bmatrix} \quad \text{and} \quad \phi^0 = \begin{bmatrix} \mu^0 \\ \alpha_1^0 \\ \alpha_2^0 \end{bmatrix}.$$

The equivalence exists since

$$\hat{\mathbf{Y}} = \mathbf{X}\phi^0$$

$$= \begin{bmatrix} 1 & 1 & 0 & 0 \\ \cdot & \cdot & \cdot & \cdot \\ \cdot & \cdot & \cdot & \cdot \\ \cdot & \cdot & \cdot & \cdot \\ 1 & 0 & 1 & 0 \\ \cdot & \cdot & \cdot & \cdot \\ \cdot & \cdot & \cdot & \cdot \\ \cdot\cdot & \cdot & \cdot & \cdot \\ 1 & 0 & 0 & 1 \end{bmatrix} \times \begin{bmatrix} \mu \\ \alpha_1 \\ \alpha_2 \\ 0 \end{bmatrix} = \begin{bmatrix} 1 & 1 & 0 \\ \cdot & \cdot & \cdot \\ \cdot & \cdot & \cdot \\ \cdot & \cdot & \cdot \\ 1 & 0 & 1 \\ \cdot & \cdot & \cdot \\ \cdot & \cdot & \cdot \\ 1 & 0 & 0 \end{bmatrix} \times \begin{bmatrix} \mu^0 \\ \alpha_1^0 \\ \alpha_2^0 \end{bmatrix}$$

$$\textbf{(15.7)} \qquad \mathbf{Y} = \begin{bmatrix} \mu^0 + \alpha_1^0 \\ \cdot \\ \cdot \\ \cdot \\ \mu^0 + \alpha_2^0 \\ \cdot \\ \cdot \\ \cdot \\ \mu^0 \end{bmatrix}$$

With the redefined $\mathbf{X}$ matrix of Equation (15.6),

$$\textbf{(15.8)} \qquad \mathbf{X'X} = \begin{bmatrix} n_. & n_1 & n_2 \\ n_1 & n_1 & 0 \\ n_2 & 0 & n_2 \end{bmatrix} \quad \text{and} \quad \mathbf{X'Y} = \begin{bmatrix} T_{..} \\ T_{.1} \\ T_{.2} \end{bmatrix}$$

To get numerical values for the elements of $\boldsymbol{\phi}^0$, we must solve $\mathbf{X'X}\boldsymbol{\phi}^0 = \mathbf{X'Y}$, or

$$\textbf{(15.9)} \qquad \begin{aligned} n_.\mu^0 + n_1\alpha_1^0 + n_2\alpha_2^0 &= T_{..} \\ n_1\mu^0 + n_1\alpha_1^0 \qquad\quad &= T_{.1} \\ n_2\mu^0 \qquad\quad + n_2\alpha_2^0 &= T_{.2} \end{aligned}$$

These can be solved either by ordinary algebra or by calculating $\boldsymbol{\phi}^0 = (\mathbf{X'X})^{-1}\mathbf{X'Y}$. Either way, the result is

$$\textbf{(15.10)} \qquad \boldsymbol{\phi}^0 = \begin{bmatrix} \mu^0 \\ \alpha_1^0 \\ \alpha_2^0 \end{bmatrix} = \begin{bmatrix} \bar{Y}_{.3} \\ \bar{Y}_{.1} - \bar{Y}_{.3} \\ \bar{Y}_{.2} - \bar{Y}_{.3} \end{bmatrix}$$

Earlier, we commented that the elements of $\boldsymbol{\phi}^0$ are not necessarily estimates of the parameters of the original design model of Equation (15.2); they are instead merely one of many sets of values that satisfy Equation (15.5). The meaning of this may be more apparent after considering Equation (15.10). The vector on the far right does not look much like any set of estimates of μ, α_1, and α_2 that we have encountered so far.

From equations (15.7) and (15.10), we find, as we might expect, that the least-squares prediction for any subject is the appropriate group mean:

$$\begin{aligned} \hat{Y}_1 &= \mu^0 + \alpha_1^0 = \bar{Y}_{.1} \\ \hat{Y}_2 &= \mu^0 + \alpha_2^0 = \bar{Y}_{.2} \\ \hat{Y}_3 &= \mu^0 \qquad\; = \bar{Y}_{.3} \end{aligned}$$

From Chapter 14, we know that

$$\textbf{(15.11)} \qquad SS_Y = \boldsymbol{\phi}^0 \mathbf{X'Y} - \mathbf{C}$$

$$= \bar{Y}_{.3}T_{..} + (\bar{Y}_{.1} - \bar{Y}_{.3})T_{.1} + (\bar{Y}_{.2} - \bar{Y}_{.3})T_{.2} - \frac{T_{..}^2}{n_.}$$

Some algebraic manipulation would show that this is identical to SS_A.

Table 15–1   Data and analysis of variance for a one-factor study

GROUP	RAW SCORES	MEANS
A_1	4, 8, 7, 9, 7	$\bar{Y}_{.1} = \frac{35}{5} = 7.00$
A_2	2, 5, 6	$\bar{Y}_{.2} = \frac{13}{3} = 4.33$
A_3	4, 1, 3, 1	$\bar{Y}_{.3} = \frac{9}{4} = 2.25$

SV	df	SS	MS	F
Total	11	80.250		
A	2	50.833	25.417	7.776[a]
S/A	9	29.417	3.269	

[a] $p < .025$

For many readers, the developments of the section may be best clarified by a numerical example. Table 15–1 presents a data set, together with a summary of an analysis of variance based on the formulas of Chapter 4. With **X** coded as in Equation (15.6), we arrive at the normal equations, (15.9):

(15.12)
$$12\mu^0 + 5\alpha_1^0 + 3\alpha_2^0 = 57$$
$$5\mu^0 + 5\alpha_1^0 \qquad\quad = 35$$
$$3\mu^0 \qquad\quad 3\alpha_2^0 = 13$$

Solving, we obtain

(15.13)
$$\boldsymbol{\phi}^0 = \begin{bmatrix} \mu^0 \\ \alpha_1^0 \\ \alpha_2^0 \end{bmatrix} = \begin{bmatrix} 2.250 \\ 4.750 \\ 2.08\bar{3} \end{bmatrix}$$

which is consistent with Equation (15.10). Next, when Equation (15.11) is applied, $SS_{\hat{Y}}$ is calculated and found identical to SS_A as calculated by the usual methods.

If the inverse of **X'X** is calculated, tests of contrasts are easily obtained within the regression framework. For the numerical example (with **X'X** defined by Equation (15.8)),

(15.14)
$$(\mathbf{X'X})^{-1} = \begin{bmatrix} .25 & -.25 & -.25 \\ -.25 & .45 & .25 \\ -.25 & .25 & .583 \end{bmatrix}$$

In Chapter 13, we proved that $\hat{\sigma}_{b_k}^2 = c_{kk}\hat{\sigma}_e^2$; c_{kk} is the entry in row k and column k of $(\mathbf{X'X})^{-1}$. Therefore, to test $H_0: \mu_1 - \mu_3 = 0$, note that α_1^0 estimates $\mu_1 - \mu_3$, and calculate

$$F = \frac{(\alpha_1^0)^2}{\hat{\sigma}_{\alpha_1^0}^2} = \frac{(\alpha_1^0)^2}{c_{11}MS_{S/A}} = \frac{(4.75)^2}{(.45)(3.269)} = 15.34$$

which is significant at the .01 level on one and nine *df*. Note that the result is exactly what we should obtain if we had computed the square of the usual *t* statistic:

$$\frac{(\bar{Y}_1 - \bar{Y}_3)^2}{[(1/n_1)+(1/n_3)]MS_{S/A}}$$

Suppose that we wanted to test $H_0 : \mu_1 - \mu_2 = 0$. Although the contrast of interest is not estimated by any single element of $\boldsymbol{\phi}^0$, we find that $\alpha_1^0 - \alpha_2^0$ provides an estimate; this can be seen by referring to Equation (15.10). This statistic is a linear combination of two random variables. The general form is

(15.15) $$L = w_0\phi_0^0 + w_1\phi_1^0 + \cdots + w_k\phi_k^0$$

and

(15.16) $$\hat{\sigma}_L^2 = w_0^2\hat{\sigma}_{\phi_0^0}^2 + \cdots + w_k^2\hat{\sigma}_{\phi_k^0}^2 + 2w_1w_2 \text{ cov } (\phi_0^0, \phi_1^0)$$
$$+ \cdots + 2w_{k-1}w_k \text{ cov } (\phi_{k-1}^0, \phi_k^0)$$

where cov represents the estimate of the population covariance and ϕ_k^0 is the kth (from zero to k) element in $\boldsymbol{\phi}^0$. We previously noted that

$$\hat{\sigma}_{\phi_k^0}^2 = c_{kk}MS_{\text{error}}$$

Also,

$$\text{cov } (\phi_k^0, \phi_{k'}^0) = c_{kk'}MS_{\text{error}}$$

In testing $H_0 : \mu_1 - \mu_2 = 0$, we get

$$L = (0)\mu^0 + (1)\alpha_1^0 + (-1)\alpha_2^0 = 2.\bar{6}$$

and

$$\hat{\sigma}_L^2 = (c_{11} + c_{22} - 2c_{12})MS_{\text{error}}$$
$$= (.45 + .58\bar{3} - .50)(3.269) = 1.743$$

The test of H_0 is $F = (2.\bar{6})^2/1.743 = 4.079$.

$\sum_{j=1}^a \alpha_j^0 = 0$. In Section 15.2.1, we argued that there were many possible solution sets for the experimental design model represented by Equations (15.2) and (15.3) because those equations represent a system of deficient rank. We have just obtained one possible solution set by imposing the constraint $\alpha_3^0 = 0$. Let us consider a different constraint. This should help us understand in what ways different solution sets are equivalent and in what ways they are not. This time, we require that $\sum \alpha_i^0 = 0$; then, $\alpha_3^0 = -(\alpha_1^0 + \alpha_2^0)$. We have

$$\mathbf{Y} = \mathbf{X}\boldsymbol{\phi}^0 = \begin{bmatrix} 1 & 1 & 0 & 0 \\ \cdot & \cdot & \cdot & \cdot \\ \cdot & \cdot & \cdot & \cdot \\ \cdot & \cdot & \cdot & \cdot \\ 1 & 0 & 1 & 0 \\ \cdot & \cdot & \cdot & \cdot \\ \cdot & \cdot & \cdot & \cdot \\ \cdot & \cdot & \cdot & \cdot \\ \cdot & \cdot & \cdot & \cdot \\ 1 & 0 & 0 & 1 \\ \cdot & \cdot & \cdot & \cdot \\ \cdot & \cdot & \cdot & \cdot \\ \cdot & \cdot & \cdot & \cdot \end{bmatrix} \times \begin{bmatrix} \mu^0 \\ \alpha_1^0 \\ \alpha_2^0 \\ -(\alpha_1^0 + \alpha_2^0) \end{bmatrix}$$

$$= \begin{bmatrix} 1 & 1 & 0 \\ \cdot & \cdot & \cdot \\ \cdot & \cdot & \cdot \\ \cdot & \cdot & \cdot \\ 1 & 0 & 1 \\ \cdot & \cdot & \cdot \\ \cdot & \cdot & \cdot \\ \cdot & \cdot & \cdot \\ 1 & -1 & -1 \\ \cdot & \cdot & \cdot \\ \cdot & \cdot & \cdot \\ \cdot & \cdot & \cdot \end{bmatrix} \times \begin{bmatrix} \mu^0 \\ \alpha_1^0 \\ \alpha_2^0 \end{bmatrix}$$

(15.17)

$$= \begin{bmatrix} \mu^0 + \alpha_1^0 \\ \cdot \\ \cdot \\ \mu^0 + \alpha_2^0 \\ \cdot \\ \cdot \\ \mu^0 \\ \cdot \\ \cdot \end{bmatrix}$$

Under the redefined code of Equation (15.17),

(15.18) $$\mathbf{X'X} = \begin{bmatrix} n_{\cdot} & n_1 - n_3 & n_2 - n_3 \\ n_1 - n_3 & n_1 + n_3 & n_3 \\ n_2 - n_3 & n_3 & n_2 + n_3 \end{bmatrix} \qquad \mathbf{X'Y} = \begin{bmatrix} T_{\cdot\cdot} \\ T_{\cdot 1} - T_{\cdot 3} \\ T_{\cdot 2} - T_{\cdot 3} \end{bmatrix}$$

To estimate parameters and test $H_0: \mu_1 = \mu_2 = \mu_3$, we must solve the normal equations, $\mathbf{X'X}\boldsymbol{\phi}^0 = \mathbf{X'Y}$:

$$\begin{aligned}
n_{.}\mu^0 + (n_1 - n_3)\alpha_1^0 + (n_2 - n_3)\alpha_2^0 &= T_{..} \\
(n_1 - n_3)\mu^0 + (n_1 + n_3)\alpha_1^0 + n_3\alpha_2^0 &= T_{.1} - T_{.3} \\
(n_2 - n_3)\mu^0 + n_3\alpha_1^0 + (n_2 + n_3)\alpha_2^0 &= T_{.2} - T_{.3}
\end{aligned}$$

(15.19)

The solution is

(15.20)
$$\boldsymbol{\phi}^0 = \begin{bmatrix} \mu^0 \\ \alpha_1^0 \\ \alpha_2^0 \end{bmatrix} = \begin{bmatrix} \dfrac{\sum_{j=1}^3 \bar{Y}_{.j}}{3} \\ \bar{Y}_{.1} - \mu^0 \\ \bar{Y}_{.2} - \mu^0 \end{bmatrix}$$

From Equations (15.17) and (15.20), we again find that the group mean is the best predictor of an individual's score:

(15.21)
$$\begin{aligned}
\hat{Y}_1 &= \mu^0 + \alpha_1^0 = \bar{Y}_{.1} \\
\hat{Y}_2 &= \mu^0 + \alpha_2^0 = \bar{Y}_{.2} \\
\hat{Y}_3 &= \mu^0 - \alpha_1^0 - \alpha_2^0 = \bar{Y}_{.3}
\end{aligned}$$

With this outcome, it follows that the variability of the predicted scores about their mean will equal SS_A:

$$\begin{aligned}
SS_{\hat{Y}} &= (\boldsymbol{\phi}^0)'\mathbf{X'Y} - \mathbf{C} \\
&= \sum_{j=1}^3 n_j(\hat{Y}_j - \bar{Y}_{..})^2 \\
&= \sum_{j=1}^3 n_j(\bar{Y}_{.j} - \bar{Y}_{..})^2 \\
&= SS_A
\end{aligned}$$

You can verify these results by the data of Table 15–1.

The two coding systems we have considered, and many others, will yield the same set of $\hat{Y}$ and consequently the same value of SS_A. Different constraints on the solution set (or equivalently, different codings of the $\mathbf{X}$ matrix) will differ only in the regression coefficients. For example, the two constraints investigated in this section resulted in different $\boldsymbol{\phi}^0$ vectors, as you can readily see by comparing Equations (15.10) and (15.20). Even this difference between the two solution sets is not critical. Suppose that we wished to test $H_0: \mu_1 - \mu_3 = 0$. Setting α_3^0 equal to zero provides a direct test; the calculations were illustrated in this section. The hypothesis can also be tested, however, with the constraint that $\sum_{j=1}^3 \alpha_j^0 = 0$. Under this system, $2\alpha_1^0 + \alpha_2^0$ provides an estimate of $\mu_1 - \mu_3$; this follows from subtracting the expression for $\bar{Y}_{.3}$ from that for $\bar{Y}_{.1}$ in Equation (15.21). Equation (15.16) provides the formula for the variance of such linear combinations of regression coefficients, and after the inverse of the $\mathbf{X'X}$ defined by Equation (15.18) is calculated, the appropriate F test can be computed. In general, such linear combinations of ϕ_k^0 will provide estimates and tests for parameters that have not been estimated under the constraint imposed.

15.2.3 CONTRAST MATRICES.   In Chapter 11, we provided an expression for the sum of squares due to a contrast among the groups means. From Equation (11.16), we know that the sum of squares for such contrasts is

$$(\mathbf{15.22}) \qquad SS_{\hat{\psi}} = \frac{(\sum_{j=1}^{a} w_j \bar{Y}_{.j})^2}{\sum_{j=1}^{a} (w_j^2/n_j)}$$

The above term is distributed on one *df*. Suppose that we wanted to test $H_0 : 2\mu_1 - \mu_2 - \mu_3 = 0$. Given the data of Table 15–1, we have

$$SS_{\hat{\psi}} = \frac{(14 - 4.\bar{3} - 2.25)^2}{\frac{4}{5} + \frac{1}{3} + \frac{1}{4}}$$

$$= 39.764$$

The approach can be extended to yield a sum of squares due to a set of K contrasts; this sum of squares is on K *df*. The value K must be less than or equal to $a - 1$, or in general, the number of cell means minus 1. For the example presented in Table 15–1, we shall get SS_A by calculations based on two linearly independent contrasts of the set of three group means.

Before proceeding, we must define some terms. First, let $\mathbf{W}$ be a $K \times a$ contrast matrix; each row consists of a set of a weights w_{jk} such that $\sum_j w_{jk} = 0$. The contrasts are linearly independent, which is another way of saying that the matrix has rank K. For our three-group example, we might have

$$(\mathbf{15.23}) \qquad \mathbf{W} = \begin{bmatrix} 1 & -1 & 0 \\ 2 & -1 & -1 \end{bmatrix}$$

The matrix $\mathbf{X}$ will be an $n \times a$ *incidence matrix* with ones representing presence in a group and zeros representing absence. In our example,

$$(\mathbf{15.24}) \qquad \mathbf{X} = \begin{bmatrix} 1 & 0 & 0 \\ 1 & 0 & 0 \\ 1 & 0 & 0 \\ 1 & 0 & 0 \\ 1 & 0 & 0 \\ 0 & 1 & 0 \\ 0 & 1 & 0 \\ 0 & 1 & 0 \\ 0 & 0 & 1 \\ 0 & 0 & 1 \\ 0 & 0 & 1 \\ 0 & 0 & 1 \end{bmatrix}$$

For such matrices,

$$(15.25) \quad \mathbf{X'X} = \begin{bmatrix} n_1 & 0 & \cdots & 0 \\ 0 & n_2 & \cdots & 0 \\ & & \cdots & \\ 0 & 0 & \cdots & n_a \end{bmatrix} \quad \text{and} \quad (\mathbf{X'X})^{-1} = \begin{bmatrix} n_1^{-1} & 0 & \cdots & 0 \\ 0 & n_2^{-1} & \cdots & 0 \\ & & \cdots & \\ 0 & 0 & \cdots & n_a^{-1} \end{bmatrix}$$

Given the n_j of Table 15–1, we get

$$\mathbf{X'X} = \begin{bmatrix} 5 & 0 & 0 \\ 0 & 3 & 0 \\ 0 & 0 & 4 \end{bmatrix} \quad \text{and} \quad (\mathbf{X'X})^{1-} = \begin{bmatrix} \frac{1}{5} & 0 & 0 \\ 0 & \frac{1}{3} & 0 \\ 0 & 0 & \frac{1}{4} \end{bmatrix}$$

Then,

$$(15.26) \quad SS_{\hat{\psi}} = (\mathbf{W\bar{Y}})' \, [\mathbf{W}(\mathbf{X'X})^{-1} \, \mathbf{W'}]^{-1} \, (\mathbf{W\bar{Y}})$$

where

$$\mathbf{\bar{Y}}' = [\bar{Y}_{.1} \cdots \bar{Y}_{.j} \cdots \bar{Y}_{.a}]$$

If $\mathbf{W}$ has only one row, Equations (15.22) and (15.26) are equivalent. We can see this by noting that

$$\mathbf{W\bar{Y}} = \sum_j w_j \bar{Y}_j$$

and

$$[\mathbf{W}(\mathbf{X'X})^{-1} \, \mathbf{W'}]^{-1} = \sum_j \frac{w_j^2}{n_j}$$

If $\mathbf{W}$ has $a - 1$ linearly independent rows, $SS_{\hat{\psi}}$ can be proved equal to SS_A for the one-factor study. Assume, for example, that $\mathbf{W}$ is defined by Equation (15.23) and $\mathbf{X}$ by (15.24). For the data of Table 15–1,

$$\mathbf{W\bar{Y}} = \begin{bmatrix} \frac{8}{3} \\ \frac{89}{12} \end{bmatrix}$$

and

$$[\mathbf{W}(\mathbf{X'X})^{-1} \, \mathbf{W'}] = \begin{bmatrix} 1 & -1 & 0 \\ 2 & -1 & -1 \end{bmatrix} \begin{bmatrix} \frac{1}{5} & 0 & 0 \\ 0 & \frac{1}{3} & 0 \\ 0 & 0 & \frac{1}{4} \end{bmatrix} \begin{bmatrix} 1 & 2 \\ -1 & -1 \\ 0 & -1 \end{bmatrix}$$

$$= \begin{bmatrix} \frac{1}{5} & -\frac{1}{3} & 0 \\ \frac{2}{5} & -\frac{1}{3} & -\frac{1}{4} \end{bmatrix} \begin{bmatrix} 1 & 2 \\ -1 & -1 \\ 0 & -1 \end{bmatrix}$$

$$= \begin{bmatrix} \frac{8}{15} & \frac{11}{15} \\ \frac{11}{15} & \frac{83}{60} \end{bmatrix}$$

The inverse is

$$\begin{bmatrix} \frac{83}{12} & -\frac{44}{12} \\ -\frac{44}{12} & \frac{32}{12} \end{bmatrix}$$

and from Equation (15.26),

$$
\begin{aligned}
SS_{\hat{\psi}} &= \begin{bmatrix} \frac{8}{3} & \frac{89}{12} \end{bmatrix} \begin{bmatrix} \frac{83}{12} & -\frac{44}{12} \\ -\frac{44}{12} & \frac{32}{12} \end{bmatrix} \begin{bmatrix} \frac{8}{3} \\ \frac{89}{12} \end{bmatrix} \\
&= \begin{bmatrix} \frac{8}{3} & \frac{89}{12} \end{bmatrix} \begin{bmatrix} -\frac{35}{4} \\ 10 \end{bmatrix} \\
&= 50.8\bar{3}
\end{aligned}
$$

a result that has appeared frequently in this chapter; we again have calculated SS_A.

   We now have seen two approaches to computing the SS_A. In Section 5.3.2, a constraint was placed on the design matrix, resulting in an X matrix of full rank. The sum of squares for regression was then computed as in ordinary multiple regression problems. Many constraints yield the same sum of squares; the difference is only in the elements of $\boldsymbol{\phi}^0$. In the present section, we began with a specific set of contrasts in mind; $\mathbf{W}\bar{\mathbf{Y}}$, the vector of $a-1$ contrasts, is that set. As long as the sets of weights in $\mathbf{W}$ are linearly independent, Equation (15.26) yields SS_A directly. The relative merits of the two procedures will be clearer in the context of a multifactor design. Therefore, we reserve further discussion of these computing approaches to Section 15.3, in which a two-factor example will be considered.

15.2.4 SELECTING WEIGHTS FOR CONTRASTS.   In the example of Table 15–1, assume that A_1 represents a sample from a population of control subjects C, and A_2 and A_3 represent samples from two different experimental populations E_1 and E_2. Suppose that we want to test $H_0: \mu_C - \mu_E = 0$, where μ_E is the average of the combined experimental populations. The weights for this contrast will depend on our assumptions about the relative sizes of the populations. We shall consider two cases.

   Populations of equal size. Assume that the differences among the n_js are due to chance and do not reflect any systematic differences in size of populations. Then,

$$\mu_E = \tfrac{1}{2}(\mu_2 + \mu_3) \quad \text{and} \quad \hat{\mu}_E = \tfrac{1}{2}(\bar{Y}_{.2} + \bar{Y}_{.3})$$

Furthermore, $H_0: \mu_C - \mu_E = 0$ is equivalent to $H_0: \mu_1 - \tfrac{1}{2}(\mu_2 + \mu_3) = 0$. The appropriate weights for testing this null hypothesis are

$$\mathbf{W} = \begin{bmatrix} 1 & -\tfrac{1}{2} & -\tfrac{1}{2} \end{bmatrix}$$

or any linear transformation of these values. Therefore, 2, -1, and -1 are allowable weights, and as we calculated immediately following Equation (15.22), $SS_{\hat{\psi}} = 39.764$.

Picturing the populations equal in size implies a restriction on the a_j, namely, $\sum_j \alpha_j = 0$. If the populations are of equal size, $\mu = (1/a) \sum \mu_j$; it follows algebraically that $\sum_j (\mu_j - \mu) = 0$. Note that this is a *restriction on the parameters;* it should be distinguished from the constraints on the solution set introduced in Section 15.2.2. The restrictions are statements about the populations from which we have sampled and form an important component of the model. Constraints, on the other hand, are merely mathematical devices introduced to generate a system of normal equations of full rank and therefore one that can be solved.

Populations of unequal sizes. Sometimes, groups represent samples from populations of different sizes. We might have several ethnic groups, personality types, levels of education, or levels of income. In such cases, equally weighting the group means does not provide an unbiased estimate of the mean of the combined populations. Let p_j be the proportion of the combined populations that is in the jth population. Then, $\mu = \sum_j p_j \mu_j$ and the $\bar{Y}_{.j}$ should be weighted in the same way to estimate μ. With this in mind, reconsider our example in which A_1 is a control condition and A_2 and A_3 are experimental conditions. Estimating μ_E, the mean of the combined experimental populations, we get

$$\mu_E = \frac{p_2 \mu_2 + p_3 \mu_3}{p_2 + p_3}$$

Therefore, $H_0: \mu_C - \mu_E = 0$ is equivalent to the null hypothesis that

$$\mu_1 - \left[\left(\frac{p_2}{p_2 + p_3} \right) \mu_2 + \left(\frac{p_3}{p_2 + p_3} \right) \mu_3 \right] = 0$$

Multiplying by $p_2 + p_3$ simplifies the expression.

Assume that the n_j of Table 15–1 are proportional to the p_j; that is, $n_j/n_. = p_j$. Then, from Equation (15.22),

$$SS_{\hat{\psi}} = \frac{[(n_2 + n_3)\bar{Y}_{.1} - n_2 \bar{Y}_{.2} - n_3 \bar{Y}_{.3}]^2}{(n_2 + n_3)^2/n_1 + n_2^2/n_2 + n_3^2/n_3}$$

$$= \frac{[(7)(7) - 3(4.\bar{3}) - 4(2.25)]^2}{7^2/5 + 3^2/3 + 4^2/4}$$

$$= 43.393$$

Note that this is not the result we obtained when we assumed that the populations were of equal size and accordingly used weights of 1, $-\frac{1}{2}$, and $-\frac{1}{2}$.

Weighting the means of the treatment population implies the following restriction on the treatment-effect parameters: $\sum w_j \alpha_j = 0$, where as usual, $\alpha_j = \mu_j - \mu$. The restriction follows from

$$\sum w_j \alpha_j = \sum w_j (\mu_j - \mu)$$
$$= \sum w_j \mu_j - \mu \sum w_j$$
$$= \mu - \mu = 0$$

since $\mu = \sum w_j \mu_j$ and $\sum w_j = 1$ (the w_j are proportions).

The developments of this section illustrate an important point: two tests of the "same" null hypothesis (for example, $\mu_C - \mu_E = 0$) will yield different results, depending on the restrictions placed on population parameters. This point will arise again in our discussion of two-factor designs. Assume, for example, a 2×3 design with unequal cell frequencies. If the six treatment populations are assumed to be of equal size (that is, variation in cell frequency is a chance occurrence), $\mu_{j.}$ is best estimated by $\sum_k \bar{Y}_{.jk}/3$. If the cell frequencies are assumed proportional to population sizes, the best estimate of $\mu_{j.}$ would be $T_{.j.}/n_{j.}$. Calculations of SS_A will depend on which assumptions are held. (Alternative computational approaches will be presented in Section 15.3.)

15.3   THE TWO-FACTOR DESIGN

15.3.1   A NUMERICAL EXAMPLE.   The developments of Section 15.3 will be easier to follow if we have a numerical example in front of us. Table 15–2 presents a data set for 14 subjects. If the A and B classifications are ignored and the design treated as though it represented one factor with six levels, the variability among the cell means readily can be computed:

$$SS_{\text{cells}} = \sum_j \sum_k \frac{T_{.jk}^2}{n_{jk}} - C$$

$$= 337.\overline{3} - 292.571429$$

$$= 44.761905$$

Within the multiple regression framework, it will be useful to relabel this quantity $SS_{A,B,AB}$. This will indicate variability about the grand mean that is accounted for by estimating the main and interaction effects. Then, we can use notation such as $SS_{A,B}$ to indicate variability due to A and B when the interaction is ignored; and

Table 15–2   Data for a two-factor study with disproportionate cell frequencies

GROUP	B	B_2	B_3	$T_{.j.}$
A_1	4, 5	3, 1, 5, 1	6, 7	32
A_2	6	2, 4	6, 6, 8	32
$T_{..k} =$	15	16	33	$T_{...} = 64$

$$SS_{\text{cells}} = \frac{9^2}{2} + \frac{10^2}{4} + \cdots + \frac{20^2}{3} - \frac{64^2}{14}$$

$$= 337.\overline{3} - 292.571429 = 44.761905$$

$$= SS_{A,B,AB}$$

$$SS_{\text{error}} = SS_{\text{total}} - SS_{\text{cells}}$$

$$= 61.428571 - 44.761905$$

$$= 16.\overline{6}$$

we can use $SS_{A|B,AB}$ to represent the variability due to A after adjusting for variability attributable to B and to the interaction. Consistent with the developments of Chapter 14, this last term would equal $SS_{A,B,AB} - SS_{B,AB}$.

Error variability is calculated as the difference between the total variability in the data set and variability attributable to A, B, and AB; that is,

$$SS_{\text{error}} = SS_{\text{total}} - SS_{A,B,AB}$$
$$= 61.428571 - 44.761905$$
$$= 16.\overline{6}$$

The error sum of squares is on $n.. - ab\ df$, $14 - 6 = 8$, in the example of Table 15–2.

The experimental design model has twelve parameters—μ, two α_j, three β_k, and six $(\alpha\beta)_{jk}$. Various sets of constraints like those given in Section 15.2.2 will yield a set of six linearly independent simultaneous equations, that is, a system of normal equations of full rank. Then, $\boldsymbol{\phi}^{0\prime}\mathbf{X}'\mathbf{Y} - \mathbf{C}$ will yield the same value of $SS_{A,\,B,\,AB}$ that was computed earlier. This outcome might suggest that the choice of constraints is not important. This is not so, however, since certain sums of squares will be different under different sets of constraints. (We shall return to this point; in the material immediately following, calculations will be developed with one system of constraints.)

15.3.2   −1, 0, 1 CODING.   We choose to constrain the solution set in the following way:

$$\text{(15.27)} \qquad\qquad \sum_j \alpha_j^0 = 0$$

$$\text{(15.28)} \qquad\qquad \sum_k \beta_k^0 = 0$$

$$\text{(15.29)} \qquad \sum_k (\alpha\beta)_{jk}^0 = 0, \qquad \sum_j (\alpha\beta)_{jk}^0 = 0$$

Given the 2×3 design of Table 15–2, Equations (15.27) through (15.29) imply

$$\text{(15.27}')\qquad\qquad \alpha_2^0 = -\alpha_1^0$$
$$\text{(15.28}')\qquad\qquad \beta_3^0 = -\beta_1^0 - \beta_2^0$$
$$(\alpha\beta)_{13}^0 = -(\alpha\beta)_{11}^0 - (\alpha\beta)_{12}^0$$
$$(\alpha\beta)_{21}^0 = -(\alpha\beta)_{11}^0$$
$$\text{(15.29}')\qquad\qquad (\alpha\beta)_{22}^0 = -(\alpha\beta)_{12}^0$$
$$(\alpha\beta)_{23}^0 = (\alpha\beta)_{11}^0 + (\alpha\beta)_{12}^0$$

The solution set can now be expressed by six elements:

$$\text{(15.30)} \qquad \boldsymbol{\phi}^{0\prime} = [\mu^0 \quad \alpha_1^0 \quad \beta_1^0 \quad \beta_2^0 \quad (\alpha\beta)_{11}^0 \quad (\alpha\beta)_{12}^0]$$

We can now construct a 14×6 (in general, $n.. \times ab$) $\mathbf{X}$ matrix such that $\mathbf{X}\boldsymbol{\phi}$ will yield predicted scores and $\boldsymbol{\phi}^{0\prime}\mathbf{X}'\mathbf{Y}$ will yield $SS_{A,B,AB}$:

$$(\mathbf{15.31}) \qquad \mathbf{X} = \begin{bmatrix} 1 & 1 & 1 & 0 & 1 & 0 \\ \cdots & \cdots & \cdots & \cdots & \cdots & \cdots \\ 1 & 1 & 0 & 1 & 0 & 1 \\ \cdots & \cdots & \cdots & \cdots & \cdots & \cdots \\ 1 & 1 & -1 & -1 & -1 & -1 \\ \cdots & \cdots & \cdots & \cdots & \cdots & \cdots \\ 1 & -1 & 1 & 0 & -1 & 0 \\ \cdots & \cdots & \cdots & \cdots & \cdots & \cdots \\ 1 & -1 & 0 & 1 & 0 & -1 \\ \cdots & \cdots & \cdots & \cdots & \cdots & \cdots \\ 1 & -1 & -1 & -1 & 1 & 1 \end{bmatrix}$$

Note that the rows are in the order $A_1B_1, A_1B_2, A_1B_3, A_2B_1, A_2B_2, A_2B_3$.

We may better understand the rationale underlying the pattern of -1s, 0s, and 1s if we consider the product matrix $\hat{\mathbf{Y}} = \mathbf{X}\boldsymbol{\phi}^0$:

$$(\mathbf{15.32}) \qquad \hat{\mathbf{Y}} = \begin{bmatrix} \mu^0 + \alpha_1^0 + \beta_1^0 + (\alpha\beta)_{11}^0 \\ \cdots \cdots \cdots \cdots \cdots \cdots \\ \mu^0 + \alpha_1^0 + \beta_2^0 + (\alpha\beta)_{12}^0 \\ \cdots \cdots \cdots \cdots \cdots \cdots \\ \mu^0 + \alpha_1^0 - \beta_1^0 - \beta_2^0 - (\alpha\beta)_{11}^0 - (\alpha\beta)_{12}^0 \\ \cdots \cdots \cdots \cdots \cdots \cdots \\ \mu^0 - \alpha_1^0 + \beta_1^0 - (\alpha\beta)_{11}^0 \\ \cdots \cdots \cdots \cdots \cdots \cdots \\ \mu^0 - \alpha_1^0 + \beta_2^0 - (\alpha\beta)_{12}^0 \\ \cdots \cdots \cdots \cdots \cdots \cdots \\ \mu^0 - \alpha_1^0 - \beta_1^0 - \beta_2^0 + (\alpha\beta)_{11}^0 + (\alpha\beta)_{12}^0 \end{bmatrix}$$

In view of the relations expressed in Equations (15.27') through (15.29'), each row of Equation (15.32) is of the general form

$$\hat{Y} = \mu^0 + \alpha_j^0 + \beta_k^0 + (\alpha\beta)_{jk}^0$$

which is analogous to the general model for the two-factor design.

It will be useful to have a standard procedure for generating the elements of $\boldsymbol{\phi}^0$ and $\mathbf{X}$ for any size of factorial design. The solution vector contains $\mu^0, \alpha_1^0, \ldots,$ $\alpha_{a-1}^0, \beta_1^0, \ldots, \beta_{b-1}^0$; interaction elements result from crossing the preceding elements. The matrix $\mathbf{X}$ is constructed by viewing each column as though it corresponded to an element of $\boldsymbol{\phi}^0$. In Equation (15.31), the first column corresponds to μ^0 and contains all ones because μ^0 contributes to all predicted scores. The next column corresponds to α_1^0; all cells at A_1 (the first three rows in that column) receive an entry of 1 and all cells at A_2 receive a -1. The next two columns represent β_1^0 and β_2^0. Enter a 1 for the B_1 cells of the first of these columns and for the B_2 cells for the second of the columns. Enter a -1 for the B_3 cells in both columns because β_3^0 has been eliminated from the solution set by the constraints placed on it. Zeros are entered in all other cells in these columns. The next column corresponds to $(\alpha\beta)_{11}^0$; its entries are generated by multiplying the values in the α_1^0 and β_1^0 columns. The same holds true for any interaction column.

15.3.3   CALCULATIONS.   With methods developed in Chapter 13, we can solve the set of six normal equations $\mathbf{X'X\phi^0} = \mathbf{X'Y}$ for numerical values of the elements of $\mathbf{\phi^0}$. We first calculate

$$(\textbf{15.33}) \quad \mathbf{X'X} = \begin{bmatrix} 14 & 2 & -2 & 1 & 2 & 3 \\ 2 & 14 & 2 & 3 & -2 & 1 \\ -2 & 2 & 8 & 5 & 0 & -1 \\ 1 & 3 & 5 & 11 & -1 & 1 \\ 2 & -2 & 0 & -1 & 8 & 5 \\ 3 & 1 & -1 & 1 & 5 & 11 \end{bmatrix} \quad \text{and} \quad \mathbf{X'Y} = \begin{bmatrix} 64 \\ 0 \\ -18 \\ -17 \\ 10 \\ 11 \end{bmatrix}$$

Solving the normal equations for the elements of $\mathbf{\phi^0}$ yields

$$(\textbf{15.34}) \qquad \mathbf{\phi^0} = \begin{bmatrix} \mu^0 = \dfrac{\sum_j \sum_k \bar{Y}_{.jk}}{6} \\[2mm] \alpha_1^0 = \dfrac{\sum_k \bar{Y}_{.1k}}{3} - \mu^0 \\[2mm] \beta_1^0 = \dfrac{\sum_j \bar{Y}_{.j1}}{2} - \mu^0 \\[2mm] \beta_2^0 = \dfrac{\sum_j \bar{Y}_{.j2}}{2} - \mu^0 \\[2mm] (\alpha\beta)_{11}^0 = \bar{Y}_{.11} - \mu^0 - \alpha_1^0 - \beta_1^0 \\[2mm] (\alpha\beta)_{12}^0 = \bar{Y}_{.12} - \mu^0 - \alpha_1^0 - \beta_2^0 \end{bmatrix} = \begin{bmatrix} 4.86\bar{1} \\ -.36\bar{1} \\ .3\bar{8} \\ -2.\bar{1} \\ -.3\bar{8} \\ .\bar{1} \end{bmatrix}$$

It might be a good idea to pause to think about what the elements of $\mathbf{\phi^0}$ estimate. It should be apparent that μ^0 is an unbiased estimate of the unweighted average of the μ_{jk}; that is,

$$\mu^0 \triangleq \frac{\sum\sum \mu_{jk}}{ab}$$

It is also an unbiased estimate of μ, the grand mean calculated for the set of populations *if, and only if,* we assume that the ab populations are of equal size, that the fluctuations among the n_{jk} are a chance occurrence. In that case, the unweighted average of the treatment population means equals μ, and accordingly, the unbiased estimate of that unweighted average must also be an unbiased estimate of μ. If the n_{jk} are assumed to represent true discrepancies in population sizes, the μ_{jk} must be weighted when μ is calculated; then, μ^0 is not an unbiased estimate of μ. The same qualifications hold when one is trying to interpret the other elements of $\mathbf{\phi^0}$. They estimate population main and interaction effects only under appropriate restrictions on the parameters of the design model. You might refer to Section 15.2.4 for additional discussion of these matters.

The value $SS_{A,B,AB}$ can be calculated by $\phi^{0\prime}X'Y - C$. As we should expect, the result is identical to that in Table 15–2, which was obtained in the usual manner. Partitioning this variability into A, B, and AB sources of variance is another matter. As we noted in Chapter 5, the usual computational formulas are inappropriate when cell frequencies are disproportionate. Instead, our approach must follow the one used in multiple regression analysis. When we considered such analyses in Chapter 14, a fundamental issue was the order in which variables were to be entered into the regression equation. The same problem arises in the present context. We have, for example, several choices in trying to assess the effects of A. We can measure the variability due to A when all other main and interaction effects are ignored; in this case, we calculate SS_A, the usual sum of squares. Alternatively, we can calculate the variability due to A after adjusting for the effects of all other main effects; in our running example, this would be $SS_{A|B}$. Still another possibility is to attribute to A that variability remaining after adjustment for all other main and interaction effects; this would be $SS_{A|B,AB}$. (We shall consider these alternatives in the pages that follow.)

Adjusting for all other main and interaction effects (*Method* 1). We first calculate

(15.35) $$SS_{A|B,AB} = SS_{A,B,AB} - SS_{B,AB}$$

To compute $SS_{B,AB}$, delete the second column from the X matrix of Equation (15.31) and also delete α_1^0 from ϕ^0. These reduced matrices will be referred to as X^* and ϕ^{0*}. The deletion operation on X results in deleting the second row and column from $X'X$ to obtain $X^{*\prime}X^*$ and the second element in $X'Y$ to obtain $X^{*\prime}Y$. Therefore, from Equation (15.33),

$$X^{*\prime}X^* = \begin{bmatrix} 14 & -2 & 1 & 2 & 3 \\ -2 & 8 & 5 & 0 & -1 \\ 1 & 5 & 11 & -1 & 1 \\ 2 & 0 & -1 & 8 & 5 \\ 3 & -1 & 1 & 5 & 11 \end{bmatrix} \quad \text{and} \quad X^{*\prime}Y = \begin{bmatrix} 64 \\ -18 \\ -17 \\ 10 \\ 11 \end{bmatrix}$$

We can now solve the system of five normal equations, $X^{*\prime}X^*(\phi^{0*})' = X^{*\prime}Y$, for the values of ϕ^{0*}:

$$\phi^{0*} = \begin{bmatrix} 4.7\overline{92} \\ .28\overline{153} \\ -2.1306 \\ -.222\overline{306} \\ -.0\overline{135} \end{bmatrix}$$

The sum of squares is calculated by the procedure first developed in Chapter 14:

$$SS_{B,AB} = (\phi^{0*})'X^{*\prime}Y - C$$
$$= 335.\overline{810} - 292.571429$$
$$= 43.239382$$

Substituting in Equation (15.35) yields

$$SS_{A|B,AB} = 44.761905 - 43.239382 = 1.\overline{522}$$

To calculate $SS_{B|A,AB}$ and $SS_{AB|A,B}$ we must first obtain numerical values for $SS_{A,AB}$ and for $SS_{A,B}$. For $SS_{A,AB}$, delete the third and fourth columns of X (that is, those corresponding to β_1^0 and β_2^0) to get the appropriate X^* matrix. The calculations proceed as above and yield

$$\phi^{0*} = \begin{bmatrix} 4.708345 \\ -.643502 \\ .074145 \\ -.259370 \end{bmatrix}$$

$$SS_{A,AB} = 299.222434 - 292.571429 = 6.651006$$

and

$$SS_{B|A,AB} = 44.761905 - 6.651006 = 38.110899$$

To compute $SS_{A,B}$, delete the last two columns of X (the columns corresponding to the $(\alpha\beta)^0$ elements), thus obtaining the appropriate X^*. Then, again solving $X^{*\prime}X^*(\phi^{0*})' = X^{*\prime}Y$, we get

$$\phi^{0*} = \begin{bmatrix} 4.800\overline{925} \\ -.291\overline{6} \\ .\overline{296} \\ -2.\overline{037} \end{bmatrix}$$

$$SS_{A,B} = 336.\overline{5} - 292.571429 = 43.984127$$

and

$$SS_{AB|A,B} = 44.761905 - 43.984127 = .\overline{7}$$

The results of the analysis are summarized in Table 15–3.

Adjusting for effects of the same and a lower order (*Method 2*). In this approach, main effects are adjusted only for all other main effects; the first-order interaction is adjusted for all main effects. If the design includes more factors, the approach is easily extended. In the present example, the sources of variance would be $SS_{A|B}$, $SS_{B|A}$, and $SS_{AB|A,B}$. Since while applying Method 1 we have already calculated $SS_{A,B}$ and $SS_{AB|A,B}$, we now require only SS_A and SS_B to complete the Method 2 analysis. These two sums of squares can be calculated

Table 15–3   Three different analyses of the data of Table 15–2

SV	df	SS	MS	F
		Preliminary Calculations		
A,B,AB	$ab-1=5$	44.762		
A,B	$a+b-2=3$	43.984		
A,AB	$ab-b=3$	6.651		
B,AB	$ab-a=4$	43.239		
A	$a-1=1$	6.095		
B	$b-1=2$	42.895		
Total	$n_{..}-1=63$	61.429		
		Method 1		
A|B,AB	1	$SS_{A,B,AB}-SS_{B,AB}=\ \ 1.522$	1.522	.731
B|A,AB	2	$SS_{A,B,AB}-SS_{A,AB}=38.111$	19.055	9.148[a]
AB|A,B	2	$SS_{A,B,AB}-SS_{A,B}=\ \ .778$	.389	.187
Error	8	$SS_{total}-SS_{A,B,AB}=16.667$	2.083	
		Method 2		
A|B	1	$SS_{A,B}-SS_{B}=\ \ 1.089$	1.089	.523
B|A	2	$SS_{A,B}-SS_{A}=37.889$	18.944	9.095[a]
AB|A,B	2	$SS_{A,B,AB}-SS_{A,B}=\ \ .778$	.389	.187
Error	8	$SS_{total}-SS_{A,B,AB}=16.667$	2.083	
		Method 3		
A	1	$SS_{A}=\ \ 6.095$	6.095	2.926
B|A	2	$SS_{A,B}-SS_{A}=37.889$	18.944	9.095[a]
AB|A,B	2	$SS_{A,B,AB}-SS_{A,B}=\ \ .778$	.389	.187
Error	8	$SS_{total}-SS_{A,B,AB}=16.667$	2.083	

[a] $p<0.01$

exactly as in Chapter 4 for one-factor designs with unequal ns. Thus,

$$SS_A = \frac{T_{.1.}^2}{n_{1.}} + \frac{T_{.2.}^2}{n_{2.}} - C$$

$$= \frac{(32)^2}{8} + \frac{(32)^2}{6} - \frac{(64)^2}{14}$$

$$= 298.\bar{6} - 292.571429$$

$$= 6.095238$$

and

$$SS_{B|A} = SS_{A,B} - SS_A$$

$$= 43.984127 - 6.094238$$

$$= 37.\bar{8}$$

Note that we could also calculate SS_A by deleting the last four columns from the

X matrix of Equation (15.31); then, as usual, $SS_A = (\boldsymbol{\phi}^{0*})'\mathbf{X}^{*'}\mathbf{Y} - \mathbf{C}$, where $\mathbf{X}^*$ is the reduced (14×2) matrix and $\boldsymbol{\phi}^{0*}$ is a two-element vector.

By either of the two procedures just described, $SS_B = 42.895238$ and

$$SS_{A|B} = 43.984127 - 42.895238 = 1.0\overline{8}$$

The results of the Method 2 analysis are also summarized in Table 15–3.

Prior orderings (*Method* 3). It was pointed out in Chapter 14 that a logical (or theoretical) analysis of the relations among independent variables, before the data are viewed, may dictate the order in which those variables are to be entered into the regression equation. A similar approach is sometimes possible with categorical variables. Suppose A is race, the two conditions being black and white. The three levels of B will, furthermore, be assumed to be levels of education: college degree, high school degree, less than high school education. In this case, it is reasonable to assume nonorthogonality in the population; race and educational level are probably correlated, with a greater proportion of whites having college degrees. It is likely that furthermore, one's race is to some degree responsible for the level of education obtained. With these assumptions, it seems appropriate to calculate a sum of squares for race, unadjusted for other sources of variance (that is, SS_A). Because the effects of education are presumed to be partly due to influences of race on education, the contribution of education will be adjusted for race ($SS_{B|A}$). Finally, the variability due to interaction will be adjusted for both main effects ($SS_{AB|A,B}$). All of the necessary terms have been calculated in completing the analyses by Method 1 and by Method 2; the last part of Table 15–3 reorganizes the relevant information for the Method 3 analysis.

In the examples we have considered, only the variability due to B proved significant, and this source was significant in all three analyses. Under some patterns of data and cell frequencies, nevertheless, the three methods of analysis can differ substantially in their results. Only when cell frequencies are identical do the three methods test the same hypotheses about population parameters, and only then will the analyses yield identical results. In such equal-n designs,

and

$$SS_A = SS_{A|B} = SS_{A|B,AB}$$

$$SS_B = SS_{B|A} = SS_{B|A,AB}$$

Beginning with a paper by Overall and Spiegel (1969), the question which method to use has received much discussion—and somewhat less clarification. A good source of references, and a clearer, more coherent statement than most of the papers it refers to, is the article by Overall, Spiegel, and Cohen (1975). Another useful article is one by Carlson and Timm (1974). These authors clarify what hypotheses are tested by the different methods. This seems to be a necessary prerequisite to choosing among methods, and therefore we shall delay our own recommendations (which are close to the ones of Overall, Spiegel, and Cohen, and Carlson and Timm) to Section 15.3.6, in which we consider the relations between methods of analyses and hypotheses about population parameters. First, let us examine constraints and coding methods.

15.3.4   OTHER CONSTRAINTS AND CODES.   In Section 15.2, we demonstrated that different constraints on the solution set, each of which generates a different coding of the **X** matrix, yielded the same SS_A for the one-factor design. You may have wondered what would happen if we imposed other constraints, and therefore coded **X** differently, in multifactor studies. Generally, other codes will yield the same value of $SS_{A,B,AB}$ that was reported in Tables 15–2 and 15–3. Other statistics of interest, however, may vary from those previously obtained. If we for example impose the constraints that $\alpha_2^0 = 0$ and $\beta_3^0 = 0$, we get an **X** matrix consisting of zeros and ones. Following the same process of deleting columns used in Section 15.3.3, we should get the values of SS_A, SS_B, $SS_{A|B}$, $SS_{B|A}$, and $SS_{AB|A,B}$ obtained earlier. In other words, Methods 2 and 3 again will yield the results presented in Table 15–3. On the other hand,

$$SS_{A|B,AB} = \frac{T_{.13}^2}{n_{13}} + \frac{T_{.23}^2}{n_{23}} - \frac{T_{..3}^2}{n_{.3}}$$

and

$$SS_{B|A,AB} = \frac{T_{.21}^2}{n_{21}} + \frac{T_{.22}^2}{n_{22}} + \frac{T_{.23}^2}{n_{23}} - \frac{T_{.2.}^2}{n_{2.}}$$

With this coding, Method 1 yields a sum of squares for A at B_3 and a sum of squares for B at A_2, a very different result from the one obtained with $-1, 0, 1$ coding.

    Other constraints provide still other results under Method 1. Searle (1971) set μ^0 and β_3^0 equal to zero and concluded that $SS_{A|B,AB}$ and $SS_{B|A,AB}$ equal zero. Carlson and Timm (1974) considered constraints such as $\sum v_j \alpha_j^0 = 0$ (the v_j are functions of the cell frequencies) and like Searle and us, have found that most sums of squares were the same as under the $-1, 0, 1$ coding used in Section 15.3.3. Again, however, $SS_{A|B,AB}$ and $SS_{B|A,AB}$ did not correspond to the values computed with the $-1, 0, 1$ coding; nor did the result correspond to tests of any of several null hypotheses about population means that Carlson and Timm wrote about.

    It appears that we must choose not only among methods of adjusting sums of squares, but also among sets of constraints on the solution set, or equivalently, codings of the **X** matrix. We prefer the $-1, 0, 1$ coding engendered by the restriction that the sums of the unweighted α^0, β^0, and $(\alpha\beta)^0$ elements be zero. As we shall see, the results under that coding scheme, particularly those obtained with Method 1, correspond to hypotheses of general interest about population parameters.

15.3.5   TESTING HYPOTHESES.   The computational approach of Section 15.3.3 is highly compatible with the multiple regression analysis of Chapter 14. Within this framework, the analysis of variance is regarded as a way of choosing between models. For example, in computing $SS_{A|B,AB}$ by Method 1, we try to decide between

$$Y_{ij} = \mu + \beta_k + (\alpha\beta)_{jk} + \varepsilon_{ijk}$$

and

$$Y_{ijk} = \mu + \alpha_j + \beta_k + (\alpha\beta)_{jk} + \varepsilon_{ijk}$$

Using this method, we ask whether adding α_j to a model that already includes β_k and $(\alpha\beta)_{jk}$ significantly improves our ability to account for the variance in the data. In contrast, Method 2 asks whether the fit of the data is significantly improved by adding α_j to a model that already includes β_k; in this case, the interaction term is ignored (that is, $(\alpha\beta)_{jk}$ is omitted from both models). The problem with this approach is that it is not always clear what null hypothesis is being tested.

In this section, we begin with a null hypothesis about the μ_{jk}. The statement of the hypothesis generates $\boldsymbol{\psi}$, a matrix of K contrasts among the ab μ_{jk}; here K equals the df associated with the source of variance under investigation. The computations are those originally presented as Equation (15.26):

(15.26)
$$SS_{\hat{\boldsymbol{\psi}}} = (\mathbf{W}\bar{\mathbf{Y}})' \, [\mathbf{W}(\mathbf{X}'\mathbf{X})^{-1}\mathbf{W}']^{-1} \, (\mathbf{W}\bar{\mathbf{Y}})$$

where $\bar{\mathbf{Y}}$ is the column vector of ab group means, $\mathbf{W}$ is a $K \times ab$ matrix of weights that is derived from the statement of the null hypothesis, and $(\mathbf{X}'\mathbf{X})^{-1}$ is an $ab \times ab$ matrix with $1/n_{jk}$ on the diagonal and zeros elsewhere. In what follows, we shall derive these quantities for the example of Table 15–2.

Equally weighting the μ_{jk}. Assume that the six populations in our example are of equal size; variation among the n_{jk} is presumed due to chance. We first consider the B source of variance, in particular, the null hypothesis that $\mu_{.1} = \mu_{.2} = \mu_{.3}$. With equal population sizes assumed, the hypothesis is equivalent to

(15.36)
$$\tfrac{1}{2}(\mu_{11}+\mu_{21}) = \tfrac{1}{2}(\mu_{12}+\mu_{22}) = \tfrac{1}{2}(\mu_{13}+\mu_{23})$$

This implies that

$$[\mu_{11}+\mu_{21}]-[\mu_{12}+\mu_{22}]=0$$

and

$$[\mu_{12}+\mu_{22}]-[\mu_{13}+\mu_{23}]=0$$

or

$$(1)\mu_{11}+(-1)\mu_{12}+(0)\mu_{23}+(1)\mu_{21}+(-1)\mu_{22}+(0)\mu_{23}=0$$

and

$$(0)\mu_{11}+(1)\mu_{12}+(-1)\mu_{13}+(0)\mu_{21}+(1)\mu_{22}+(-1)\mu_{23}=0$$

In matrix terms, the null hypothesis now can be stated as $\boldsymbol{\psi}_B = \mathbf{W}_B \boldsymbol{\mu} = \mathbf{0}$, where

(15.37)
$$\mathbf{W}_B = \begin{bmatrix} 1 & -1 & 0 & 1 & -1 & 0 \\ 0 & 1 & -1 & 0 & 1 & -1 \end{bmatrix}$$

There are many other possible W matrices that test the same null hypothesis and give the same results when substituted in Equation (15.26). Equation (15.36)

implies the general form

$$(15.38) \qquad \mathbf{W}_B = \begin{bmatrix} v_1 & v_2 & -(v_1+v_2) & v_1 & v_2 & -(v_1+v_2) \\ z_1 & z_2 & -(z_1+z_2) & z_1 & z_2 & -(z_1+z_2) \end{bmatrix}$$

and all $\mathbf{W}$ matrices of this form will give the same value of $SS_{\hat{\psi}_B}$.

Having obtained the $\mathbf{W}$ matrix, we can proceed. Again using the data of Table 15–2, we calculate the components of the expression in Equation (15.26). We compute

$$\mathbf{W}\bar{\mathbf{Y}} = \begin{bmatrix} 5 \\ -\frac{23}{3} \end{bmatrix}$$

and

$$(\mathbf{X}'\mathbf{X})^{-1} = \begin{bmatrix} \frac{1}{2} & & & & \\ & \frac{1}{4} & & \mathbf{0} & \\ & & \frac{1}{2} & & \\ & & & 1 & \\ & \mathbf{0} & & \frac{1}{2} & \\ & & & & \frac{1}{3} \end{bmatrix}$$

Carrying out the calculations required by Equation (15.26), we get

$$\mathbf{W}(\mathbf{X}'\mathbf{X})^{-1}\mathbf{W}' = \begin{bmatrix} \frac{9}{4} & -\frac{3}{4} \\ -\frac{3}{4} & \frac{19}{12} \end{bmatrix}$$

which has as its inverse

$$[\mathbf{W}(\mathbf{X}'\mathbf{X})^{-1}\mathbf{W}']^{-1} = \left(\frac{1}{36}\right)\begin{bmatrix} 19 & 9 \\ 9 & 27 \end{bmatrix}$$

and $SS_{\hat{\psi}_B} = 38.\overline{1}$, which is exactly the result obtained under Method 1 in Table 15–3.

It is no coincidence that $SS_{\hat{\psi}_B}$ is identical to $SS_{B|A,AB}$. Method 1 tests null hypotheses of the form expressed in Equation (15.36); when identical weights are given to all the means at any level of B, it is equivalent to adjusting for all other main and interaction effects. This assumes that the $-1, 0, 1$ coding was used with Method 1 (see Section 15.3.4 for a discussion of the role of coding) and that $\mathbf{W}$ is of rank equal to the df for the source of variance under consideration ($b-1$, in this case).

According to the usual design model,

$$\mu_{jk} = \alpha_j + \beta_k + (\alpha\beta)_{jk}$$

If these terms are substituted into Equation (15.36), the null hypothesis can be restated in terms of main and interaction effects:

$$(15.39) \quad \beta_1 + \tfrac{1}{2}[(\alpha\beta)_{11} + (\alpha\beta)_{21}] = \beta_2 + \tfrac{1}{2}[(\alpha\beta)_{12} + (\alpha\beta)_{22}] = \beta_3 + \tfrac{1}{2}[(\alpha\beta)_{13} + (\alpha\beta)_{23}]$$

The important implication of Equation (15.39) is that a significant B effect under Method 1, or under the equivalent hypothesis testing approach just calculated, may reflect interaction effects in the population.[1] These tests do reflect only the main effects of interest, however, under certain restrictions on the parameters. If the populations are equal in size

$$\mu = \frac{1}{a} \sum_j \mu_j = \frac{1}{b} \sum_k \mu_k$$

(15.40)
$$\mu_j = \frac{1}{b} \sum_k \mu_{jk}$$

$$\mu_k = \frac{1}{a} \sum_k \mu_{jk}$$

It follows algebraically from Equation (15.40) that

$$\sum_j \alpha_j = 0$$

(15.41)
$$\sum_k \beta_k = 0$$

$$\sum_j (\alpha\beta)_{jk} = \sum_k (\alpha\beta)_{jk} = 0$$

Then, Equation (15.39) states that $\beta_1 = \beta_2 = \beta_3$.

The hypothesis testing approach can also be used to compute $SS_{\hat{\psi}_A}$ and $SS_{\hat{\psi}_{AB}}$. The calculations follow those for $SS_{\hat{\psi}_B}$, except that

$$\mathbf{W}_A = [1 \quad 1 \quad 1 \quad -1 \quad -1 \quad -1]$$

and the interaction weights are easily obtained by multiplying the entries in the $\mathbf{W}_A$ and $\mathbf{W}_B$ matrices:

$$\mathbf{W}_{AB} = \begin{bmatrix} 1 & -1 & 0 & -1 & 1 & 0 \\ 0 & 1 & -1 & 0 & -1 & 1 \end{bmatrix}$$

You can verify that the results of applying Equation (15.26) will be identical to $SS_{A|B,AB}$ and $SS_{AB|A,B}$.

Unequal weighting. Assume that the n_{jk} are proportional to the sizes of the treatment populations. Then, once again considering the B source of variance, you will find that a reasonable statement of the null hypothesis ($\mu_{.1} = \mu_{.2} = \mu_{.3}$) is

(15.42)
$$\frac{n_{11}}{n_{.1}} \mu_{11} + \frac{n_{21}}{n_{.1}} \mu_{21} = \frac{n_{12}}{n_{.2}} \mu_{12} + \frac{n_{22}}{n_{.2}} \mu_{22} = \frac{n_{13}}{n_{.3}} \mu_{13} + \frac{n_{23}}{n_{.3}} \mu_{23}$$

---

[1] There is an apparent contradiction in this statement. After all, the term $SS_{B|A,AB}$ implies an adjustment for interaction effects. In what sense, then, would the variance of the $\mu_{.k}$ reflect interaction effects? The answer lies in the distinction between constraints on elements of the solution set and restrictions on parameters. The adjustment that our sum of squares terminology refers to is an adjustment for the α_j^0 and the $(\alpha\beta)_{jk}^0$. These elements represent estimates of population parameters only under certain restrictions specified in the next part of the main text. Under those restrictions, variability in the $\mu_{.k}$ reflects variability in only the β_k.

First substituting the values of n_{jk} from Table 15–2, we can rewrite Equation (15.43) in a form that will generate the **W** matrix:

(15.43)
$$[\tfrac{2}{3}\mu_{11}+\tfrac{1}{3}\mu_{21}]-[\tfrac{4}{6}\mu_{12}+\tfrac{2}{6}\mu_{22}]=0$$
$$[\tfrac{4}{6}\mu_{12}+\tfrac{2}{6}\mu_{22}]-[\tfrac{2}{5}\mu_{13}+\tfrac{3}{5}\mu_{23}]=0$$

Multiplying by a constant will not affect the sum of squares calculations and will simplify them; therefore, multiply the first line by 3 and the second line by 15. We can now write

(15.44)
$$\mathbf{W}_B = \begin{bmatrix} 2 & -2 & 0 & 1 & -1 & 0 \\ 0 & 10 & -6 & 0 & 5 & -9 \end{bmatrix}$$

As usual, the cells are in the order A_1B_1, A_1B_2, ..., A_2B_3.

The calculations proceed just as for the equally weighted means. First compute

$$\mathbf{W}_B\bar{Y}=\begin{bmatrix} 7 \\ -59 \end{bmatrix}$$

and

$$\mathbf{W(X'X)^{-1}W'}=\frac{1}{2}\begin{bmatrix} 9 & -15 \\ -15 & 165 \end{bmatrix}$$

which has the inverse

$$[\mathbf{W(X'X)^{-1}W'}]^{-1}=\frac{1}{620}\begin{bmatrix} 165 & 15 \\ 15 & 9 \end{bmatrix}$$

Applying Equation (15.26) gives $SS_{\hat{\psi}_B} = 42.895$, which is identical to the value of SS_B reported in Table 15–3. In general, proportional weighting is equivalent to computing sums of squares for a variable, ignoring other main and interaction effects.

To apply the unequal weighting approach to the A variable, we begin with the null hypothesis

$$\psi_A = (\tfrac{2}{8}\mu_{11}+\tfrac{4}{8}\mu_{12}+\tfrac{2}{8}\mu_{13})-(\tfrac{1}{6}\mu_{21}+\tfrac{2}{6}\mu_{22}+\tfrac{3}{6}\mu_{23})=0$$

The proportions are based on the cell frequencies of Table 15–2. After multiplying the above weights by 12, we have

$$\mathbf{W}_A =[3 \quad 6 \quad 3 \quad -2 \quad -4 \quad -6]$$

and

$$\mathbf{W(X'X)^{-1}W'}=42$$

We also find $\mathbf{W}_A\bar{Y}=-16$. Then,

$$SS_{\hat{\psi}_A}=\frac{(-16)^2}{42}=6.095$$

which is the result listed as SS_A in Table 15–3.

Hypothesis-testing approach equivalent to Method 2. Carlson and Timm (1974) have presented **W** matrices that lead to sums of squares equivalent to $SS_{A|B}$ and $SS_{B|A}$. The null hypothesis for A is

$$\psi_A = \left[\left(n_{11}-\frac{n_{11}^2}{n_{.1}}\right)\mu_{11}+\left(n_{12}-\frac{n_{12}^2}{n_{.2}}\right)\mu_{12}+\left(n_{13}-\frac{n_{13}^2}{n_{.3}}\right)\mu_{13}\right]$$
$$-\left[\frac{n_{11}n_{21}}{n_{.1}}\mu_{21}+\frac{n_{12}n_{22}}{n_{.2}}\mu_{22}+\frac{n_{13}n_{23}}{n_{.3}}\mu_{23}\right]=0$$

These weights lack any intuitive rationale. Indeed, one is sympathetic to Carlson and Timm's conclusion that people may calculate $SS_{A|B}$ only because they fail to understand what hypothesis is being tested.

15.3.6   WHICH METHOD?   In the usual experiment, it is reasonable to assume that treatment populations are equal in size. The assumption will hold for studies in which the variables are manipulated by the investigator and there are no combinations of variables that result in systematic losses of subjects. Under these conditions, $\mu_{j.}$ is an average obtained by equal weighting of the μ_{jk} at A_j, and the F test for A should reflect the variation among the $\mu_{j.}$ so defined. Method 1 (assuming -1, 0, 1 coding) provides such a test and so does the algebraically equivalent hypothesis-testing approach described under "*Equally weighting the μ_{jk}*" in Section 15.3.5. This is the method we recommend whenever variations in cell frequency can reasonably be assumed due to chance.

Often, nonorthogonality in a study represents nonorthogonality in the population. This is particularly true when factors are attributes of subjects, like socioeconomic level, race, or personality type. Such studies do not differ basically from studies in which the independent variables are continuous and correlated. The same problems of interpretation and analysis arise that were met in the discussion of multiple regression (Chapter 14). When causal relations among factors can be assumed, Method 3 makes sense. If the investigator lacks a hypothesis about priorities within a set of correlated variables, it is not clear which analysis should be pursued. It is also unclear whether such a study should be carried out when no more precise motivating theory exists.

15.4   CONCLUDING REMARKS

When the logic of the problem permits, equal cell frequencies are desirable. This not only yields a computationally simple, easily interpreted, orthogonal partitioning of the variability among cell means but also may result in a more robust set of F tests. We know, for example, that the consequences of heterogeneity of variance are greatly exaggerated when cell frequencies are markedly disparate.

When cell frequencies are disproportionate, there are many possible analyses available to the investigator. Some, like Method 1 with -1, 0, 1 coding and the first hypothesis-testing approach presented, are algebraically equivalent.

Others, like Method 1 under two different sets of constraints (codes), may be equivalent for some terms but not for others. Section 5.3 has focused on these relations among data analyses. There is really no one correct analyses of data when cell frequencies are disproportionate. The investigator has a decision to make, one that should be based on assumptions about the populations sampled and on the relations among variables in that population.

EXERCISES

15.1   Consider the following data:

$$A_1:   2, 7, 2, 6, 3$$
$$A_2:   8, 9, 7$$

(a) Calculate SS_A as in the usual analysis of variance.
(b) Find $\mathbf{X}$, $\boldsymbol{\phi}^0$, and SS_A for each of the following constraints: (i) $\mu^0 = 0$; (ii) $\sum_j \alpha_j^0 = 0$; (iii) $\sum_j n_j \alpha_j^0 = 0$.

15.2   The data set is presented in Table 15–1 of Section 15.2.
(a) Under the constraint that $\mu^0 = 0$, find $\mathbf{X}$, $\boldsymbol{\phi}^0$, and SS_A.
(b) Assuming that the populations are of equal size, test the null hypothesis that $\mu = 0$.
(c) Assuming that the n_j represent the population sizes, again test the null hypothesis of part (b). Comment on any discrepancy in results.
(d) In parts (b) and (c), the answers required that we calculate linear combinations of the solution set elements and the variances of these combinations. What constraint (code) would provide the estimate of μ required by (b) as a single element of $\boldsymbol{\phi}^0$? What constraint would do the same for (c)? Do the calculations.

15.3   Using the data of Table 15–1, test the null hypotheses that $\mu_1 = \mu_2$ and that the average of μ_1 and μ_2 equals μ_3 (a) assuming equal-size populations, and (b) assuming that the n_j are proportional to population sizes. In which case, if either, are the tests orthogonal?

15.4   Present the appropriate $\mathbf{X}$ matrix for a 3×4 design using -1, 0, 1 coding.

15.5   The following contrast matrix conforms to Equation (15.39), and therefore should yield $SS_{B|A,AB}$ when applied to the data of Table 15–2; the value should be $38.\overline{1}$. The matrix is

$$\mathbf{W}_B = \begin{bmatrix} 3 & 2 & -5 & 3 & 2 & -5 \\ 1 & -3 & -4 & 1 & -3 & -4 \end{bmatrix}$$

Carry out the calculations. It will help to know that the inverse of

$$\begin{bmatrix} a & b \\ c & d \end{bmatrix} \quad \text{is} \quad (ad - bc)^{-1} \begin{bmatrix} d & -b \\ -c & a \end{bmatrix}$$

REFERENCES

Carlson, J. E., and Timm, N. H. Analysis of nonorthogonal fixed-effect designs. *Psychological Bulletin* 81:563–70 (1974).

Overall, J. E., and Spiegel, D. K. Concerning least squares analysis of experimental data. *Psychological Bulletin* 72:311–22 (1969).

Overall, J. E., Spiegel, D. K., and Cohen, J. Equivalence of orthogonal and nonorthogonal analysis of variance. *Psychological Bulletin* 82:182–86 (1975).

SUPPLEMENTARY READING

The textbooks by Graybill and Searle, referred to in Chapter 14, present mathematical treatments of the material in this chapter. Searle's presentation includes numerical examples.

A good general reference on multiple regression is the book by Cohen and Cohen; the full reference also appears at the close of Chapter 14. Their development of analyses for factorial studies emphasizes the relation of the analyses to multiple and partial correlations and should give you additional insight.

16 ANALYSIS OF COVARIANCE

16.1 INTRODUCTION

Much of the error in experimentation can be traced to those characteristics of individual subjects that correlate highly with the dependent variable. Evaluation of the effects of different instructions on concept formation performance, for example, is made more difficult by the relation between this performance and the intelligence of the subject. Variability in intelligence among subjects increases variability in performance within groups. Several experimental designs have been presented, each of which provides an alternative approach to this problem of experimental error. Now we take a different tack; we consider a statistical adjustment of each dependent measure (for example, number of errors in a concept formation task) for the contribution to that measure of a concomitant variable (for example, intelligence test score). The technique that will be examined is the analysis of covariance.

This approach assumes that some portion of the usual error component is predictable if we know the individual's score on some related measure X, the covariate. Removing the variance of these predicted error components leaves a smaller error variance and thus a more efficient, more powerful test of treatment effects. The simplest set of computations, and the one most frequently used, rests on the assumption that Y is a linear function of X. Chapter 12, or its equivalent, represents the minimum knowledge about linear regression that you should have for studying analysis of covariance.

Analysis of covariance has been used for other purposes besides reducing error variance. In many studies, investigators have used the covariance procedure as a way of adjusting for systematic differences among groups in X, a concomitant variable correlated with the performance measure Y. This may happen because subjects are not randomly assigned to treatment groups. Different methods of teaching, for example, may be applied to two schools that differ markedly in aptitude test scores of students. In such a study, analysis of covariance can be used to determine whether the methods differ in effects on learning when initial differences in aptitude are considered.

Groups may also differ in a concomitant variable because that variable is affected by the experimental treatment. In studying performance as a function of method of instruction, the researcher may note that study time is influenced by the method and may analyze covariance to determine whether performance differences among experimental groups remain after adjusting for variation in study time.

The results of analyzing covariance to adjust for systematic differences in X often have been misinterpreted. Section 16.4 provides a discussion of the interpretation of such analyses. Consideration of that material should help you to avoid some all too common inferential pitfalls.

Although we have written about analysis of covariance under the assumption that you understand simple linear regression, we do not count on your knowing either matrix algebra or multiple regression. We have done this, recognizing that traditional experimental design courses often omit such material. For those who are comfortable with the multiple regression approach, however (the contents of Chapters 13–15 in this book), Section 16.2.5 presents an alternative view of the covariance calculations. Although that material should provide some additional insight and should integrate analysis of covariance into the broad regression framework, it is not essential to understanding the rest of the chapter.

16.2 THE COMPLETELY RANDOMIZED ONE-FACTOR DESIGN

16.2.1 THE COVARIANCE MODEL.   The simple one-factor design has the following structural equation (Chapter 4):

(16.1)
$$Y_{ij} = \mu + \alpha_j + \varepsilon_{ij}$$

We are used to thinking of ε_{ij} as the sum of two components: error due to variation in measurements and error due to individual differences. Individuals differ on many dimensions that may contribute to the difference in their scores. Analysis of covariance provides a way of adjusting for differences on one or more of those dimensions. We assume

(16.2)
$$Y_{ij} = \mu + \alpha_j + \varepsilon'_{ij} + \beta(X_{ij} - \bar{X}_{..})$$

where β is the regression coefficient that characterizes the rate of change of Y relative to X in the population sampled; $\beta(X_{ij} - \bar{X}_{..})$ is therefore the error in Y that would be predicted on the basis of knowledge of X, and ε'_{ij} is the residual error unaccounted for by X. For example, individuals may differ from the treatment population average because they differ from other individuals in intelligence, motivation, amount of previous experience with related tasks, or any one of numerous other factors. Each of these factors contributes a positive or a negative component to the individual's score Y_{ij}; the sum of these components plus errors in measurements is ε_{ij}. If we have an intelligence test score for the individual, we can estimate that component of the individual's total deviation from μ_j due to intelligence. Removing components of error predicted on the basis of knowledge of the covariate reduces the variance of the residual error component. Since

$$\varepsilon_{ij} = \varepsilon'_{ij} + \beta(X_{ij} - \bar{X}_{..})$$

and since we assume that the predicted and residual components are independent,

$$\sigma_e^2 = \sigma_{e'}^2 + \sigma_{\text{predicted}}^2$$

which demonstrates that $\sigma_{e'}^2$, the residual error variance remaining after adjustment, is less than σ_e^2, the error variance estimated under the usual analysis of variance procedure (Chapter 4). Thus, efficiency can be increased through the covariance procedure.

If the ratio of mean squares computed in analyzing the covariance is to be distributed as *F*, we require assumptions analogous to those usually used in analysis of variance. Specifically, we assume that for any value of the covariate X_{ij} within each treatment population, there exists a population of independently and normally distributed scores with variance $\sigma_{e'}^2$. Furthermore, as in the earlier treatment of linear regression (Chapter 12), we assume that *X* is fixed over replications of the study and measured without error. Finally, we assume that within each treatment population, *Y* varies as a linear function of *X* and that the *a* regression lines characterizing the treatment populations have the same slopes.

To review briefly, we assume that the adjusted error components ε'_{ij} (sometimes referred to as the residuals) are distributed within each treatment population:

1. Independently.
2. Normally.
3. With mean zero and homogeneous variances $\sigma_{e'}^2$.

Furthermore, we assume:

4. *X* is fixed and measured without error.
5. Linear regression of *Y* on *X* within each treatment population.
6. Homogeneous regression coefficients, that is,

$$\beta_1 = \beta_2 = \cdots = \beta_{j...} = \cdots = \beta_a$$

Consequences of violating assumptions. Relative to the analysis of variance, little is known about what effects violating assumptions has on inferences based on the covariance procedure. The available evidence suggests that like analysis of variance, analysis of covariance is reasonably robust with respect to the normality and the homogeneity of variance assumptions (Atiqullah, 1964).[1] It has also been shown that most results of interest still hold when *X* varies over samples although errors in measuring *X* can result in biased estimates of β and loss of precision. These consequences are slight provided that the variance of the measurement error is small relative to the variance in *X* itself (Snedecor and Cochran, 1967).

---

[1] Good reviews of covariance assumptions, with additional references to studies of robustness, can be found in Elashoff (1969) and Glass, Peckham, and Sanders (1972).

Nonlinearity results in biased estimates of treatment effects; the magnitude of the bias depends on the true form of the function relating X and Y, and is least severe when subjects are randomly assigned to groups and the dependent variable is normally distributed. Departure from linearity also tends to reduce the efficiency of the covariance analysis.

Heterogeneity of the β_j has been found to result in a loss of power (Atiqullah, 1964). Furthermore, the departure from parallelism of the treatment population regression lines makes it difficult to interpret the results of the test of main effects. Such heterogeneity of regression is essentially an interaction of A with the covariate; it represents variation in the effect of the treatment variable as a function of the covariate score. When this occurs, it would seem of most interest to investigate the treatment variable within regions of X; a treatment by blocks design might be more illuminating than the covariance approach.

16.2.2 PARTITIONING THE ADJUSTED SUM OF SQUARES. Equation (16.2), which defines Y in terms of population parameters, provides a starting point for our computations. Let $y_{ij} = Y_{ij} - \bar{Y}_{..}$ and $x_{ij} = X_{ij} - \bar{X}_{..}$. Then, the deviation of the adjusted score from the grand mean of the Y values ($\bar{Y}_{..}$) can be expressed as

$$(\mathbf{16.3}) \qquad Y_{ij} - b_{\text{tot}}(X_{ij} - \bar{X}_{..}) - \bar{Y}_{..} = y_{ij} - b_{\text{tot}}x_{ij}$$

Since we are concerned with the actual data analysis, we have replaced β of Equation (12.2) with a least-squares estimate based on the total set of *an* scores. This estimate b_{tot} is computed as

$$(\mathbf{16.4}) \qquad b_{\text{tot}} = \frac{\sum_i \sum_j x_{ij} y_{ij}}{\sum_i \sum_j x_{ij}^2} = \frac{SP_{\text{tot}}}{SS_{\text{tot}(x)}}$$

Note that the regression coefficient involves (1) a sum of cross-products of deviations about $\bar{X}_{..}$ and $\bar{Y}_{..}$ (SP_{tot}), and (2) the total sum of squares with X as the variable being measured ($SS_{\text{tot}(x)}$). Squaring and summing the right-hand side of Equation (16.3) gives the $SS_{\text{tot(adj)}}$. Our present problem is to determine how this adjusted total sum of squares is to be partitioned to provide the components of the F ratio. The following identity is a step toward the desired partitioning of the $SS_{\text{tot(adj)}}$:

$$y_{ij} - b_{\text{tot}}x_{ij} = (b_j - b_{S/A})(x_{ij} - \bar{x}_{.j}) + [(y_{ij} - \bar{y}_{.j}) - b_j(x_{ij} - \bar{x}_{.j})]$$
$$(\mathbf{16.5}) \qquad \qquad + (\bar{y}_{.j} - b_A \bar{x}_{.j}) + [(b_A - b_{S/A})\bar{x}_{.j} - (b_{\text{tot}} - b_{S/A})x_{ij}]$$

where $\bar{x}_{.j} = \bar{X}_{.j} - \bar{X}_{..}$ and $\bar{y}_{.j} = \bar{Y}_{.j} - \bar{Y}_{..}$. Also

$(\mathbf{16.6})$ b_j (regression coefficient for the best-fitting line for A_j)

$$= \frac{\sum_i (x_{ij} - \bar{x}_{.j})(y_{ij} - \bar{y}_{.j})}{\sum_i (x_{ij} - \bar{x}_{.j})^2} = \frac{SP_j}{SS_{j(x)}}$$

(16.7) $b_{S/A}$ (least-squares estimate of β derived under the homogeneity of regression assumption)

$$= \frac{\sum_j SP_j}{\sum_j SS_{j(x)}}$$

(16.8) b_A (regression coefficient for the line that best fits the a pairs of group means)

$$= \frac{n \sum_j \bar{x}_{.j} \bar{y}_{.j}}{n \sum_j \bar{x}_{.j}^2} = \frac{SP_A}{SS_{A(x)}}$$

Squaring both sides of Equation (16.5) and summing yield

$$\underset{SS_{\text{tot(adj)}}}{\sum_j \sum_i (y_{ij} - b_{\text{tot}} x_{ij})^2} = \underset{SS_1}{\sum_j [(b_j - b_{S/A}) \sum_i (x_{ij} - \bar{x}_{.j})]^2}$$

$$+ \underset{SS_2}{\sum_j \sum_i [(y_{ij} - \bar{y}_{.j}) - b_j (x_{ij} - \bar{x}_{.j})]^2}$$

(16.9)

$$+ \underset{SS_3}{n \sum_j (\bar{y}_{.j} - b_A \bar{x}_{.j})^2}$$

$$+ \underset{SS_4}{\sum_j \sum_i [(b_A - b_{S/A}) \bar{x}_{.j} - (b_{\text{tot}} - b_{S/A}) x_{ij}]^2}$$

The value SS_1 is a measure of the variability of the b_j defined by Equation (16.6). If the slopes were identical for all a groups, all the b_j would equal their average, $b_{S/A}$, and SS_1 would equal zero. Obviously, the group regression coefficients will vary to some extent due to sampling error; the question is whether they vary enough to indicate that the β_j, the regression coefficients for the treatment populations, actually differ. To answer this question, we require a measure of error variance. Such a measure is provided by SS_2. This term reduces to a sum of squared deviations of scores (values of Y) about the best-fitting straight line for their group. In essence, this is a within-group sum of squares, analogous to $SS_{S/A}$ in the usual analysis of variance; the difference is that we are now taking deviations from the best-fitting straight line rather than from the mean for each group.

The quantities SS_1 and SS_2 provide a basis for testing the assumption that the best-fitting regression line has the same slope in all treatment populations. If the F ratio, MS_1/MS_2, is not significant, then it may be concluded that the β_j are essentially homogeneous and that the observed variability in the b_j is attributable to sampling error. In this case, MS_1 and MS_2 are both estimates of error variance, and the two components can be pooled to provide an error term for the test of the adjusted treatment effects.

We next consider the interpretation of SS_3 and SS_4. If the adjusted treatment effects were all zero, then the plot of $\bar{Y}_{.j}$ as a function of $\bar{X}_{.j}$ should resemble the plot of the Y_{ij} as a function of the X_{ij}; in both cases error variance is the only factor contributing to the variability of data points about the best-fitting linear function. In deciding that the two plots resemble each other, we are particularly concerned with two aspects of the plots. If the null hypothesis is true, (1) departures of the $\bar{Y}_{.j}$ from the line that best fits the group means should be no greater than one would expect by chance; and (2) the regression coefficient of that line b_A should not differ significantly from $b_{S/A}$, which represents the regression of individual scores about the average group regression line. The term SS_3 clearly represents the deviation of group means from the best fitting line. And although SS_4 is a more complicated expression, it is definitely a function of $b_A - b_{S/A}$. In fact, it can be shown that

$$SS_4 = \frac{(b_A - b_{S/A})^2}{(1/SS_{A(x)}) + (1/SS_{S/A(x)})}$$

Generally, SS_3 and SS_4 are pooled and divided by their pooled df to provide the numerator of the F test for the adjusted treatment effects. The denominator is the pooled error term based on the first two components of the adjusted total sum of squares.

An artificial numerical example may help clarify relations among the various regression coefficients and also the interpretation of SS_3 and SS_4 as measures of treatment effects. Consider the parent population consisting of nine pairs of X and Y scores in which

$$Y = 2X + 3$$

This is errorless data in the sense that all the variability in Y can be attributed to variability in X. Suppose we divide the data into three sets of three pairs each, as in the upper half of Table 16-1. Note that there is, at the stage, *no* treatment effect. The variation in the $\bar{Y}_{.j}$ is due solely to the variation in the $\bar{X}_{.j}$. This is readily showed by computing SS_3 and SS_4; both are zero. All the regression coefficients equal 2, as can be verified by applying Equations (16.4), (16.6), (16.7), and (16.8).

Suppose we now apply three treatments, one to each group. Let $\alpha_1 = -1$, $\alpha_2 = 2$, $\alpha_3 = -1$. The resulting data set is in the middle of Table 16-1. The b_j are all still 2, as $b_{S/A}$ is; adding the α_js has changed only the slope-intercepts of the within-group regression lines. Furthermore, applying Equation (16.8) yields

$$b_A = \frac{3[(-1)(-3) + (0)(2) + (1)(1)]}{3[(-1) + (0)^2 + (1)^2]} = 2$$

Furthermore, b_{tot} is still 2. Thus, SS_4 must be zero (see Equation (16.9)). However,

$$SS_3 = 3\{[(10 - 13) - 2(4 - 5)]^2 + [(15 - 13) - 2(5 - 5)]^2$$
$$+ [(14 - 13) - 2(6 - 5)]^2\} = 18$$

In this example, we chose our treatment effects so that they varied in a curvilinear manner with $\bar{X}_{.j}$. Note that $\bar{Y}_{.j}$ at first increases and then decreases with increased $\bar{X}_{.j}$. Thus, the $\bar{Y}_{.j}$ no longer fall on the best-fitting straight line with regression coefficient $b_A = 2$.

The term SS_4 was unaffected because the straight line that best describes the relation between α_j and $\bar{X}_{.j}$ has a zero regression coefficient. Thus, when the α_j are added to the original $\bar{Y}_{.j}$, the value b_A is unchanged.

Suppose we chose our α_j in such a way that they did have a linear relation to $\bar{X}_{.j}$. For example, let $\alpha_j = -1$, 0, +1, respectively. Adding these effects to the *original* data set yields the data at the bottom of Table 16–1. Computing b_A again, we now obtain $b_A = 3$. The b_j and $b_{S/A}$ are as usual unchanged by adding a constant to the scores within each group. However, b_{tot} is now $\frac{55}{40}$, or 1.375. From Equation (16.9), it is apparent without further calculation that SS_4 will be greater than zero. However, SS_3 is still zero because the $\bar{Y}_{.j}$ do fall on a straight line when plotted as a function of $\bar{X}_{.j}$.

To summarize: In analyzing variance, the total sum of squares is partitioned; in analyzing covariance, the adjusted total sum of squares is partitioned. The first

Table 16–1   An artificial example of relations in a covariance problem

$\alpha_1 = \alpha_2 = \alpha_3 = 0$

	A_1		A_2		A_3	
	X	Y	X	Y	X	Y
	3	9	2	7	4	11
	4	11	8	19	9	21
	5	13	5	13	5	13
Means =	4	11	5	13	6	15

$\alpha_1 = -1,\ \alpha_2 = +2,\ \alpha_3 = -1$

	A_1		A_2		A_3	
	X	Y	X	Y	X	Y
	3	8	2	9	4	10
	4	10	8	21	9	20
	5	12	5	15	5	12
Means =	4	10	5	15	6	14

$\alpha_1 = -1,\ \alpha_2 = 0,\ \alpha_3 = +1$

	A_1		A_2		A_3	
	X	Y	X	Y	X	Y
	3	8	2	7	4	12
	4	10	8	19	9	22
	5	12	5	13	5	14
Means =	4	10	5	13	6	16

component resulting from this partitioning, SS_1, measures the variability of the group regression coefficients about an average coefficient. The second term, SS_2, measures the variability of scores about each group regression line. The hypothesis of homogeneity of regression can be tested by a ratio of mean squares based on the two terms just described. If the F statistic is not significant, the two terms can be pooled to form a single estimate of error, which will be subsequently used in testing treatment effects. The third component of $SS_{tot(y')}$, which is SS_3, reflects the variability of treatment means about the line that gives the predicted value of $\bar{Y}_{.j}$. The fourth component, SS_4, measures the difference between the slope of that line and the slope of the average within-group regression line. If variability of the means about this best-fitting line is significant, then variation among the $\bar{Y}_{.j}$ is attributable to something more than the variation in $\bar{X}_{.j}$ and error variance; the "something more" is presumably treatment effects. If SS_4 is significantly large, the rate of change of $\bar{Y}_{.j}$ as a function of $\bar{X}_{.j}$ is not the same as the rate of change within groups, and again we conclude that the treatments are playing a role. The point is that if either SS_3 or SS_4 is significantly large, then the same function that describes the plot within a group does not adequately describe the plot of group means, and we conclude that the difference can be attributed to the presence of treatment effects.

16.2.3 COMPUTATIONAL FORMULAS FOR THE ANALYSIS OF COVARIANCE. To analyze covariance as efficiently as possible, it is desirable to provide raw score computational formulas for the four components of $SS_{tot(y')}$ that we have just discussed. These four expressions can be obtained by appropriately combining entries in Table 16.2. The entries in the first two columns are simply the usual formulas for the analysis of variance of a completely randomized one-factor design with the addition of formulas for group j; their role will shortly become

Table 16–2 Computational formulas for the analysis of covariance

	SUMS OF SQUARES (SS)		SUMS OF PRODUCTS (SP)
	Y	X	
Total	$\sum_i \sum_j Y_{ij}^2 - C$	$\sum_i \sum_j X_{ij}^2 - C_x$	$\sum_i \sum_j X_{ij} Y_{ij} - C_{xy}$
A	$\dfrac{\sum_j T_{.j(y)}^2}{n} - C_y$	$\dfrac{\sum_j T_{.j(x)}^2}{n} - C_x$	$\dfrac{\sum_j T_{.j(x)} T_{.j(y)}}{n} - C_{xy}$
S/A	$SS_{tot} - SS_A$	$SS_{tot} - SS_A$	$SP_{tot} - SP_A$
Group j	$\sum_i Y_{ij}^2 - \dfrac{T_{.j(y)}^2}{n}$	$\sum_i X_{ij}^2 - \dfrac{T_{.j(x)}^2}{n}$	$\sum_i X_{ij} Y_{ij} - \dfrac{T_{.j(x)} T_{.j(y)}}{n}$
	$C_y = \dfrac{T_{..(y)}^2}{an}$	$C_x = \dfrac{T_{..x}^2}{an}$	$C_{xy} = \dfrac{T_{..(x)} T_{..(y)}}{an}$

clearer. The *SP* (sums of cross-products) terms can be arrived at by analogy to the *SS* terms. For example,

$$SS_{A(x)} = \frac{\sum_j (\sum_i X_{ij})^2}{n} - \frac{(\sum_i \sum_j X_{ij})^2}{an}$$

$$= \frac{\sum_j (\sum_i X_{ij})(\sum_i X_{ij})}{n} - \frac{(\sum_i \sum_j X_{ij})(\sum_i \sum_j X_{ij})}{an}$$

Then, if one *X* is replaced by *Y*,

$$SP_A = \frac{\sum_j (\sum_i X_{ij})(\sum_i Y_{ij})}{n} - \frac{(\sum_i \sum_j X_{ij})(\sum_i \sum_j Y_{ij})}{an}$$

Computing the entries in Table 16–2 is the first step in analysis of covariance. They will be combined to yield raw score expressions for the four components of $SS_{tot(y)}$. The terms SS_1 and SS_2 will then be used to provide a test of the homogeneity of regression assumption (Section 16.2.1). If this assumption seems tenable, SS_1 and SS_2 will provide a pooled error term against which treatment effects SS_3 and SS_4 are tested.

Raw score formulas for the four sums of squares components are readily derived. The terms are defined in Equation (16.9). The regression coefficients are expressed in terms of sums of squares *SS* and sums of cross-products *SP* in Equations (16.4), (16.6), (16.7) and (16.8). Raw score formulas for the *SS* and *SP* terms are obtained as indicated earlier. The final equations are:

(16.10)
$$SS_1 = \sum_j \frac{SP_j^2}{SS_{j(x)}} - \frac{SP_{S/A}^2}{SS_{S/A(x)}}$$

(16.11)
$$SS_2 = SS_{S/A(y)} - \sum_j \frac{SP_j^2}{SS_{j(x)}}$$

(16.12)
$$SS_3 = SS_{A(y)} - \frac{SP_A^2}{SS_{A(x)}}$$

(16.13)
$$SS_4 = \frac{SP_A^2}{SS_{A(x)}} + \frac{SP_{S/A}^2}{SS_{S/A(x)}} - \frac{SP_{tot}^2}{SS_{tot(x)}}$$

To complete the analysis of covariance it is also helpful to note that

(16.14)
$$SS_{tot(y')} = SS_1 + SS_2 + SS_3 + SS_4$$
$$= SS_{tot(y)} - SS_{lin}$$
$$= SS_{tot(y)} - \frac{SP_{tot}^2}{SS_{tot(x)}}$$

The SS_1 is distributed on $(a-1)$ *df*, since we are concerned with the variability of *a* regression coefficients about the pooled coefficient, $b_{S/A}$. Note that this is consistent with Equation (16.10), in which one squared quantity is subtracted from the sum of *a* squared quantities. The *df* for SS_2 are $a(n-2)$; the explanation lies in a closer examination of the meaning of this variability. In each group, the variance of n scores is taken about a group regression line. Estimating this line involves losing two df, one for estimating β_j, the regression coefficient, and the other for estimating $\bar{Y} - \beta\bar{X}$, the slope intercept. Thus there are $(n-2)$ df pooled over a groups. Referring to Equation (16.11), we note that the relation of squared quantities to df still holds; we have $a(n-1) - a$ $[= a(n-2)]$ squared quantities on the right-hand side. The SS_3 measures the variability of a means about a regression line; again the estimation of the line involves losing two df, so that SS_3 is distributed on $(a-2)$ df. Alternatively, we refer to Equation (16.12) and note that we have $(a-1)$ df for $SS_{A(y)}$, and that one more is lost for the squared cross-product term. Since SS_4 measures the difference between two regression coefficients, it is on 1 df. The computational formula is again consistent with the conclusion.

Before testing treatment effects, we must consider the null hypothesis that $\beta_1 = \beta_2 = \cdots = \beta_a$. The appropriate statistic is

$$F = \frac{SS_1/(a-1)}{SS_2/a(n-2)}$$

The logic of this F test should be apparent. The question is whether the variability among treatment regression coefficients is significantly greater than the pooled variability about the group regression lines.

We are now ready to test the adjusted treatment effects. The null hypothesis is that $\alpha_1 = \alpha_2 = \cdots = \alpha_a$; that is, the adjusted treatment effects are homogeneous. The appropriate statistic is

$$F = \frac{(SS_3 + SS_4)/(a-1)}{(SS_1 + SS_2)/[a(n-1) - 1]}$$

The pool of SS_1 and SS_2 requires that the test of homogeneity of regression coefficients does not have a significant result. Table 16-3 summarizes the analysis of covariance and includes the adjusted (adj) sources, their df, the SS formulas for the adjusted terms, the EMS and the F ratios. Note that we obtain the sums of squares for the adjusted A source by subtraction: $SS_{tot(y')} - SS_{S/A(y')}$. This is

Table 16–3 Analysis of covariance for a completely randomized one-factor design

SV	df	SS	EMS	F
Total (adj)	$an-2$	$SS_{tot(y')} = SS_{tot(y)} - \dfrac{SP_{tot}^2}{SS_{tot(x)}}$		
A(adj)	$a-1$	$SS_{A(y')} = SS_{tot(y')} - SS_{S/A(y')}$	$\sigma_{e'}^2 + n\theta_{A'}^2$	$\dfrac{MS_{A(adj)}}{MS_{S/A(adj)}}$
S/A(adj)	$a(n-1)-1$	$SS_{S/A(y')} = SS_{S/A(y)} - \dfrac{SP_{S/A}^2}{SS_{S/A(x)}}$	$\sigma_{e'}^2$	

equivalent to adding $SS_3 + SS_4$, since

$$SS_{A(y')} = SS_{\text{tot}(y')} - SS_{S/A(y')}$$

$$= \left(SS_{\text{tot}(y)} - \frac{SP_{\text{tot}}^2}{SS_{\text{tot}(x)}}\right) - \left(SS_{S/A(y)} - \frac{SP_{S/A}^2}{SS_{S/A(x)}}\right)$$

$$= SS_{\text{tot}(y)} - SS_{S/A(y)} - \frac{SP_{\text{tot}}^2}{SS_{\text{tot}(x)}} + \frac{SP_{S/A}^2}{SS_{S/A(x)}}$$

And since $SS_{A(y)} = SS_{\text{tot}(y)} - SS_{S/A(y)}$,

(16.15)
$$SS_{A(y')} = SS_{A(y)} - \frac{SP_{\text{tot}}^2}{SS_{\text{tot}(x)}} + \frac{SP_{S/A}^2}{SS_{S/A(x)}}$$

$$= SS_3 + SS_4$$

summing Equations (16.12) and (16.13).

To summarize the computations briefly:

1. The total sums of squares for the X and for the Y data are separately analyzed as in the usual analysis of variance; the cross-products sums are similarly treated. The formulas are in Table 16–2.
2. Substituting into Equations (16.10) and (16.11), a test of homogeneity of regression is carried out.
3. Assuming the population regression coefficients to be homogeneous, the adjusted sums of squares of Table 16–3 can be calculated, and the F tests of that table carried through.

Relations between the adjusted sums of squares and correlation coefficients are illuminating; it is important to note that a squared correlation coefficient can be interpreted as a proportion of variance. Consider the adjusted error sum of squares:

$$SS_{S/A(y')} = SS_1 + SS_2$$

$$= SS_{S/A(y)} - \frac{SP_{S/A}^2}{SS_{S/A(x)}}$$

(16.16)
$$= SS_{S/A(y)} - \frac{SP_{S/A}^2}{SS_{S/A(x)}}\left(\frac{SS_{S/A(y)}}{SS_{S/A(y)}}\right)$$

$$= SS_{S/A(y)}\left(1 - \frac{SP_{S/A}^2}{SS_{S/A(x)}SS_{S/A(y)}}\right)$$

$$= SS_{S/A(y)}(1 - r_{xy/A}^2)$$

where $r_{xy/A}$ is the correlation of X and Y scores, pooled over the a treatment groups. Therefore, our adjusted error sum of squares is that proportion of the unadjusted error sum of squares not attributable to the linear relation between X and Y. It is clear from Equation (16.17) that the efficiency of covariance relative to the usual analysis will depend on the magnitude of the correlation between X and Y; the larger the correlation, the smaller the adjusted error term and the greater the profit from performing the covariance analysis.

Manipulations like those performed above show that

(16.17) $$SS_{A(y')} = SS_{A(y)} - [SS_{\text{tot}(y)}r^2_{xy/\text{tot}} - SS_{S/A(y)}r^2_{xy/A}]$$

where $r_{xy/\text{tot}}$ is the correlation of the *an* X and Y scores, with treatment classifications disregarded. Equation (16.17) thus provides an interpretation of the adjusted treatment variability. The adjustment is the difference between the total variability predicted from X and the variability within groups predicted from X.

16.2.4 A NUMERICAL EXAMPLE. An analysis of covariance will now be applied to the data of Table 16–4. We first partition the X variability. The total is

$$SS_{\text{tot}(x)} = (12)^2 + (10)^2 + \cdots + (7)^2 + (9)^2$$
$$- \frac{(12 + 10 + 7 + \cdots + 7 + 9)^2}{18}$$
$$= 2{,}385 - 2{,}200.06 = 184.94$$

Then,

$$SS_{A(x)} = \frac{(12 + 10 + \cdots + 11)^2 + (11 + 12 + \cdots + 11)^2}{6}$$
$$+ \frac{(6 + 13 + \cdots + 9)^2}{6} - C_x$$
$$= 2{,}200.83 - 2{,}200.06 = .77$$

and

$$SS_{S/A(x)} = 184.94 - .77 = 184.17$$

We next turn to the Y data:

$$SS_{\text{tot}(y)} = (26)^2 + (22)^2 + \cdots + (30)^2 - \frac{(26 + 22 + \cdots + 30)^2}{18}$$
$$= 17{,}099 - 16{,}260.06 = 838.94$$

Table 16–4 Data for the analysis of covariance for a one-factor design

	A_1		A_2		A_3
X	Y	X	Y	X	Y
12	26	11	32	6	23
10	22	12	31	13	35
7	20	6	20	15	44
14	34	18	41	15	41
12	28	10	29	7	28
11	26	11	31	9	30

The components are

$$SS_{A(y)} = \frac{(23+22+\cdots+26)^2+(2+31+\cdots+31)^2}{6}$$

$$+ \frac{(23+35+\cdots+30)^2}{6} - C_y$$

$$= 16{,}432.17 - 16{,}260.06 = 172.11$$

and

$$SS_{S/A(y)} = 838.94 - 172.11 = 666.83$$

Next we obtain the cross-product terms. The total is

$$SP_{\text{tot}} = (12)(26)+(10)(22)+\cdots+(9)(30)$$

$$- \frac{(12+10+\cdots+9)(26+22+\cdots+30)}{18}$$

$$= 6{,}317 - 5{,}981.06 = 335.94$$

The treatment term is

$$SP_A = \frac{(12+10+\cdots+11)(26+22+\cdots+26)}{6}$$

$$+ \frac{(11+12+\cdots+11)(32+31+\cdots+31)}{6}$$

$$+ \frac{(6+13+\cdots+9)(23+35+\cdots+30)}{6} - C_{xy}$$

$$= 5{,}978.83 - 5{,}981.06 = -2.23$$

Note that it is possible to obtain negative SP terms (but not SS), since the SP term is the numerator of a correlation coefficient. The $SP_{S/A}$ is

$$SP_{S/A} = SP_{\text{tot}} - SP_A = 335.94 - (-2.23) = 338.17$$

Table 16–5 summarizes the analysis thus far. The remainder of the analysis of covariance involves manipulating the quantities in Table 16–5 according to the

Table 16–5   Preliminary computations for the covariance analysis for a one-factor design

	$SS_{(x)}$	$SS_{(y)}$	SP
Total	184.94	838.94	335.94
A	.77	172.11	−2.23
S/A	184.17	666.83	338.17

Table 16–6   Analysis of covariance for a one-factor design

SV	df	SS	MS	F
Total	16	228.71		
A	2	182.82	91.41	27.87[a]
S/A	14	45.89	3.28	

[a] $p < .001$

formulas of Table 16–3. The adjusted total variability is

$$SS_{tot(y')} = SS_{tot(y)} - \frac{SP^2_{tot}}{SS_{tot(x)}}$$

$$= 838.94 - \frac{(335.94)^2}{184.94}$$

$$= 228.71$$

The error variability is computed next:

$$SS_{S/A(y')} = SS_{S/A(y)} - \frac{SP^2_{S/A}}{SS_{S/A(x)}}$$

$$= 666.83 - \frac{(338.17)^2}{184.17}$$

$$= 45.89$$

The residual variability accounts for the treatment effects:

$$SS_{A(y')} = SS_{tot(y')} - SS_{S/A(y')}$$

$$= 228.71 - 45.89$$

$$= 182.82$$

Table 16–6 presents the final results of the analysis of covariance. The A main effect is a highly significant source of variance. If the covariance adjustment had not been made, F would have had a lower value:

$$F = \frac{SS_{A(y)}/2}{SS_{S/A(y)}/15} = \frac{172.11/2}{666.83/15} = \frac{86.06}{44.46} = 1.94$$

It is clear that the covariance adjustment has resulted in a marked change in the results of the F test.

16.2.5   CALCULATING WITHIN A MULTIPLE REGRESSION FRAMEWORK.   The results obtained in Section 16.2.4 can also be calculated by formulas developed earlier (Chapters 14 and 15). Equation (16.2) can be rewritten in matrix notation:

(16.18) $$\mathbf{Y} = \boldsymbol{\phi}\mathbf{X} + \boldsymbol{\varepsilon}$$

Assuming the restriction on the parameters, $\sum_j \alpha_j = 0$, we find that X is an $an \times (a+1)$ matrix (15×4 for the example of the preceding section) of the general form

(**16.19**)
$$\mathbf{X} = \begin{bmatrix} 1 & 1 & 0 & X_{11}-\bar{X}_{..} \\ \cdot & \cdot & \cdot & \cdot \\ \cdot & \cdot & \cdot & \cdot \\ \cdot & \cdot & \cdot & \cdot \\ 1 & 0 & 1 & X_{12}-\bar{X}_{..} \\ \cdot & \cdot & \cdot & \cdot \\ \cdot & \cdot & \cdot & \cdot \\ \cdot & \cdot & \cdot & \cdot \\ 1 & -1 & -1 & X_{13}-\bar{X}_{..} \\ \cdot & \cdot & \cdot & \cdot \\ \cdot & \cdot & \cdot & \cdot \\ \cdot & \cdot & \cdot & \cdot \end{bmatrix}$$

and

(**16.20**)
$$\boldsymbol{\phi} = \begin{bmatrix} \mu \\ \alpha_1 \\ \alpha_2 \\ \beta \end{bmatrix}$$

The adjusted sums of squares for treatments is then calculated as in the two preceding chapters. We have

(**16.21**) $$SS_A(y') = SS_{A|X} = SS_{A,X} - SS_X = \hat{\boldsymbol{\phi}}'\mathbf{X}'\mathbf{Y} - \hat{\boldsymbol{\phi}}^{*\prime}\mathbf{X}^{*\prime}\mathbf{Y}$$

where $\mathbf{X}$ and $\boldsymbol{\phi}$ are defined by Equations (16.19) and (16.20), $\mathbf{X}^*$ consists of only the first and last columns of $\mathbf{X}$, and $\boldsymbol{\phi}^*$ contains only μ and β. By the data of Table 16.4,

$$\mathbf{X}'\mathbf{X} = \begin{bmatrix} 18 & 0 & 0 & 0 \\ 0 & 12 & 6 & 1 \\ 0 & 6 & 12 & 3 \\ 0 & 1 & 3 & 184.9\bar{4} \end{bmatrix} \quad \text{and} \quad \mathbf{X}'\mathbf{Y} = \begin{bmatrix} 541 \\ -45 \\ -17 \\ 335.9\bar{4} \end{bmatrix}$$

Solving for $\hat{\boldsymbol{\phi}}$ by the methods presented in Chapter 13 gives

$$\hat{\boldsymbol{\phi}} = \begin{bmatrix} \bar{Y}_{..} \\ \bar{Y}'_{.1} - \bar{Y}_{..} \\ \bar{Y}'_{.2} - \bar{Y}_{..} \\ b_{S/A} \end{bmatrix} = \begin{bmatrix} 30.0\bar{5} \\ -3.953545 \\ .101056 \\ 1.836199 \end{bmatrix}$$

Note that

(16.22)
$$\bar{Y}'_{.j} = \bar{Y}_{.j} - b_{S/A}(\bar{X}_{.j} - \bar{X}_{..})$$

This is the adjusted group mean; the second and third entries in $\hat{\boldsymbol{\phi}}$ are estimates of adjusted treatment effects.

We can now calculate

$$\boldsymbol{\phi}'\mathbf{X}'\mathbf{Y} = 17{,}053.107950$$

and

$$SS_{A,X} = \boldsymbol{\phi}'\mathbf{X}'\mathbf{Y} - \mathbf{C}$$
$$= 17{,}053.11 - 16{,}260.06 = 793.05$$

This is the variability about $\bar{Y}_{..}$ accounted for by the simultaneous estimation of the α_j and the regression coefficient β. To get the adjusted treatment sum of squares, we must subtract the sum of squares that is accounted for when the α_j are ignored and only β is estimated. For this purpose, we calculate

$$\mathbf{X}^{*\prime}\mathbf{X}^* = \begin{bmatrix} 1/18 & 0 \\ 0 & 1/184.9\bar{4} \end{bmatrix} \quad \text{and} \quad \mathbf{X}^{*\prime}\mathbf{Y} = \begin{bmatrix} 541 \\ 335.9\bar{4} \end{bmatrix}$$

Then,

$$\hat{\boldsymbol{\phi}}^* = \begin{bmatrix} \bar{Y}_{..} \\ b_{\text{tot}} \end{bmatrix} = \begin{bmatrix} 30.0\bar{5} \\ 1.816461 \end{bmatrix}$$

and

$$SS_X = \boldsymbol{\phi}^{*\prime}\mathbf{X}^{*\prime}\mathbf{Y} - \mathbf{C} = 16{,}870.29 - 16{,}260.06 = 610.23$$

Finally, the adjusted treatment sum of squares is

$$SS_{A(y')} = SS_{A,X} - SS_X = 793.05 - 610.23 = 182.82$$

the result obtained in the preceding section.

The calculations of this section are essentially the calculations of Method 2 of Chapter 15. Under that method, the sum of squares for any main effect was adjusted only for other main effects and not for interaction effects; such interactions were assumed to be nonexistent in the population. In the present case, the analogous assumption is homogeneity of regression; the effects of A are presumed not to vary as a function of the "level" of X. That assumption can be tested as follows. Add two columns to the X matrix defined by Equation (16.19); these columns contain the cross-products of the second and fourth columns ($A_1 \times X$) and the third and fourth columns ($A_2 \times X$), respectively. Also, the elements β_1 and β_2 are added to the ϕ vector defined by Equation (16.20). The sum of squares based on these matrices, reduced by $SS_{A,X'}$ is the SS_1 of Section 16.2.4. From this perspective, SS_1 is a measure of the additional variability accounted for when the β_j are estimated separately.

16.2.6   COMPARING THE ADJUSTED GROUP MEANS.   Equation (16.22) provides an expression for the group mean, adjusted on the basis of $\bar{X}_{.j}$. It is frequently desirable to test for a difference between some pair of means. The appropriate F statistic is

(16.23) $$F = \frac{(\bar{Y}'_{.j} - \bar{Y}'_{.j'})^2}{MS_{S/A(y)}\{2/n + [(\bar{X}_{.j} - \bar{X}_{.j'})^2/SS_{S/A(x)}]\}}$$

and is distributed on one and $a(n-1)-1$ df. Note that the precision of the test increases with increased variability of the covariate within groups (that is, $SS_{S/A(x)}$). Precision is reduced, however, by increasing differences between groups on the covariate measure.

16.3   THE ANALYSIS OF COVARIANCE FOR MULTIFACTOR DESIGNS

16.3.1   THE COMPLETELY RANDOMIZED TWO-FACTOR DESIGN.   We shall first consider a completely randomized two-factor design. The techniques of this section and Section 16.3.2 generalize readily to designs involving more treatment variables.

    Calculations of the usual sums of squares for both X and Y are presented in Chapter 4, and therefore $SS_{A(x)}$, $SS_{A(y)}, \ldots, SS_{AB(y)}$, $SS_{S/AB(x)}$, and $SS_{S/AB(y)}$ require no further comment. We shall need calculations of sums of cross-products. The appropriate formulas are

(16.24) $$C_{xy} = \frac{T_{\ldots(x)}T_{\ldots(y)}}{abn}$$

(16.25) $$SP_{tot} = \sum_i^n \sum_j^a \sum_k^b X_{ijk} Y_{ijk} - C_{xy}$$

(16.26) $$SP_A = \frac{\sum_j^a T_{.j.(x)}T_{.j.(y)}}{bn} - C_{xy}$$

(16.27) $$SP_B = \frac{\sum_k^b T_{..k(x)}T_{..k(y)}}{an} - C_{xy}$$

(16.28) $$SP_{AB} = \frac{\sum_j^a \sum_k^b T_{.jk(x)}T_{.jk(y)}}{n} - C_{xy} - SP_A - SP_B$$

(16.29) $$SP_{S/AB} = SP_{tot} - SP_A - SP_B - SP_{AB}$$

Again note the correspondence between the SP and SS terms. For example,

$$SS_{A(y)} = \frac{\sum_j T_{.j.(y)}^2}{bn} - C = \frac{\sum_j T_{.j.(y)}T_{.j.(y)}}{bn} - C$$

and

$$SP_A = \frac{\sum_j T_{.j.(x)}T_{.j.(y)}}{bn} - C_{xy}$$

We are now ready to present exact computational formulas for the adjusted sums of squares. Considering $SS_{A(y')}$, we begin by computing an adjusted pooled sum of squares for A and S/AB:

(16.30)
$$SS_{(A+S/AB)(y')} = SS_{A(y)} + SS_{S/AB(y)} - \frac{(SP_A + SP_{S/AB})^2}{SS_{A(x)} + SS_{S/AB(x)}}$$

Next, we obtain the adjusted error sum of squares:

(16.31)
$$SS_{S/AB(y')} = SS_{S/AB(y)} - \frac{SP^2_{S/AB}}{SS_{S/AB(x)}}$$

The adjusted sum of squares for the treatment variable A is

(16.32)
$$SS_{A(y')} = SS_{(A+S/AB)(y')} - SS_{S/AB(y')}$$
$$= SS_{A(y)} - \frac{(SP_A + SP_{S/AB})^2}{SS_{A(x)} + SS_{S/AB(x)}} + \frac{SP^2_{S/AB}}{SS_{S/AB(x)}}$$

Corresponding to Equation (16.32), we have

$$df_{A(adj)} = (a-1) - 1 + 1 = a - 1$$

The above calculations can be better understood by comparing Equation (16.26) with Equation (16.15). Rewrite Equation (16.15), making the following substitutions:

$$SS_{tot(x)} = SS_{A(x)} + SS_{S/A(x)}$$

and

$$SP_{tot} = SP_A + SP_{S/A}$$

It is now clear that Equation (16.15) and Equation (16.32) are of identical form. Equation (16.32) also describes the pool of two sums of squares quantities, analogous to SS_3 and SS_4 of the preceding section.

The $SS_{B(y')}$ and $SS_{AB(y')}$ are computed in similar manner to $SS_{A(y')}$. We have

(16.33)
$$SS_{B(y')} = SS_{B(y)} - \frac{(SP_B + SP_{S/AB})^2}{SS_{B(x)} + SS_{S/AB(x)}} + \frac{SP^2_{S/AB}}{SS_{S/AB(x)}}$$

and

(16.34)
$$SS_{AB(y')} = SS_{AB(y)} - \frac{(SP_{AB} + SP_{S/AB})^2}{SS_{AB(x)} + SS_{S/AB(x)}} + \frac{SP^2_{S/AB}}{SS_{S/AB(x)}}$$

The test for homogeneity of regression coefficients also follows that for the one-factor design. We compute

(16.35)
$$SS_1 = \sum_j \sum_k \frac{SP^2_{jk}}{SS_{jk(x)}} - \frac{SP^2_{S/AB}}{SS_{S/AB(x)}}$$

and

(16.36)
$$SS_2 = SS_{S/AB(y)} - \sum_j \sum_k \frac{SP^2_{jk}}{SS_{jk(x)}}$$

To test the null hypothesis that $\beta_{11} = \beta_{12} = \cdots = \beta_{ab}$, we compute

(16.37)
$$F = \frac{SS_1/(ab-1)}{SS_2/ab(n-2)}$$

The procedures of this section are readily extended to designs involving additional factors and to other designs besides the completely randomized design. All that is necessary is that the transition from SS formulas to SP formulas be understood, and that the general form for the sum of squares for any main or interaction effect be recognized:

(16.38)
$$SS_{\text{effect}(y')} = SS_{\text{effect}(y)} - \frac{(SP_{\text{effect}} + SP_{\text{error}})^2}{SS_{\text{effect}(x)} + SS_{\text{error}(x)}} + \frac{SP_{\text{error}}^2}{SS_{\text{error}(x)}}$$

In the next section, the application of Equation (16.38) to a design involving between-subject and within-subjects variability will be illustrated. Following that is a numerical example.

16.3.2   A MIXED DESIGN.   Consider a groups of n subjects who are given b trials on a paired-associate task. The groups differ relative to the meaningfulness of the material. All an subjects have previously been tested for b trials on one list of associates that is not included among the a experimental lists. Thus, there are b pretest scores X and also b dependent measures Y for each of the an subjects. We can readily apply Equation (16.38) to this experimental design, which involves one between-subjects and one within-subject variable. The "error" of Equation (16.38) depends on which treatment effect the adjusted sum of squares is being computed for. Thus, in our example,

$$SS_{A(y')} = SS_{A(y)} - \frac{(SP_A + SP_{S/A})^2}{SS_{A(x)} + SS_{S/A(x)}} + \frac{SP_A^2}{SS_{S/A(x)}}$$

$$SS_{B(y')} = SS_{B(y)} - \frac{(SP_B + SP_{SB/A})^2}{SS_{B(x)} + SS_{SB/A(x)}} + \frac{SP_{SB/A}^2}{SS_{SB/A(x)}}$$

$$SS_{AB(y')} = SS_{AB(y)} - \frac{(SP_{AB} + SP_{SB/A})^2}{SS_{AB(x)} + SS_{SB/A(x)}} + \frac{SP_{SB/A}^2}{SS_{SB/A(x)}}$$

The error term calculations are

$$SS_{S/A(y')} = SS_{S/A(y)} - \frac{SP_{S/A}^2}{SS_{S/A(x)}}$$

and

$$SS_{SB/A(y')} = SS_{SB/A(y)} - \frac{SP_{SB/A}^2}{SS_{SB/A(x)}}$$

The sum of squares calculations have been previously presented in Chapter

Table 16–7 Data for the covariance of a mixed design

		X Data					Y Data		
		B_1	B_2	B_3			B_1	B_2	B_3
A_1	S_{11}	22	23	20	A_1	S_{11}	14	17	23
	S_{21}	23	18	26		S_{21}	16	20	18
A_2	S_{12}	22	18	21	A_2	S_{12}	6	23	18
	S_{22}	19	26	28		S_{22}	8	27	26

8. The *SP* calculations follow readily as in the past. For example,

$$SS_{S/A(y)} = \frac{\sum_{j=1}^{a}\sum_{i=1}^{n} T_{ij.(y)}^2}{b} - \frac{\sum_{j=1}^{a} T_{.j.(y)}^2}{bn}$$

$$= \frac{\sum_j \sum_i T_{ij.(y)} T_{ij.(y)}}{b} - \frac{\sum_j T_{.j.(y)} T_{.j.(y)}}{bn}$$

and

$$SP_{S/A} = \frac{\sum_j \sum_i T_{ij.(x)} T_{ij.(y)}}{b} - \frac{\sum_j T_{.j.(x)} T_{.j.(y)}}{bn}$$

16.3.3 A NUMERICAL EXAMPLE.   Table 16–7 presents *X* and *Y* data for an experiment involving four subjects, two at A_1 and two at A_2; all four are tested at all levels of *B*. The sums of squares for *X* and for *Y* are computed as in Chapter 8. Therefore, computational details are omitted and results are merely listed in Table 16–8. The cross-product (*SP*) calculations parallel those for sums of

Table 16–8 Preliminary computations for the analysis of covariance for a mixed design

	$SS_{(x)}$	$SS_{(y)}$	SP
Total	115.67	870.92	133.17
Between S	25.00	48.92	21.50
A	.33	30.08	3.17
S/A	24.67	18.84	18.33
Within S	90.67	822.00	111.67
B	15.17	623.17	71.42
AB	8.17	197.16	39.08
SB/A	67.33	1.67	1.17

squares. The total is given by

$$SP_{tot} = (22)(14) + (23)(17) + \cdots + (28)(35)$$
$$- \frac{(22 + 23 + \cdots + 28)(14 + 17 + \cdots + 35)}{12}$$
$$= 5,564.00 - 5,430.83 = 133.17$$

As with the sums of squares, the total can be partitioned into a between-subjects and a within-subject component. Thus, we have

$$SP_{B.S} = \frac{(22 + 23 + 20)(14 + 17 + 22)}{3} + \cdots$$
$$+ \frac{(19 + 26 + 28)(8 + 27 + 35)}{3} - C_{xy}$$
$$= 5,452.33 - 5,430.83 = 21.50$$

This, in turn can be partitioned:

$$SP_A = \frac{(22 + 23 + \cdots + 26)(14 + 16 + \cdots + 24)}{6}$$
$$+ \frac{(22 + 19 + \cdots + 28)(6 + 8 + \cdots + 35)}{6} - C_{xy}$$
$$= 5,434.00 - 5,430.83 = 3.17$$

and
$$SP_{S/A} = SP_{B.S} - SP_A = 21.50 - 3.17 = 18.33$$

The within-subject component is obtained by subtraction:

$$SP_{WS} = SP_{tot} - SP_{B.S}$$
$$= 133.17 - 21.50 = 111.67$$

This is now partitioned into SP_B, SP_{AB}, and $SP_{SB/A}$. First,

$$SP_B = \frac{(22 + \cdots + 19)(14 + \cdots + 8)}{4} + \cdots$$
$$+ \frac{(20 + \cdots + 28)(22 + \cdots + 35)}{4} - C_{xy}$$
$$= 5,502.25 - 5,430.83 = 71.42$$

Next,

$$SP_{AB} = \frac{(22 + 23)(14 + 16) + \cdots + (21 + 28)(33 + 35)}{2}$$
$$- C_{xy} - SP_A - SP_B$$
$$= 5,544.40 - 5,430.83 - 3.17 - 71.42 = 39.08$$

Finally,

$$SP_{SB/A} = SP_{W.S} - SP_B - SP_{AB}$$
$$= 111.67 - 71.42 - 39.08 = 1.17$$

These results are also included in Table 16–8.

Using the entries in Table 16–8, we can now obtain the adjusted sums of squares. The key is the correct application of Equation (16.38). Thus, we have

$$SS_{A(y')} = SS_{A(y)} - \frac{(SP_A + SP_{S/A})^2}{SS_{A(x)} + SS_{S/A(x)}} + \frac{SP_{S/A}^2}{SS_{S/A(x)}}$$

$$= 30.08 - \frac{(3.17 + 18.33)^2}{.33 + 24.67} + \frac{(18.33)^2}{24.67}$$

$$= 25.21$$

The error term is straightforward:

$$SS_{S/A(y')} = SS_{S/A(y)} - \frac{SP_{S/A}^2}{SS_{S/A(x)}}$$

$$= 18.83 - \frac{(18.33)^2}{24.66} = 5.22$$

Turning to the within-subject effects, we have

$$SS_{B(y')} = SS_{B(y)} - \frac{(SP_B + SP_{SB/A})^2}{SS_{B(x)} + SS_{SB/A(x)}} + \frac{SP_{SB/A}^2}{SS_{SB/A(x)}}$$

$$= 623.17 - \frac{(71.42 + 1.17)^2}{15.17 + 67.33} + \frac{(1.17)^2}{67.33}$$

$$= 559.33$$

and

$$SS_{AB(y')} = SS_{AB(y)} - \frac{(SP_{AB} + SP_{SB/A})^2}{SS_{AB(x)} + SS_{SB/A(x)}} + \frac{SP_{SB/A}^2}{SS_{SB/A(x)}}$$

$$= 197.16 - \frac{(39.08 + 1.17)^2}{8.17 + 67.33} + \frac{(1.17)^2}{67.33}$$

$$= 175.72$$

For the error term, we have

$$SS_{SB/A(y')} = SS_{SB/A(y)} - \frac{SP_{SB/A}^2}{SS_{SB/A(x)}}$$

$$= 1.67 - \frac{(1.17)^2}{67.33}$$

$$= 1.65$$

The final analysis is summarized in Table 16–9.

Table 16–9   Analysis of covariance for a mixed design

SV	df	SS	MS	F
Between S	2	30.43		
A	1	25.21	25.22	4.83
S/A	1	5.22	5.21	
Within S	7	736.70		
B	2	559.33	279.67	527.68[a]
AB	2	175.72	87.86	165.77[a]
SB/A	3	1.65	.53	

[a] $p < .001$

16.4   INTERPRETING THE RESULTS OF ANALYSIS OF COVARIANCE

If treatment groups do not differ systematically in X, analysis of covariance can be an extremely useful tool for improving the precision of a study. Variability about the group regression line will be less than variability about the group mean; at the cost of one error *df*, the error mean square will be reduced relative to that computed in the standard analysis of variance. Furthermore, if we assume homogeneity of regression, interpreting the analysis of covariance is straightforward. The adjustment has little effect on estimation of treatment means; consequently, we are testing much the same treatment effects in analyzing covariance as in analyzing variance, but against a reduced error term. This situation is shown in Figure 16–1, in which the primary effect of the treatment is on the intercept of the group regression functions.

In Figures 16–2 and 16–3, estimates of treatment effects are different after covariance adjustment from what they are before such adjustments are imposed.

Figure 16–1   A data set showing homogeneity of regression with similar values of X for the two groups.

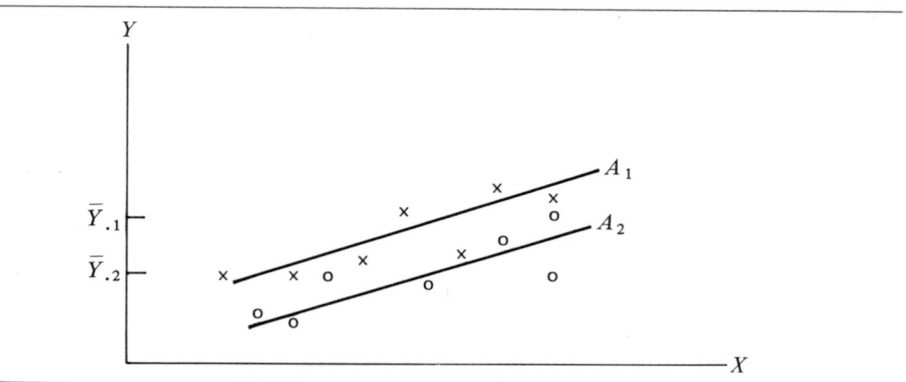

Figure 16–2   A data set in which the performance means are identical but group intercepts differ.

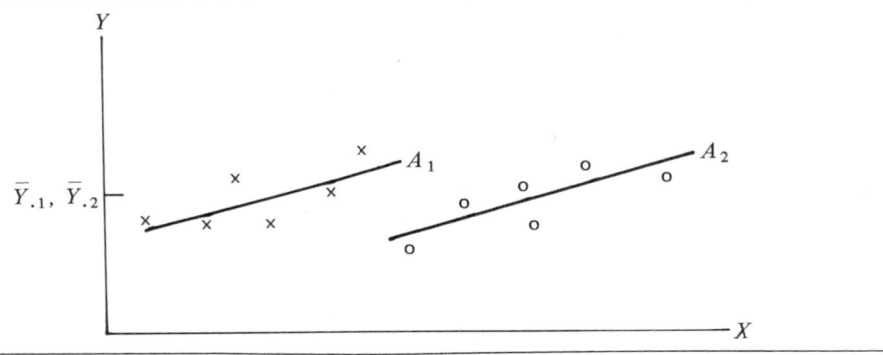

Both figures contain apparent paradoxes. In Figure 16–2, the $\bar{Y}_{.j}$ are identical; the usual analysis of variance leads us to conclude that there are no treatment effects. Analysis of covariance of these data well may prove significant, however. To see why, recall that the adjusted mean for group A_j is

(16.39)
$$\bar{Y}'_{.j} = \bar{Y}_{.j} - \beta(\bar{X}_{.j} - \bar{X}_{..})$$
$$= (\bar{Y}_{.j} - \beta\bar{X}_{.j}) + \beta\bar{X}_{..}$$

The first term to the right of the equals sign is the expression for the intercept for A_j. It is evident in Figure 16–2 that the group intercepts differ and therefore the adjusted means will differ, even though the unadjusted means are identical. This happens because the $\bar{X}_{.j}$ differ; if they did not, the intercepts would be identical.

Figure 16–3 presents the complementary situation. The group means clearly differ but the group intercepts are identical and analysis of covariance will

Figure 16–3   A data set in which performance means differ but group intercepts are identical.

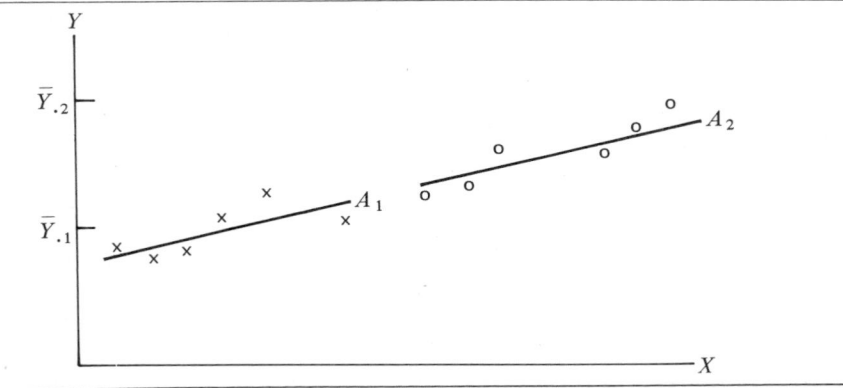

produce nonsignificant results. Again referring to Equation (16.39), it should be evident that if the intercepts are identical, the adjusted means also will be, even though the unadjusted means differ greatly.

Let us summarize the three situations depicted in Figures 16–1 through 16–3. When groups are similar in the covariate, adjusted treatment effects are similar to unadjusted effects; the only important consequence of the analysis of covariance is a reduction of error variance that may produce significant Fs in cases in which analyzing variance failed to yield significance. When groups differ significantly in the covariate, adjusted and unadjusted treatment effects can appear very different. It is often unclear how the results of the analysis of covariance are to be interpreted in such situations. Frequently, the investigator makes statements either that do not follow from the research or that can be easily misinterpreted by the reader.

Suppose three methods of teaching arithmetic are compared; Y is a performance measure and X is amount of study time. Method 1 is significantly superior to the other two when an analysis of variance test is carried out. After adjustment for study time, however, the effect due to methods is no longer significant. The experimenter concludes that the originally obtained difference in performance was due to differences in study time, and that when study time is held constant, the three methods are equally effective. This interpretation is not necessarily correct. Variation in performance may not be due to variation in study time; variability in both measures may instead be due to a third factor, for example, differences in motivation resulting from the three methods. If this were the case, and if study time were actually held constant (all subjects were required to study for a set time period), methods might have a significant effect, since the degree of motivation is still free to vary. Another reason why the experimental control of study time may have different results from the statistical control is that the experimental environment has changed. For example, subjects instructed to study for a given amount of time may have a different set from subjects in the covariance experiment, who were given no particular instructions about study time. The point is that *statistically* adjusting for study time is not the same as *experimentally* holding study time constant.

Furthermore, suppose that our interest lies solely in deciding which method is best. How this method achieves its success is irrelevant. In this case, it is of no interest to adjust the mean performance for variability among the mean study times. On the other hand, suppose the interest lies in the influence of study time on performance. In this case, study time should be systematically manipulated as an independent variable in a factorial design.

In the above example, group differences in the covariate were the result of treatment effects on the covariate. Group differences in X can also be found when existing groups are compared on some measure. Suppose a researcher compares, for example, attitudes of blacks and of whites toward birth control and finds that the whites show greater approval. The researcher notes that the white sample has more years of formal education, hypothesizes that a positive attitude toward birth control is a product of the educational system, and therefore that the difference between the two groups in the study is largely due to differences in education. The data are reanalyzed, this time by analysis of covariance with years of schooling as a covariate. The difference between the adjusted means is not significant. The investigator concludes that "adjusting for educational level, there is no effect of race." There is some sense in which the statement is true; the result of analyzing covariance implies acceptance of the model

$$Y_{ij} = \mu + \beta(X_{ij} - \bar{X}_{..}) + \varepsilon_{ij}$$

in preference to

$$Y_{ij} = \mu + \alpha_j + \beta(X_{ij} - \bar{X}_{..}) + \varepsilon_{ij}$$

It is not clear, however, that increasing the years of education of blacks would result in markedly greater approval of birth control. The kind of education might be very different from that of whites and other differences in cultural factors would still remain. This point may seem obvious, but unfortunately, conclusions like the one objected to here permeate the literature.

It is all too easy to draw inferences that go beyond the data when the groups differ in the covariate. In an excellent treatment of problems of interpretation in analysis of covariance, Smith (1957) has argued that adjusted means might better be referred to as fictitious means. Smith also noted that many applications of analysis of covariance are like the analysis of the heights of mountains with air density as a covariate. Such an analysis might not yield a significant difference between the Rockies and the Himalayas; however, it is doubtful that the Rockies would suddenly shoot up if we could find a way to thin out the surrounding air.

In addition to the possible misinterpretations that arise when the groups differ in X, there is a potential loss of power. Consider Equation (16.23) and note that the variance of the difference between two adjusted means increases as $\bar{X}_{.j} - \bar{X}_{.j'}$ increases. We lose precision, and thus power, when groups differ in the covariate.

16.5   COMPARISON WITH THE TREATMENTS × BLOCKS DESIGN

Section 16.4 suggests that one should generally be interested in covariance as a technique for reducing error variance. In this regard, it is important to compare the analysis with using the treatments × blocks design, since both use a concomitant variable to increase precision. Three advantages of the covariance approach are discernible:

1. The concomitant data can be used after the fact if the covariance analysis is applied. For example, if problem-solving scores prove highly variable, intelligence test data can be collected and an analysis of covariance carried out even though this had not been planned before the data were collected.
2. Establishing blocks is often impractical. For example, we are interested in the problem-solving behavior of high and low socially cohesive groups. It is most

efficient, and possibly more interesting, to work with already established groups, such as Boy Scout troops. It may be impossible, however, to find an equal number of high and low intelligence groups at each level of cohesiveness. It is more practical merely to measure intelligence and use it as a covariate instead of as a factor in the design.

3. The analysis of covariance is more precise than the treatments×blocks design when the true correlation between X and Y is greater than .6 (Feldt, 1958).

Although the covariance approach has the advantages just cited, the treatments× blocks design is superior in several other respects:

1. The computational labor involved in an analysis of variance performed on treatments×blocks data is approximately one-third to one-half the labor involved in the covariance analysis, thus partly compensating for the increased experimental labor resulting from establishing blocks of subjects.

2. The treatments×blocks approach (assuming the optimal number of levels) is more precise than the covariance approach when the true correlation between X and Y is less than .4. This is important, since correlations of less than .4 are more frequent in psychological research than correlations greater than .6.

3. The treatments×blocks interaction may be of interest. Furthermore, if there is reason to expect such an interaction to be significant, covariance should be avoided. If the block means differ more at one treatment level than at another, then the values of Y_{ij} are changing more rapidly as a function of X_{ij} at one treatment level than at another. In short, if a treatments×blocks interaction is significant, it is not correct to assume homogeneity of regression coefficients, and the covariance model presented earlier in this chapter is not appropriate to the data.

4. Perhaps the most important advantage of the experimental over the statistical approach lies in the relative complexity of the covariance model. With it, there are more assumptions, and more things that can go wrong, and statisticians have not yet adequately assessed the consequences of violating the model.

Add to the points just cited the inferential problems raised in the preceding section, and one begins to understand why most biometricians and statisticians recommend the experimental over the statistical approach. Covariance is frequently useful but the experimenter should consider the estimated correlation of X and Y, the probable validity of the covariance model for the data to be collected, and any possible inferential problems. If the choice is approached this way, more often than not the treatments×blocks design will be used.

EXERCISES

16.1 Do the following analysis of covariance:

		A_1		A_2	
		X	Y	X	Y
		23.8	7.9	28.5	25.1
		23.8	7.1	18.5	20.7
		22.6	7.7	20.3	20.3
B_1		22.8	11.2	26.6	18.9
		22.0	6.4	21.2	25.4
		19.6	10.0	24.0	30.0
		27.5	20.1	22.9	19.9
		28.1	17.7	25.2	28.2
		35.7	16.8	20.8	18.1
B_2		27.7	30.5	13.5	13.5
		25.9	21.0	19.1	19.3
		27.9	29.3	32.2	35.1

16.2 The following 3×3 Latin square has one S in each sequence and an X (covariate) and Y (dependent) measure in each cell. Compute the adjusted F test for treatments.

	A_1	A_2	A_3
X	4	1	5
Y	8	9	21

	A_2	A_3	A_1
X	3	2	4
Y	5	18	2

	A_3	A_1	A_2
X	1	3	4
Y	10	2	4

16.3 Each of three of eight Ss is run through four problem-solving tasks. Time to solve is recorded. IQ scores are also available.

		Y data (time to solve)				X data (IQ)
		P_1	P_2	P_3	P_4	
	S_{11}	34	46	48	64	108
A_1	S_{21}	36	41	40	60	112
	S_{31}	28	37	35	52	124
	S_{12}	46	60	63	84	116
A_2	S_{22}	40	51	48	74	127
	S_{32}	55	72	73	96	103
	S_{13}	45	70	74	88	106
A_3	S_{23}	41	63	62	70	135
	S_{33}	49	71	70	85	112

　　(a) Set up the analysis of covariance SS computations.

　　(b) Comment on the effects of using covariance in the present case.

16.4 The relation between Y and X for each treatment group can reasonably be described by the nonlinear function

$$Y_{ij} = K_i + 5X_{ij}^2$$

How would you perform an analysis of covariance?

REFERENCES

Atiqullah, M. The robustness of the covariance analysis of a one-way classification. *Biometrika* 51: 365–72 (1964).

Elashoff, J. D. Analysis of covariance: A delicate instrument. *American Educational Research Journal* 6: 383–401 (1969).

Feldt, L. S. A comparison of the precision of three experimental designs employing a concomitant variable. *Psychometrika* 23: 335–53 (1958).

Glass, G. V., Peckham, P. D., and Sanders, J. R. Consequences of failure to meet assumptions underlying the analysis of variance and covariance. *Review of Educational Research* 42: 237–88 (1972).

Smith, H. F. Interpretation of adjusted treatment means and regression in analysis of covariance. *Biometrics* 13: 282–308 (1957).

Snedecor, G. W., and Cochran, W. G. *Statistical Methods* 6th ed. Ames: Iowa State University, 1967.

SUPPLEMENTARY READING

An excellent summary of the covariance model and its application can be found in

Biometrics 13, no. 3 (September 1957).

　　In that issue, readers should find particularly useful

Cochran, W. G. Analysis of covariance: Its nature and uses, pp. 261–81.

The Smith article, referred to above, is also in that issue of *Biometrics*.

　　Many psychologists have studied how to interpret the results of the analysis of covariance:

Lord, F. M. A paradox in the interpretation of group comparisons. *Psychological Bulletin* 68: 304–5 (1967).

Lord, F. M. Statistical adjustments when comparing preexisting groups. *Psychological Bulletin* 72: 336–37 (1969).

Maxwell, S., and Cramer, E. M. A note on analysis of covariance. *Psychological Bulletin* 82: 187–90 (1975).

This last paper lists several others on the same topic that preceded it in the *Psychological Bulletin*.

17 FURTHER DATA ANALYSES: QUANTITATIVE INDEPENDENT VARIABLES

17.1 INTRODUCTION

In Chapter 11 various procedures for comparing the means of treatment groups were discussed. The primary concern was answering questions that seem most appropriate when the independent variable is qualitative. Certainly it is possible to investigate the contrasts of Chapter 11 even when the independent variable is quantitative; with quantitative variables, however, it will usually be more interesting to consider the overall trend in the treatment group means rather than to make the specific comparisons among means, which was the chief concern in Chapter 11. Since the formulas are basically the same for quantitative and qualitative contrasts, the emphasis in the present chapter will be the rationale for the evaluation of trend.

Consider an experimental study of generalization. A mild shock is paired with a 1,000-Hz tone. Subjects are then divided into several groups, each of which experiences a different tonal frequency but no shock. Galvanic skin responses (GSR) are measured in this test phase for each individual. Let us assume that there are two independent processes at work that sum together to produce the individual scores. One process is the consequence of stimulus frequency; the effect on GSR increases linearly as the frequency of the test tone increases. A second process is generalization; the effect on GSR decreases symmetrically as distance of the test tone from the training stimulus of 1,000 Hz increases. Figure 17–1 depicts this hypothetical situation. The dashed lines represent the effects of the two processes underlying the data. Summing these yields the α_j, represented by the solid line.

In actual fact, all the experimenter really has is a set of observed deviations from the means, the xs of Figure 17–1. Assuming that the F test of the *frequency* main effect is significant, we are left with the inferential problem of determining whether there is a stimulus frequency process, a generalization process, or both, or other processes contributing to the significant variation within the observed set of means. On other possible processes, note the slight upturn in the last data point. If this is more than chance variability, the straight line and inverted U representing stimulus frequency and generalization could not account for it. There is the possibility that some S-shaped process is also at work.

Figure 17–1   Assumed effects of two processes, the sum of effects, and observed data points.

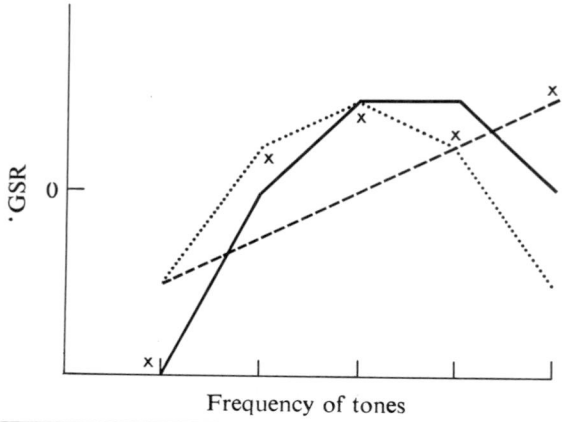

Frequency of tones

The procedures involved in drawing inferences about the true shape of the function relating treatment population means and levels of the independent variable are usually referred to as trend, or orthogonal polynomial, analysis. Such procedures can attack additional questions also. Suppose, for example, we carried out the experiment with several different groups of subjects drawn from populations defined by various degrees of severity of schizophrenia. Several personality theorists have considered the possibility that schizophrenics generalize more than normal subjects. We should therefore expect that the underlying generalization gradient would become flatter as severity increased, the highly schizophrenic subject showing little variation in *GSR* over a wide range of test tones. If this hypothesis is true, we should have a personality×frequency interaction. That in itself does not prove the case, however, because such an interaction could be due to other variations in underlying processes. For example, if the linear function relating *GSR* and intensity varied in regression coefficient over clinical categories, this too would give rise to a significant interaction. We shall consider analyses that attack questions about variations in the slope and curvature of several population functions.

Before developing the rationale and computations involved in trend analysis, it might help to consider a second instance of its application, this time an experiment on signal detection. A subject sits a fixed distance from a 25-sq-ft screen. Every 10 seconds, a small circle of light appears on the screen, and the subject must report the location of this target. Detection time is recorded by the experimenter. One independent variable is the intensity of the target illumination. A second variable is the structure of the screen, which may be one open area, or an area divided by a vertical line into two equal segments or by two vertical lines into three equal segments, and so on. The experimenter hypothesizes that a certain amount of segmenting of the screen facilitates the search. There will be

Figure 17–2 Hypothesized functions for a signal detection experiment.

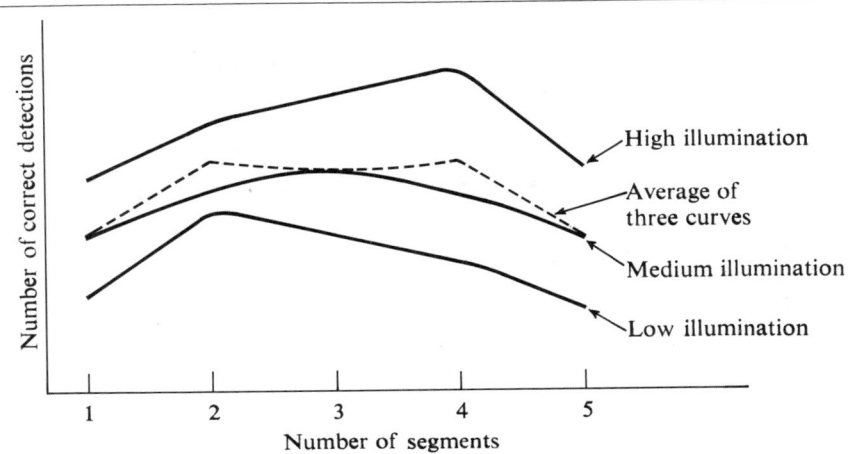

some point, however, at which further segmenting will result in deterioration of performance; the subject will become confused and will repeatedly search the same segment, thus losing time. It is further hypothesized that the optimal number of segments will be a function of intensity of illumination. The higher the illumination, the higher the optimal number of segments. Figure 17–2 illustrates one way the data might look if the experimenter's hypotheses were correct.

In Figure 17–2, the behavioral hypotheses mentioned above have been translated into hypotheses about performance trends over numbers of segments. On the average curve—the plot of the main effect of segments—one might note that (1) if a straight line were fitted to this curve, its slope would be zero, and (2) the function appears to be better fitted by a curved than by a straight line. The behavioral hypotheses also lead to definite inferences about the three intensity curves. If a straight line were fitted to each, the three lines would have different slopes. It appears, furthermore, that the three functions differ in their shapes.

In view of the initial hypotheses, and of Figure 17–2, which provides one plausible representation of the hypotheses, it seems that tests are required to answer such questions as:

1. Does the plot of mean detection time as a function of number of segments have a slope different from zero?
2. Is the plot of mean detection time as a function of segments adequately fitted by a straight line?
3. Do the three plots of detection time against segments differ in slope?
4. Do the three plots of detection time against segments differ in shape?

The computations involved in answering questions such as these about the shapes and slopes of functions come under the general heading of trend. As in the example of the generalization experiment, the answer to our questions lies in trend analysis.

17.2   ORTHOGONAL POLYNOMIALS

17.2.1   RATIONALE FOR TESTING NULL HYPOTHESES.

We begin our discussion of the complete trend analysis by noting that any a data points can be described by an equation having the general form

$$(17.1) \qquad Y = b_0 + b_1 X + b_2 X^2 + \cdots + b_p X^p + \cdots + b_{a-1} X^{a-1}$$

Equations of this form are referred to as *polynomial functions of order $a-1$*. Figure 17–3 presents several such functions, each labeled by the appropriate equation. Note the restriction that if there are a points, the order of the polynomial is at most $a-1$ (it can be less, since b_{a-1}, b_{a-2}, . . . , can be zero). To understand why this is so, consider a first-order polynomial, the straight line

$$(17.2) \qquad Y = b_0 + b_1 X$$

At least two data points are required to estimate the parameters, b_0 and b_1. Similarly, three data points are required to fit a second-order polynomial, otherwise known as a quadratic function,

$$(17.3) \qquad Y = b_0 + b_1 X + b_2 X^2$$

Figure 17–3   Some sample polynomial functions.

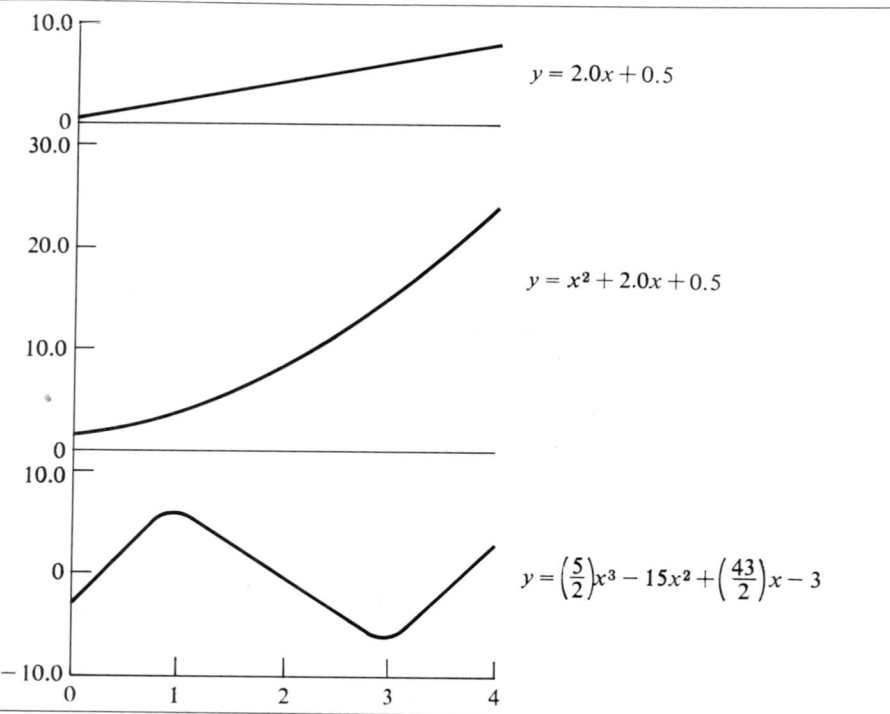

since three parameters are to be estimated. A polynomial of order $a-1$ generally involves estimating a parameters, and at least that many data points are required.

Suppose that the a data points are treatment population means and the values of X are levels of some quantitative independent variable, for example, stimulus intensity, magnitude of reward, or amount of practice. In this context, Equation (17.1) becomes

(17.1′) $$\mu_j = \beta_0 + \beta_1 X_j + \beta_2 X_j^2 + \cdots + \beta_p X_j^p + \cdots + \beta_{a-1} X_j^{a-1}$$

which can be rewritten

(17.4) $$\mu_j = \beta_0' + \beta_1'(a_1 + X_j) + \beta_2'(a_2 + b_2 X_j + X_j^2) + \cdots$$
$$+ \beta_{a-1}'(a_{a-1} + b_{a-1} X_j + c_{a-1} X_j^2 + d_{a-1} X_j^3 + \cdots + X_j^{a-1})$$

where

$$\beta_0 = \beta_0' + \beta_1' a_1 + \beta_2' a_2 + \cdots + \beta_{a-1}' a_{a-1}$$
$$\beta_1 = \beta_1' + \beta_2' b_2 + \cdots + \beta_{a-1}' b_{a-1}$$
$$\cdots$$
$$\beta_{a-1} = \beta_{a-1}'$$

By the above definitions of β_p', Equation (17.1′) and (17.4) are algebraically identical. The presentation will now be considerably simplified if we rewrite Equation (17.4)

(17.5) $$\mu_j = \beta_0' + \beta_1' \xi_{j1} + \beta_2' \xi_{j2} + \cdots + \beta' \xi_{j,a-1}$$

where the definitions of the ξ_{jp} follow from comparing Equations (17.4) and (17.5). We have taken the trouble to recast Equation (17.1′) as (17.5), because if the ξ_{jp} are properly chosen, the terms being summed are independent. The term $\beta_1' \xi_{j1}$ represents an underlying linear process. The term $\beta_2' \xi_{j2}$ represents an underlying independent quadratic process, and so on; the critical point is that in this form, with ξ_{jp} properly chosen, the $a-1$ functions are independent of each other. We generally refer to *orthogonal* (independent) *polynomial* functions.

Let us now reconsider some of the questions raised in Section 17.1. In the generalization example, we wanted to determine whether a linear process due to stimulus frequency contributed to the overall main effect. In the detection example, the experimenter's hypotheses led to the expectation that the underlying linear component would have slope zero. In terms of Equation (17.5), the relevant null hypothesis is

$$H_0 : \beta_1' = 0$$

In both experiments, there is reason to believe that the plot of the main effect against levels of the independent variable (frequency, number of segments) would have a quadratic component. The relevant null hypothesis is

$$H_0 : \beta_2' = 0$$

Tests of hypotheses about polynomial components that may contribute to main effects generally have the form

$$H_0 : \beta'_p = 0$$

In short, Equation (17.5) states that the plot of the μ_j against the levels of the independent variable can be held to result from the summing of several independent polynomial functions of different orders; Figure 17.1 presented a graphic analogue to Equation (17.5). The statistical problem is to determine which of the possible $a - 1$ polynomial functions contribute significantly to the plot of the μ_j, which of the regression coefficients β' are greater than zero.

How do we test null hypotheses of the sort described? First, note that β'_p is the coefficient for the regression of μ_j on ξ_{jp}. We have pointed out (Chapter 12) that in dealing with the regression of Y on X the appropriate expression for the regression coefficient is

$$\beta = \frac{\sum_j (Y_j - \bar{Y})(X_j - \bar{X})}{\sum_j (X_j - \bar{X})^2}$$

By analogy,

(17.6)
$$\beta'_p = \frac{\sum_j (\mu_j - \mu)(\xi_{jp} - \bar{\xi}_p)}{\sum_j \xi_{jp}^2}$$

The ξ_{jp} can be selected so that $\bar{\xi}_p = (\sum_j \xi_{jp})/a = 0$. Therefore, from Equation (17.6) it follows that the null hypothesis $\beta'_p = 0$, is true if $\sum_j (\mu_j - \mu)\xi_{jp} = 0$. However,

$$\sum_j (\mu_j - \mu)\xi_{jp} = \sum_j \xi_{jp}\mu_j - \mu \sum_j \xi_{jp} = \sum_j \xi_{jp}\mu_j$$

This is the exact form of the null hypotheses introduced in Section 11.1 except that there we used w_j instead of ξ_j. It therefore follows that the computations to test for contributions of orthogonal polynomial functions are identical to those used to test contrasts in Chapter 11, and that furthermore, each null hypothesis of the class now under discussion will be tested on a single *df*.

Since Chapter 11 provides the computational formulas for sums of squares, all we really require is a method for obtaining the weights—in the case of trend, the ξ_{jp}. As preparation, let us briefly reconsider the application of trend analysis.

We apply the analyses of this chapter when we are interested in testing the contributions of various polynomial functions to the variability among the treatment effects. We can, as in the generalization example, imagine the data as the sum of several independent processes represented by different orders of polynomial functions. Alternatively, as in the signal detection example, we may be unable to label separate component processes although we have reason to believe that the observed data function will be of a particular nature; this belief is consistent with viewing the data as though they composed the sum of certain underlying components and proceeding to test for these components. In both the generalization and detection examples, we have reason to be interested in the linear and quadratic components. Each of these should be tested. Since the components are othogonal, these tests will have accounted for two of the

$a - 1$ *df*; we should then test the residual, $SS_A - SS_{lin} - SS_{quad}$, on $(a-1) - 2$ *df*. Significance of the residual term would suggest the presence of processes that had not been expected before the experiment. If the residual is significant, it may be desirable to partition it into its polynomial components, testing them separately, the better to specify the unlooked-for process.

Frequently, the investigator's sole interest is in possible deviations from linearity. A common paradigm in the study of memory involves presenting subjects with displays of digits and then measuring the latency with which a subject reports whether a test digit was or was not in the original display. The most common theoretical position holds that the subject has a representation of the display in memory and scans that representation one digit at a time, comparing each with the test digit. If the test digit had not been in the display, and if it requires t milliseconds to scan each digit in memory, latency is a linear function of K, the display size. In this example, we should perform two tests: linearity on 1 *df* and residual (deviations from linearity) on $(a-1) - 1$ *df*. Significance of the latter term would pose a problem for the theory.

17.2.2 DERIVING VALUES OF ξ.   Table A–11 presents values of ξ' for various values of a under the assumptions that there are equal numbers of measures for each level of A and that the levels of A are equally spaced. The derivational technique frequently gives noninteger values of ξ. These values were multiplied by λ to give the integers ξ' of Table A–11; remember that multiplying a set of weights by a constant λ will not change the sum of squares based on those weights.

For several reasons, it is worth while to show how the tabled polynomial coefficients are derived. We shall probably be more comfortable with the analyses if we can sense where the coefficients come from. More important, there may be occasions in which ns are not equal, or more frequently, in which the levels of A are not equally spaced. Suppose we have a study of delay of reward with four delay intervals. The values of X, delay magnitude, and numbers of subjects are:

X_j	0	1	3	4
n_j	4	5	5	4

From Equations (17.4) and (17.5), we know that the linear coefficients ξ_{j1} are of the form

$$\xi_{j1} = a_1 + X_j$$

Substituting the actual delay intervals, we have

$$\xi_{11} = a_1 + 0$$
$$\xi_{21} = a_1 + 1$$
$$\xi_{31} = a_1 + 3$$
$$\xi_{41} = a_1 + 4$$

We now impose the restriction $\sum_j n_j \xi_{jp} = 0$. Then,

$$(4)(a_1) + (5)(a_1 + 1) + 5(a_1 + 3) + 4(a_1 + 4) = 0$$

and on solving this, we have

$$18a_1 + 36 = 0$$
$$a_1 = -2$$

Then,

$$\xi_{11} = -2$$
$$\xi_{21} = -2 + 1 = -1$$
$$\xi_{31} = -2 + 3 = 1$$
$$\xi_{41} = -2 + 4 = 2$$

The quadratic coefficients are of the form

$$\xi_{j2} = a_2 + b_2 X_j + X_j^2$$

Then,

$$\xi_{12} = a_2 + (0)(b_2) + 0^2$$
$$\xi_{22} = a_2 + b_2 + 1$$
$$\xi_{32} = a_2 + 3b_2 + 9$$
$$\xi_{42} = a_2 + 4b_2 + 16$$

We have two unknowns, a_2 and b_2. Because of the orthogonality requirement, we also have two simultaneous equations:

$$\sum_j n_j \xi_{j2} = 0$$

$$\sum_j n_j \xi_{j1} \xi_{j2} = 0$$

The second equation represents the orthogonality requirement. Applying these restrictions yields

$$(4)(a_2) + (5)(a_2 + b_2 + 1) + (5)(a_2 + 3b_2 + 9) + (4)(a_2 + 4b_2 + 16) = 0$$
$$(4)(-2)(a_2) + (5)(-1)(a_2 + b_2 + 1) + (5)(1)(a_2 + 3b_2 + 9)$$
$$+ (4)(2)(a_2 + 4b_2 + 16) = 0$$

Simplifying, we have

$$18a_2 + 36b_2 + 114 = 0$$
$$42b_2 + 168 = 0$$

Then,

$$b_2 = \frac{-168}{42} = -4$$

and

$$18a_2 + (36)(-4) + 114 = 0$$

$$a_2 = \tfrac{5}{3}$$

Substituting into the original expressions for ξ_{j2} gives

$$\xi_{12} = \tfrac{5}{3}$$
$$\xi_{22} = \tfrac{5}{3} + (-4) + 1 = -\tfrac{4}{3}$$
$$\xi_{32} = \tfrac{5}{3} + (3)(-4) + 9 = -\tfrac{4}{3}$$
$$\xi_{42} = \tfrac{5}{3} + (4)(-4) + 16 = \tfrac{5}{3}$$

Let $\lambda = 3$. Then, we have integer values:

$$\xi'_{12} = 5$$
$$\xi'_{22} = -4$$
$$\xi'_{32} = -4$$
$$\xi'_{42} = 5$$

The cubic coefficients will involve three unknowns: a_3, b_3, c_3. We also have three simultaneous equations,

$$\sum n_j \xi_{j3} = 0$$
$$\sum n_j \xi_{j1} \xi_{j3} = 0$$
$$\sum n_j \xi_{j2} \xi_{j3} = 0$$

We leave the solution as an exercise.

Usually, the values of X are equally spaced or can be transformed to a scale on which they are equally spaced; if $A = 1, 2, 4, 8$, for example, then $\log_2 X = 0, 1, 2, 3$. When spacing is equal, and observations at each value of X occur equally frequently, Table A–11 immediately provides the desired coefficients.

17.2.3 A NUMERICAL EXAMPLE. Assume, as an example of the calculations, that we have the following treatment totals:

$$\begin{array}{cccc} A_1 & A_2 & A_3 & A_4 \\ 22 & 22 & 20 & 36 \end{array}$$

Further assume that $n = 10$ and $MS_{S/A} = 1.2$. We have

$$SS_A = \frac{(22)^2 + \cdots + (36)^2}{10} - \frac{(22 + \cdots + 36)^2}{40}$$

$$= 16.4$$

Applying Equation (11.17) with w_{ip} first equal to the ξ'_{j1}, we have

$$SS_{\text{lin (A)}} = \frac{[(-3)(22) + (-1)(22) + (1)(20) + (3)(36)]^2}{(10)(20)}$$

$$= 8.0$$

Using the ξ_2' next, we have

$$SS_{\text{quad (A)}} = \frac{(+22 - 22 - 20 + 36)^2}{(10)(4)}$$

$$= 6.4$$

The cubic contrast yields

$$SS_{\text{cub (A)}} = \frac{[-22 + (3)(22) + (-3)(20) + 36]^2}{(10)(20)}$$

$$= 2.0$$

Note that

$$SS_A = SS_{\text{lin (A)}} + SS_{\text{quad (A)}} + SS_{\text{cub (A)}}$$

which must be true if our calculations are correct. Only the linear and quadratic components are significant; we conclude that the cubic sum of squares reflects chance variability.

Consider the data set just analyzed. The parameter b_0' is merely $\bar{Y}_{..}$; other coefficients are obtained from the general expression

(17.6)
$$b_p' = \frac{\sum_j \xi_{jp} \bar{Y}_{.j}}{\sum_j \xi_{jp}^2}$$

Our numerical coefficients are

$$b_0' = \frac{22 + 22 + 20 + 36}{40} = 2.5$$

$$b_1' = \frac{(-3)(2.2) - 2.2 + 2.0 + (3)(3.6)}{20} = .20$$

$$b_2' = \frac{2.2 - 2.2 - 2.0 + 3.6}{4} = .40$$

Then, we have

	A_1	A_2	A_3	A_4
$b_0' =$	2.5	2.5	2.5	2.5
$b_1'\xi_1' =$	-.6	-.2	.2	.6
$b_2'\xi_2' =$	.4	-.4	-.4	.4
	2.3	1.9	2.3	3.5

Each of the rows is plotted in the upper panel of Figure 17–4. The column totals fall on the solid line in the lower panel. This function is not the function that best fits the data points; a perfect fit could be obtained by adding in $b_3'\xi_3'$. We do not wish to fit the data points (the $\bar{Y}_{.j}$), however; our goal is to describe the plot of the population values (the μ_j). Since our analysis has led us to conclude that the cubic component is not significant, the curve in the lower panel of Figure 17–4 is our

Figure 17–4   Trend analysis for a data set.

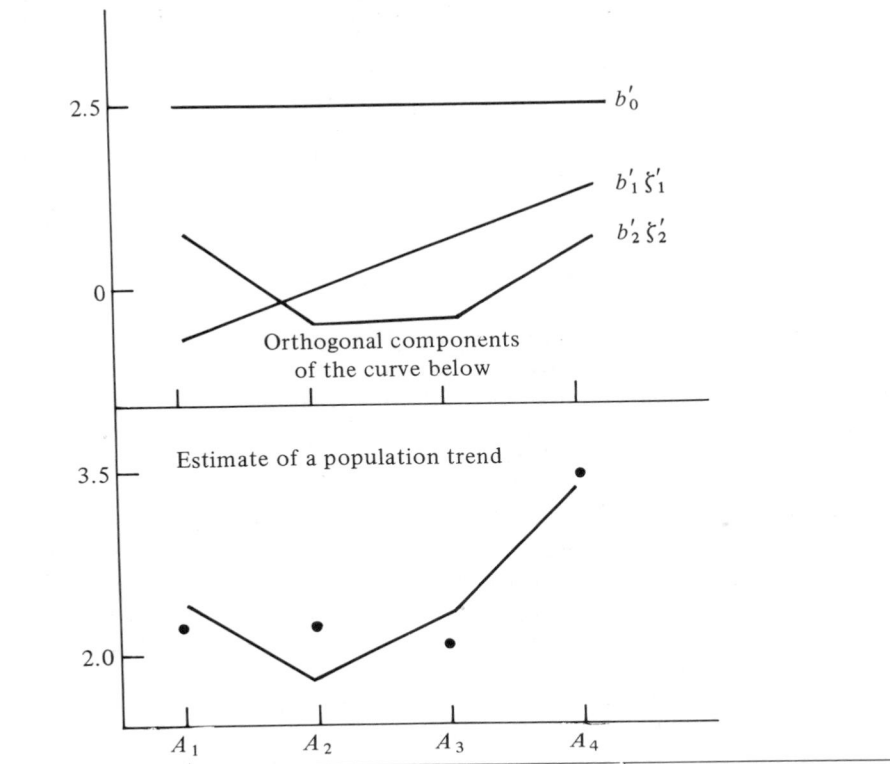

estimate of the population function. The variability of data points about that function is assumed to be error variability rather than the result of a cubic component in the population function.

17.3   MULTIFACTOR DESIGNS

17.3.1   THE ANALYSIS OF INTERACTION.   We have thus far considered the orthogonal components of a main effect. This analysis enables us to answer several specific questions, such as, Is the slope of the best-fitting straight line significantly different from zero? Is there quadratic curvature? Other questions arise when the design involves more than one independent variable. If there is a significant interaction between the variables A and B, and if A is a quantitative variable, we might wish to investigate further the source of the interaction. Why are the b curves not parallel? Do they differ in their slopes? in their quadratic components?

in their cubic components? These equations imply tests of the null hypothesis that the β'_p are the same for all values of k (levels of B). For example, the comparison of slopes involves testing the null hypothesis

$$H_0 : \beta'_{11} = \beta'_{21} = \cdots = \beta'_{k1} = \cdots = \beta'_{b1}$$

Let us consider some examples. In the generalization experiment, A might be the intensity of the test stimulus and B might represent clinical populations. Our theory is that generalization increases with degree of severity of schizophrenic symptoms. In terms of trend analysis, we believe that the quadratic coefficient β'_{k2} decreases (the quadratic component becomes flatter) as schizophrenia becomes more pronounced. The corresponding null hypothesis is that β'_{k2} is the same at all levels of B.

Consider a second example. In the memory experiment described earlier, it was hypothesized that response time would increase linearly with an increase in size of display, each added item in the display resulting in a constant increment in time to scan memory of the display. It has been further hypothesized that the time to scan each item would decrease with amount of prior practice with the experimental task. Let B represent several groups having had different numbers of previous trials and let A still be display size. Now consider a plot of response time as a function of display size with a separate curve for each level of practice. Our theory holds that the linear coefficient β'_{k1} will decrease as practice increases; that is, response time will change less rapidly with display size for more practiced subjects. The null hypothesis is that β'_{k1} will be the same at all levels of B.

Consider carefully what is implied. There are b curves, each plotted over the levels of A. We want to compare the b linear components of A, then the b quadratic components of A, and so on. There are $a - 1$ possible comparisons of this type, that is, as many as there are orthogonal components of the A effect. In general, we are interested in testing the interaction of the pth component of A with the levels of B, in determining whether the pth component of A is a function of the level of B. The relevant source of variance is labeled $p(A) \times B$, following our practice in Section 11.5. The calculations and df are exactly those presented in Chapter 11, with the w_{jp} of the sum of squares formula replaced by the appropriate ξ'_{jp}. The error terms are also those presented in Section 11.5 and are dictated by the design and analysis of variance model.

In Section 11.5 we considered the complete analysis of interaction into $(a - b)(b - 1)$ components, each distributed on a single df. If we replace the ws by ξs, the same breakdowns are possible in trend analysis. Suppose that there are four performance curves, each obtained under a different training method A, plotted over five blocks of trials B. In our analysis of the interaction of trend components, we should first test the source, $\text{lin}(B) \times A$, which represents the variability of the slopes of the four curves. If this is significant, the slopes are a function of training method. Additional hypotheses can now be considered. The average slope, for example, of the A_1 and A_2 curves may differ from the average slope of the A_3 and A_4 curves. In this case, we are concerned with $\text{lin}(B) \times p(A)$ (p represents the contrast of A_1 and A_2 against A_3 and A_4), whose sum of squares is distributed on a single *df*. Two sets of weights are involved:

$$\xi_{11} = -2 \qquad w_{1p} = +1$$
$$\xi_{21} = -1 \qquad w_{2p} = +1$$
$$\xi_{31} = 0 \qquad \text{and} \qquad w_{3p} = -1$$
$$\xi_{41} = +1 \qquad w_{4p} = -1$$
$$\xi_{51} = +2$$

Given these weights, we calculate $SS_{\text{lin}(B) \times p(A)}$ as we calculated $SS_{q(B) \times p(A)}$ in Section 11.5. Calculating error terms again follows directly from the developments in Chapter 11.

Suppose that in the experiment just described, A is also a quantitative variable, for example, amount of practice. Then, there are two sets of ξs, enabling us to obtain such terms as $\text{lin}(A) \times \text{lin}(B)$ and $\text{quad}(A) \times \text{cub}(B)$. The calculations are the same as for any single *df* component of interaction. The interpretation is more complicated than before.

Reconsider the example of the memory experiment in which A is display size and B is amount of practice. Assume that we plot the data as in the left-hand panel of Figure 17–5; reaction time is plotted as a function of display size A, with a separate function at each level of practice B. Assume that there is very little error variance; then, $\text{lin}(A) \times B$ is significant because the slopes of the three functions are quite different. Those slopes ($\text{lin}(A)$) are plotted as a function of B in the right-hand panel. If a best-fitted straight line is fitted to the function plotted in that panel, it would have a slope different from zero. This implies a $\text{lin}(A) \times \text{lin}(B)$ component. The right-hand function is also curved, implying a significant $\text{lin}(A) \times \text{quad}(B)$ component to the AB sum of squares.

Before proceeding to a numerical example, it may be wise to summarize the analysis of the interaction source. Assuming that B is a quantitative scaled

Figure 17–5 Cell means and linear coefficients for a two-factor experiment.

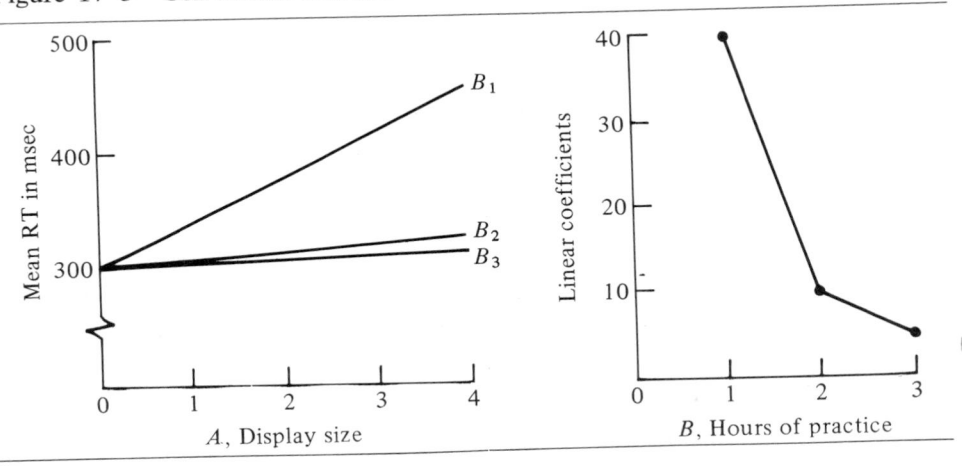

variable, we can examine the qth component of B as a function of the levels of A. Therefore, our first breakdown of the AB source is as shown in the accompanying chart.

SV	df
AB	$(a-1)(b-1)$
$\text{lin}(B) \times A$	$a-1$
$\text{quad}(B) \times A$	$a-1$
$\vdots$	$\vdots$
$q(B) \times A$	$a-1$
$\vdots$	$\vdots$
$(b-1)(B) \times A$	$a-1$

Each of these sources can be further analyzed by assigning weights to the levels of A. If A is also a quantitative variable, then we get a second set of data, as shown.

SV	df
AB	$(a-1)(b-1)$
$\text{lin}(B) \times A$	$a-1$
$\text{lin}(B) \times \text{lin}(A)$	1
$\text{lin}(B) \times \text{quad}(A)$	1
$\vdots$	$\vdots$
$\text{lin}(B) \times p(A)$	1
$\vdots$	$\vdots$
$\text{lin}(B) \times (a-1)(A)$	1
$\text{quad}(B) \times A$	$a-1$
$\text{quad}(B) \times \text{lin}(A)$	1
$\vdots$	$\vdots$
$\text{quad}(B) \times p(A)$	1
$\vdots$	$\vdots$
$\text{quad}(B) \times (a-1)(A)$	1
$q(B) \times A$	$a-1$
$q(B) \times \text{lin}(A)$	1
$\vdots$	$\vdots$
$q(B) \times p(A)$	1
$\vdots$	$\vdots$
$q(B) \times (a-1)(A)$	1
$(b-1)(B) \times A$	$a-1$
$(b-1)(B) \times \text{lin}(A)$	1
$\vdots$	$\vdots$
$(b-1)(B) \times p(A)$	1
$\vdots$	$\vdots$
$(b-1)(B) \times (a-1)(A)$	1

The sum of squares formulas and error terms follow from Chapter 11. To ensure that the application is clear, however, we proceed to a numerical example involving two quantitative variables.

17.3.2 A NUMERICAL EXAMPLE.   We assume three equally spaced levels of the independent variable A and five equally spaced levels of the independent variable B with four scores in each cell. The data, together with various totals and subtotals, are presented in Table 17–1. The appropriate sets of orthogonal coefficients, taken from Table A–11 are also present. The total variability is calculated in the usual way:

$$SS_{tot} = (20)^2 + (18)^2 + \cdots + (7)^2 - \frac{(802)^2}{60}$$

$$= 11,792.000 - 10,720.067$$

$$= 1,071.933$$

Table 17–1   Data for a trend analysis

		A_1	A_2	A_3	$T_{i\cdot k}$	ξ'_{k1}	ξ'_{k2}	ξ'_{k3}	ξ'_{k4}
		20	18	16	54				
		18	17	15	50				
B_1		19	18	17	54	-2	2	-1	1
		16	16	15	47				
	$T_{\cdot j1} = 73$		69	63	$T_{\cdot\cdot 1} = 205$				
		18	15	14	47				
		18	16	13	47				
B_2		17	14	14	45	-1	-1	2	-4
		16	13	13	42				
	$T_{\cdot j2} = 69$		58	54	$T_{\cdot\cdot 2} = 181$				
		16	12	11	39				
		18	13	14	45				
B_3		17	12	13	42	0	-2	0	6
		16	10	12	38				
	$T_{\cdot j3} = 67$		47	50	$T_{\cdot\cdot 3} = 164$				
		15	5	6	26				
		18	8	8	34				
B_4		17	7	9	33	1	-1	-2	-4
		17	5	5	27				
	$T_{\cdot j4} = 67$		25	28	$T_{\cdot\cdot 4} = 120$				
		17	7	6	30				
		18	10	9	37				
B_5		18	9	8	35	2	2	1	1
		15	8	7	30				
	$T_{\cdot j5} = 68$		34	30	$T_{\cdot\cdot 5} = 132$				
	$T_{\cdot j\cdot} = 344$		233	225	$T_{\cdots} = 802$				
	ξ_{j1}	-1	0	1					
	ξ_{j2}	1	-2	1					

For the A main effect, we have

$$SS_A = \frac{(344)^2 + (233)^2 + (225)^2}{20} - 10,720.067$$

$$= 442.433$$

and for B,

$$SS_B = \frac{(205)^2 + \cdots + (132)^2}{12} - 10,720.067$$

$$= 405.433$$

We next compute the variability due to polynomial components according to Equation (11.16):

$$SS_{\text{lin }(A)} = \frac{[-344 + (0)(233) + 225]^2}{(20)(2)}$$

$$= 354.025$$

and

$$SS_{\text{quad }(A)} = \frac{[344 + (-2)(233) + 225]^2}{(20)(6)}$$

$$= 88.408$$

The sum of the above two terms is 442.433, which equals SS_A, as it should. Partitioning SS_B, we obtain

$$SS_{\text{lin }(B)} = \frac{[(-2)(205) - 181 + (0)(164) + 120 + (2)(132)]^2}{(12)(10)}$$

$$= \frac{(207)^2}{120} = 357.075$$

$$SS_{\text{quad }(B)} = \frac{[(2)(205) - 181 + (-2)(164) - 120 + (2)(132)]^2}{(12)(14)}$$

$$= \frac{(45)^2}{168} = 12.054$$

$$SS_{\text{cub }(B)} = \frac{[-205 + (2)(181) + (0)(164) + (-2)(120) + 132]^2}{(12)(10)}$$

$$= \frac{(49)^2}{120} = 20.008$$

and

$$SS_{\text{quart }(B)} = \frac{[205 + (-4)(181) + (6)(164) + (-4)(120) + 132]^2}{(12)(70)}$$

$$= \frac{(117)^2}{840} = 16.296$$

Summing the four components, we obtain SS_B.

The interaction sum of squares is obtained as previously; thus

$$SS_{AB} = \frac{(73)^2 + (69)^2 + \cdots + (30)^2}{4} - C - SS_A - SS_B$$

$$= 151.067$$

There are two ways to partition the AB variability; the choice depends on the questions that are asked of the data. We can compare the five curves (one at each level of B) in linear and quadratic components of A; then we are concerned with $SS_{p(A) \times B}$. Alternatively, we can compare the three curves (one at each level of A) in each of the four components of B; then we are concerned with $SS_{q(B) \times A}$. The choice of approach does not affect the subsequent further analysis of interaction into $(a-1)(b-1)$ single df components of the form $p(A) \times q(B)$. For simplicity, we assume that our primary interest lies in plotting a curve for each level of A over the five B_k data points and therefore in assessing $q(B) \times A$. Then, according to Equation (11.21),

$$SS_{\text{lin}(B) \times A} = \frac{[(-2)(73) - 69 + (0)(67) + 67 + (2)(68)]^2}{(4)(10)}$$

$$+ \frac{[(-2)(69) - 58 + (0)(47) + 25 + (2)(34)]^2}{(4)(10)}$$

$$+ \frac{[(-2)(63) - 54 + (0)(50) + 28 + (2)(30)]^2}{(4)(10)} - SS_{\text{lin}(B)}$$

$$= 123.350$$

Calculations of $SS_{\text{quad}(B) \times A}$, $SS_{\text{cub}(B) \times A}$, and $SS_{\text{quart}(B) \times A}$ are parallel to the above. We merely use a different set of ξ'. For example,

$$SS_{\text{quad}(B) \times A} = \frac{[(2)(73) - 69 + (-2)(67) - 67 + (2)(68)]^2}{(4)(14)}$$

$$+ \frac{[(2)(69) - 58 + (-2)(47) - 25 + (2)(34)]^2}{(4)(14)}$$

$$+ \frac{[(2)(63) - 54 + (-2)(50) - 28 + (2)(30)]^2}{(4)(14)} - SS_{\text{quad}(B)}$$

$$= 5.821$$

Similar manipulations yield

$$SS_{\text{cub}(B) \times A} = 13.067 \quad \text{and} \quad SS_{\text{quart}(B) \times A} = 8.829$$

Each of the above four sums of squares components is distributed on $2\ df$, indicating the possibility of further partitioning. For example,

$$SS_{\text{lin}(B) \times A} = SS_{\text{lin}(B) \times \text{lin}(A)} + SS_{\text{lin}(B) \times \text{quad}(A)}$$

We compute these single df components as follows:

$$SS_{\text{lin}(B)\times\text{lin}(A)} = \frac{[(-2)(-1)(73)+(-2)(0)(69)+ \cdots +(2)(1)(30)]^2}{(4)(20)}$$

$$= 80$$

and

$$SS_{\text{lin}(B)\times\text{quad}(A)} = \frac{[(-2)(1)(73)+(-2)(-2)(69)+ \cdots +(2)(1)(30)]^2}{(4)(60)}$$

$$= 43.350$$

The complete breakdown of the interaction sum of squares is presented in Table 17–2. We shall now consider appropriate error terms for three experimental designs: completely randomized, repeated measurements $(S \times A \times B)$, and a mixed design in which B is a between-subjects variable.

Completely randomized. Assume that each of the 60 scores in Table 17–1 represents a different subject. Then, all terms in the analysis of variance are tested against $MS_{S/AB}$. The results are presented in Table 17–2.

Repeated measurements. Assume that the first row at each level of B in Table 17–1 represents the performance of a single subject, that the second row represents a second subject, and so on. Then we have 4 subjects going through 15 combinations of levels of A and B. With a nonadditive model, the selection of error terms follows the developments of Chapters 7 and 11. The A and B main effects are tested against the $S \times A$ and $S \times B$ terms, respectively. The $p(A)$ term is tested against the $S \times p(A)$ term. If we wish to test, for example, whether there is a significant linear component in the plot of the A main effect, we compute $SS_{S\times\text{lin}(A)}$, which is distributed on 3 df. Calculating this error term is aided if we pool the data over levels of B, establishing a table containing totals for each $S \times A$ combination. Table 17–3 accomplishes this for the data of Table 17–1. Then we calculate

$$SS_{S\times\text{lin}(A)} = \frac{[-86+(0)(57)+53]^2+ \cdots +[-80+(0)(52)+52]^2}{(5)(2)}$$

$$- SS_{\text{lin}(A)}$$

$$= 356.300-354.025 = 2.275$$

The calculations of $SS_{S\times\text{quad}(A)}$ and $SS_{S\times q(B)}$ ($q=$ lin, quad, cub, quart) are left as an exercise for the reader. The results that you should get are presented in Table 17–2.

We next consider an error term for the F test of the significance of the general interaction component $q(B) \times A$. The appropriate error sum of squares, $SS_{S\times q(B)\times A}$, which is distributed on $(n-1)(a-1)$ df, will be most easily calculated if the data are regrouped into four (in general, n) $S \times A$ tables, as in Table 17–3.

Table 17-2 Trend analyses for three two-factor designs

SV	df	SS
A	2	442.433
lin (A)	1	354.025
quad (A)	1	88.408
B	4	405.433
lin (B)	1	357.075
quad (B)	1	12.054
cub (B)	1	20.008
quart (B)	1	16.296
AB	8	151.067
lin (B)×A	2	123.350
lin (B)×lin (A)	1	80.000
lin (B)×quad (A)	1	43.350
quad (B)×A	2	5.821
quad (B)×lin (A)	1	.571
quad (B)×quad (A)	1	5.250
cub (B)×A	2	13.067
cub (B)×lin (A)	1	5.000
cub (B)×quad (A)	1	8.067
quart (B)×A	2	8.829
quart (B)×lin (A)	1	7.779
quart (B)×quad (A)	1	1.050

NOTE: Table 17–2 is continued on the following page.

Table 17-2 (*Continued*)

ERROR TERMS

DESIGN	SV	df	SS	ERROR TERM FOR
Completely randomized				
Repeated measurements	S/AB	45	73.000	all sources
	$S \times A$	6	3.567	A
	$S \times \text{lin}(A)$	3	2.275	$\text{lin}(A)$
	$S \times \text{quad}(A)$	3	1.292	$\text{quad}(A)$
	$S \times B$	12	21.767	B
	$S \times \text{lin}(B)$	3	15.692	$\text{lin}(B)$
	$S \times \text{quad } B$	3	3.065	$\text{quad}(B)$
	$S \times \text{cub}(B)$	3	2.892	$\text{cub}(B)$
	$S \times \text{quart}(B)$	3	.118	$\text{quart}(B)$
	$S \times A \times B$	24	12.912	AB
	$S \times A \times \text{lin}(B)$	6	1.183	$A \times \text{lin}(B)$
	$S \times \text{lin}(A) \times \text{lin}(B)$	3	1.000	$\text{lin}(A) \times \text{lin}(B)$
	$S \times \text{quad}(A) \times \text{lin}(B)$	3	.183	$\text{quad}(A) \times \text{lin}(B)$
	$S \times A \times \text{quad}(B)$	6	5.418	$A \times \text{quad}(B)$
	$S \times \text{lin}(A) \times \text{quad}(B)$	3	3.000	$\text{lin}(A) \times \text{quad}(B)$
	$S \times \text{quad}(A) \times \text{quad}(B)$	3	2.418	$\text{quad}(A) \times \text{quad}(B)$
	$S \times A \times \text{cub}(B)$	6	2.734	$A \times \text{cub}(B)$
	$S \times \text{lin}(A) \times \text{cub}(B)$	3	2.500	$\text{lin}(A) \times \text{cub}(B)$
	$S \times \text{quad}(A) \times \text{cub}(B)$	3	.234	$\text{quad}(A) \times \text{cub}(B)$
	$S \times A \times \text{quart}(B)$	6	3.600	$A \times \text{quart}(B)$
	$S \times \text{lin}(A) \times \text{quart}(B)$	3	2.349	$\text{lin}(A) \times \text{quart}(B)$
	$S \times \text{quad}(A) \times \text{quart}(B)$	3	1.251	$\text{quad}(A) \times \text{quart}(B)$
Mixed	S/B	15	56.500	B and all $q(B)$
	$S \times A/B$	30	16.500	AB, all $q(B) \times A$
	$S \times \text{lin}(A)/B$	15	11.130	all $q(B) \times \text{lin}(A)$
	$S \times \text{quad}(A)/B$	15	5.360	all $q(B) \times \text{quad}(A)$

Table 17–3   Reorganization of Table 17–1 to facilitate calculation of interaction effects involving subjects

		A_1	A_2	A_3	$\sum_j \xi'_{j1} Y_{ijk}$	$\sum_j \xi'_{j2} Y_{ijk}$
	B_1	20	18	16	-4	0
	B_2	18	15	14	-4	2
S_1	B_3	16	12	11	-5	3
	B_4	15	5	6	-9	11
	B_5	17	7	6	-11	9
	$T_{1j\cdot} =$	86	57	53	$\sum_k \sum_j \xi'_{j1} Y_{1jk} = -33$	$\sum_k \sum_j \xi'_{j2} Y_{1jk} = 25$
	B_1	18	17	15	-3	-1
	B_2	18	16	13	-5	-1
S_2	B_3	18	13	14	-4	6
	B_4	18	8	8	-10	10
	B_5	18	10	9	-9	7
	$T_{2j\cdot} =$	90	64	59	$\sum_k \sum_j \xi'_{j1} Y_{2jk} = -31$	$\sum_k \sum_j \xi'_{j2} Y_{2jk} = 21$
	B_1	19	18	17	-2	0
	B_2	17	14	14	-3	3
S_3	B_3	17	12	13	-4	6
	B_4	17	7	9	-8	12
	B_5	18	9	8	-10	8
	$T_{3j\cdot} =$	88	60	61	$\sum_k \sum_j \xi'_{j1} Y_{3jk} = -27$	$\sum_k \sum_j \xi'_{j2} Y_{3jk} = 29$
	B_1	16	16	15	-1	-1
	B_2	16	13	13	-3	3
S_4	B_3	16	10	12	-4	8
	B_4	17	5	5	-12	12
	B_5	15	8	7	-8	6
	$T_{4j\cdot} =$	80	52	52	$\sum_k \sum_j \xi'_{j1} Y_{4jk} = -28$	$\sum_k \sum_j \xi'_{j2} Y_{4jk} = 28$

Then, for the specific test of lin $(B) \times A$, we require

$$SS_{S \times \text{lin} (B) \times (A)} = \frac{[(-2)(20) - 18 + (0)(16) + 15 + (2)(17)]^2}{10} + \cdots$$

$$+ \frac{[(-2)(15) - 13 + (0)(12) + 5 + (2)(7)]^2}{10}$$

$$- SS_{\text{lin} (B)} - SS_{S \times \text{lin} (B)} - SS_{A \times \text{lin} (B)}$$

$$= 1.182$$

Values for sums of squares for other error terms of the form $S \times q(B) \times A$ are given in Table 17–2.

To test $p(A) \times q(B)$, we require $SS_{S \times p(A) \times q(B)}$, which is distributed on

$n-1\ df$. If we are interested, for example, in testing $\lin(A)\times\lin(B)$, we compute

$$SS_{S\times\lin(A)\times\lin(B)} = \frac{[(-2)(-1)(20)+(-2)(0)(18)+\cdots+(2)(0)(7)+(2)(1)(6)]^2}{20} + \cdots$$

$$+\frac{[(-2)(-1)(16)+(-2)(0)(16)+\cdots+(2)(0)(8)+(2)(1)(7)]^2}{20}$$

$$-SS_{\lin(A)\times\lin(B)}$$

$$= 1.00$$

Mixed design. Suppose that there are different subjects at each level of B, but all subjects are tested at all levels of A. Then each row of scores in Table 17–1 represents a different subject. The qth component of the B main effect is tested against $MS_{S/B}$, the between-subjects error term. To test the pth component of A, we require $SS_{S\times p(A)/B}$, which is distributed on $b(n-1)\ df$. For example, if we wish to test $\lin(A)$, we compute

$$SS_{S\times\lin(A)/B} = \frac{[-20+(0)(18)+16]^2+\cdots+[(-15)+(0)(8)+7]^2}{2}$$

$$-SS_{\lin(A)} - SS_{\lin(A)\times B}$$

$$= 11.130$$

The $MS_{S\times p(A)/B}$ is also the appropriate error term for testing all terms of the form $p(A)\times q(B)$. Thus $S\times\lin(A)/B$ would be the appropriate component for a test of $\lin(A)\times\lin(B)$ or $\lin(A)\times\quad(B)$ or $\lin(A)\times\cub(B)$ or $\lin(A)\times\quart(B)$. This is analogous to the test of the interaction of a between-subjects and a within-subject variable; the error term is the error term for testing the within-subject main effect. In the present instance, the error term for the interaction of a between-subjects and a within-subject polynomial component is the error term for testing the within-subject component.

17.4 CONCLUDING REMARKS

Trend, or orthogonal polynomial, analyses should never be routinely applied whenever one or more independent variables are quantitative. Any set of a data points can be fitted by a polynomial of order $a-1$, but if the population function is not polynomial (a sine curve, for example), the polynomial analysis will be misleading. It is also dangerous to identify statistical components freely with psychological processes. It is one thing to postulate a cubic component of A, to test for it, and to find it significant, thus substantiating the theory. It is another matter to assign psychological meaning to a significant component that has not been postulated on a priori grounds. An unexpected significant component would be of interest and should alert the experimenter to the possible need for revising the behavioral hypotheses. Since calculating several polynomial F tests will increase the overall Type I error rate, however, significant results established on an a posteriori basis should require subsequent experimental validation before they are drawn into the body of scientific conclusions. Furthermore, when several trend tests are carried out on a set of data, inflation of the error rate for the family of tests can be a problem. For planned tests, the approach developed in Section 11.42 applies. For post hoc trend tests, EF can be controlled by setting the significance level for the individual tests equal to the EF divided by the total number of orthogonal tests. With these caveats in mind, trend analysis can be a powerful tool for establishing the true shapes of data functions. As such, these methods of analyses should go hand in hand with developing precise quantitative behavioral theories.

EXERCISES

17.1 In this problem assume that the levels of A are equally spaced. Test the orthogonal polynomial components of A and AB (including single df components of AB). Assume that $n = 5$ and $MS_{S/AB} = .40$

	A_1	A_2	A_3	A_4
B_1	15	5	5	15
B_2	24	18	12	26

17.2 A conflict theorist has scaled TAT cards and chosen seven that are equally spaced along a sexual content continuum. He predicts that low-guilt subjects will give increasing numbers of sexual responses as sexual content increases, and that high-guilt subjects will show the same number of responses to the low-sex-content cards but will inhibit responses to the high-sex-content cards. Assuming 20 subjects in each group (no counterbalance required), give the SV, df, and error terms, stating explicitly what terms should be significant according to the hypotheses and why.

17.3 There are three training methods T and four amounts of practice P in a completely randomized design. There are five subjects in each cell. The numbers given below represent the total number of errors for each of the 12 groups.

	P_1	P_2	P_3	P_4
T_1	9	5	3	4
T_2	6	8	5	4
T_3	5	4	6	3

The levels of P are so chosen as to be equally spaced (that is, 1, 2, 3, and 4 hr).
(a) Find the sum of squares for the linear components of P and $P \times T$.
(b) We are interested in comparing the rate of learning under T_1 with the average of T_2 and T_3. Compute the appropriate sum of squares.
(c) Suppose the subjects at each level of T went through all levels of P (P might be the stage of practice). Give computational formulas for the error terms for the effects to be tested in (a) and (b).

17.4 Three classes of 20 patients each undergo group therapy over a 2-year period. They are individually rated by the clinical staff every 6 months (5 times). According to one theory of personality (by Freud out of Skinner), one group A should show improvement, then regression, then improvement but with overall trend to improve; A_2 should improve and then regress to the initial level; A_3 should show a steady rate of improvement. List the terms that should be significant.

17.5 In a study of short-term memory, 12 groups of 10 Ss are run. They are all required to tell whether a comparison tone is the same as a standard presented several seconds earlier or different from it. The groups differ in D, duration of the interstimulus interval: $D = 1, 2,$ or 4. They also differ in I, the stimulus presented during the interval: blank, noise, tone 15 Hz above standard, tone 30 Hz above standard. It is hypothesized that

H_1: D will not influence memory when the interval is blank.

H_2: All other I levels will result in a nonlinear (exponential drop) with increased D.

H_3: The rate of decay (drop in memory over D) will be most pronounced if $I = $ noise, next when $I = +30$ Hz, next when $I = +15$ Hz.

Describe the statistical tests that you would make to test these Hs.

17.6 A large-scale study of programmed instruction is carried out with three variables: method (linear program, branching program, material is just read); material (math, English, social studies); and time per day (15, 30, or 45 min). There are 20 students in each cell. The hypotheses are:

H_1: Math scores are higher than scores on the other materials.

H_2: Programmed instruction is superior to nonprogrammed instruction.

H_3: Performance improves generally with instruction time but there is a leveling off between 30 and 45 min.

H_4: The superiority of math performance is more marked under the two programming methods than under the straight reading method.

H_5: The superiority of math is more marked under branching than under linear programming.

H_6: The superiority of mathematics performance increases as instructional time increases.

Set up the appropriate contrasts for each hypothesis.

17.7 The levels of A are

$$X_1 = 0 \qquad X_2 = 1 \qquad X_3 = 2$$

The corresponding means are

$$\bar{Y}_1 = 2 \qquad \bar{Y}_2 = 20 \qquad \bar{Y}_3 = 8$$

(a) Find the numerical values of b'_0, b'_1, and b'_2 for the equation

$$\bar{Y}_j = b'_0 + b'_1 \xi'_1 + b'_2 \xi'_2$$

(b) Derive the values of ξ that are tabled in A–11 for the three groups.

(c) Using the values obtained in the course of doing (b), find the numerical values of b_0, b_1, and b_2 for $\bar{Y}_j = b_0 + b_i X_j + b_2 X_j^2$.

18 AN INTRODUCTION TO MULTIVARIATE STATISTICS

18.1 INTRODUCTION

So far, in examining the analysis of variance of repeated measurements, we have treated the lack of independence among the scores at the least as a minor hindrance to the inferential process, and at the most as a severe impediment to a complete analysis of data. In again considering designs in which several measures are obtained from each subject, we consider analyses that require no restrictions on the patterns of variances and covariances. We do not require homogeneity of variance and covariance (also referred to as compound symmetry) or homogeneity of variance of difference scores. By the terminology already developed (Chapter 7), the general variance-covariance matrix provides a sufficient basis for our analysis together with the following assumptions:

1. The distribution of the population of measures is multivariate normal. While much less in known about the robustness of multivariate statistics (that is, their lack of sensitivity to violations of assumptions) than about the robustness of the univariate statistics (presented earlier in this text), the best guess is that violating the multivariate normality assumption has the same minor consequences that violating the normality assumption has in the univariate case.
2. The vector of measurements for any one individual is stochastically independent of the vector for any other individual. If we conceive of a matrix in which each row represents a subject and each column a measure, this assumption implies zero covariances among the rows. The departure from the univariate cases is that we need place no restriction on the covariances among the columns.
3. When several populations are being compared on a series of measures, we assume the same variance-covariance matrix for all populations. Again, there is little available information on the consequences of violating this assumption; we can only conjecture that the situation is similar to that for homogeneity of variance in the univariate case (Section 4.32).

Entire textbooks have been written on multivariate analysis; in one chapter, we can only introduce the topic. We shall neglect certain approaches to hypothesis testing, not because they are necessarily inferior to the approaches we do present, but because they involve advanced aspects of matrix algebra. For the same reason, some of the most interesting applications of multivariate statistics—in particular, discriminant analysis—will not be covered. (References to more advanced and more complete treatments will be presented at the end of the chapter.)

Although our coverage is introductory and necessarily limited, the several analyses are of interest and should provide the basis for further reading on multivariate statistics. The material covered includes alternatives to the tests of main and interaction effects that were earlier presented (Chapters 7 and 8). In contrast to the univariate F tests of within-subject effects, the multivariate procedures have the advantage of not requiring any constraints on the variance-covariance matrix. The distributions of the multivariate test statistics to be considered do not depend on assuming homogeneity of variances of difference scores nor even on the weaker condition, compound symmetry. The existence of alternative, multivariate test statistics raises the question whether to perform univariate or multivariate analyses on the data from any repeated-measurement design. (See Section 18.6 for this issue.)

Some multivariate tests demand knowledge of aspects of matrix algebra that were not presented in Chapter 13. The determinant of a matrix is a particularly useful concept because several common multivariate statistics are expressed as a function of the determinant of the variance-covariance matrix. Therefore, we shall first define and provide numerical examples of the determinant of a matrix.

18.2 DETERMINANTS

Every square matrix has associated with it a number, a scalar, that provides information about the pattern of values in the matrix. This index of a square matrix is called its *determinant*. Rather than attempt a general definition of a determinant, we illustrate the concept through a series of examples, using progressively larger matrices until the general idea appears clear. First consider a matrix consisting of only one element. If we let

$$\mathbf{A} = [a_{11}]$$

then the determinant of $\mathbf{A}$ is

$$|\mathbf{A}| = a_{11'}$$

the element itself. Note the vertical lines ($|\mathbf{A}|$), which denote "determinant of."

In the case of a 2×2 matrix, the determinant is the difference between the diagonal products. That is, if

$$\mathbf{A} = \begin{bmatrix} a_{11} & a_{12} \\ a_{21} & a_{22} \end{bmatrix}$$

then

(18.1) $$|\mathbf{A}| = a_{11}a_{22} - a_{12}a_{21}$$

Applying Equation (18.1) to a set of numerical values gives

$$\mathbf{A} = \begin{bmatrix} 4 & 12 \\ 3 & -5 \end{bmatrix}$$

and

$$|\mathbf{A}| = (4)(-5) - (12)(3) = -56$$

Next, we consider a 3×3 matrix. We define m_{ij}, *the minor of* a_{ij}, as the determinant of the 2×2 matrix that remains when we delete the ith row and jth column of the full matrix. Then, selecting any row i of the matrix, we calculate the determinant as

(18.2) $$|\mathbf{A}| = \sum_{j=1}^{3} (-1)^{i+j} a_{ij} m_{ij}$$

The same result can be reached by selecting a particular column j, and summing over rows:

(18.3) $$|\mathbf{A}| = \sum_{i=1}^{3} (-1)^{i+j} a_{ij} m_{ij}$$

An example may help. Let

$$\mathbf{A} = \begin{bmatrix} 2 & -8 & 6 \\ 3 & 5 & -2 \\ 4 & -2 & -7 \end{bmatrix}$$

We shall use the first row to calculate the determinant. Applying Equation (18.2), we get

$$|\mathbf{A}| = (2)[(5)(-7) - (-2)(-2)] - (-8)[(3)(-7) - (-2)(4)]$$
$$+ (6)[(3)(-2) - (5)(4)]$$
$$= (2)(-39) - (-8)(-13) + (6)(-26)$$
$$= -338$$

Now, let us use the second column as the base for our calculations. Applying Equation (18.3), we get

$$|\mathbf{A}| = -(-8)[(3)(-7) - (-2)(4)] + (5)[(2)(-7) - (6)(4)]$$
$$- (-2)[(2)(-2) - (6)(3)]$$
$$= -(-8)(-13) + (5)(-38) - (-2)(-22)$$
$$= -338$$

Expansion based on any other row or column will produce the same result. Furthermore, Equations (18.2) and (18.3) apply to a square matrix of any size. If a rows and columns are assumed, m_{ij} is the determinant of the $(a-1) \times (a-1)$ matrix that remains when row i and column j are deleted. The only change required in Equations (18.2) and (18.3) is that the upper limit of summation should be a instead of 3.

Simpler procedures for finding determinants of large matrices exist. They have as a basis that the determinant of a *triangular matrix*, a matrix with zeros in all cells below (or above) the diagonal, is the product of the diagonal entries. For example,

$$\begin{vmatrix} 7 & -5 & 4 \\ 0 & 6 & -2 \\ 0 & 0 & 3 \end{vmatrix} = (7)(6)(3) = 126$$

Let us reconsider the matrix whose determinant was shown to be -338. We shall transform the original matrix to a triangular matrix. Remember that adding a multiple of any row (column) to any other row (column) of the matrix does not change the value of the determinant. Therefore,

$$|\mathbf{A}| = \begin{vmatrix} 2 & -8 & 6 \\ 3 & 5 & -2 \\ 4 & -2 & -7 \end{vmatrix}$$

$$= \begin{vmatrix} 2 & -8 & 6 \\ 3-(\frac{3}{2})(2) & 5-(\frac{3}{2})(-8) & -2-(\frac{3}{2})(6) \\ 4-(2)(2) & -2-(2)(-8) & -7-(2)(6) \end{vmatrix}$$

$$= \begin{vmatrix} 2 & -8 & 6 \\ 0 & 17 & -11 \\ 0 & 14 & -19 \end{vmatrix}$$

To achieve a triangular matrix, we now need only subtract $\frac{14}{17}$ of row 2 from row 3:

$$|\mathbf{A}| = \begin{vmatrix} 2 & -8 & 6 \\ 0 & 17 & -11 \\ 0 & 0 & -\frac{169}{17} \end{vmatrix}$$

$$= (2)(17)(-\tfrac{169}{17}) = -338$$

A distinction is often made between *nonsingular* and *singular* matrices; nonsingular matrices have an inverse whereas singular matrices do not. The distinction can also be made by the determinant. If a matrix is singular, its determinant will be zero. As we noted earlier, a matrix of deficient rank will be singular. Therefore, the value of the determinant provides a test of whether a matrix is of full rank.

There are many other properties of determinants. Since they are not all useful for our limited purposes, we shall omit their description. (Helpful sources are listed at the end of Chapter 13.)

18.3 HOTELLING'S T^2: ONE-SAMPLE CASE

18.3.1 TESTING $H_0: \mu = \mu_0$.   Assume that we have a group of n subjects, each of whom provides a set of a measures. A null hypothesis of possible interest is

(18.4)
$$H_0: \begin{bmatrix} \mu_1 \\ \cdot \\ \cdot \\ \cdot \\ \mu_a \end{bmatrix} = \begin{bmatrix} \mu_{01} \\ \cdot \\ \cdot \\ \cdot \\ \mu_{0a} \end{bmatrix}$$

That is, μ, the vector of a population means is identical to μ_0, some vector that has been specified before the data are viewed. We can test this null hypothesis by carrying out a set of a univariate t (or F) tests; if one or more of such tests proves significant, the overall null hypothesis of Equation (18.4) can be rejected. When a tests are carried out, the procedure might be modified to the extent of using Bonferroni statistics (Table A–12).

It would be helpful if we could calculate a single statistic to test the null hypothesis of Equation (18.4). This should be a more direct way of evaluating the null hypothesis than what a series of tests provides. Furthermore, a multivariate test statistic (because it takes into account the correlations between measures) incorporates information that is not available when univariate tests are carried out as though they are independent of each other. As a consequence, the multivariate test may lead to a different conclusion from what a series of univariate tests gives. A numerical example will clarify this last point.

Hotelling (1931) developed a multivariate extension of the univariate t to test the null hypothesis of Equation (18.4). Let

$$\mathbf{Y}_i = \begin{bmatrix} Y_{i1} \\ \cdot \\ \cdot \\ \cdot \\ Y_{ia} \end{bmatrix} \quad \text{and} \quad \bar{\mathbf{Y}} = \begin{bmatrix} \bar{Y}_{.1} \\ \cdot \\ \cdot \\ \cdot \\ \bar{Y}_{.a} \end{bmatrix}$$

and let

$$\mathbf{S} = \left(\frac{1}{n-1} \right) \begin{bmatrix} \sum_i y_{i1}^2 & \sum_i y_{i1}y_{i2} & \cdots & \sum_i y_{i1}y_{ia} \\ \sum_i y_{i1}y_{i2} & \sum_i y_{i2}^2 & \cdots & \sum_i y_{i2}y_{ia} \\ \vdots & \vdots & & \vdots \\ \sum_i y_{i1}y_{ia} & \sum_i y_{i2}y_{ia} & \cdots & \sum_i y_{ia}^2 \end{bmatrix}$$

where $y_{ij} = Y_{ij} - \bar{Y}_{.j}$. It is assumed that the a measures have a multivariate normal distribution and that the vectors for the subjects (the $\mathbf{Y}_i$) are independently distributed. The matrix $\mathbf{S}$ is the $a \times a$ variance-covariance matrix based on the data and estimates the population matrix $\mathbf{\Sigma}$ (defined in Chapter 7). The only restriction on $\mathbf{\Sigma}$ in the present context is that it must be nonsingular.

Following Hotelling, we define

(18.5)
$$T^2 = n(\bar{\mathbf{Y}} - \boldsymbol{\mu}_0)' \mathbf{S}^{-1} (\bar{\mathbf{Y}} - \boldsymbol{\mu}_0)$$

and

(18.6)
$$F = \frac{n - a}{a(n - 1)} T^2$$

is distributed as F on a and $n - a$ df.

If $a = 1$, then $\mathbf{S}^{-1}$ reduces to $1/\hat{\sigma}^2$; under these conditions, $F = T^2 = t^2$, the square of the usual univariate t statistic. Another important point about the df is that whenever multivariate analyses are planned, n must exceed a. Otherwise, neither the numerator of Equation (18.6) nor df_2 will be positive. A related point is that the greater the number of subjects relative to the number of measures per subject, the greater the df_2, $n - a$, and accordingly, the power of the T^2 test.

Ordinarily, it is easier to compute determinants of matrices than inverses of matrices. Thus, it is useful to note that an expression equivalent to Equation (18.5) is

(18.7)
$$T^2 = \frac{|\mathbf{S} + n(\bar{\mathbf{Y}} - \boldsymbol{\mu}_0)(\bar{\mathbf{Y}} - \boldsymbol{\mu}_0)'|}{|\mathbf{S}|} - 1$$

Before presenting a numerical example, let us consider one view of T^2. Assume a vector of a nonzero weights $\mathbf{w}$. For any subject, we can construct a single score U_i by calculating

(18.8)
$$U_i = \mathbf{w}' \mathbf{Y}_i = w_1 Y_{i1} + w_2 Y_{i2} + \cdots + w_a Y_{ia}$$

We could then calculate a univariate t statistic to test $H_0 : \mathbf{w}' \boldsymbol{\mu} = \mathbf{w}' \boldsymbol{\mu}_0$. The square of this statistic is

(18.9)
$$t^2(\mathbf{w}) = \frac{n[(\bar{U} - \mathbf{w}' \boldsymbol{\mu}_0)]^2}{\hat{\sigma}_U^2} = \frac{n[\mathbf{w}'(\bar{\mathbf{Y}} - \boldsymbol{\mu}_0)]^2}{\mathbf{w}' \mathbf{S} \mathbf{w}}$$

It can be demonstrated (for example, see Morrison, 1976, Chapter 4) that Hotelling's T^2 is equivalent to the t^2 of Equation (18.9), with $\mathbf{w}$ chosen so as to maximize the size of $t^2(\mathbf{w})$. The statistic T^2 (or max $t^2(\mathbf{w})$) is distributed on a and $n - a$ df, and not the univariate df of 1 and $n - 1$, because a population means are estimated. Furthermore, the fraction in Equation (18.6), $(n - a)/a(n - 1)$, adjusts to accommodate that T^2 is equivalent to a post hoc search for an optimal set of weights.

A numerical example. Consider a group of 20 subjects who are tested on 2 discrimination tasks; one is a 2-choice task and the other is a 4-choice task. There are 24 trials on each task. It follows that if subjects are responding on a chance basis on both tasks, the expected number of correct responses is 12 for Task 1 and 6 for Task 2. In other words, the null hypothesis is

$$H_0 : \begin{bmatrix} \mu_1 \\ \mu_2 \end{bmatrix} = \begin{bmatrix} 12 \\ 6 \end{bmatrix}$$

The summary statistics necessary to calculate T^2 are presented in Table 18–1.

Table 18–1 Summary statistics for a simple repeated-measurement design

n	$\bar{Y}_{.1}$	$\bar{Y}_{.2}$
20	13.2	5.2

$$\mathbf{S} = \begin{bmatrix} 8 & 5.1 \\ 5.1 & 5 \end{bmatrix} \quad \mathbf{S}^{-1} = \left(\frac{1}{13.99}\right)\begin{bmatrix} 5 & -5.1 \\ -5.1 & 8 \end{bmatrix}$$

Applying Equation (18.5), we have

$$T^2 = (20)[1.2 - .8]\left\{\frac{1}{40 - 5.1^2}\begin{bmatrix} 5 & -5.1 \\ -5.1 & 8 \end{bmatrix}\right\}\begin{bmatrix} 1.2 \\ -.8 \end{bmatrix}$$

$$= 31.61$$

and substituting into Equation (18.6) gives

$$F = \tfrac{18}{38}(31.61) = 14.97$$

which exceeds the value required for significance at the .001 level with 2 and 18 *df*.

Equation (18.7) yields the same value of T^2:

$$\mathbf{S} + n(\bar{\mathbf{Y}} - \boldsymbol{\mu}_0)(\bar{\mathbf{Y}} - \boldsymbol{\mu}_0)' = \begin{bmatrix} 8 & 5.1 \\ 5.1 & 5 \end{bmatrix} + (20)\begin{bmatrix} 1.44 & -.96 \\ -.96 & .64 \end{bmatrix}$$

$$= \begin{bmatrix} 36.8 & -14.1 \\ -14.1 & 17.8 \end{bmatrix}$$

The determinant of this last matrix is

$$(36.8)(17.8) - (-14.1)^2 = 456.23$$

and the determinant of **S** is

$$(8)(5) - (5.1)^2 = 13.99$$

Therefore, from Equation (18.7),

$$T^2 = \frac{456.23}{13.99} - 1 = 31.61$$

the result obtained earlier by applying Equation (18.5).

Just as confidence intervals can be calculated for a single population mean, confidence regions can be constructed about the vector of p population means. Such a vector of means is often referred to as a *centroid*; it is a point in a multidimensional space. We shall use the data of Table 18–1 to construct a 95 percent confidence region for $\boldsymbol{\mu}$, the true population centroid for the two tasks. From Equations (18.5) and (18.6), it follows that

(18.10)
$$P\left[n(\bar{\mathbf{Y}}_. - \boldsymbol{\mu})'\mathbf{S}^{-1}(\bar{\mathbf{Y}}_. - \boldsymbol{\mu}) \leq \frac{(n-1)a}{n-a} F_{.05;a,n-a}\right] = .95$$

That is, in 95 percent of the replications of the study, the stated inequality would be satisfied. Substituting values from Table 18–1, the inequality becomes

$$(20)[13.2-\mu_1 \quad 5.2-\mu_2]\left(\frac{1}{13.99}\right)\begin{bmatrix} 5 & -5:1 \\ -5.1 & 8 \end{bmatrix}\begin{bmatrix} 13.2-\mu_1 \\ 5.2-\mu_2 \end{bmatrix}$$
$$\leq \tfrac{38}{18}(3.55)$$

or more simply,

(18.11) $(7.15)(13.2-\mu_1)^2-(14.58)(13.2-\mu_1)(5.2-\mu_2)$
$$+(11.44)(5.2-\mu_2)^2\leq 7.49$$

Equation (18.11) describes an ellipse centered at a point whose coordinates are $\bar{Y}_{.1}$ and $\bar{Y}_{.2}$. That ellipse is plotted in Figure 18–1. The multivariate F statistic based on the data of Table 18–1 will fail to result in rejection (at the .05 level) of any null hypothesis that specifies a population centroid lying within the ellipse. On the other hand, any null hypothesis that specifies a centroid lying outside the ellipse will be rejected at the .05 level on the basis of the current data set. Note that the theoretical vector (12, 6) falls well outside the ellipse; consistent with the results of the F test calculated earlier, the confidence region approach leads us to reject this particular null hypothesis.

The multivariate testing procedure can be put into broader perspective by considering what would have happened had we carried out a pair of univariate t tests. The horizontal line that goes through the theoretical centroid extends from 11.88 to 14.52; these values were obtained by considering only the first-task data

Figure 18–1   Showing a 95 percent confidence interval based on the data of Table 18–1.

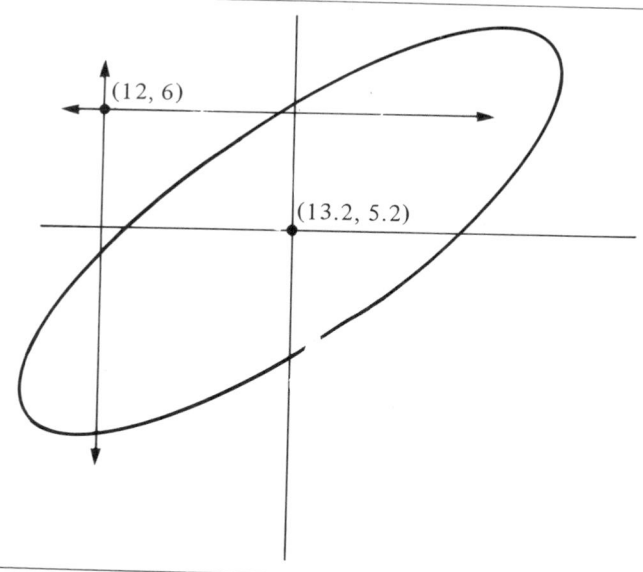

and calculating the univariate confidence interval:

$$CI = \bar{Y}_{.1} \pm \sqrt{(\hat{\sigma}/n)} t_{.05;19}$$

$$= 13.2 \pm \sqrt{\tfrac{8}{20}}(2.093)$$

The Y_1 coordinate specified under the null hypothesis (that is, $\mu_1 = 12$) falls within this confidence interval; a univariate t test would not have rejected $H_0 : \mu_1 = 12$ at the .05 level of significance. Similarly, considering the vertical line that runs through the theoretical centroid, it is clear that a univariate t test based on the Y_2 data would not reject $H_0 : \mu_2 = 6$.

There is an apparent contradiction here. Two univariate t tests are not significant at the .05 level but a single multivariate test statistic is significant at well beyond that level. We can resolve this paradox, and better understand the nature of the multivariate test if we note that the correlation of scores on the two tests is .81 $(=5.1/\sqrt{(8)(5)})$. From this high positive correlation, we should expect that if the null hypothesis were true, both sample means would be high, or both low, relative to the hypothesized values. In fact, this is not the case; $\bar{Y}_{.1}$ is high relative to μ_{01} and $\bar{Y}_{.2}$ low relative to μ_{02}. Thus, when we consider the correlation between the two measures, the sample centroid is unlikely, according to the centroid specified under the null hypothesis. The univariate tests are totally unresponsive to the size of the correlation. The T^2 does take the correlation into consideration and for that reason is more sensitive to a departure from the null hypothesis in the present data set.

We can also use the data of Table 18–1 to illustrate the relation between T^2 and a squared univariate t statistic based on a linear combination of the a measures. As we indicated earlier in this section, the two measures in the example could be combined into a single score:

$$U_i = w_1 Y_{i1} + w_2 Y_{i2}$$

The n U scores could then be the basis for a univariate t test of

$$H_0: w_1 \mu_1 + w_2 \mu_2 = w_1 \mu_{01} + w_2 \mu_{02}$$

We noted that T^2 was equivalent to the t^2 that had the largest possible value as a function of the vector $\mathbf{w}$. For the present numerical example, the weights that maximize t^2 are[1]

$$\mathbf{w}' = [-.63 \quad .78]$$

Substituting these values into the numerator of t^2 $(\mathbf{w})$, defined by Equation (18.9),

---

[1] We shall not derive these values; such a derivation involves a lengthy digression into aspects of matrix algebra that we have decided to omit. The reader wishing to pursue the matter should know that $\mathbf{w}$ is referred to as the *characteristic*, or *eigen*, vector of $\mathbf{S}$; its derivation can be found in any matrix algebra or multivariate statistics textbook (for example, Morrison, 1976, Section 2.10). There is really an infinite set of linearly related vectors that will maximize t^2. We have followed the usual practice of dividing each of the originally derived weights by the square root of the sum of the squared weights. The resulting values have the nice property that the sum of their squares is 1; for example, $(-.63)^2 + (.78)^2 = 1.0$.

we get

$$n[\mathbf{w}'(\bar{\mathbf{Y}} - \boldsymbol{\mu}_0)]^2 = (20)[(-.63)(1.2) + (.78)(-.8)]^2$$
$$= 38.088$$

The denominator is

$$\mathbf{w}'\mathbf{Sw} = [-.63 \quad .78]\begin{bmatrix} 8 & 5.1 \\ 5.1 & 5 \end{bmatrix}\begin{bmatrix} -.63 \\ .78 \end{bmatrix}$$
$$= 1.20492$$

and the ratio, $t^2(\mathbf{w})$, is the by-now-familiar value 31.61. You can substitute other weights into Equation (18.9) to convince yourself that this is the largest possible value of t^2.

Viewing T^2 as a maximum value of a t^2 based on a linear combination of the measures provides another way of understanding why univariate t tests on the individual measures may not prove significant even though the multivariate test is. The univariate test based on a single measure is equivalent to a test of a linear combination in which the combining weights are zero and one; for example, one null hypothesis is

$$(1)\mu_1 + (0)\mu_2 = (1)\mu_{01} + (0)\mu_{02}$$

Since 1 and 0 are far from the optimal weights of $-.63$ and $.78$, it is not too surprising that the tests on the separate measures were not significant.

18.3.2 TESTING $H_0: \mu_1 = \mu_2 = \cdots = \mu_a$: There is a multivariate alternative to the univariate test of treatments (presented in Chapter 7) that deals with simple repeated-measurement designs. In contrast to the univariate test, no restrictions (except for nonsingularity) are imposed on the variance-covariance matrix. The precise calculations involve a small modification of those presented in the immediately preceding section. The approach rests on the following observation: If $\mu_1 = \mu_2 = \cdots = \mu_a$, then

(18.12)
$$\begin{bmatrix} \mu_1 - \mu_2 \\ \mu_2 - \mu_3 \\ \cdot \\ \cdot \\ \cdot \\ \mu_{a-1} - \mu_a \end{bmatrix} = \mathbf{0}$$

It is also true that

(18.13)
$$\begin{bmatrix} \mu_1 - \mu_a \\ \mu_2 - \mu_a \\ \cdot \\ \cdot \\ \cdot \\ \mu_{a-1} - \mu_a \end{bmatrix} = \mathbf{0}$$

In fact, the overall null hypothesis of no treatment effects implies that any one of a number of vectors of $a-1$ contrasts will be identically equal to zero. Therefore, to carry out a multivariate test of treatment effects in the simple repeated-measurement design, we need only convert the set of a measures to $a-1$ contrast scores of the sort shown in Equation (18.12) or (18.13) and then apply Equations (18.5) and (18.6) to a variance-covariance matrix based on the contrast scores.

We can formulate this procedure more precisely. Let $\mathbf{Y}_i$ be an a element column vector consisting of the scores for the ith subject and let $\mathbf{C}$ be the $(a-1)\times a$ matrix,

(18.14)
$$\mathbf{C} = \begin{bmatrix} 1 & -1 & 0 \cdots 0 & 0 \\ 0 & 1 & -1 \cdots 0 & 0 \\ & & \vdots & \vdots \\ & & \vdots & \vdots \\ 0 & 0 & 0 \quad 1 & -1 \end{bmatrix}$$

Then,

(18.15)
$$\mathbf{d}_i = \mathbf{C}\mathbf{Y}_i = \begin{bmatrix} Y_{i1} - Y_{i2} \\ Y_{i2} - Y_{i3} \\ \vdots \\ \vdots \\ \vdots \\ Y_{i,a-1} - Y_{ia} \end{bmatrix}$$

It is not necessary actually to calculate the $a-1$ difference scores for each of the n subjects. We need only note that $\bar{\mathbf{d}}$, the vector of $a-1$ means, can be obtained as

(18.16)
$$\bar{\mathbf{d}} = \mathbf{C}\bar{\mathbf{Y}}$$

and that $\mathbf{S}_d$, the $(a-1)\times(a-1)$ variance-covariance matrix for the variable defined by Equation (18.15) is

(18.17)
$$\mathbf{S}_d = \mathbf{C}\mathbf{S}\mathbf{C}'$$

Under H_0, we get $E(\bar{\mathbf{d}})=0$; therefore, Equation (18.5) becomes

(18.18)
$$T^2 = n(\mathbf{C}\bar{\mathbf{Y}})'(\mathbf{C}'\mathbf{S}\mathbf{C})^{-1}(\mathbf{C}\bar{\mathbf{Y}})$$

and because we now have $a-1$ instead of a measures, Equation (18.6) becomes

(18.19)
$$F = \frac{n-a+1}{(n-1)(a-1)} T^2$$

This statistic has the F distribution on $a-1$ and $n-a+1$ df. Note that the multivariate test, unlike the univariate, requires that n be greater than $a-1$. Also, we might observe that when $a=2$, the univariate and multivariate tests are identical. Both are equivalent to calculating a squared univariate t statistic for matched pairs.

18.4   HOTELLING'S T^2: TWO-SAMPLE CASE

18.4.1   COMPARISON OF GROUP CENTROIDS.   Consider two groups of subjects, A_1 and A_2. There are n_1 and n_2 subjects in the two groups and b measures are obtained from each subject. These measures may have been obtained under b different treatments or at b stages of practice, or may represent scores on b different scales. Particularly in this last case, it is often of interest to ask whether the two population centroids are the same. Do two ethnic populations, for example, or two populations taught by different methods, produce the same vector of means for a set of b achievement measures? The question is reminiscent of one frequently raised when we have one measure for each subject: Do the two population means differ? That question can be handled by calculating the ordinary univariate t for two groups; its square is

(18.20)
$$t^2 = \frac{(\bar{Y}_{.1} - \bar{Y}_{.2})^2}{MS_{S/A}[(1/n_1) + (1/n_2)]}$$
$$= \left(\frac{n_1 n_2}{n_1 + n_2}\right)(MS_{S/A})^{-1}(\bar{Y}_{.1} - \bar{Y}_{.2})^2$$

As in the one-sample case, T^2 can be formed by analogy to the univariate t^2. In the univariate case, we used Equation (18.20) to test $H_0: \mu_1 = \mu_2$. In the multivariate case, we test $H_0: \boldsymbol{\mu}_1 = \boldsymbol{\mu}_2$, where $\boldsymbol{\mu}_j$ is a vector of b means; the test statistic is

(18.21)
$$T^2 = \left(\frac{n_1 n_2}{n_1 + n_2}\right)(\bar{\mathbf{Y}}_1 - \bar{\mathbf{Y}}_2)'(\mathbf{S}^{-1})(\bar{\mathbf{Y}}_1 - \bar{\mathbf{Y}}_2)$$

Note the similarity between Equations (18.20) and (18.21). We have replaced the group means of Equation (18.20) by group vectors with the latter defined as

(18.22)
$$\bar{\mathbf{Y}}_j = \begin{bmatrix} \bar{Y}_{j1} \\ \bar{Y}_{j2} \\ \cdot \\ \cdot \\ \cdot \\ \bar{Y}_{jb} \end{bmatrix}$$

The $MS_{S/A}$ of Equation (18.20) has been replaced by $\mathbf{S}$, a $b \times b$ variance-covariance matrix. The entries on the diagonal of this matrix are estimates of population variances. More precisely, the entry in the kth row and kth column is the error mean square for a one-factor (A) univariate analysis of variance using only the kth score for each subject:

(18.23)
$$S_{kk} = \frac{\sum_{j=1}^2 \sum_{i=1}^{n_j} (Y_{ijk} - \bar{Y}_{.jk})^2}{n_1 + n_2 - 2}$$

The off-diagonal entries are covariances based on pooling cross-product terms for the two groups:

(18.24)
$$S_{kk'} = \frac{\sum_{j=1}^{2} \sum_{i=1}^{n_j} (Y_{ijk} - \bar{Y}_{.jk})(Y_{ijk'} - \bar{Y}_{.jk'})}{n_1 + n_2 - 2}$$

To test the null hypothesis that the two vectors of population means (the centroids) are identical, calculate

(18.25)
$$F = \frac{N - b - 1}{b(N - 2)} T^2$$

where $N = n_1 + n_2$. This statistic has the F distribution on b and $N - b - 1$ *df*, it being assumed that (1) the measures have a multivariate normal distribution within each population, (2) the two populations have the same variance-covariance matrix Σ, and (3) the subject vectors are independently distributed.

As in the one-sample case, T^2 can be viewed as the largest possible t^2 that can be computed through varying a vector **w** that is used to form a linear combination of the scores. It is as though we were to calculate for the ith subject in group A_j:

$$U_{ij} = w_1 Y_{ij1} + w_2 Y_{ij2} + \cdots + w_b Y_{ijb}$$

We should then calculate a squared t to test the null hypothesis that the expected values of U are the same for the two populations. The t^2 that had the largest possible value as a function of the choice of **w** would equal T^2.

There is still another parallel with the one-sample case in that T^2 can be calculated without the inverse being calculated. We have the following, analogous to Equation (18.7) for one sample, and algebraically equivalent to Equation (18.21),

(18.26)
$$T^2 = \frac{|\mathbf{S} + (n_1 n_2 / N)(\bar{\mathbf{Y}}_1 - \bar{\mathbf{Y}}_2)(\bar{\mathbf{Y}}_1 - \bar{\mathbf{Y}}_2)'|}{|\mathbf{S}|} - 1$$

A numerical example. Table 18–2 presents some artificial data for a hypothetical study of performance on three measures in two age groups. The question is whether the populations perform identically on the three tasks; we shall calculate Hotelling's T^2 to try to get the answer.

We assume that there are 51 subjects in each group. Although the 102 sets of three scores are not presented, those summary statistics necessary for calculating T^2 are included in Table 18–2. The matrices labelled *SP* contain numerators of variances and covariances—sums of squares and sums of products—for each group. The value of 357 in the first row and second column of $\mathbf{SP}_1$, for example, is calculated from the raw data as $\sum_i (Y_{i11} - \bar{Y}_{.11})(Y_{i12} - \bar{Y}_{.12})$. The first diagonal entry is the numerator of the variance of the fifty-one A_1 scores on the arithmetic measure:

$$\sum_i (Y_{i11} - \bar{Y}_{.11})^2$$

The variance-covariance matrix **S** is the result of summing the two **SP** matrices and dividing by the pooled *df*, 100 in this case. The inverse has been calculated by

Table 18–2   Summary statistics for a two-group study, 51 subjects in each group

		TEST MEANS	
	ARITHMETIC (Y_1)	INFORMATION (Y_2)	PERCEPTION (Y_3)
A_1 (21–30 years)	16.3	14.5	10.6
A_2 (51–60 years)	14.5	14.3	10.3
$\bar{Y}_{..k} =$	15.4	14.4	10.45
$\bar{Y}_{.1k} - \bar{Y}_{.2k} =$	1.8	.2	.3

$$SP_1 = \begin{bmatrix} 583 & 357 & 130 \\ & 567 & 165 \\ & & 273 \end{bmatrix}$$

$$SP_2 = \begin{bmatrix} 620 & 374 & 187 \\ & 592 & 146 \\ & & 324 \end{bmatrix}$$

$$S = \begin{bmatrix} 12.03 & 7.31 & 3.17 \\ & 11.59 & 3.11 \\ & & 5.97 \end{bmatrix}$$

$$S^{-1} = \begin{bmatrix} .140152 & -.079549 & -.032980 \\ & .145451 & -.033532 \\ & & .202485 \end{bmatrix}$$

the methods of Chapter 13. Note that all these matrices are symmetric, and that therefore we have omitted the entries below the diagonal.

Substituting into Equation (18.21) gives

$$T^2 = \frac{(51)^2}{102} [1.8 \quad .2 \quad .3] \begin{bmatrix} .140152 & -.079549 & -.032980 \\ — & .145451 & -.033532 \\ — & — & .202485 \end{bmatrix} \begin{bmatrix} 1.8 \\ .2 \\ .3 \end{bmatrix}$$

$$= 9.72$$

and from Equation (18.25),

$$F = \frac{98}{300} T^2 = 3.18$$

which, on 3 and 98 df, is significant at the .05 level.

At this point, the investigator might want to pinpoint more closely the source of the differences between the groups. A simple approach is to perform univariate t tests on each of the b measures, controlling the Type I error rate for the family of b tests by the Bonferroni procedure (Chapter 11). In brief review, if each of a set of k planned significance tests is carried out at the α/k level, the probability of at least one Type I error (the error rate per family) is less than α or equal to it. Table A–12 is useful for carrying out such tests. The Bonferroni approach can be taken regardless of whether or not the multivariate F is significant. Remember, however, that significance of the multivariate F does not ensure that any test on a single measure will prove significant (a point demonstrated and discussed in Section 18.3).

18.4.2 TWO-SAMPLE CASE: FURTHER ANALYSES.   The two-group study is a special case of the two-factor mixed design of Section 8.2 with $a = 2$. In such designs, the univariate analysis involved testing three sources of variance, (1) the between-subjects variable A, (2) the within-subject variable B, and (3) the AB interaction term. The corresponding multivariate tests are often referred to as tests of (1) *levels* (When all measures are averaged over, are the groups at the same level?); (2) *flatness* (When the groups are averaged over, is the function plotted relative to the b measures flat?); and *profile* (When the profiles of the two groups are plotted over measures, are the profiles parallel?). The entire set of tests is often referred to as *profile analysis*. The analysis makes sense only if the measures are on comparable scales, or if they represent the same score taken under b different conditions. Otherwise, significant flatness and profile effects might reflect scale variation rather than any interesting psychological phenomena.

The test of level is identical to the test of the A main effect in Section 8.2. We calculate $MS_A/MS_{S/A}$ and evaluate it against an F required for significance on 1 and $n-1$ df. The flatness test is a slight modification of the test of equality of measures (Section 18.3.2). We define the $(b-1) \times b$ matrix $\mathbf{C}$ as in Equation (18.14). We also define a b element vector of population response means for group A_j:

(18.27)
$$\boldsymbol{\mu}_j = \begin{bmatrix} \boldsymbol{\mu}_{j1} \\ \boldsymbol{\mu}_{j2} \\ \cdot \\ \cdot \\ \cdot \\ \boldsymbol{\mu}_{jb} \end{bmatrix}$$

Then, the null hypothesis that there is no B main effect (that is, that $\mu_{.1} = \mu_{.2} = \cdots = \mu_{.b}$) is equivalent to

(18.28)
$$H_0 : \mathbf{C}(\boldsymbol{\mu}_1 + \boldsymbol{\mu}_2) = \mathbf{0}$$

where

(18.29)
$$\mathbf{C}(\boldsymbol{\mu}_1 + \boldsymbol{\mu}_2) = \begin{bmatrix} (\mu_{11} + \mu_{21}) - (\mu_{12} + \mu_{22}) \\ (\mu_{12} + \mu_{22}) - (\mu_{13} + \mu_{23}) \\ \cdot \\ \cdot \\ \cdot \\ (\mu_{1,b-1} + \mu_{2,b-1}) - (\mu_{1b} + \mu_{2b}) \end{bmatrix}$$

The test of the null hypothesis of Equation (18.28) follows readily from the developments of Section 18.3.2, in which we tested the equivalent null hypothesis for the one-sample case. Restating Equation (18.18) gives

(18.30) $$T^2 = (n_1 + n_2)(C\bar{Y})'(CSC')^{-1}(C\bar{Y})$$

The matrix **S** is the same variance-covariance matrix whose elements were defined by Equations (18.23) and (18.24), and $\bar{Y}$ is a weighted average of the two vectors of response means:

(18.31) $$\bar{Y} = \left(\frac{n_1}{n_1 + n_2}\right)\bar{Y}_1 + \left(\frac{n_2}{n_1 + n_2}\right)\bar{Y}_2$$

where $\bar{Y}_j$ is defined by Equation (18.22). To complete the flatness test, we compute

(18.32) $$F = \frac{n_1 + n_2 - b}{(n_1 + n_2 - 2)(b - 1)} T^2$$

which is distributed on $b - 1$ and $n_1 + n_2 - b$ *df*.

The multivariate test of the *AB* interaction—the comparison of group profiles—also uses the matrix **C**. Note that if there is no interaction, the difference between any two response means should be the same at A_1 and A_2. Thus, the null hypothesis of no interaction is equivalent to

(18.33) $$\begin{bmatrix} \mu_{11} - \mu_{12} \\ \cdot \\ \cdot \\ \cdot \\ \mu_{1,b-1} - \mu_{1,b} \end{bmatrix} = \begin{bmatrix} \mu_{21} - \mu_{22} \\ \cdot \\ \cdot \\ \cdot \\ \mu_{2,b-1} - \mu_{2,b} \end{bmatrix}$$

or in matrix form,

(18.34) $$C\mu_1 = C\mu_2$$

where **C** is again the $(b-1) \times b$ matrix defined by Equation (18.14) and μ_j is defined by Equation (18.27). Then

(18.35) $$T^2 = \left(\frac{n_1 n_2}{n_1 + n_2}\right)[C(\bar{Y}_1 - \bar{Y}_2)]'(CSC')^{-1}[C(\bar{Y}_1 - \bar{Y}_2)]$$

and

(18.36) $$F = \frac{n_1 + n_2 - b}{(n_1 + n_2 - 2)(b - 1)} T^2$$

with *df* of $b - 1$ and $n_1 + n_2 - b$.

A numerical example. We shall again use the data of Table 18–2 for this example. The level, or *A* main effect, test is the usual analysis of variance test. Since there have already been several examples, we shall not bother with it here. The flatness, or *B* main effect test, proceeds as follows. First calculate the vector of differences among the three adjacent response means:

$$C\bar{Y} = \begin{bmatrix} 1 & -1 & 0 \\ 0 & 1 & -1 \end{bmatrix} \begin{bmatrix} 15.40 \\ 14.40 \\ 10.45 \end{bmatrix} = \begin{bmatrix} 1.00 \\ 3.95 \end{bmatrix}$$

The variance-covariance matrix for the difference scores is

$$\mathbf{CSC'} = \begin{bmatrix} 1 & -1 & 0 \\ 0 & 1 & -1 \end{bmatrix} \begin{bmatrix} 12.03 & 7.31 & 3.17 \\ 7.31 & 11.59 & 3.11 \\ 3.17 & 3.11 & 5.97 \end{bmatrix} \begin{bmatrix} 1 & 0 \\ -1 & 1 \\ 0 & -1 \end{bmatrix}$$

$$= \begin{bmatrix} 9.00 & -4.34 \\ -4.34 & 11.34 \end{bmatrix}$$

The inverse is

$$(\mathbf{CSC'})^{-1} = \begin{bmatrix} .136258 & .052148 \\ .052148 & .108141 \end{bmatrix}$$

Substituting into Equation (18.30) yields

$$T^2 = (102)[1 \quad 3.95] \begin{bmatrix} .136258 & .052148 \\ .052148 & .108141 \end{bmatrix} \begin{bmatrix} 1 \\ 3.95 \end{bmatrix} = 228.04$$

and the F on 2 and 99 df is

$$F = \frac{99}{200} T^2 = 112.88$$

which is clearly significant at any of the usual levels.

The test of parallel profiles, or of no interaction, is equally straightforward. We find

$$\mathbf{C}(\bar{\mathbf{Y}}_1 - \bar{\mathbf{Y}}_2) = \begin{bmatrix} 1 & -1 & 0 \\ 0 & 1 & -1 \end{bmatrix} \begin{bmatrix} 1.8 \\ .2 \\ .3 \end{bmatrix} = \begin{bmatrix} 1.6 \\ -.1 \end{bmatrix}$$

The inverse of **CSC'** has already been computed so that we need only substitute into Equation 18.35 to find that

$$T^2 = 8.50 \quad \text{and} \quad F = \frac{99}{200} T^2 = 4.20$$

which, with 2 and 99 df, is significant at the .05 level.

18.5 MULTIVARIATE ANALYSIS FOR *a* GROUPS

We next consider the general case of the mixed design (Section 8.2). There are a groups of subjects, n_j subjects in group A_j, and b measures for each subject. Several analyses, which are not mathematically equivalent, have been proposed for such studies. We focus on Wilks's (1932) Λ (*lambda*) test because it is computationally simpler than the alternatives and does not require comprehension of the characteristic roots (also known as eigen values or latent roots) of the variance-covariance matrix, nor the ability to compute them.

18.5.1 COMPARING *a* CENTROIDS. Recall that a population centroid is a point in a b-dimensional space, the vector of the b means of the b measurements for the population in question. The null hypothesis we shall test is that the population centroids are identical; that is,

(18.37) $$H_0: \mu_1 = \mu_2 = \cdots = \mu_a$$

Wilks's Λ provides one way of testing this null hypothesis. It is defined as

(18.38) $$\Lambda = \frac{|\mathbf{W}|}{|\mathbf{W} + \mathbf{B}|}$$

The matrices $\mathbf{W}$ and $\mathbf{B}$ contain sums of squares and sums of products. Let us consider the within-group matrix $\mathbf{W}$ first. Suppose that we were to analyze variance using only the kth measure. Then $SS_{S/A}$ for this analysis would provide the entry in the kth diagonal cell of $\mathbf{W}$. In notational terms, the kth diagonal entry is

(18.39) $$W_{kk} = \sum_{j=1}^{a} \sum_{i=1}^{n_j} (Y_{ijk} - \bar{Y}_{.jk})^2$$

The entry in the kth row and k'th column is the numerator of a covariance. It takes the computational form

(18.40) $$W_{kk'} = \sum_{j=1}^{a} \sum_{i=1}^{n_j} (Y_{ijk} - \bar{Y}_{.jk})(Y_{ijk'} - \bar{Y}_{.jk'})$$

Those who have followed closely the developments of Section 18.4 may think of calculating the variance-covariance matrix $\mathbf{S}$ and then multiplying all entries by the *df*, which are $N - a$, where $N = \sum n_j$, in order to obtain $\mathbf{W}$.

The diagonal entries in the between-groups matrix $\mathbf{B}$ are also sums of squares. It is as though we were to calculate SS_A for each of the b measures:

(18.41) $$B_{kk} = \sum_{j=1}^{a} n_j (\bar{Y}_{.jk} - \bar{Y}_{..k})^2$$

The off-diagonal entries are sums of cross-products for the group means of the various measures:

(18.42) $$B_{kk'} = \sum_{j=1}^{a} n_j (\bar{Y}_{.jk} - \bar{Y}_{..k})(\bar{Y}_{.jk'} - \bar{Y}_{..k'})$$

Multiplying all elements of Equation (18.38) by $\mathbf{W}^{-1}$ yields an alternative way of calculating Λ.

(18.43) $$\Lambda = \frac{1}{|\mathbf{I} + \mathbf{W}^{-1}\mathbf{B}|}$$

If there is only one measure in the study ($b = 1$), $\mathbf{W}$ and $\mathbf{B}$ reduce each to a single element; these are $SS_{S/A}$ and SS_A, respectively. Since the determinant of a matrix containing one element is just the element, we should have

$$\Lambda = \frac{SS_{S/A}}{SS_{S/A} + SS_A}$$

It should be evident that decreasing values of Λ correspond to increasing values of F. In the general multivariate case with b measures, smaller values of Λ cast greater doubt on the validity of the null hypothesis.

Several approaches to evaluating the significance of Λ exist. Various approximations to the chi square distribution have been considered (for example, Bartlett, 1947; Box, 1949; Schatzoff, 1966). These differ in the adequacy of the approximation, particularly when N is small. The more accurate approximations tend to be more complex; the Schatzoff paper provides a table of transformations to exact chi square distributions. We take a different approach, a statistic presented by Rao (1952), which has approximately an F distribution. The statistic in question is

(18.44)
$$R = \left[\left(\frac{1}{\Lambda}\right)^{1/s} - 1\right]\frac{ms+1-b(a-1)/2}{b(a-1)}$$

where

(18.45)
$$m = N-1-(\tfrac{1}{2})(a+b)$$

and

(18.46)
$$s = \sqrt{\frac{b^2(a-1)^2-4}{b^2+(a-1)^2-5}}$$

Note that $(1/\Lambda)^{1/s}$ can be evaluated by finding $(1/s)\log\Lambda$ and then finding its antilog. Many inexpensive electronic calculators provide the exponential result directly.

The degrees of freedom for evaluating R are

$$df_1 = b(a-1) \qquad \text{and} \qquad df_2 = ms - \tfrac{1}{2}b(a-1)$$

The test is the usual univariate F test when $b=1$. The statistic is also exactly distributed like F when $b=2$ or when $a=2$ or 3. When $a=2$, then R reduces algebraically to the F of Equation (18.25); that is,

$$R = F = \frac{N-b-1}{b(N-2)} T^2$$

In all other cases, the F distribution is an approximation to the distribution of R.

A numerical example. We shall again use the data of Table 18–2. Although it is true that we should ordinarily compute T^2 for a two-group problem, Wilks's Λ can also be calculated. It is convenient to use these data since many of the calculations have already been carried out in earlier sections and doing so provides a demonstration of the equivalence of R and F when $a=2$. Furthermore, if the calculations of $\mathbf{W}$ and $\mathbf{B}$ are understood in the two-group case, there should be no problem in making similar calculations for more than two groups.

As pointed out earlier, $\mathbf{W}=(N-a)\mathbf{S}$. Therefore,

$$\mathbf{W}^{-1} = \left(\frac{1}{N-a}\right)\mathbf{S}^{-1}$$

The matrix $\mathbf{S}^{-1}$ is presented in Table 18–2. Also, with a total of 102 subjects in 2 groups, $N-a = 100$.

If we had not already calculated the variance-covariance matrix, the approach would still be straightforward. We should compute an *SP* matrix for each treatment group, as in Table 18–2. The matrix $\mathbf{W}$ is just the sum of the *a SP* matrices.

The matrix $\mathbf{B}$ can be computed from the six means in Table 18–2. For a single measure,

$$SS_A = n \sum_j (\bar{Y}_{.jk} - \bar{Y}_{..k})^2 = n \sum_j \bar{Y}^2_{.jk} - 2n \sum_j \bar{Y}^2_{..k}$$

The *SP* for k and k' has a similar form:

$$SP_A = n \sum_j \bar{Y}_{.jk} \bar{Y}_{.jk'} - 2n \bar{Y}_{..k} \bar{Y}_{..k'}$$

Therefore, the first diagonal entry in $\mathbf{B}$ is

$$(51)[(16.3)^2 + (14.5)^2] - (102)(15.4)^2 = 82.62$$

and the entry in the first row and second column is

$$(51)[(16.3)(14.5) + (14.5)(14.3)] - (102)(15.4)(14.4) = 9.18$$

The entire matrix is

$$\mathbf{B} = \begin{bmatrix} 82.62 & 9.18 & 13.77 \\ & 1.02 & 1.53 \\ & & 2.295 \end{bmatrix}$$

and

$$\mathbf{W}^{-1}\mathbf{B}+\mathbf{I} = \begin{bmatrix} 1.103950 & .001550 & .017325 \\ -.056988 & .993668 & -.009498 \\ -.002444 & -.000272 & .999593 \end{bmatrix}$$

which has the determinant

$$|\mathbf{W}^{-1}\mathbf{B}+\mathbf{1}| = 1.097211$$

Substituting into Equation (18.43), we get $\Lambda = .9911402$. When $a = 2$, it can be shown that

$$T^2 = (N-2)\left(\frac{1}{\Lambda} - 1\right)$$

or 9.72 in the present case. This agrees with the value of T^2 obtained in Section 18.4.1 when we first tested the null hypothesis of equal centroids with these data.

The other way of evaluating the significance of Λ is to calculate R. When $a = 2$, substitutions into Equations (18.44) through (18.46) reveal that

$$R = \left(\frac{1}{\Lambda} - 1\right)\left(\frac{N-b-1}{b}\right)$$

With $N = 102$ and $b = 3$, and substituting in the calculated value of Λ, $R = 3.18$, exactly the value of F obtained in Section 18.4.1.

18.5.2   PROFILE ANALYSIS.   Once again we are concerned with the multivariate analog to the analysis of variance for the mixed design. We require tests of the A main effect (the "levels" test), the B main effect (the "flatness" test), and the AB interaction (the "profiles" test). Because the nature of the variance-covariance matrix is irrelevant to the test of the A main effect, we use the usual ratio, $MS_A/MS_{S/A}$. Regardless of the number of levels of A, the T^2 test can be used to evaluate the null hypothesis that the b response means are identical. The general form of the T^2 statistic for this flatness hypothesis is

$$(\textbf{18.47}) \qquad\qquad T^2 = N(\mathbf{C\bar{Y}})'(\mathbf{CSC}')^{-1}(\mathbf{C\bar{Y}})$$

where $\mathbf{C}$ is defined by Equation (18.14),

$$(\textbf{18.48}) \qquad\qquad \mathbf{\bar{Y}} = \sum_j \left(\frac{n_j}{N}\right)\mathbf{\bar{Y}}_j$$

and we evaluate for significance

$$(\textbf{18.49}) \qquad\qquad F = \frac{N-b}{(N-2)(b-1)}\, T^2$$

which is distributed on $b-1$ and $N-b\,df$. Note that Equation (18.18) for the one-sample case and (18.30) for the two-sample case are special instances of Equation (18.47).

As we noted in Section 18.4.2, the null hypothesis of parallel profiles (no AB interaction) can be restated as

$$H_0: \begin{bmatrix} \mu_{11} - \mu_{12} \\ \vdots \\ \mu_{1,b-1} - \mu_{1b} \end{bmatrix} = \begin{bmatrix} \mu_{21} - \mu_{22} \\ \vdots \\ \mu_{2,b-1} - \mu_{2b} \end{bmatrix} = \cdots = \begin{bmatrix} \mu_{a1} - \mu_{a2} \\ \vdots \\ \mu_{a,b-1} - \mu_{ab} \end{bmatrix}$$

This suggests transforming the b measures into $b-1$ difference scores, and then computing the R statistic presented in the preceding section. An equivalent approach is to transform the matrices $\mathbf{W}$ and $\mathbf{B}$ in the same way that we transformed the variance-covariance matrix $\mathbf{S}$ in making the profile analysis for two groups. That is, calculate $\mathbf{CWC}'$ and $\mathbf{CBC}'$. Then compute Wilks's Λ, using the transformed matrices. The only change from the formulation of Section 18.5.1 is that b is replaced by $b-1$ in Equations (18.44) through (18.46) and in calculating df. This is because the present analysis is equivalent to an analysis performed on $b-1$ difference scores.

A numerical example. Again using the data of Table 18–2, we shall illustrate the calculation of Wilks's Λ for a test of AB interaction. We require

$$\Lambda = \frac{1}{|(\mathbf{CWC}')^{-1}\mathbf{CBC}' + \mathbf{I}|}$$

Because $\mathbf{W} = (1/(N-2))\mathbf{S}$,

$$(\mathbf{CWC}')^{-1} = \tfrac{1}{100}(\mathbf{CSC}')^{-1} = \begin{bmatrix} .00136258 & .00052148 \\ .00052148 & .00108141 \end{bmatrix}$$

We calculated **B** in Section 18.5.1. With

$$\mathbf{C} = \begin{bmatrix} 1 & -1 & 0 \\ 0 & 1 & -1 \end{bmatrix}$$

we find that

$$\mathbf{CBC'} = \begin{bmatrix} 65.28 & -4.08 \\ -4.08 & .255 \end{bmatrix}$$

Then,

$$(\mathbf{CWC'})^{-1}(\mathbf{CBC'}) + \mathbf{I} = \begin{bmatrix} 1.086822 & -.005426 \\ .029630 & .998148 \end{bmatrix}$$

Its determinant is the difference between the diagonal products, or 1.084970. The reciprocal of this value is Λ and equals .921685. Substituting into Equations (18.44) through (18.46) (and replacing b by 2, not 3), we find that R equals 4.20, the value of F obtained by the T^2 approach in Section 18.4.2.

18.6   MULTIVARIATE VERSUS UNIVARIATE TESTS

Some applications clearly dictate the use of multivariate analyses. Of particular utility is extending the developments of this chapter to problems of classifying individuals—for example, for job placement or into diagnostic categories. The general topic of discriminant analysis is treated in all standard textbooks on multivariate statistics (for example, Morrison, 1976); and Tatsuoka (1971) has presented a particularly readable introduction, giving the material from both a geometric and an algebraic viewpoint.

Our primary concern here is to compare multivariate and univariate procedures in problems to which both can be applied. We refer to tests of treatment effects in the simple repeated-measurement designs and to the tests of within-subject main and interaction effects in mixed designs.

The first, although least important, consideration in comparing the two approaches is computational simplicity. It is clear that the univariate approach has a marked advantage. If ε is to be estimated to adjust df for heterogeneity of variances and covariances (the adjustment is presented in Chapter 7), the labor involved in the univariate approach is greatly increased because the variance-covariance matrix is required. Nevertheless, it is still somewhat less than in the multivariate case; neither determinants nor inverses are required. Too much should not be made of this computational difference in an age of high-speed computers and canned programs. On the other hand, it is feasible to perform the univariate tests on small, inexpensive electronic calculators. More time and effort would be required to do the multivariate analyses on such instruments.

A more important consideration is the assumptions underlying the two approaches. As first noted in Chapter 7, MS_A/MS_{SA} is distributed exactly like F if and only if the variances of difference scores are homogeneous. More precisely, it is assumed that for the population from which the data are sampled, the variances of $Y_{ij} - Y_{ij'}$ are constant for all pairs of j and j'. Failure to meet this condition results in an inflation of the Type I error rate. The multivariate procedures impose no other restrictions on **S** but nonsingularity. Although this seems to weigh strongly in favor of the procedures presented in this chapter, things are not quite that simple. It is possible to adjust the *df* associated with the univariate test. In what follows, we shall frequently refer to the ε-adjusted *F* test. By this, we mean that the univariate *F* has been calculated and evaluated against an *F* whose associated *df* are the usual ones multiplied by the fraction ε. Let us consider how the ε-adjusted univariate *F* and the multivariate *F* compare in Type I and Type II error rates.

Collier, Baker, Mandeville, and Hayes (1967) ran computer simulations of a mixed design with 4 groups and 3 measures; one set of simulations had 5 subjects in each group and another set had 15 subjects per group. Population variance-covariance matrices were selected to ensure a wide range of variation in the degree of heterogeneity of covariance as evidenced by ε values ranging from .45 to 1. The important consequence of this research is the finding that even under conditions in which the unadjusted univariate *F* yielded empirical error rates much larger than the nominal, the ε-adjusted procedure resulted in close agreement between empirical and nominal α values. With ε of .45, about .05 of the unadjusted *F*s exceeded the .01 criterion *F* and about .10 exceeded the .05 criterion; with the ε adjustment, the empirical rate was usually within a fraction of a percentage point of the nominal α.

Stoloff (1967) has provided additional support for concluding that the ε-adjusted *F* has an honest α rate. He simulated simple subjects $\times$ treatments designs involving 3 measures and either 4, 6, 10, or 15 subjects and 5 measures with 6, 7, 10, or 15 subjects. The values of ε ranged from .5 to 1. Again, under maximum heterogeneity of covariance, error rates were markedly inflated for the univariate *F* with no adjustment of *df*. When *df* were adjusted, empirical error rates were again very close to nominal rates. In fact, the degree of agreement was as good as it was for Hotelling's T^2, which is presumably unaffected by the pattern of variances and covariances.

The relative power of univariate and multivariate testing procedures is the next consideration. When the assumptions underlying the univariate *F* test are met (that is, when $\varepsilon = 1$), it should be more powerful than the multivariate test. To see why this is so, consider a simple repeated-measurement design with *n* subjects and *a* measures. The value df_1 will be the same for both the *F* test and T^2; however, df_2 will be greater by $a(n-2)$ for the univariate test. Since the univariate assumptions will not often be met, the interesting question is the relative power of the ε-adjusted *F* test and T^2. Stoloff (1967) found the adjusted univariate test generally to be more powerful. Davidson (1972), however, has noted that the relative power of the two procedures will depend on aspects of the data that are not necessarily reflected in the variance-covariance matrix. He has compared the powers of the ε-adjusted univariate *F* and T^2 for three different patterns of data. The variance-covariance matrices were identical in all three cases and had maximum heterogeneity of variance and covariance; therefore, the univariate *F* test used $1/(n-1)$ *df*.

The univariate approach was more powerful when the data could be divided into two subsets of treatment levels—(1) there were high correlations within subsets but low correlations between subsets; and (2) the means within each subset were similar but the two subsets differed in average performance. When the first held and (2′) the two subsets had the same mean but a small difference within one subset, the multivariate procedure was more powerful. In fact, the multivariate procedure had the same power as in the first case (because the variance-covariance matrix was the same) and the univariate test had no power. With a third data pattern somewhere between the first two, but still with the same variance-covariance matrix, relative power depended on degrees of freedom. When $n = a + 1$, both tests have lower power but the univariate test has the advantage. For large n, the situation is reversed; for large-treatment effects, the multivariate approach has an advantage of approximately .40.

We need to know considerably more about the power of univariate and multivariate tests, and about the effect of adjusting degrees of freedom on that power. Davidson's conclusions about the univariate F test, although interesting, are conservative. It is likely that with less heterogeneity of the variance-covariance matrix (greater ε) and with degrees of freedom adjusted for ε, the power of the univariate test would increase. More variations in experimental design parameters and data patterns must be investigated. With the gaps in our knowledge, any recommendations about data-analysis strategies must be tentative.

We believe that in most cases the repeated-measurement design strategy we have described (Chapter 7) will combine computational simplicity, an honest Type I error rate, and reasonable power against multivariate alternatives. The computed F ratio is first evaluated against a criterion F distributed on 1 and $n - 1$ df; if it is significant by this conservative criterion, the null hypothesis is rejected. If the result is not significant, the computed F is evaluated against a criterion F distributed on the usual univariate df, $a - 1$ and $(a - 1)(n - 1)$. If this proves nonsignificant, we fail to reject the null hypothesis. If this sequence of tests fails to yield significance by the first, conservative criterion but does yield significance by the second, more liberal criterion, an adjusted set of df can be computed. The adjustment ε is based on the variance-covariance matrix $\mathbf{S}$, and details can be found in Section 7.4.

One other point about testing within-subject effects is worth considering. Any single df term, if tested against its own error term is the basis for a proper univariate F test. (This is described in Chapters 11 and 17; for example, in trend analysis, the linear component would be tested against $S \times \text{lin}\,(A)$, not $S \times A$.) It is essentially the same as when there are only two measures for each subject; the variance-covariance matrix is not at issue. In many instances, it might be best to decide on a family of single df contrasts in advance of the data collection and to carry out those tests.

EXERCISES

18.1   Label the elements of a 2×2 matrix a through d. Then prove that (a) if all elements in a row are multiplied by k, the determinant is also; (b) if all elements in the matrix are multiplied by k, the determinant increases by the factor k^2; (c) the determinant of the matrix is unchanged if a multiple of one row is added to another; (d) the determinant of two matrices of the same size is the product of their determinants.

18.2 Suppose that the sign of the covariance in Table 18–1 is reversed. Calculate the new value of T^2 and plot the .95 confidence ellipse. Compare the result to that plotted in Figure 18–1.

18.3 In the chapter, we tested the null hypothesis of no B effect for the data of Table 18–2. We should obtain the same value of T^2 using any of a number of other $\mathbf{C}$ matrices. Recalculate T^2 with

$$\mathbf{C} = \begin{bmatrix} 1 & 1 & -2 \\ 1 & -1 & 0 \end{bmatrix}$$

18.4 For the comparison of centroids for the data set of Table 18–2, calculate T^2 using Equation (18.26).

18.5 Compute both Hotelling's T^2 and Wilks's Λ to test the null hypothesis that the two population centroids are identical; use the following data set. Note that $T^2 = (2n - 2)(1/\Lambda - 1)$ and use that to check your calculations.

		B_1	B_2	B_3
	S_{11}	12	8	9
A_1	S_{21}	15	7	10
	S_{31}	5	9	12
	S_{41}	16	8	14
	S_{12}	5	10	6
A_2	S_{22}	9	9	4
	S_{32}	3	12	7
	S_{32}	8	11	8

18.6 Make a complete profile analysis on the preceding data set.

REFERENCES

Bartlett, M. S. Multivariate analysis. *Journal of the Royal Statistical Society, Series B,* 9:176–97 (1947).

Box, G. E. P. A general distribution theory for a class of likelihood criteria. *Biometrika* 36:317–46 (1949).

Collier, R. O., Baker, F. D., Mandeville, G. K., and Hayes, T. F. Estimates of test size for several test procedures based on conventional variance ratios in the repeated measure design. *Psychometrika* 32:339–53 (1967).

Davidson, M. L. Univariate vs. multivariate tests in repeated-measures experiments. *Psychological Bulletin* 77:446–52 (1972).

Hotelling, H. The generalization of Student's ratio. *Annals of Mathematical Statistics* 2:360–78 (1931).

Morrison, D. F. *Multivariate Statistical Methods*, 2nd ed. New York: McGraw-Hill, 1976.

Schatzoff, M. Exact distribution of Wilks' likelihood ratio criterion. *Biometrika* 53:347–58 (1966).

Stoloff, P. H. An empirical evaluation of the effects of violating the assumption of homogeneity of covariance for the repeated measures design of the analysis of variance. University of Maryland, Technical Report, TR-66-28, NSG-398, May 1966.

Tatsuoka, M. M. *Discriminant Analysis: The Study of Group Differences* (Booklet No. 6). Champaign, Ill.: Institute for Personality and Ability Testing, 1970.

Wilks, S. S. Certain generalizations in the analysis of variance. *Biometrika* 24:471–94 (1932).

SUPPLEMENTARY READING

Alternatives to Wilks's Λ are presented in several sources, with discussion of the relative merits of the different test procedures. The reader should consult both Morrison's text and

Harris, R. J. *A Primer of Multivariate Statistics.* New York: Academic Press, 1975.

There is also a good discussion of alternative testing procedures in

Olson, C. L. On choosing a test statistic in multivariate analysis of variance. *Psychological Bulletin* 83:579–86 (1976).

Good presentations of discriminant analysis, an important topic not covered in the present chapter, can be found in the Harris and Morrison books, and in

Tatsuoka, M. M. *Multivariate Analysis: Techniques for Educational and Psychological Research.* New York: John Wiley & Sons, 1971.

ANSWERS TO EXERCISES

CHAPTER 2

2.1 $\sum\limits_{i=3}^{6} Y_i$

2.2 (a)

$$\sum_{i=1}^{n} (Y_i - \bar{Y}) = \sum_{i=1}^{n} Y_i - \sum_{i=1}^{n} \bar{Y} \qquad \text{(Rule 3)}$$

$$= \sum Y_i - n\bar{Y} \qquad \text{(Rule 2)}$$

$$= \sum Y_i - n\frac{\sum Y_i}{n} \qquad \text{(Substitution)}$$

$$= 0$$

(b)

$$\sum_{i=1}^{n} (Y_i + k - n) = \sum_{i=1}^{n} Y_i + nk - n^2$$

2.3

$$\frac{1}{k} \sum_{i=1}^{k} (k - kY_i) = \frac{1}{k}\left(\sum_{i=1}^{k} k + \sum_{i=1}^{k} kY_i \right) \qquad \text{(Rule 3)}$$

$$= \frac{1}{k}\left(k^2 + k \sum_{i=1}^{k} Y_i \right) \qquad \text{(Rules 1 and 2)}$$

$$= k + \sum_{i=1}^{k} Y_i$$

2.4

$$S_{y+c}^2 = \frac{\sum_{i=1}^{n} [(Y_i + C) - (\overline{Y+C})]^2}{n-1} \qquad \text{(by definition of a variance)}$$

$$(\overline{Y+C}) = \frac{\sum_{i=1}^{n} (Y_i + C)}{n}$$

$$= \frac{\sum Y_i}{n} + \frac{\sum C}{n}$$

$$= \bar{Y} + C$$

Then,

$$S^2_{y+c} = \frac{\sum_{i=1}^n [(Y_i + C) - (\bar{Y} + C)]^2}{n-1}$$

$$= \frac{\sum (Y_i - \bar{Y})^2}{n-1} = S^2_y$$

2.5

$$S^2_{cy} = \frac{\sum_{i=1}^n [CY_i - (\overline{CY})]^2}{n-1}$$

$$(\overline{CY}) = \frac{\sum_{i=1}^n CY_i}{n}$$

$$= \frac{C \sum Y_i}{n}$$

$$= C\bar{Y}$$

Then,

$$S^2_{cy} = \sum_{i=1}^n \frac{(CY_i - C\bar{Y})^2}{n-1}$$

$$= \frac{\sum [C(Y_i - \bar{Y})]^2}{n-1}$$

$$= C^2 \frac{\sum (Y_i - \bar{Y})^2}{n-1} = C^2 S^2_y$$

2.6 (a)

$$\bar{z} = \sum_{i=1}^n \frac{(Y_i - \bar{Y})/n}{S_y}$$

$$= \frac{1}{S_y} \frac{\sum (Y_i - \bar{Y})}{n} \qquad \text{(since } S_y \text{ is a constant)}$$

$$= \frac{1}{S_y} \cdot 0 \qquad \text{(see Exercise 2.2)}$$

(b)

$$S^2_z = \frac{\sum_{i=1}^n (z_i - \bar{z})^2}{n-1} \qquad \text{(by definition)}$$

$$\bar{z} = 0$$

Therefore,

$$S^2_z = \frac{\sum_{i=1}^n z^2_i}{n-1}$$

$$z^2_i = \frac{(Y_i - \bar{Y})^2}{S^2_y}$$

Then,

$$\sum_{i=1}^n z^2_i = \frac{\sum_{i=1}^n (Y_i - \bar{Y})^2}{S^2_y}$$

Dividing by $n-1$,

$$S_z^2 = \frac{\sum (Y_i - \bar{Y})^2/(n-1)}{S_y^2}$$

$$= \frac{S_y^2}{S_y^2}$$

$$= 1$$

2.7 (a)

$$\bar{D} = \frac{\sum_{i=1}^n (X_i - Y_i)}{n}$$

$$= \frac{\sum X_i - \sum Y_i}{n}$$

$$= \bar{X} - \bar{Y}$$

(b)

$$S_d^2 = \frac{\sum_{i=1}^n (D_i - \bar{D})^2}{n-1}$$

$$= \frac{\sum [(X_i - Y_i) - (\bar{X} - \bar{Y})]^2}{n-1}$$

Rearrange terms inside the brackets. Then

$$S_d^2 = \frac{\sum [(X_i - \bar{X}) - (Y_i - \bar{Y})]^2}{n-1}$$

$$= \frac{\sum [(X_i - \bar{X})^2 + (Y_{i.} - \bar{Y})^2 - 2(X_i - \bar{X})(Y_i - \bar{Y})]}{n-1}$$

$$= \frac{\sum (X_i - \bar{X})^2}{n-1} + \frac{\sum (Y_i - \bar{Y})^2}{n-1} - \frac{2\sum (X_i - \bar{X})(Y_i - \bar{Y})}{n-1}$$

$$= S_x^2 + S_y^2 - \frac{2\sum (X_i - \bar{X})(Y_i - \bar{Y})}{n-1}$$

Note:

$$r_{xy} = \frac{\sum_{i=1}^n (X_i - \bar{X})(Y_i - \bar{Y})/(n-1)}{S_x S_y}$$

Then

$$\frac{\sum (X_i - \bar{X})(Y_i - \bar{Y})}{n-1} = r_{xy} S_x S_y$$

Therefore,

$$S_d^2 = S_x^2 + S_y^2 - 2r_{xy} S_x S_y$$

2.8 $i = 1, 2, \ldots, n$ (subjects)
$j = 1, 2, \ldots, a$ (A)
$k = 1, 2, \ldots, d$ (D)
$m = 1, 2, \ldots, t$ (T)

(a)
$$\sum_{j=1}^{a} \left[\sum_{i=1}^{n} \sum_{k=1}^{d} \sum_{m=1}^{t} Y_{ijkm} \right]^2$$

(b)
$$\sum_{k=1}^{d} \sum_{m=1}^{t} \left[\sum_{i=1}^{n} \sum_{j=1}^{a} Y_{ijkm} \right]^2$$

2.9
$$\sum_{j=1}^{a} \sum_{i=1}^{n} (\bar{Y}_{i.} - \bar{Y}_{..})^2 = \sum_{j} \sum_{i} (\bar{Y}_{i.}^2 + \bar{Y}_{..}^2 - 2\bar{Y}_{i.}\bar{Y}_{..})$$

$$= \sum_{i} (a\bar{Y}_{i.}^2 + an\bar{Y}_{..}^2 - 2a\bar{Y}_{i.}\bar{Y}_{..})$$

$$= a \sum_{i} \bar{Y}_{i.}^2 + an\bar{Y}_{..}^2 - 2a\bar{Y}_{..} \sum_{i} \bar{Y}_{i.}$$

$$= a \sum_{i} \frac{(T_{i.}^2)}{a^2} + an\frac{T_{..}^2}{a^2 n^2} - 2a\frac{T_{..}}{an} \sum_{i} \frac{T_{i.}}{a}$$

$$= \frac{\sum_{i} T_{i.}^2}{a} + \frac{T_{..}^2}{an} - \frac{2T_{..}^2}{an}$$

$$= \frac{\sum_{i} T_{i.}^2}{a} - \frac{T_{..}^2}{an}$$

2.10
$$na \sum_{k=1}^{b} (\bar{Y}_{.k.} - \bar{Y}_{...})^2 = na \sum_{k} (\bar{Y}_{.k.}^2 + \bar{Y}_{...}^2 - 2\bar{Y}_{.k.}\bar{Y}_{...})$$

$$= na \sum_{k} \bar{Y}_{.k.}^2 + nab\bar{Y}_{...}^2 - 2na\bar{Y}_{...} \sum_{k} \bar{Y}_{.k.}$$

$$= na \sum_{k} \frac{T_{..k}^2}{n^2 a^2} + nab \frac{T_{...}^2}{n^2 a^2 b^2}$$

$$- 2na \frac{T_{...}}{nab} \sum_{k} \frac{T_{..k}}{na}$$

$$= \sum_{k} \frac{T_{..k}^2}{na} - \frac{T_{...}^2}{nab}$$

2.11   (a) $(14+12+2+8)^2 + (3+8+1+9)^2 + \cdots + (2+3+6+3)^2$

(b) $(14+2)^2 + (12+8)^2 + \cdots + (3+3)^2$

(c) $\bar{Y}_{...2} = \dfrac{T_{...2}}{18} = \dfrac{2+8+1+9+\cdots+7+6+3}{18}$

(d) $\bar{Y}_{..2.} = \dfrac{T_{..2.}}{12} = \dfrac{3+6+4+7+\cdots+7+6}{12}$

2.12 (a) $(42+93)(70+80)+(60+99)(77+91)$
(b) $(27+30)^2+(19+20)^2+(24+30)^2+\cdots+(33+27)^2$
(c) $(11+21)(27+22)+(15+16)(19+22)+\cdots+(27+32)(30+27)$
(d) $(4+1+8+6+\cdots+9+14)^2+(3+9+\cdots+6+7)^2$
$$+(4+5+\cdots+13+11)^2$$

CHAPTER 3

3.2 (b) The result is not in conflict with the statement that these are 95 percent confidence intervals. *In the long run,* we expect 95 percent of the intervals computed to contain μ. Alternatively, if we compute many sets of 100 intervals, the average proportion of intervals containing μ should be .95.

3.3 From Equation (3.20)

$$\left[\bar{Y}+\frac{(1.96)(15)}{\sqrt{n}}\right]-\left[\bar{Y}-\frac{(1.96)(15)}{\sqrt{n}}\right]=10$$

Then
$$\frac{(2)(1.96)(15)}{\sqrt{n}}=10$$

$$\sqrt{n}\approx\tfrac{60}{10}$$

$$n\approx 36$$

3.4 With this unusual critical region, the area under the H_1 distribution corresponding to power will be a small segment in the tail of the distribution. In fact, as the distributions become more separate, power will actually decrease.

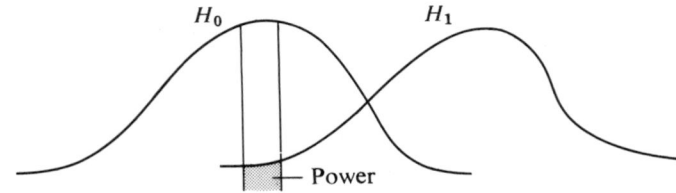

3.5 (a) The critical region is $z\geq 1.645$.

$$\sigma_{\bar{x}}=\frac{\sigma}{\sqrt{n}}=4$$

$$\frac{\mu_1-\mu_0}{\sigma_{\bar{x}}}=\frac{10}{4}=2.5$$

Thus μ_1 lies 2.5 standard deviation units above μ_0. The critical value of z lies $2.5-1.645=.855\sigma$ below μ_1. Then power is approximately .80, the proportion of the μ_1 distribution above $-.855$.

(b) The critical region consists of values of $z \geq 1.96$ or ≤ -1.96. The upper value, 1.96, lies $.54\sigma$ below μ_1 $(2.5 - 1.96 = .54)$ and this is exceeded approximately 71 percent of the time under H_1. The lower critical value, -1.96, is at 4.46σ below μ_1. Power is the proportion of the H_1 distribution $> -.54$ or < -4.46 under H_1, or .71 plus (essentially) .00. Note that we have lost power against one alternative in switching from a one- to a two-tailed test.

3.6 $Y - \mu = Y - \bar{Y} + \bar{Y} - \mu$

$$\frac{\sum (Y - \mu)^2}{n} = \frac{\sum (Y - \bar{Y})^2}{n} + \frac{\sum (\bar{Y} - \mu)^2}{n}$$

Noting that $\bar{Y} - \mu$ is a constant,

$$\frac{\sum (Y - \bar{Y})^2}{n} = \frac{\sum (Y - \mu)^2}{n} - (\bar{Y} - \mu)^2$$

$$E\left(\frac{\sum (Y - \bar{Y})^2}{n}\right) = \sum \frac{E(Y - \mu)^2}{n} - E(\bar{Y} - \mu)^2$$

$$= \frac{n\sigma^2}{n} - \frac{\sigma^2}{n}$$

$$= \left(\frac{n-1}{n}\right)\sigma^2$$

CHAPTER 4

4.1 Since ε_{ij} is a random variable, and a given experimental group contains only n of the infinite number of values in the treatment population, its sum will not generally be zero. The α_j, on the other hand, are fixed-effect variables; all deviations are represented in the experiment, and the sum will therefore be zero. If the α_j were a random sample from a larger population of such values, they would no longer sum to zero, although $E(\sum_j \alpha_j) = 0$.

4.2 (a)

SV	df	SS	MS	F
A	2	3,886.9	1,943.5	31.53
S/A	27	1,664.1	61.6	

(b)

A	4	3,541.9	885.5	1.30
S/A	26	17,727.5	681.8	

4.3 (a)

SV	df	SS	MS	F
A	2	81.67	40.84	9.42
S/A	27	117.00	4.33	

(b)

A	2	17.08	8.54	1.87
S/A	21	96.00	4.57	

4.4 Hal clearly has a more powerful F test. In both studies, however, $\theta_A^2 = 1.5$ and $\omega_A^2 = 1/(1 + 20)$. The variable A accounts for less than 5 percent of the population variance. It appears to be of little practical interest.

4.5 (a) (i) .5.   (ii) .5.

(b) If θ_A^2 and σ_e^2 do not vary, changes in n will not affect ω_A^2.

(c) If $F = 1$, then θ_A^2 must be zero. But this means that ω_A^2 must also be zero.

(d) (i) $E(SS_A) = (a-1)\sigma_e^2 + n \sum_j \alpha_j^2$

(ii) $E(SS_{S/A}) = a(n-1)\sigma_e^2$

(iii) $\hat{\eta}_A^2 = \dfrac{E(SS_A)}{E(SS_{S/A}) + E(SS_A)}$

(e) (i) $\hat{\eta}_A^2 = 160/(135+160) = .54$

(ii) $\hat{\eta}_A^2 = 85/(60+85) = .59$

(f) $\hat{\eta}_A^2 = (a-1)/(an-1)$

(g) ω_A^2 reflects the total proportion of population variance due to A. Unlike η_A^2, it is not affected by sample size and it equals zero when it should. It is preferable as a measure of the importance of a variable.

4.6 (a) Disagree. $\theta_{A,6}^2 = 12$; $\theta_{A,10}^2 = 10$

(b) Disagree. $\theta_{A,6}^2 = 10$; $\theta_{A,6}^2 = 12$

(c) Disagree. $\theta_{A,6}^2 = 10$; $\theta_{A,6}^2 = 9$

(d) The F ratios should not be compared because the size of the F depends on σ_e^2 and n, as well as the effects of A.

4.7
$$\frac{\sum (\mu_j - \mu)^2}{3} = \frac{200}{3}$$

$$\phi' = \sqrt{\frac{200}{(3)(225)}} \, n \approx .55\sqrt{n}$$

if $n = 16$, $df_2 = 45$, and $\phi = 2.2$. Then, Power $\approx .90$.

4.8 (a) $\phi = 1.22$ and power is less than .50.

(b) $\hat{\theta}_A^2 = 3$; $\delta_A^2 = \left(\dfrac{a-1}{a}\right)\hat{\theta}_A^2 = (\tfrac{1}{2})(3) = 1.5$

CHAPTER 5

5.1 B, AB, and AC.

5.2 $\hat{\delta}_A^2 = \left(\dfrac{a-1}{a}\right)\hat{\theta}_A^2$ $\qquad$ $\hat{\theta}_A^2 = \left(\dfrac{1}{bn}\right)(MS_A - MS_{S/AB})$

$\hat{\delta}_B^2 = \left(\dfrac{b-1}{b}\right)\hat{\theta}_B^2$ $\qquad$ $\hat{\theta}_B^2 = \left(\dfrac{1}{an}\right)(MS_B - MS_{S/AB})$

$\hat{\delta}_{AB}^2 = \left(\dfrac{(a-1)(b-1)}{ab}\right)\hat{\theta}_{AB}^2$ $\qquad$ $\theta_{AB}^2 = \left(\dfrac{1}{n}\right)(MS_{AB} - MS_{S/AB})$

$\hat{\sigma}_e^2 = MS_{S/AB}$

$\hat{\omega}_A^2 = \dfrac{\hat{\delta}_A^2}{\hat{\sigma}_e^2 + \hat{\delta}_A^2 + \hat{\delta}_B^2 + \hat{\delta}_{AB}^2}$

5.3 (a) $MS_B = 35.00$ $\quad MS_{S/AB} = 18.00$ $\qquad F = 1.94$
(b) $MS_{A/B_3} = 180$ $\quad F = 180/18 = 10$

Since the variances do not differ significantly, $MS_{S/AB}$ is justified. This error term provides more df, and therefore more power, than $MS_{S/A/B_3}$.

(c) 23.33
(d) $df = 2$. In general, $df = b(a-1) - (a-1)$.
(e) The df in (d) $= (a-1)(b-1)$, suggesting that SS_{AB} was calculated in (c). We have taken the variability of b simple effects $(SS_{A/B})$ about their average (SS_A).

5.4 The significant SV should be therapy and therapy $\times$ socio-economic level.

	S-E Level		
	Low	Middle	High
Psychotherapy	$\begin{bmatrix} 10 \\ 20 \end{bmatrix}$	$\begin{matrix} 12 \\ 18 \end{matrix}$	$\begin{bmatrix} 14 \\ 16 \end{bmatrix}$
Behavior Therapy			

5.5 Let $t_c = 40$, 80, and 120 for the three list lengths. Let $t_e = 100$ for the fuzzy probe and 50 for the clear probe. The "data" are:

		L		
		2	3	4
Q	Fuzzy	$\begin{bmatrix} 480 \\ 430 \end{bmatrix}$	$\begin{matrix} 520 \\ 470 \end{matrix}$	$\begin{bmatrix} 560 \\ 510 \end{bmatrix}$
	Clear			

The L and Q terms should be significant but not $L \times Q$.

5.6 (a) The terms D, P, and $D \times P$ should all be significant.
(b) The terms A and I should be significant but $A \times I$ should not.
(c) This theory predicts D, O, DO, AO, and DOA effects. There is no clear prediction for A or $A \times D$.

5.7 The terms A, D, AD, and CD should be significant.

5.8 $E\left(\sum_i \sum_j \sum_k \hat{\varepsilon}_{ijk}^2\right) = E\left[\sum_i \sum_j \sum_k (\varepsilon_{ijk} - \bar{\varepsilon}_{.j.} - \bar{\varepsilon}_{..k} + \bar{\varepsilon}_{...})^2\right]$

Expanding the right-hand side and assuming the ε_{ijk} are independently and homogeneously distributed, we get the following terms.

(a) $E\left(\sum_i \sum_j \sum_k \varepsilon_{ijk}^2\right) = abn\sigma_e^2$

(b) $E\left(\sum_i \sum_j \sum_k \bar{\varepsilon}_{.j.}^2\right) = abn(\sigma_e^2/bn) = a\sigma_e^2$

(c) $E\left(\sum_i \sum_j \sum_k \bar{\varepsilon}_{..k}^2\right) = abn(\sigma_e^2/an) = b\sigma_e^2$

(d) $E\left(\sum_i \sum_j \sum_k \bar{\varepsilon}_{...}^2\right) = abn(\sigma_e^2/abn) = \sigma_e^2$

(e) $-2E\left(\sum_i \sum_j \sum_k \varepsilon_{ijk} \bar{\varepsilon}_{\cdot j\cdot}\right) = -2E\sum_j (bn\bar{\varepsilon}_{\cdot j\cdot})\bar{\varepsilon}_{\cdot j\cdot} = -2a\sigma_e^2$

(f) $-2E\left(\sum_i \sum_j \sum_k \varepsilon_{ijk} \bar{\varepsilon}_{\cdot\cdot k}\right) = -2E\sum_k (an\bar{\varepsilon}_{\cdot\cdot k})\bar{\varepsilon}_{\cdot\cdot k} = -2b\sigma_e^2$

(g) $2E\left(\sum_i \sum_j \sum_k \varepsilon_{ijk} \bar{\varepsilon}_{\cdots}\right) = -2E(abn\bar{\varepsilon}_{\cdots})\bar{\varepsilon}_{\cdots} = 2\sigma_e^2$

(h) $2E\left(\sum_i \sum_j \sum_k \bar{\varepsilon}_{\cdot j\cdot}\bar{\varepsilon}_{\cdot\cdot k}\right) = 2E\left(\sum_i (a\bar{\varepsilon}_{\cdots})(b\bar{\varepsilon}_{\cdots})\right) = 2\sigma_e^2$

(i) $-2E\left(\sum_i \sum_j \sum_k \bar{\varepsilon}_{\cdot j\cdot}\bar{\varepsilon}_{\cdots}\right) = -2E\left(\sum_i \sum_k (a\bar{\varepsilon}_{\cdots})\bar{\varepsilon}_{\cdots}\right) = -2\sigma_e^2$

(j) $-2E\left(\sum_i \sum_j \sum_k \bar{\varepsilon}_{\cdot\cdot k}\bar{\varepsilon}_{\cdots}\right) = -2E\left(\sum_i \sum_j (b\bar{\varepsilon}_{\cdots})(\bar{\varepsilon}_{\cdots})\right) = -2\sigma_e^2$

Combining terms, we get the desired result. The potential gain from the pooling procedure is an increase in error df from $ab(n-1)$ to $abn - a - b + 1$. If $\theta_{AB}^2 > 0$, however, the pooled error MS will estimate more than σ_e^2. The ratios of mean squares used to test A and B will not have the F distribution.

5.9 (a) $df_{pool} = (abcen - 1) - (b-1) - (c-1) - (b-1)(c-1) - abce(n-1)$

$$= bc(ae-1)$$

Note that this result can be obtained more directly. Within each BC cell, there are ae treatment combinations on $ae - 1 df$. Summing this variability over BC combinations, $bc(ae-1)df$.

Expanding the df, we get the SS:

$$abce - bc \rightarrow \sum_{}^{a}\sum_{}^{b}\sum_{}^{c}\sum_{}^{e} \frac{(\sum^n Y)^2}{n} - \sum_{}^{b}\sum_{}^{c} \frac{(\sum^a \sum^e \sum^n Y)^2}{aen}$$

(b) If any of the effects involving A or E are not negligible (e.g., $\theta_{AB}^2 > 0$ or $\theta_E^2 > 0$), the residual error term will be spuriously inflated (see Exercise 5.8). The df are $abce(n-1) + bc(ae-1)$.

5.10 $SS_D = \dfrac{[(52+80+84+81) - (107+58+60+115)]^2}{(8)(24)}$

$SS_{AC} = \dfrac{[(52+107) + (81+115) - (80+58) - (84+60)]^2}{(8)(24)}$

$SS_{ACD} = \dfrac{[(52+81) - (84+80) - (107+115) + (60+58)]^2}{(8)(24)}$

5.11 $\omega_A^2 = 10/(62+10+20+8) = .10$

$\omega_B^2 = .20$

$\omega_{AB}^2 = .08$

5.12 (a) $\boldsymbol{\alpha} = \begin{bmatrix} 4 \\ -1 \\ -3 \end{bmatrix}$ $\boldsymbol{\beta} = \begin{bmatrix} -3 \\ -4 \\ 7 \end{bmatrix}$ $\boldsymbol{\alpha\beta} = \begin{bmatrix} 1 & -2 & 1 \\ 2 & 1 & -3 \\ -3 & 1 & 2 \end{bmatrix}$

(b) $\begin{bmatrix} 8 & 4 & 18 \\ 9 & 7 & 14 \\ 4 & 7 & 19 \end{bmatrix}$

The column (B) means are unchanged.

(c) $\boldsymbol{\alpha} = \begin{bmatrix} 2.9583 \\ -2.8750 \\ -3.2083 \end{bmatrix}$ $\boldsymbol{\beta} = \begin{bmatrix} -3.5833 \\ -5.9583 \\ 5.1667 \end{bmatrix}$

Note that $\sum_j n_{j.}\alpha_j = \sum_k n_{.k}\beta_k = 0$. The column (B) means are unchanged.

(d) $\boldsymbol{\alpha}_j = [3.1667 \quad .6667 \quad -3.8333]$

After adjustment for the α_j, the column means change. The B_1 mean was originally 110/20, or 5.5. But

$$[5(12-3.17)+5(8-.67)+10(1+3.83)]/20 = 6.46$$

The other means also change.

(e) When cell frequencies are disproportional the A and B sums of squares are correlated.

CHAPTER 6

6.1 Let C designate blocks. To equate the sizes of the two designs, we will have cn scores in each of the ab cells of the completely randomized (c.r) design. For that design,

$$E(SS_{tot}) = (a-1)(\sigma^2_{e_{c.r.}} + nbc\theta^2_A) + (b-1)(\sigma^2_{e_{c.r.}} + nac\theta^2_B)$$
$$+ (a-1)(b-1)(\sigma^2_{e_{c.r.}} + nc\theta^2_{AB}) + ab(cn-1)\sigma^2_{e_{c.r.}}$$

For the blocks design $(t \times b)$, we have

$$E(SS_{tot}) = (a-1)(\sigma^2_{e_{t\times b}} + nbc\theta^2_A) + (b-1)(\sigma^2_{e_{t\times b}} + nac\theta^2_B)$$
$$+ (a-1)(b-1)(\sigma^2_{e_{t\times b}} + nc\theta^2_{AB}) + E(SS_C) + E(SS_{AC})$$
$$+ E(SS_{BC}) + E(SS_{ABC}) + abc(n-1)\sigma^2_{e_{t\times b}}$$

Equating the two $E(SS_{tot})$ and cancelling the θ^2_A, θ^2_B, and θ^2_{AB} terms, we have

$$(abcn-1)\sigma^2_{e_{c.r.}} = [abc(n-1)+(ab-1)]\sigma^2_{e_{t\times b}} + E(SS_C)$$
$$+ E(SS_{AC}) + E(SS_{BC}) + E(SS_{ABC})$$

Relative efficiency of the $t \times b$ design to c.r. design is

$$RE = \frac{[abc(n-1)+(ab-1)]MS_{S/ABC} + SS_{AC} + SS_{BC} + SS_{ABC} + SS_C}{(abcn-1)MS_{S/ABC}}$$

$$= 1 - \frac{ab(c-1)}{abcn-1} + \frac{SS_{AC} + SS_{BC} + SS_{ABC} + SS_C}{(abcn-1)MS_{S/ABC}}$$

6.2

SV	df	SS	MS
A	2	136.8	68.4
B	3	25.5	8.5
AB	6	153.9	25.7
S/AB	12	62.5	5.2

$$RE = \frac{(13)(24)}{(15)(22)}\left[1 - \frac{(3)(3)}{23} + \frac{25.5 + 153.9}{(23)(5.2)}\right] = 1.99$$

from Equations (6.6) and (6.7)

The AB variability is considerable.

6.3 Consider $\rho = .2$, $N = 70$. Then $a = 4$, by linear interpolation, yields $b = 3.7$. If $N = 100$, $b = 5$ then, for $N = 80$, we want a value of b one-third the distance between the values at $N = 70$ ($b = 3.7$) and $N = 100$ ($b = 5$); $b = 4$ is a reasonable approximation. Similarly, for $\rho = .4$, we arrive at $b = 7$. Our estimates of ρ range from .25 to .35. Taking $\rho = .30$, its squared value, .09, lies a little less than midway between $(.2)^2$ and $(.4)^2$; $b = 5$ would seem a reasonable approximation.

6.4 (a) Taking $MS_{S/AB}$ as σ_e^2,

$$\phi^2 = \frac{(12)(32)/3}{16} = 8$$

Then $\phi = 2.8$ and, with 2 and 18 df,

$$Power \approx .91$$

(b) We must re-estimate σ_e^2. From Equation (6.5),

$$MS_{S/A} = \left(1 - \frac{15}{35}\right)(16) + \frac{800 + 220}{35} = 9.1 + 29.1 \approx 38$$

Now

$$\phi^2 = \frac{(12)(32)/3}{38}$$

and

$$\phi \approx 1.84$$

with $df = 2$ and 33

$$Power \approx .52$$

6.5 (a) Blocking will result in reduced error variance because the range of Y scores will be smaller in each block than in a level of A.

(b) Feldt's Tables are inappropriate because they are derived assuming a linear relation between X and Y.

CHAPTER 7

7.1 The interaction term is now the appropriate error term. Note that in this case the presence of an AB interaction poses the problems generally associated with nonadditivity.

7.2 $N = 2985.17$ $D = (250.17)(854.08)$ $SS_{nonadd} = \dfrac{N^2}{D} = 41.71$

$SS_{AS} = 44.00$ $F = \dfrac{41.71}{2.29/5} = 91.07$

Nonadditivity is significant. To transform the data, we calculate

$$\hat{p} = 1 - \left(\frac{N}{D}\right)\bar{Y}_{..} = .48$$

Good results are obtained by letting $p = .5$; this means that each of the original scores should be transformed by $Y' = \sqrt{Y}$.

7.3 Only set 3 should be sensitive to the Tukey test.

7.4
$$(\eta\alpha)_{ij} = \mu_{ij} - \mu_i - \mu_j + \mu$$

$$\sum_j (\eta\alpha)_{ij} = \sum_j \mu_{ij} - a\mu_i - \sum_j \mu_j + a\mu$$

$$= a\mu_i - a\mu_i - a\mu + a\mu$$

$$= 0.$$

7.5 The variance-covariance matrix is

$$\begin{bmatrix} 2.917 & 4.083 & .750 \\ 4.083 & 10.917 & 3.583 \\ .750 & 3.583 & 4.250 \end{bmatrix}$$

$$\bar{S}_{..} = 3.88$$

$$\bar{S}_{jj} = 6.03$$

$$\sum_j \sum_k \bar{S}_{jk}^2 = 205.89$$

$$\sum_j \bar{S}_j^2 = 53.23$$

$$\hat{\varepsilon} = \frac{a(\bar{S}_{jj} - \bar{S}_{..})^2}{(a-1)(\sum_j \sum_k S_{jk}^2 - 2a_j \sum \bar{S}_j^2 + a^2\bar{S}_{..}^2)}$$

$$= \frac{9(6.03 - 3.88)^2}{2[205.89 - (6)(53.23) + (9)3.88^2]}$$

$$= .94$$

7.6 (a) If words are viewed as having fixed effects, this procedure yields the same F in a test of A as in a full analysis using all abn scores. This is because σ^2_{AB} makes no contribution to $E(MS_A)$ and it is therefore proper to average over levels of B.

(b) If words are viewed as having random effects, the proposed procedure is improper. The variance σ^2_{AB} will contribute to $E(MS_A)$ but not to the SA error term, the only error term available if we average over levels of B.

7.7 Under the independence assumption, $\rho = 0$. Therefore, $\Sigma_{..} = \sigma^2/a$ (see Equation (7.20)). From Equations (7.21) and (7.22), $\sigma^2_S = \sigma^2/a$ and $\sigma^2_{SA} = \sigma^2$. Independence is not the same as additivity.

7.8 The variances and covariances are heterogeneous. The variances of difference scores, however, are homogeneous. To prove this, first remember that the variance of a difference σ^2_d is given by

$$\sigma^2_d = \sigma^2_j + \sigma^2_{j'} - 2cov(jj')$$

Then,

$$\sigma^2_{d_{12}} = 1.0 + 3.0 - 2(.5) = 3.0$$
$$\sigma^2_{d_{13}} = 1.0 + 5.0 - 2(1.5) = 3.0$$
$$\sigma^2_{d_{23}} = 3.0 + 5.0 - 2(2.5) = 3.0$$

The adjustment $\hat{\varepsilon}$ equals 1.0 as it should when we have homogeneous variances of difference scores.

CHAPTER 8

8.1

SV	df	SS	MS	F
A	2	90.70	45.35	5.94
S/A	6	45.78	7.63	
B	1	93.35	93.35	10.73
AB	2	84.93	42.46	4.88
SB/A	6	52.22	8.70	
C	2	667.70	333.85	53.25
AC	4	10.07	2.52	.40
SC/A	12	75.22	6.27	
BC	2	89.93	44.96	5.77
ABC	4	13.63	3.41	.44
SBC/A	12	93.44	7.79	

8.2 (i) $SS_A = \dfrac{(33-30)^2}{12} = .75$

(ii) $SS_{S/A} = \dfrac{(15-18)^2}{6} + \dfrac{(12-18)^2}{6} = 7.5$

(iii) $SS_{A'} = .25$ $SS_{S/A'} = 2.5$

Generally, division by a constant (b, or 3 in this case) changes the sum of squares by that constant. The F ratio stays the same.

(iv) $SS_{SB/A} = SS_{SB/A_1} + SS_{SB/A_2} = 14.00$

$MS_{SB/A} = 14/4 = 3.75$

8.3

SV	EM_S
A	$\sigma_e^2 + n\sigma_{AB}^2 + b\sigma_{S/A}^2 + \sigma_{SB/A}^2 + nb\theta_A^2$
S/A	$\sigma_e^2 \qquad\qquad + b\sigma_{S/A}^2 + \sigma_{SB/A}^2$
B	$\sigma_e^2 \qquad\qquad\qquad\qquad + \sigma_{SB/A}^2 + na\sigma_B^2$
AB	$\sigma_e^2 + n\sigma_{AB}^2 \qquad\qquad + \sigma_{SB/A}^2$
$S_{B/A}$	$\sigma_e^2 \qquad\qquad\qquad\qquad + \sigma_{SB/A}^2$

$$MS_{error} = MS_{S/A} + MS_{AB} - MS_{SB/A}$$

$$df_{error} = \frac{(MS_{error})^2}{\dfrac{MS_{S/A}^2}{a(n-1)} + \dfrac{MS_{AB}^2}{(a-1)(b-1)} + \dfrac{MS_{SB/A}^2}{a(n-1)(b-1)}}$$

If $\sigma_{AB}^2 = 0$, A may be tested against S/A. If $\sigma_{S/A}^2 = 0$, A may be tested against AB.

8.4 Two runs through the computer suffice. *First run:* First factor = Ss (an levels). Second factor = B (b levels). We obtain:

Between Ss	$an - 1$
B	$b - 1$
Between Ss × B	$(an-1)(b-1)$

Second run: First factor = A. Second factor = B. We obtain A, B, and AB sums of squares. Subtracting by hand:

$$SS_{B.Ss} - SS_A = SS_{S/A}$$

$$SS_{B.Ss} \times B - SS_{AB} = SS_{SB/A}$$

The analysis is complete. The approach is readily extended to other variables and to other computer limitations (e.g., if there is a limit on the total number of scores to be read in, much of the analysis can still be carried out automatically by analyzing in parts, say first at A_1, then at A_2 in the next run, etc. This would yield S/A, SB/A. Further reduction in number of observations might be obtained by computing SS' for cell means, then multiplying by appropriate number of observations to obtain SS.)

8.5 If we assume extreme heterogeneity of variances of difference scores, the true F distribution has 1 and $a(n-1)$ df, or 1 and 27 (instead of 3 and 81). From Table A-5, we find that $p < .20$. Note that on the usual df, $p \approx .10$.

CHAPTER 9

9.1

SV	df	SS	MS	F
Between G	5	1,422.25		
A	2	1,324.67	662.33	20.36
G/A	3	97.58	32.53	.57
S/G/A	30	1,706.50	56.88	

9.2

SV	df	SS	MS	F
Between S	15	1,277.25		
Between G	3	1,210.42		
A	1	1,160.33	1,160.33	46.34
G/A	2	50.08	25.04	4.49
S/G/A	12	66.83	5.57	
Within S	32	1,006.00		
B	2	919.63	459.81	50.86
AB	2	17.04	8.52	0.94
GB/A	4	36.16	9.04	6.55
SB/G/A	24	33.26	1.38	

9.3

SV	df	MS
S	4	.72
A	1	456.01
SA	4	.92
B	1	.31
SB	4	.59
AB	1	37.81
SAB	4	1.16
P/AB	12	1.62
SP/AB	48	1.75

Ordinarily, quasi F ratios would be required to test effects of interest. The F test for A, for example, would be $MS_A/(MS_{SA} + MS_{P/AB} - MS_{SP/AB})$. In the present example, however, preliminary F tests against SP/AB reveal that several terms estimate the same variance and can therefore be pooled. This gives the revised table below.

SV	df	MS	F
S	4	.72	.45
A	1	456.01	287.70
B	1	.31	.20
AB	1	37.81	23.85
Residual	72	1.59	

9.4

SV	df	EMS
Between S	71	
Between G	35	
Probability (P)	2	$\sigma_e^2 + 16\sigma_{G/P}^2 + 192\theta_P^2$
G/P	33	$\sigma_e^2 + 16\sigma_{G/P}^2$
Within G	36	
Role (R)	1	$\sigma_e^2 + 8\sigma_{GR/P}^2 + 288\theta_P^2$
PR	2	$\sigma_e^2 + 8\sigma_{GR/P}^2 + 96\theta_{PR}^2$
GR/P	33	$\sigma_e^2 + 8\sigma_{GR/P}^2$
Within S	504	
Trial Type (T)	1	$\sigma_e^2 + 8\sigma_{GT/P}^2 + 288\theta_T^2$
PT	2	$\sigma_e^2 + 8\sigma_{GT/P}^2 + 96\theta_{PT}^2$
GT/P	33	$\sigma_e^2 + 8\sigma_{GT/P}^2$
RT	1	$\sigma_e^2 + 4\sigma_{GRT/P}^1 + 144\theta_{RT}^2$
PRT	2	$\sigma_e^2 + 4\sigma_{GRT/P} + 8\theta_{PRT}^2$
GRT/P	33	$\sigma_e^2 + 4\sigma_{GRT/P}^2$
Blocks (B)	3	$\sigma_e^2 + 4\sigma_{GB/P}^2 + 144\theta_B^2$
PB	6	$\sigma_e^2 + 4\sigma_{GB/P}^2 + 48\theta_{PB}^2$
GB/P	99	$\sigma_e^2 + 4\sigma_{GB/P}^2$
RB	3	$\sigma_e^2 + 2\sigma_{GRB/P}^2 + 72\theta_{RB}^2$
PRB	6	$\sigma_e^2 + 2\sigma_{GRB/P}^2 + 24\theta_{PRB}^2$
GRB/P	99	$\sigma_e^2 + 2\sigma_{GRB/P}^2$
TB	3	$\sigma_e^2 + 2\sigma_{GTB/P}^2 + 72\theta_{TB}^2$
PTB	6	$\sigma_e^2 + 2\sigma_{GTB/P}^2 + 24\theta_{PTB}^2$
GTB/P	99	$\sigma_e^2 + 2\sigma_{GTB/P}^2$
RTB	3	$\sigma_e^2 + \sigma_{GRTB/P}^2 + 36\theta_{RTB}^2$
PRTB	6	$\sigma_e^2 + \sigma_{GRTB/P} + 12\theta_{PRTB}^2$
GRTB/P	99	$\sigma_e^2 + \sigma_{GRTB/P}^2$

9.5

SV	df
D	2
Sc/D	27
I	1
DI	2
$Sc \times I/D$	27
$S/Sc \times I/D$	540

We now can test hypotheses about variability due to schools (Sc) and their interaction with I. Furthermore, if these sources do contribute variability the earlier pooled error term yielded a biased F test.

9.6

SV	df	EMS
Between G	7	
Orientation (O)	1	$\sigma_e^2+6\sigma_{G/O}^2+24\theta_O^2$
G/O	6	$\sigma_e^2+6\sigma_{G/O}^2$
Within G	40	
Between Problems (P)	5	
Stress (St)	1	$\sigma_e^2+3\sigma_{GO/St}^2+24\theta_{St}^2$
P/St	4	$\sigma_e^2+\sigma_{GP/St}^2+8\theta_{P/St}^2$
$O\times St$	1	$\sigma_e^2+3\sigma_{GO/St}^2+12\theta_{O\times St}^2$
OP/St	4	$\sigma_e^2+\sigma_{GP/St}^2+8\theta_{OP/St}^2$
$G\times St/O$	6	$\sigma_e^2+\sigma_{G\times St/O}^2$
$GP/St\times O$	24	$\sigma_e^2+\sigma_{GP/St\times O}^2$

Note: P is treated here as a fixed-effect variable. In an experiment of the sort described, the problems might well be selected at random from a larger pool of items. In that case, the EMS are changed and quasi F ratios are required.

9.7 (a) The random variables are $S/GX/T$ and G/T. The fixed variables are T and X.

(b)

SV	df	EMS	F
T	1	$\sigma_e^2+6\sigma_{G/T}^2+36\theta_T^2$	$MS_T/MS_{G/T}$
G/T	10	$\sigma_e^2+6\sigma_{G/T}^2$	
X	1	$\sigma_e^2+3\sigma_{GX/T}^2+36\theta_X^2$	$MS_X/MS_{GX/T}$
TX	1	$\sigma_e^2+3\sigma_{GX/T}^2+18\theta_{TX}^2$	$MS_{TX}/MS_{GX/T}$
GX/T	10	$\sigma_e^2+3\sigma_{GX/T}^2$	
S/GX/T	48	σ_e^2	

(c) Because of the small number of error df, the tests of these effects lack power. If the G/T terms were clearly not significant when tested against $S/GX/T$, however, these three sources could be pooled to get an error term on 68 df.

9.8 (a) Let A represent associative strength, T represent truth value, and s represent statements.

SV	df	EMS	Error Term
S	23	$\sigma_e^2+120\sigma_S^2$	Ss/TA
T	1	$\sigma_e^2+60\sigma_{ST}^2+24\sigma_{s/TA}^2+1440\theta_T^2$	$ST+s/TA-Ss/TA$
ST	23	$\sigma_e^2+60\sigma_{ST}^2$	Ss/TA
A	2	$\sigma_e^2+40\sigma_{SA}^2+24\sigma_{s/TA}^2+960\theta_A^2$	$SA+s/TA-Ss/TA$
SA	46	$\sigma_e^2+40\sigma_{SA}^2$	Ss/TA
TA	2	$\sigma_e^2+20\sigma_{STA}^2+24\sigma_{s/TA}^2+480\theta_{TA}^2$	$STA+s/TA-Ss/TA$
STA	46	$\sigma_e^2+20\sigma_{STA}^2$	Ss/TA
s/TA	114	$\sigma_e^2+24\sigma_{s/TA}^2$	Ss/TA
Ss/TA	2622	σ_e^2	

(b)

SV	df	EMS	Error Term
A	2	$\sigma_e^2 + 60\sigma_{S/A}^2 + 48\sigma_{s/TA}^2 + 960\theta_A^2$	$S/A + s/TA - Ss/TA$
S/A	45	$\sigma_e^2 + 60\sigma_{S/A}^2$	Ss/TA
T	1	$\sigma_e^2 + 30\sigma_{ST/A}^2 + 48\sigma_{s/TA}^2 + 1440\theta_T^2$	$ST/A + s/TA - Ss/Ta$
TA	2	$\sigma_e^2 + 30\sigma_{ST/A}^2 + 48\sigma_{s/TA}^2 + 480\theta_{TA}^2$	$ST/A + s/TA - Ss/TA$
ST/A	45	$\sigma_e^2 + 30\sigma_{ST/A}^2$	Ss/TA
s/TA	174	$\sigma_e^2 + 48\sigma_{s/TA}^2$	Ss/TA
Ss/TA	2610	σ_e^2	

Design (b) requires fewer statements but more subjects. It probably will yield a less precise test of A because variability due to individual differences contributes to the error term. Both designs involve quasi-F tests.

(c)

SV	df	EMS	Error Term
S	23	$\sigma_e^2 + 120\sigma_S^2$	s/STA
T	1	$\sigma_e^2 + 60\sigma_{ST}^2 + 1440\theta_T^2$	ST
ST	23	$\sigma_e^2 + 60\sigma_{ST}^2$	s/STA
A	2	$\sigma_e^2 + 40\sigma_{SA}^2 + 960\theta_A^2$	SA
SA	46	$\sigma_e^2 + 40\sigma_{SA}^2$	s/STA
TA	2	$\sigma_e^2 + 20\sigma_{STA}^2 + 480\theta_{TA}^2$	STA
STA	46	$\sigma_e^2 + 20\sigma_{STA}^2$	s/STA
s/STA	2736	σ_e^2	

Design (c) requires no quasi-F tests. The design has the disadvantage, however, of requiring 2,880 statements. For the experiment under consideration it is probably impossible to generate this many stimuli. Nevertheless, the principle of assigning each subject to a different stimulus set may prove a useful way of avoiding quasi-F tests in some studies.

CHAPTER 10

10.1 We assume that

$$Y_{ijkm} = \mu + \alpha_j + \beta_k + \gamma_m + \varepsilon_{ijkm}$$

SV	df	MS	F
A	2	8.01	8.90
B	2	20.08	22.31
C	2	1.14	1.27
Between cells residual	2	1.12	1.24
S/cells	81	.90	

If the test of the between-cells residual terms had a significant result, it would imply that interactions were contributing to the data. In such a case, the F's may be spuriously inflated.

10.2

SV	df
Between S	35
Sequences (R)	3
Instruction (I)	2
RI	6
S/RI	24
Within S	108
Location (L)	3
Column (C)	3
IL	6
IC	6
Between-cells res.	6
$I \times B$. cells res.	12
Within-cells res.	72

10.3

SV	df
Between S	59
Alcohol (A)	2
S/A	57
Within S	180
Mode of presentation (M)	1
Tasks (T)	1
$M \times T$	1
Columns (C)	3
AM	2
AT	2
AMT	2
AC	6
Residual	162

10.4

SV	df
Between cells	24
S	4
C	4
Lists (L)	4
Between-cells res.	12
Within cells	150
Position (P)	6
LP	24
CP	24
SP	24
Residual	72

10.5

SV	df
Between S	47
Between G	15
Sequences (R)	3
G/R	12
S/G/R	32
Within S	144
C	3
Problems (P)	3
Between-cells res.	6
Within-cells res.	132
GP/R(GC/R)	36
SP/GR(SC/GR)	96

10.6 (a) 9×9 Latin square with each subject tested under all combinations of two of the variables (e.g., payoff and probability) and 27 subjects at each level of initial stake. We could use nine different squares or replicate the same square nine times. This approach affords precise tests of two of the three main effects (we would have less precision in testing the between-subjects effect) and of all interactions of interest. However, it requires nine measurements/subject.

(b) 3×3 Latin square with each subject tested under all levels of one variable and nine subjects at each combination of levels of the other two variables. The design saves measurement/subject but yields less precise tests of the two between-S variables and their interaction. Nonadditivity may also be more of a problem in the smaller square.

(c) No repeated measurements. Using the Latin square principle, we could have nine cells with nine subjects in each. But all interaction information is lost and, if the three factors interact, F tests against W. cells res. will be positively biased while those against B. cells res. will be negatively biased.

(d) Complete factorial design, 27 cells with three subjects in each. There are 54 error df, a reasonable number, no need to hold subjects for several measures, and no additivity problems. On the other hand, (a) and (b) will generally be more efficient.

10.7

$$SS_A = SS_{(E_1+E_3)-(E_2+E_4)}$$
$$SS_{BC} = SS_{(E_1+E_2)-(E_3+E_4)}$$
$$SS_{ABC} = SS_{(E_1+E_4)-(E_2+E_3)}$$

CHAPTER 11

11.1 *The Tukey test.* Use $q_{.05}$ for $df = 60$; this is 3.98. The statistic $\sigma_{\bar{Y}}^2$ is $\sqrt{297.44/75} = 1.99$. The critical difference is therefore 7.93. The results can be represented as

A   B   C   D   E

The Neuman-Keuls Test. The result is

<u>A</u>   <u>B   C</u>   <u>D   E</u>

The Bonferroni test. For 10 comparisons, $t_{.05,60} = 2.91$. The standard error of $\bar{Y}_{.j'} - \bar{Y}_{.j}$ is $\sqrt{\dfrac{(247.44)^2}{75}} = 2.82$. The critical difference is therefore 8.18. The result is

<u>A</u>   <u>B   C   D</u>   <u>E</u>

11.2 (a)

$$SS_\psi = \frac{[(3)(5) + (-2)(5) + (-2)(7) + (-2)(9) + (3)(2)]^2}{30/6}$$

$$= \frac{(21)^2}{5} = 88.2$$

$$F = \frac{88.2}{4} = 22.05$$

$$F_{.05;1,25} = 4.24$$

The result is significant. Note that we have a single a priori hypothesis. (b) *Scheffé.* Let the weights be $-\frac{1}{2}, +\frac{1}{3}, +\frac{1}{3}, +\frac{1}{3}, -\frac{1}{2}$ so that the results are on the original scale. Then the limits are

$$3.5 \pm \frac{\sqrt{(4)(2.76)(4)(5)}}{(6)(6)} = 3.5 \pm 2.47$$

11.3 (a)

$$SS_\psi = \frac{[13 + 22 - (2)(11)]^2}{(4)(6)} = 7.04$$

(b)

$$SS_\psi = \frac{[5 + 11 - (2)(5)]^2}{(2)(6)} + \frac{[8 + 11 - (2)(6)]^2}{(2)(6)} - 7.04 = .04$$

The error term for both (a) and (b) is

$$\frac{[4 + 5 - (2)(3)]^2}{6} + \cdots + \frac{[3 + 7 - (2)(4)]^2}{6} - \frac{[5 + 11 - (2)(5)]^2}{(2)(6)} - \frac{[8 + 11 - (2(6)]^2}{(2)(6)}$$

$$= 7.83 - 7.08 = .75$$

11.4 If we assign the weights $+1, -1$ to B_1 and B_2 and $+2, -1, -1$ to the A levels, we have part of AB on 1 *df*. Since $0, +1, -1$ gives an orthogonal A

contrast, the solution follows.

$$SS_{\hat{\psi}_1} = \frac{[(1)(2)(21)+(1)(-1)(16)+(1)(-1)(9)+(-1)(2)(4) \atop +(-1)(-1)(12)+(-1)(-1)(7)]^2}{(5)(12)}$$

$$SS_{\hat{\psi}_2} = \frac{[(1)(1)(16)+(1)(-1)(9)+(-1)(1)(12)+(-1)(-1)(7)]^2}{(5)(4)}$$

$$SS_{AB} = SS_{\hat{\psi}_1} + SS_{\hat{\psi}_2}$$

11.5 *Tukey.* To be significant, $\hat{\psi} > q_{.05,5,20}\sqrt{12/5}$; this critical difference is $4.23\sqrt{11/5} = 6.27$. No drug groups differ significantly from the control. *Dunnett.* We require $\hat{\psi} > 2.70\sqrt{22/5} = 5.66$. D_1 and D_4 differ significantly from C. *Bonferroni.* We require $\hat{\psi} > 2.74\sqrt{22/5} = 5.75$. D_1 and D_4 differ significantly from C.

  The Dunnett and Bonferroni approaches have more power for comparisons with C than the Tukey does because the former control the EF for a smaller family.

11.6 The six schools form a 3×2 array:

Public

East	Midwest	West
Penn State	Minnesota	California

Private

East	Midwest	West
Yale	Northwestern	Stanford

Thus, the overall design is a $3 \times 3 \times 2$.
Let Public, Private $= A_1, A_2$. East, Midwest, West $= B_1, B_2, B_3$.
Clinicians, Experimentalists, Social—C_1, C_2, C_3 and let

$$T_{jkm} = \text{total for } A_j B_k C_m$$

$$T_j = \text{total for } A_j = \sum_k \sum_m T_{jkm}$$

$$T_{jk} = \text{total for } A_j B_k = \sum_m T_{jkm}$$

etc.

H_1: $(2)(T_{.2.}) - T_{.1.} - T_{.3.}$
H_2: $[(2)(T_{12.}) - T_{11.} - T_{13.}] - [(2)(T_{22.}) - T_{21.} - T_{23.}]$
H_3: $(2)[(2)(T_{.21}) - T_{.11} - T_{.31}] + (-1)[(2)(T_{.22}) - T_{.12} - T_{.32}]$
     $+ (-1)[(2)(T_{.23}) - T_{.13} - T_{.33}]$
H_4: $(5)(T_{121}) - T_{111} - T_{131} - T_{211} - T_{221} - T_{231}$
H_5: $(2)[(5)(T_{121}) - T_{111} - T_{131} - T_{211} - T_{221} - T_{231}] + (-1)$
     $\times [(5)(T_{122}) - (T_{112}) - T_{132} - T_{222} - T_{232}] + (-1)$
     $\times [(5)(T_{123}) - T_{113} - T_{213} - T_{223} - T_{233}]$

11.7 (a) $F = \dfrac{25}{(9.6)(.8)} = 3.25$ which is not significant.

(b) $F = \dfrac{25}{(6.60)(.4)} = 9.47$ which is significant.

(c) The procedure in (b) is improper because it neglects the sampling variance of the control group means. Even if 11.40 were used in (b) as an estimate of σ_e^2, the procedure is improper because it is multiplied by $2/n$ (.4) rather than $4/n$ (.8). The procedure is proper only if the subtracted control group values are determined prior to data collection on the basis of some theory. Then, the values are hypothesized parameters and would have no sampling variability.

11.8 (a) $SS_{AB} = 62.50$. The weights for A and AB are

	A_1B_1	A_1B_2	A_2B_1	A_2B_2
A	1	1	-1	-1
AB	1	-1	-1	1

After adjustment for effects of A, SS_{AB} is still 62.50. The point is that if contrasts are orthogonal, adjusting the data for one contrast does not affect variability of the others.

(b) $SS_{\hat{\psi}_1} = 245 \qquad SS_{\hat{\psi}_2} = 481.67 \qquad SS_{\hat{\psi}_3} = 720.00$

Adjusting for ψ_1, we get the means 14.5, 14.5, and 6.0. The $SS_{\hat{\psi}_2}$ is still the same but $SS_{\hat{\psi}_3}$ is now 361.25. This is what we expect on the basis of summing the crossproducts of weights.

11.9 H_0: $\dfrac{\mu_{F_1} + \mu_{F_2}}{2} - \mu_c = \mu_c - \dfrac{\mu_{I_1} + \mu_{I_2}}{2}$

Multiplying by 2 and rearranging terms, we get

H_0: $\mu_{F_1} + \mu_{F_2} + \mu_{I_1} + \mu_{I_2} - 4\mu_c = 0$

Choice (iii) is correct.

CHAPTER 12

12.1 It is useful to calculate several terms. We get

$$\sum y^2 = \sum Y^2 - \frac{(\sum Y)^2}{n} = 9{,}084 - 7{,}728.4 = 1{,}355.6$$

$$\sum x^2 = \sum X^2 - \frac{(\sum X)^2}{n} = 324 - 270.4 = 53.6$$

$$\sum xy^2 = \sum XY - \frac{(\sum X)(\sum Y)}{n} = 1{,}713 - 1{,}445.6 = 267.4$$

(a) $b_1 = \sum xy / \sum x^2 = 4.99$

(b) $\hat{\sigma}_e^2 = [\sum y^2 - (\sum xy)^2 / \sum x^2]/(n-2) = 2.70$

(c) $\hat{\sigma}^2_{b_1} = \hat{\sigma}^2_e / \sum x^2 = .05$

$$t = \frac{4.99}{\sqrt{.05}} = 22.23$$

This is significant at the .01 level.

(d) $t_{8,.05} = 2.306$ $\qquad \hat{Y} = 1.85 + (4.99)(6) = 31.79$

$$\hat{\sigma}^2_{\hat{Y}} = \hat{\sigma}^2_e \left(\frac{1}{n} + \frac{x}{\sum x^2} \right) = .31$$

The confidence interval is $31.79 \pm 2.306\sqrt{.31}$

12.2 Extrapolation involves a greater danger that the function will not be linear in the region of interest. Furthermore, $\hat{\sigma}^2_{\hat{Y}}$ is larger the further X is from $\bar{X}$ (see Equation 12.28).

12.3 (a) $MS_A = 8,000$ $\qquad MS_{S/A} = 333$ $\qquad F = 24.02$

(b) No. Joe is hypothesizing any list length effect while Jim is looking for a particular pattern of effects. If Jim is correct in predicting linear variation as a function of list length, then Joe must be correct (there is some effect of L). The converse is not necessarily true.

(c) The list length deviations are $x_{.j} = -3$, -1, 1, and 3. Therefore, $\sum\sum x^2 = 10(9+1+1+9) = 200$. Also, from (a), $\sum\sum y^2 = 24,000 + 11,988 = 35,988$. To obtain $\sum_j \sum_i x_{ij} y_{ij}$, note that the x_{i_1} are all -3, the x_{i_2} are all -1, and so on. Also, $\sum_i y_{ij} = \sum_i (Y_{ij} - \bar{Y}_{..}) = [n\bar{Y}_{.j} - \bar{Y}_{..}]$. Therefore, $\sum\sum xy = 10[(-3)(-40) + (-1)(0) + (1)(20) + (1)(20)] = 1600$.

(d) $SS_{\hat{Y}} = \dfrac{(\sum\sum xy)^2}{\sum x^2} = 12,800.$ $\qquad SS_{error} = 35,988 - 12,800$
$$= 23,188.$$

(e) $F = \dfrac{(b_1 - 0)^2}{\hat{\sigma}^2_e / \sum\sum x^2} = \dfrac{8^2}{\left(\dfrac{12,800}{38}\right) \Big/ 200} = 38.00$

12.4 (a) SS_A includes $SS_{\hat{Y}}$ plus any other variation and must always be the larger quantity.

(b) We could have a model that postulated a higher (than linear) relation. A possibility would be to have squared x_{ij} terms.

CHAPTER 13

13.1 (a) No. (b) No. (c) Yes. (d) Yes. (e) Yes.

$$\mathbf{X'Y} = \begin{bmatrix} 34 & 22 \\ 22 & 12 \\ 86 & 78 \end{bmatrix}$$

$$\mathbf{YX} = \begin{bmatrix} 22 & 16 & 38 \\ 46 & 30 & 114 \end{bmatrix}$$

$$\mathbf{Y'X} = \begin{bmatrix} 34 & 22 & 86 \\ 22 & 12 & 78 \end{bmatrix}$$

Note that $\mathbf{Y'X} = (\mathbf{X'Y})'$. This is always true.

13.2 The entry in row i and column j of $\mathbf{XY}$ is $X_{i_1}Y_{1_j}+X_{i_2}Y_{2_j}$. If we take the transpose, $(\mathbf{XY})'$, this entry is in row j and column i. It is also the entry in row j and column i of $\mathbf{Y}'\mathbf{X}'$.

13.3 $S_x^2=\left(\dfrac{1}{n}\right)\mathbf{x}'\mathbf{x}$ $\qquad$ $\mathrm{Cov}=\left(\dfrac{1}{n}\right)\mathbf{x}'\mathbf{y}$

13.4 $\mathbf{X}'=\begin{bmatrix}3 & -1 & 4 & -2\end{bmatrix}$

13.5 Line $2-(\frac{2}{3})\times$Line 1 is $\begin{bmatrix}0 & \frac{10}{3} & -\frac{4}{3} & \frac{4}{3}\end{bmatrix}$. Line $3-(\frac{7}{3})\times$Line 1 is $\begin{bmatrix}0 & \frac{5}{3} & -\frac{2}{3} & \frac{2}{3}\end{bmatrix}$. These two results are in a 2:1 ratio and therefore convey the same information. Another way of expressing this is that

$$\text{Line } 2-(\tfrac{2}{3})\times\text{Line }1=2\times[\text{Line }3-(\tfrac{7}{3})\text{Line }1]$$

and, solving,

$$\text{Line }3=(\tfrac{1}{2})\times\text{Line }2+2\times\text{Line }1.$$

Since any one line can be expressed as a linear function of the other two lines, the rank is only 2.

CHAPTER 14

14.1 First calculate $\mathbf{X}'\mathbf{X}$ and $\mathbf{X}'\mathbf{Y}$. These are

$$\mathbf{X}'\mathbf{X}=\begin{bmatrix}10 & 104 & 87 & 149\\104 & 2{,}192 & 996 & 1{,}569\\87 & 996 & 825 & 1{,}263\\149 & 1{,}569 & 1{,}263 & 2{,}293\end{bmatrix} \qquad \mathbf{X}'\mathbf{Y}=\begin{bmatrix}188\\2{,}153\\1{,}677\\2{,}815\end{bmatrix}$$

Then use the Doolittle approach to solve $(\mathbf{X}'\mathbf{X})\hat{\boldsymbol{\beta}}=\mathbf{X}'\mathbf{Y}$. The solution is $\hat{Y}=4.590750+.113410X_1+.687397X_2+.473116X_3$.

14.2 (a) The SS_{total}, calculated as usual, is 209.6. We obtain $SS_{\hat{Y}(123)}$ by $\hat{\boldsymbol{\beta}}'\mathbf{X}'\mathbf{Y}-C=57.419039$ and $SS_{error}=152.180961$. The $MS_{error}=SS_{error}/(n-1-p)=25.3694$. Ignoring X_2 and X_3, $\hat{Y}=16.947406+.178134X_1$. Then the $SS_{\hat{Y}(1)}=35.234830$. The F ratio of 1.39 is not significant on 1 and 6 df.
(b) Ignoring X_1, $\hat{Y}=12.675551+.213068X_2+.286628X_3$. The $SS_{\hat{Y}(23)}=12.776491$. Subtracting from $SS_{\hat{Y}(123)}$ we get $SS_{\hat{Y}(1/23)}=44.642548$.

14.3 Substituting for $\mathbf{Y}$, we have

$$\mathbf{b}=(\mathbf{X}'\mathbf{X})^{-1}\mathbf{X}'[\mathbf{X}\boldsymbol{\beta}+\mathbf{X}_3\boldsymbol{\beta}_3+\boldsymbol{\varepsilon}]$$

Taking expectations and expanding, we have

$$E(\mathbf{b})=(\mathbf{X}'\mathbf{X})^{-1}(\mathbf{X}'\mathbf{X})\boldsymbol{\beta}+(\mathbf{X}'\mathbf{X})^{-1}\mathbf{X}'\mathbf{X}_3\boldsymbol{\beta}_3+(\mathbf{X}'\mathbf{X})^{-1}\mathbf{X}'E(\boldsymbol{\varepsilon}).$$

(Note that $E(\mathbf{X})=\mathbf{X}$ because $\mathbf{X}$ is assumed to be fixed.) Since $(\mathbf{X}'\mathbf{X})^{-1}(\mathbf{X}'\mathbf{X})=\mathbf{I}$ and $E(\boldsymbol{\varepsilon})=0$, we have the desired result.

14.4 (a) Summing crossproducts, $(2)(-2)+(-1)(-1)+\cdots+(2)(2)=0$. If $\mathbf{X}_j$ is a 5×1 column vector, we can express this as $\mathbf{X}_1\mathbf{X}_2=0$. The matrix $\mathbf{X'X}$ is diagonal; the general form is

$$\mathbf{X'X} = \begin{bmatrix} \sum X_1^2 & 0 & 0 & \cdots \\ 0 & \sum X_2^2 & 0 & \cdots \\ & & \cdots & \end{bmatrix}$$

(b) $(\mathbf{X'X})^{-1} = \begin{bmatrix} \frac{1}{14} & 0 \\ 0 & \frac{1}{10} \end{bmatrix}$

(c) $\mathbf{b'} = [5.8 \quad 1 \quad 2.2]$

(d) $SS_{\hat{Y}(12)} = \mathbf{b'X'Y} - C = 62.40$

(e) The regression coefficients are unchanged.

(f) The $SS_{\hat{Y}(1|2)} = 62.40 - 48.40 = 14.00 = SS_{\hat{Y}(1)}$.

CHAPTER 15

15.1 (a) $SS_A = 30.00$

(b) (i) $\mathbf{X'} = \begin{bmatrix} 1 & 1 & 1 & 1 & 1 & 0 & 0 & 0 \\ 0 & 0 & 0 & 0 & 0 & -1 & -1 & -1 \end{bmatrix}$

$\boldsymbol{\phi}^{0'} = [4 \quad 8]$

(ii) $\mathbf{X'} = \begin{bmatrix} 1 & 1 & 1 & 1 & 1 & 1 & 1 & 1 \\ 1 & 1 & 1 & 1 & 1 & -1 & -1 & -1 \end{bmatrix}$

$\boldsymbol{\phi}^{0'} = [6 \quad -2]$

(iii) $\mathbf{X'} = \begin{bmatrix} 1 & 1 & 1 & 1 & 1 & 1 & 1 & 1 \\ 1 & 1 & 1 & 1 & 1 & -\frac{5}{3} & -\frac{5}{3} & -\frac{5}{3} \end{bmatrix}$

$\boldsymbol{\phi}^{0'} = [5.5 \quad -1.5]$

The $SS_A = 30$ in all cases.

15.2 (a) $\mathbf{X'} = \begin{bmatrix} 1 & 1 & 1 & 1 & 1 & 0 & 0 & 0 & 0 & 0 & 0 & 0 \\ 0 & 0 & 0 & 0 & 0 & 1 & 1 & 1 & 0 & 0 & 0 & 0 \\ 0 & 0 & 0 & 0 & 0 & 0 & 0 & 0 & 1 & 1 & 1 & 1 \end{bmatrix}$

$\boldsymbol{\phi}^{0'} = [7.00 \quad 4.33 \quad 2.25]$

$SS_A = \boldsymbol{\phi}^{0'}\mathbf{X'Y} - C = 50.83$

(b)
$$SS_{\hat{\psi}} = \frac{[(\bar{Y}_1 + \bar{Y}_2 + \bar{Y}_3)/3]^2}{(\frac{1}{9})(\frac{1}{5} + \frac{1}{3} + \frac{1}{4})} = 235.54$$

$$F = SS_{\hat{\psi}}/MS_{S/A} = 72.06$$

(c) Let $p_j = n_j/n$. Then

$$SS_{\hat{\psi}} = \frac{(\sum_j p_j \bar{Y}_{.j})^2}{\sum_j p_j^2/n_j} = \frac{(5\bar{Y}_1 + 3\bar{Y}_2 + 4\bar{Y}_3)^2}{12} - C = 270.75$$

$$F = SS_{\hat{\psi}}/MS_{S/A} = 82.82$$

(d) (i) The constraint is $\alpha_3^0 = -(\alpha_1^0 + \alpha_2^0)$ (see Equation (15.20))

$$\mathbf{X'X} = \begin{bmatrix} 12 & 1 & -1 \\ 1 & 9 & 4 \\ -1 & 4 & 7 \end{bmatrix} \qquad \mathbf{X'Y} = \begin{bmatrix} 57 \\ 26 \\ -4 \end{bmatrix}$$

Solving $(\mathbf{X'X})\boldsymbol{\phi}^0 = \mathbf{X'Y}$, we get

$$\boldsymbol{\phi}^{0'} = [\mu^0 \quad \bar{Y}_{.1} - \mu^0 \quad \bar{Y}_{.2} - \mu^0] = [4.53 \quad 2.47 \quad -.20]$$

where $\mu^0 = (\frac{1}{3})(\bar{Y}_{.1} + \bar{Y}_{.2} + \bar{Y}_{.3})$. An easier way to derive these results is to note that

$$\hat{\mathbf{Y}} = \bar{\mathbf{Y}}_{.j} = \mathbf{X}\boldsymbol{\phi}^0$$

Then,

$$\bar{Y}_{.1} = \mu^0 + \alpha_1^0$$
$$\bar{Y}_{.2} = \mu^0 + \alpha_2^0$$
$$\bar{Y}_{.3} = \mu^0 - \alpha_1^0 - \alpha_2^0$$

We can now solve for the elements of $\boldsymbol{\phi}^0$ in terms of the $\bar{Y}_{.j}$.

(ii) The constraint is $5\alpha_1^0 + 3\alpha_2^0 + 4\alpha_3^0 = 0$. Then,

$$\alpha_3^0 = -\tfrac{5}{4}\alpha_1^0 - \tfrac{3}{4}\alpha_2^0$$

$$\mathbf{X'X} = \begin{bmatrix} 12 & 0 & 0 \\ 0 & \frac{45}{4} & \frac{15}{4} \\ 0 & \frac{15}{4} & \frac{21}{4} \end{bmatrix} \qquad \mathbf{X'Y} = \begin{bmatrix} 57 \\ 23.75 \\ 6.25 \end{bmatrix}$$

$$\boldsymbol{\phi}^{0'} = [4.75 \quad 2.25 \quad .42]$$

15.3 (a) $SS_{\hat{\psi}_1} = \dfrac{(7 - 4.33)}{\frac{1}{5} + \frac{1}{3}} = 13.33$

$$SS_{\hat{\psi}_2} = \frac{\left(\frac{7 + 4.33}{2} - 2.25\right)^2}{\left(\frac{1}{2}\right)^2 \left(\frac{1}{5}\right) + \left(\frac{1}{2}\right)^2 \frac{1}{3} + \frac{1}{4}} = 30.45$$

(b) $SS_{\hat{\psi}}$ is unchanged.

$$SS_{\hat{\psi}_2} = \frac{[(\frac{5}{8})\bar{Y}_1 + (\frac{3}{8})\bar{Y}_2 - \bar{Y}_3]^2}{(\frac{5}{8})^2(\frac{1}{5}) + (\frac{3}{8})^2\frac{1}{3} + \frac{1}{4}} = 37.50$$

The second set is orthogonal. Note that $SS_{\hat{\psi}_1} + SS_{\hat{\psi}_2} = SS_A$ in this case.

15.4 $\phi^{0'} = [\mu^0 \ \alpha_1^0 \ \alpha_2^0 \ \beta_1^0 \ \beta_2^0 \ \beta_3^0 \ (\alpha\beta)_{11}^0 \ (\alpha\beta)_{12}^0 \ (\alpha\beta)_{13}^0 \ (\alpha\beta)_{21}^0 \ (\alpha\beta)_{22}^0 \ (\alpha\beta)_{23}^0]$

$$\mathbf{X} \equiv \begin{array}{c} A_1B_1 \\ A_1B_2 \\ A_1B_3 \\ A_1B_4 \\ A_2B_1 \\ A_2B_2 \\ A_2B_3 \\ A_2B_4 \\ A_3B_1 \\ A_3B_2 \\ A_3B_3 \\ A_3B_4 \end{array} \begin{bmatrix} 1 & 1 & 0 & 1 & 0 & 0 & 1 & 0 & 0 & 0 & 0 & 0 \\ 1 & 1 & 0 & 0 & 1 & 0 & 0 & 1 & 0 & 0 & 0 & 0 \\ 1 & 1 & 0 & 0 & 0 & 1 & 0 & 0 & 1 & 0 & 0 & 0 \\ 1 & 1 & 0 & -1 & -1 & -1 & -1 & -1 & -1 & 0 & 0 & 0 \\ 1 & 0 & 1 & 1 & 0 & 0 & 0 & 0 & 0 & 1 & 0 & 0 \\ 1 & 0 & 1 & 0 & 1 & 0 & 0 & 0 & 0 & 0 & 1 & 0 \\ 1 & 0 & 1 & 0 & 0 & 1 & 0 & 0 & 0 & 0 & 0 & 1 \\ 1 & 0 & 1 & -1 & -1 & -1 & 0 & 0 & 0 & -1 & -1 & -1 \\ 1 & -1 & -1 & 1 & 0 & 0 & -1 & 0 & 0 & -1 & 0 & 0 \\ 1 & -1 & -1 & 0 & 1 & 0 & 0 & -1 & 0 & 0 & -1 & 0 \\ 1 & -1 & -1 & 0 & 0 & 1 & 0 & 0 & -1 & 0 & 0 & -1 \\ 1 & -1 & -1 & -1 & -1 & -1 & 1 & 1 & 1 & 1 & 1 & 1 \end{bmatrix}$$

15.5 We calculate $(\mathbf{W}\bar{\mathbf{Y}})' = [-23.\bar{3} \quad -25.\bar{6}]$ and

$$[\mathbf{W}(\mathbf{X}'\mathbf{X})^{-1}\mathbf{W}'] = \begin{bmatrix} 37.\bar{3} & 25.\bar{6} \\ 25.\bar{6} & 21.58\bar{3} \end{bmatrix}$$

Then, $SS_{\hat{\psi}} = (\mathbf{W}\bar{\mathbf{Y}})'[\mathbf{W}(\mathbf{X}'\mathbf{X})^{-1}\mathbf{W}'^{-1}(\mathbf{W}\bar{\mathbf{Y}})] = 38.\bar{1}$.

CHAPTER 16

16.1

	SS_y	SS_x	SP	$SS_{y'}$
A	328.56	49.88	−128.02	434.75
B	258.73	44.83	107.69	129.89
AB	348.08	79.21	166.04	145.56
S/AB	586.66	343.83	203.77	465.89

SV	df	MS	F
A	1	434.75	17.73
B	1	129.89	5.30
AB	1	145.56	5.94
S/AB	19	24.52	

16.2

	SS_y	SS_x	SP	$SS_{y'}$
A	262.89	2.00	14.33	268.82
C	6.22	6.67	−0.67	15.14
S	81.56	0.67	7.33	51.96
Res	14.89	4.67	7.67	2.29

SV	df	MS	F
A	2	134.41	58.60
C	2	7.57	3.30
S	2	25.98	11.32
Res	1	2.29	

16.3 The point is that we have only one covariate measure for each subject. Since there is no within-subject variability on X, covariance is meaningless for the Within-S SV. The Between-S SV can be adjusted. The simplest approach is to sum the four scores for each subject and then carry out the usual one-factor analysis of covariance.

16.4 Define a new covariate, Z:

$$Z = X^2$$

Now Y is a linear function of Z and the usual covariance may be carried out. Of course, this assumes the usual covariance model; for example, that the regression of Y on Z is linear and the same for all treatment populations.

CHAPTER 17

17.1 A(lin)

$$SS = \frac{[(-3)(39) - 25 + 17 + (3)(41)]^2}{(10)(20)}$$
$$= 0$$

A(quad)

$$SS = \frac{(39 - 25 - 17 + 41)^2}{(10)(4)}$$
$$= 40$$

A(cub)

$$SS = \frac{[-39 + 3(25) + (-3)(17) + 41]^2}{(10)(20)}$$
$$= 2$$

A(lin) $\times B$

$$SS = \frac{[(-3)(15) - 7 + 5 + (3)(15)]^2 + [(-3)(24) - 18 + 12 + (3)(26)]^2}{(5)(20)}$$
$$- SS_{A(\text{lin})} = 0$$

A(quad) $\times B$

$$SS = \frac{(15 - 7 - 5 + 15)^2 + (24 - 18 - 12 + 26)^2}{(5)(4)}$$
$$- SS_{A(\text{quad})} = 0$$

A(cub) $\times B$

$$SS = \frac{[-15 + (3)(7) - (3)(5) + 15]^2 + [-24 + (3)(18) - (3)(12) + 26]^2}{(5)(20)}$$
$$SS_{A(\text{cub})} = 2$$

17.2 Whether there is a sexual content main effect will depend upon whether or not high-guilt responses decline at the same rate as low-guilt responses increase as a function of sexual content. There should clearly be a guilt main effect, since the two groups have equal response frequency for low sexual content but the low group has a higher frequency at higher-content levels. There should be an interaction, since the two curves diverge. In particular, $G \times S(\text{lin})$ should be significant. The residual, $G \times S(\text{curv})$, should also be tested.

17.3 (a)
$$SS_{P(\text{lin})} = \frac{(30)^2}{(15)(20)} = 3.0$$

$$SS_{P(\text{lin}) \times T} = \frac{(17)^2 + (9)^2 + (4)^2}{(5)(20)} - 3.0 = .86$$

(b)
$$SS_{\hat{\psi}} = \frac{[(-2)(-17) - 9 - 4]^2}{(5)(20)(6)} = .735$$

17.4 A_1 apparently exhibits a linear and cubic trend; A_2 a quadratic trend; A_3 a linear trend. Assuming that the groups start at the same level, we would expect an A main effect, A_2 performing less well over the five tests (T). There might also be a T main effect; averaging two linear components $(A_1$ and $A_3)$ with a zero linear component should yield $T(\text{lin})$. On similar grounds we could look for $T(\text{quad})$ and $T(\text{cub})$. Most clearly, the hypothesis implies $A \times T(\text{lin})$, $A \times T(\text{quad})$, and $A \times T(\text{cub})$.

17.5 H_1: Test the simple effect of D for blank interval data.
H_2: Discarding the blank interval data, test for $D(\text{lin})$ and $D(\text{quad})$.
H_3: Test $I \times D(\text{lin})$.

17.6 Method: Linear, branching, reading $= A_1, A_2, A_3$
Material: Math, English, Social Studies $= B_1, B_2, B_3$.
Time/Day: 15, 30, 45 $= C_1, C_2, C_3$.

H_1: $2T_{.1.} - T_{.2.} - T_{.3.}$
H_2: $T_{1..} + T_{2..} - T_{3..}$
H_3: $T(\text{lin}) = T_{..1} - T_{..3}$
$T(\text{quad}) = T_{..1} - 2T_{..2} + T_{..3}$
H_4: $(2T_{11.} - T_{12.} - T_{13.}) + (2T_{21.} - T_{22.} - T_{23.}) - 2(T_{31.} - T_{32.} - T_{33.})$
H_5: $(2T_{11.} - T_{12.} - T_{13.}) - 2T_{21.} - T_{22.} - T_{23.})$
H_6: $(2T_{.11} - T_{.21} - T_{.31}) - 2T_{.13} - T_{.23} - T_{.33})$

17.7 (a)
$$b'_0 = \tfrac{30}{3} = 10$$

$$b'_1 = \frac{\sum_i \sum_{ij} \bar{Y}_j}{\sum_j \xi'^2_{1j}} = 3$$

$$b'_2 = -5$$

$$\bar{Y}_{ij} = 10 + 3\xi'_{1j} - 5\xi'_{2j}$$

To check:

$$\bar{Y}_1 = 10 + 3(-1) - 5(+1) = 2$$
$$\bar{Y}_2 = 10 + 3(0) - 5(-2) = 20$$
$$\bar{Y}_3 = 10 + 3(+1) - 5(+1) = 8$$

(b)

$$\xi_{11} = \alpha_1 \qquad\qquad \xi_{12} = \alpha_2$$
$$\xi_{21} = \alpha_1 + 1 \qquad \xi_{22} = \alpha_2 + \beta_2 + 1$$
$$\xi_{31} = \alpha_1 + 2 \qquad \xi_{32} = \alpha_2 + 2\beta_2 + 4$$

$$0 = 3\alpha_1 + 3 \qquad\qquad 0 = 3\alpha_2 + 3\beta_2 + 5$$
$$\alpha_1 = -1 \qquad\qquad 0 = -\alpha_2 + (0)(\alpha_2 + \beta_2 + 1) + (1)(\alpha_2 + 2\beta_2 + 4)$$
$$\beta_2 = -2\alpha_2 = \tfrac{1}{3}$$

(c)

$$\bar{Y}_j = 10 + 3(\alpha_1 + X) - (5)(3)(\alpha_2 + \beta_2 X + X^2)$$
$$= 2 + 33X - 15X^2$$

Note: We multiply $(\alpha_2 + \beta_2 + X^2)$ by (3) to obtain the integer ξ_{j2}. To check:

$$\bar{Y}_1 = 2 + 33(0) - 15(0) = 2$$
$$\bar{Y}_2 = 2 + 33(1) - 15(1) = 20$$
$$\bar{Y}_3 = 2 + 33(2) - 15(4) = 8$$

CHAPTER 18

18.1 (a) The determinant is $(ka)d - (kb)c = k\,|\mathbf{A}|$.
   (b) The determinant is $(ka)(kd) - (kb)(kc) = k^2\,|\mathbf{A}|$.
   (c) Let

$$\mathbf{B} = \begin{bmatrix} a & b \\ c + ka & d + kb \end{bmatrix}$$

Then $|\mathbf{B}| = |\mathbf{A}|$.
   (d) Let

$$\mathbf{A} = \begin{bmatrix} a & b \\ c & d \end{bmatrix} \qquad \mathbf{B} = \begin{bmatrix} e & f \\ g & h \end{bmatrix}$$

$$\mathbf{AB} = \begin{bmatrix} ae + bg & af + bh \\ ce + dg & cf + dh \end{bmatrix}$$

Then $|\mathbf{AB}| = aedh - adfg - bceh + bcgf = |\mathbf{A}| \cdot |\mathbf{B}|$.

18.2 $T^2 = 3.61$. Equation (18.11) describes the confidence ellipse except that the crossproduct term is now preceded by a plus sign.

18.3
$$(\mathbf{C\bar{Y}})' = [8.9 \quad 1.0]$$

$$(\mathbf{CSC'}) = \begin{bmatrix} 37.00 & .32 \\ .32 & 9.00 \end{bmatrix}$$

$$T^2 = (102)(\mathbf{C\bar{Y}})'(\mathbf{CSC'})^{-1}(\mathbf{C\bar{Y}}) = 228.04$$

18.4
$$T^2 = \frac{\left| \mathbf{S} + \dfrac{(51)^2}{102}(\bar{\mathbf{Y}}_{..1} - \bar{\mathbf{Y}}_{..2})(\bar{\mathbf{Y}}_{..1} - \bar{\mathbf{Y}}_{..2})' \right|}{|\mathbf{S}|} - 1$$

$$= \frac{4563.36}{424.91} - 1 = 9.72$$

18.5
$$\mathbf{S} = \begin{bmatrix} 16.125 & -2.91\bar{6} & -.541\bar{6} \\ -2.91\bar{6} & 1.1\bar{6} & 1.25 \\ -.541\bar{6} & 1.25 & 3.91\bar{6} \end{bmatrix} \qquad |\mathbf{S}| = 18.78$$

$$\left| \mathbf{S} + \frac{(4)^2}{8}(\bar{\mathbf{Y}}_{.1} - \bar{\mathbf{Y}}_{.2})'(\bar{\mathbf{Y}}_{.1} - \bar{\mathbf{Y}}_{.2}) \right| = 1482.97$$

$$T^2 = 77.98$$

$$|\mathbf{W} + \mathbf{B}| = 56,766.50 \qquad |\mathbf{W}| = 4,055.50$$

Note that each entry in $\mathbf{W} = 2(n-1)$ times the corresponding entry in $\mathbf{S}$. Therefore (see Exercise 18.1(b))

$$|\mathbf{W}| = [2(n-1)]^n \, |\mathbf{S}|.$$

$$\Lambda = .07$$

Checking, we get

$$T^2 = 6\left(\frac{1}{\Lambda} - 1\right) = 77.98$$

18.6 Let

$$\mathbf{C} = \begin{bmatrix} 1 & -1 & 0 \\ 0 & 1 & -1 \end{bmatrix}$$

Then,

$$(\mathbf{CWC'})^{-1} = \left(\frac{1}{1,961.5625}\right)\begin{bmatrix} 15.50 & 13.75 \\ 13.75 & 138.75 \end{bmatrix}$$

and

$$(\mathbf{CBC'}) = \begin{bmatrix} 136.125 & -123.750 \\ -123.750 & 112.500 \end{bmatrix}$$

We then calculate

$$\Lambda = [|(\mathbf{CWC'})^{-1}(\mathbf{CBC'})|]^{-1} = .120506$$

and

$$T^2 = 6\left(\frac{1}{\Lambda} - 1\right) = 43.790$$

Finally,

$$F = \frac{N-b}{(b-1)(N-2)}\, T^2 = (\tfrac{5}{12})(43.790) = 18.25,$$

which is distributed on $b - 1$ (or 2) and $N - b$ (or 5) *df*.

INDEX